The Latest Methods of Construction Design

Vojtěch Dynybyl • Ondrej Berka • Karel Petr
František Lopot • Martin Dub
Editors

The Latest Methods of Construction Design

Springer

Editors
Vojtěch Dynybyl
Faculty of Mechanical Engineering
Czech Technical University of Prague
Prague, Czech Republic

Karel Petr
Faculty of Mechanical Engineering
Czech Technical University of Prague
Prague, Czech Republic

Martin Dub
Faculty of Mechanical Engineering
Czech Technical University of Prague
Prague, Czech Republic

Ondrej Berka
Faculty of Mechanical Engineering
Czech Technical University of Prague
Prague, Czech Republic

František Lopot
Faculty of Mechanical Engineering
Czech Technical University of Prague
Prague, Czech Republic

ISBN 978-3-319-22761-0 ISBN 978-3-319-22762-7 (eBook)
DOI 10.1007/978-3-319-22762-7

Library of Congress Control Number: 2015954952

Springer Cham Heidelberg New York Dordrecht London

Printed on acid-free paper

Springer International Publishing AG Switzerland is part of Springer Science+Business Media (www.springer.com)

Preface

The book "The Latest Methods of Construction Design" has been created on the basis of contributions of the 55th International Conference of Machine Design Departments. This book is a follow-up to an earlier very successful book "Modern Methods of Construction Design."

This conference is one of the oldest central conferences, dealing with methods and applications in the machine design. The main aim of the conference is to provide an international forum where experts, researchers, engineers, and also industrial practitioners, managers, and PhD students can meet, share their experiences, and present the results of their efforts in the broad field of machine design and related fields.

In the year 2014, the Department of Designing and Machine Components, Faculty of Mechanical Engineering, Czech Technical University in Prague, whose members deal above all with machine design, experimental methods and measuring, engineering analyses, and products innovation, organized the 55th International Conference of Machine Design Departments on the occasion of 150th anniversary of Faculty of Mechanical Engineering (CTU in Prague). The mountain town of Beroun in Central Bohemia was chosen for this celebratory event.

Since 1960, when the first conference was organized in Melnik by Brno University of Technology, more than 50 years have passed. The main aim of the conference was to providing an opportunity for professional experiences sharing in the field of machine design, gears, and transmission mechanisms. The heads of Mechanical Design Departments decided to organize these conferences annually first at the national and later at the international level. The significance of the conference has grown. Historically, the conferences were organized in different places by different mechanical design departments of Czech and Slovak technical universities:

1960	**Brno University of Technology**	*Melnik*
1960	**Brno University of Technology**	*Brno*
1962	**University of West Bohemia**	*Pilsen*
1963	**Slovak University of Technology in Bratislava**	*Bratislava*
1964	**Technical University of Liberec**	*Liberec*
1965	**Technical University of Kosice**	*Kosice*
1966	**Czech Technical University in Prague**	*Prague*
1967	**VŠB-Technical University of Ostrava**	*Ostrava*
1968	**University of Zilina**	*Zilina*
1969	**University of West Bohemia**	*Pilsen*
1970	**Brno University of Technology**	*Brno*
1971	**Technical University of Zvolen**	Žvolen
1972	**Slovak University of Technology in Bratislava**	*Bratislava*
1973	**Technical University of Liberec**	*Liberec*
1974	**Technical University of Kosice**	*Kosice*
1975	**Brno University of Technology**	*Brno*
1976	**Czech Technical University in Prague**	*Prague*
1977	**Czech University of Life Sciences Prague**	*Prague*
1978	**University of Zilina**	*Kysucke Nove Mesto*
1979	**Technical University of Liberec**	*Liberec*
1980	**Slovak University of Technology in Bratislava**	*Smolenice*
1981	**Brno University of Technology**	*Brno*
1982	**Technical University in Zvolen**	*Sachticky*
1983	**Technical University of Kosice**	*Michalovce*
1984	**VŠB-Technical University of Ostrava**	*Podolanky*
1985	**Czech Technical University in Prague**	*Kladno*
1986	**University of West Bohemia**	*Pilsen*
1987	**Slovak University of Technology in Bratislava**	*Bratislava*
1988	**Brno University of Technology**	*Brno*
1989	**Technical University of Liberec**	*Liberec*
1990	**University of Zilina**	*Papradno*
1991	**Technical University of Kosice**	*Herlany*
1992	**VŠB-Technical University of Ostrava**	*Tesinske Beskydy*
1993	**Technical University in Zvolen**	*Zvolen*
1994	**Czech Technical University in Prague**	*Kladno*
1995	**Brno University of Technology**	*Brno*
1996	**University of West Bohemia**	*Pilsen*
1997	**Slovak University of Technology in Bratislava**	*Gabcikovo*
1998	**Technical University of Liberec**	*Liberec*
1999	**University of Zilina**	*Belueske Slatiny*
2000	**Technical University of Kosice**	*Herlany*
2001	**VŠB-Technical University of Ostrava**	*Ostrava*
2002	**Technical University of Zvolen**	*Bienska dolina*
2003	**Czech Technical University in Prague**	*Prague*

(continued)

2004	**Brno University of Technology**	*Blansko*
2005	**Technical University of Liberec**	*Sedmihorky*
2006	**Czech University of Life Sciences Prague**	*Prague*
2007	**Slovak University of Technology in Bratislava**	*Smolenice*
2008	**University of West Bohemia**	*Srni*
2009	**University of Zilina**	*Techova*
2010	**Technical University of Kosice**	*Slovensky Raj*
2011	**Brno University of Technology**	*Mikulov*
2012	**VŠB-Technical University of Ostrava**	*Sepetna*
2013	**Technical University of Liberec**	*Hejnice*
2014	**Czech Technical University in Prague**	*Beroun*

We firmly believe the tradition of ICMD will continue in the coming years.
Members of conference organizing committee

Prague, Czech Republic — Vojtěch Dynybyl
Prague, Czech Republic — Ondrej Berka
Prague, Czech Republic — Karel Petr
Prague, Czech Republic — František Lopot
Prague, Czech Republic — Martin Dub

Committee

International Scientific/Programme Committee of the 55th ICMD 2014

V. Dynybyl-Chairman, Czech Republic
J. Machado, Portugal
Z. Matuszak, Poland
K. Peszynski, Poland
E.N. Mohammadali, Iran
A. Ahameedi, Iraq
Y. Takeichi, Japan
H. Haberhauer, Germany
M. Bozca, Turkey
M. Rackov, Serbia
H. Mickoski, Macedonia
V. Arakelyan, France
W. Toisuta, Australia
S. Popescu, Romania
E. Coatanea, Finland
M. Abdulgadir, Libyan Arab
S. Loginov, Russia
S. Simanjuntak, Indonesia
D. Bakti, Indonesia
J. Boamah, Ghana
P. L Nbah, Cameroon
L. Yaoming, China
R. Basan, Croatia
B. Križan Croatia

S. Hrček, Slovakia
S. Hosnedl, Czech Republic
J. Hošek, Czech Republic
T. Jirout, Czech Republic
J. Křička, Czech Republic
F. Lopot, Czech Republic
M. Dub, Czech Republic
O. Berka, Czech Republic
Ś. Medvecký, Slovakia
M. Němeček Czech Republic
L. Pešík, Czech Republic
L. Ševčík, Czech Republic
P. Lepšík Czech Republic
M. Petrů, Czech Republic
M. Španiel, Czech Republic
M. Vereš, Slovakia
M. Bošanský, Slovakia
Z. Češpíro, Czech Republic
Z. Dejl, Czech Republic
M. Hartl, Czech Republic
D. Herák, Czech Republic
J. Homišin, Slovakia

Review Committee of the 55th ICMD 2014

S. Loginov, Russia
W. Toisuta, Australia
Ondřej Novák, Czech Republic
J. Petrik, Czech Republic
S. Chowdhury, USA
J. Broncek, Slovakia
S. Popescu, Romania
M. Komarek, Czech Republic
V. Arakelyan, France
M. Bozca, Turkey
B. Křižan, Croatia
T. Jirout, Czech Republic
J. Hošek, Czech Republic
J. Bečka, Czech Republic

V. Dynybyl, Czech Republic
L. Jančík, Czech Republic
F. Lopot, Czech Republic
J. Mrázek, Czech Republic
M. Španiel, Czech Republic
D. Herák, Czech Republic
S. Holý, Czech Republic

Organizing Committee of the 55th ICMD 2014

K. Petr, J. Andruš, F. Lopot, P. Malý, J. Mrázek, F. Starý, R. Uhlíř

Contents

Part I Machine Design

A New Stand for Fatigue Testing of Pillars 3
L. Barton

HCR Gears in the Industrial Gearbox 7
M. Burián, M. Trochta, and J. Havlik

Design of the Vibration Separatory Apparatus 15
M. Dub, S. Sadek, J. Hendrych, and P. Spacek

Designing of Mixing Equipment, Reactors and Bioreactors 19
T. Jirout and D. Jiroutova

Modern Trends in the Drive Wheelsets of Rail Vehicles 27
J. Kolar

Scale-Up of Equipment for Thermal-Expansionary Pretreatment of Lignocelluloses 37
L. Kratky

Should We Teach Polymeric Machine Elements as Well? 45
B. Križan

Design of Semi-active Magnetorheological Valve 51
M. Kubík and I. Mazůrek

Support Tools in the Development of Bearings for Wind Turbines 57
R. Madaj and R. Kohar

Note to an Undercut Limit of Involute Gearing 63
M. Nemecek

Assessment of Industrial Design Projects with Use of Online Evaluation System 69
M. Ondra, R. Machalek, D. Skaroupka, D. Semrad, and F. Uhlir

Vibration Isolation Pneumatic System with a Throttle Valve 75
L. Pesik, A. Skarolek, and O. Kohl

Development of the Assembly Set for the Logistic Transport Solution 81
R. Sásik, M. Haluška, R. Madaj, M. Gregor, and P. Grznár

The Design of the Device for the Manufacture of Planar Filter 87
L. Sevcik

The Structural Design of Industrial Vacuum Cleaner for Dental Laboratories 93
R. Slanina and P. Sniehotta

Reading of Operating Values of End Effector During Goods Manipulation 99
F. Stary, V. Dynybyl, and J. Mrazek

Building Platform for Identification of Important Parameters for 3D Printing of Huge Parts 103
M. Suransky and M. Zlebek

On Cardan Joint: Some Special Applications 111
J. Zicha and K. Macuchova

Part II Tribology

Experimental Analysis of Tribological Properties of Heat Treated Graphitic Cast Irons 117
J. Broncek, P. Fabian, M. Dzimko, and N. Radek

The Experimental Determination of the Grease Amount to Effective Wear Reduction in the Wheel-Rail Contact 127
J. Fryza and M. Omasta

The Effect of Friction Modifier on the Wheel-Rail Contact 133
R. Galas and M. Omasta

Measurement Methodology of Low-Shear Viscosity of Reference Liquid Under High-Pressure 139
S. Hornak

Modification of Tooth Flank as the Possibility of Increasing Resistance of Gearing to Pitting 145
B. Kopilakova, M. Bosansky, and J. Zapotocny

Including the Method of Lubricant Film Thickness Measurement with the Use of the Monochromatic Interferometry to the Education of Tribology 153
D. Kostal, I. Křupka, and M. Hartl

Frictional Properties of PVA Hydrogel 159
D. Nečas, M. Vrbka, S. Yarimitsu, K. Nakashima, Y. Sawae, P. Šperka, I. Křupka, and M. Hartl

Effect of Sanding on Adhesion Under Oil and Water Condition Between Wheel and Rail 165
D. Smejkal

Part III Hydraulics – Fluid Mechanisms

Experimental Research and Optimization of Characteristic Parameters of the Valve Plate of the Axial Piston Pump/Motor 173
R. Petrovic, M. Andjelkovic, M. Radosavljevic, and N. Todic

The Effects of the Temperatures on the Performance of the Gas Turbine 185
A. Wahab and T. Ibrahim

Part IV Engineering Analyses

Innovation of Methodological Support of Risk Analyses of Technical Products in Their Life Cycle During Their Designing 197
J. Dvorak, M. Sasiadek, and S. Hosnedl

The Mathematical Description of Sunflower Seed's Hulling Process Using Impact Huller 203
J. Hoidekr, J. Andrus, and V. Dynybyl

Roller Bearing Fatigue Life Calculation with Inclusion of the Fits Parameters 207
S. Hrcek and R. Kohar

Application of Statistical Methods to Analysis of Parallel Structure of Gearbox Test Rig 215
J. Chmelar, K. Petr, and V. Dynybyl

Driving Shaft Fracture Analysis 221
K. Kaláb

Force Conditions at an Antibacklash Designed Gear Mechanism 227
V. Kloucek

The Effect of Fits Stiffness on Distribution of Internal Loading in Rolling Bearings 235
R. Kohar and S. Hrcek

Design Calculation of Spur Gear Module 243
E. Kronerova

Application of the Bond Graphs to the Description of Technical Systems 249
Z. Matuszak

Assessment of the Validity of Technical Elements of the Complex Structure Using Flow Streams of Energy Factors 255
Z. Matuszak

Influence of Short-Circuit Valve on Hydraulic Recovery of Vehicle Kinetic Energy 263
Z. Nemec and J. Nevrly

FEM Simulations on Gears with Different Carbonized Depth 269
K. Petr, V. Dynybyl, and J. Krepela

Numerical Simulation of Ductile Fracture 277
M. Spaniel

Dynamic of Propulsion, New Challenges for Analysis 283
V. Zoul

Part V Modern Material and Technology

Corrosion in Some Iron Alloys 291
M. Khidhair and M. Hasan

Overview on the Clinched Joint Method 299
P. Maly, V. Dynybyl, J. Sojka, and F. Lopot

The Process of Preparation Prototypes Bearing Cages for the Durability Testing 305
P. Spišák and R. Madaj

Part VI Optimization and Design

Optimisation of the Geometric Design Parameters of a Five Speed Gearbox for an Automotive Transmission 313
M. Bozca

Modification of Experimental Station AXMAT 319
J. Dvoracek, L. Komenda, and L. Nohal

Basic Conditions of the Implementation of Extremal Regulation 325
J. Homišin and M. Moravič

Design of a Disc Grinder for EDM Electrodes Machining 331
J. Hosek and J. Drahokoupil

Optimized Design of the Wind Turbine's Composite Blade to Flatten the Stress Distribution in the Mounting Areas 335
P. Oganesyan, I. Zhilyaev, S. Shevtsov, and J.-K. Wu

Research of New Composites for Lightweight Construction with Low Impact on the Environment 343
M. Petrů, P. Lepsik, and Ondřej Novák

System of Low Profile Roller Conveyors 349
F. Stary and V. Hanus

Fiber Driven Tilting Mechanism 353
P. Svatoš, Z. Šika, M. Valášek, J. Zicha, V. Bauma, and V. Rada

Part VII Product Innovation

Pneumatic Dual Mass Flywheel – Damper Concept for Downspeeding 361
R. Grega and P. Baran

Rationalising the Use of Design for Assembly as DfX Knowledge to Increase Competitiveness of Designed Technical Products 369
S. Hosnedl, M. Kopecky, and M. Sasiadek

Materials and Sustainable Development 377
I. Mazinova and M. Bochinsky

Part VIII Experimental Methods and Measuring

Design of Experiment for Load Gear Solution in Planetary Transmission 385
O. Berka, F. Lopot, and M. Dub

Investigation of Tram Gearbox Hinge Behaviour 393
M. Dub, J. Kolar, F. Lopot, V. Dynybyl, and O. Berka

Mechanical Properties of Fasteners 399
V. Fliegel, A. Lufinka, and R. Martonka

Mathematical Description of Non-linear Mechanical Behaviour of Materials Under Compression Loading, Case Study: Spruce Bulk Wood Chips 405
D. Herak, A. Kabutey, and R. Sigalingging

Influence of Steel Sheet Width on Bearing Capacity of Resistance Spot Welding 411
P. Hrabe and M. Muller

Calculation of Fuel Consumption on the Vehicle Simulation Model 417
M. Kobza

Identification of Parameters of Stiffness and Damping Based on the Amplitude Frequency Characteristics 421
O. Kohl

Measurements of the Integral Characteristics of Twisted Flow Using a Five-Tube Pressure Probe 427
J. Kuzel, M. Golda, M. Kral, and M. Hraska

Method of Experimental Measurement of a Bucket in Laboratory Conditions 433
P. Kribala, V. Dynybyl, and J. Kricka

Biaxial Stress of the Textile Car Seat Cover Tested in a Climatic Chamber 437
P. Kulhavý, N. Kovalova, R. Martonka, and J. Petrik

Hexapod Control System Optimization 443
A. Lufinka and R. Martonka

Integration of Light: A Way to Measure Spectral Characteristics 449
K. Macuchova

The Equipment for Temperature Measurements Near a Gear Tooth Flank 457
M. Mazac, P. Herajn, and M. Svoboda

Determination of Power Losses Based on the Measurement in Closed Loop Test Bench for Automotive Gearboxes 465
O. Milacek and G. Achtenova

Experimental Measurements and Development of Agricultural Trailer 471
P. Mossoczy, R. Uhlir, and P. Maly

Deformable Mirror for High Power Lasers 475
S. Nemcova and J. Zicha

Identification of Failures of Technical Equipment Using Acoustic Camera 483
S. Poljak and I. Gajdac

Assessment of the Mechanical Properties the Polyurethane Foam Under Multiaxial Loading in a Different Climatic Conditions . . . 491
P. Srb, P. Kulhavý, and R. Martonka

Experimental Analysis of Prototype Rubber Suspension for Agricultural Machine 497
P. Syrovatka, F. Lopot, and M. Janda

Contribution to the Methodology of Measuring Transmission Error . . . 503
M. Trochta, Z. Folta, and M. Burián

Climate Chamber for Testing of Multiaxial Loading 509
D. Vejrych, R. Martonka, and R. Kovář

Part IX Nanotechnology

System for High-Temperature Calcination of Nanofibers 515
R. Kovář

Author Index 521

Subject Index 525

Contributors

G. Achtenova Czech Technical University in Prague, Prague, Czech Republic

M. Andjelkovic Faculty for Strategic and Operational Management Belgrade, University Union Nikola Tesla of Belgrade, Belgrade, Serbia

J. Andrus Czech Technical University in Prague, Prague, Czech Republic

P. Baran Technical University of Košice, Košice, Slovak Republic

L. Barton University of West Bohemia, Pilsen, Czech Republic

V. Bauma Faculty of Mechanical Engineering, Czech Technical University, Prague, Czech Republic

O. Berka Faculty of Mechanical Engineering, Czech Technical University of Prague, Prague, Czech Republic

M. Bochinsky University of West Bohemia, Bohemia, Czech Republic

M. Bosansky Slovak University of Technology in Bratislava, Bratislava, Slovakia

M. Bozca Yildiz Technical University, Istanbul, Turkey

J. Broncek University of Zilina, Žilina, Slovak Republic

M. Burián VŠB – Technical University of Ostrava, Ostrava, Czech Republic

J. Chmelar Czech Technical University in Prague, Prague, Czech Republic

J. Drahokoupil Czech Technical University in Prague, Prague, Czech Republic

M. Dub Faculty of Mechanical Engineering, Czech Technical University of Prague, Prague, Czech Republic

J. Dvoracek Brno University of Technology, Brno, Czech Republic

J. Dvorak University of West Bohemia, Plzeň, Czech Republic

V. Dynybyl Faculty of Mechanical Engineering, Czech Technical University of Prague, Prague, Czech Republic

M. Dzimko University of Zilina, Žilina, Slovak Republic

P. Fabian University of Zilina, Žilina, Slovak Republic

V. Fliegel Technical University of Liberec, Liberec, Czech Republic

Z. Folta VŠB-TU Ostrava, Ostrava, Czech Republic

J. Fryza Brno University of Technology, Brno, Czech Republic

I. Gajdac University of Zilina, Žilina, Slovak republic

R. Galas Brno University of Technology, Brno, Czech Republic

M. Golda Aerospace Research and Test Establishment, Prague, Czech Republic

R. Grega Technical University of Košice, Košice, Slovak Republic

M. Gregor University of Zilina, Žilina, Slovak Republic

P. Grznár University of Zilina, Žilina, Slovak Republic

M. Haluška University of Zilina, Žilina, Slovak Republic

V. Hanus Czech Technical University, Prague, Czech Republic

M. Hartl Brno University of Technology, Brno, Czech Republic

M. Hasan Al-Mustaqbal College – Babylon, Hilla, Iraq

J. Havlik VŠB – Technical University of Ostrava, Ostrava, Czech Republic

J. Hendrych Institute of Chemical Technology, Prague, Czech Republic

P. Herajn Technical University of Liberec, Liberec, Czech Republic

D. Herak Czech University of Life Sciences Prague, Prague, Czech Republic

J. Hoidekr Czech Technical University in Prague, Prague, Czech Republic

J. Homišin Technical University in Košice, Košice, Slovak Republic

S. Hornak Brno University of Technology, Brno, Czech Republic

J. Hosek Czech Technical University in Prague, Prague, Czech Republic

S. Hosnedl University of West Bohemia, Plzeň, Czech Republic

P. Hrabe Czech University of Life Sciences Prague, Prague, Czech Republic

M. Hraska Aerospace Research and Test Establishment, Prague, Czech Republic

S. Hrcek University of Zilina, Žilina, Slovak Republic

T. Ibrahim Tikrit University, Tikrit, Iraq

M. Janda Czech Technical University, Prague, Czech Republic

T. Jirout Faculty of Mechanical Engineering, Department of Process Engineering, Czech Technical University in Prague, Prague, Czech Republic

D. Jiroutova Klokner Institute, Czech Technical University in Prague, Prague, Czech Republic

A. Kabutey Czech University of Life Sciences Prague, Prague, Czech Republic

K. Kaláb VSB-Technical University Ostrava, Ostrava, Czech Republic

M. Khidhair Al-Mustaqbal College – Babylon, Hilla, Iraq

V. Kloucek VÚTS Liberec, a.s., Liberec, Czech Republic

M. Kobza Brno University of Technology, Brno, Czech Republic

R. Kohar University of Zilina, Žilina, Slovak Republic

O. Kohl Faculty of Mechanical Engineering, Technical University of Liberec, Liberec, Czech Republic

J. Kolar Czech Technical University, Prague, Czech Republic

L. Komenda Brno University of Technology, Brno, Czech Republic

M. Kopecky University of West Bohemia, Pilsen, Czech Republic

B. Kopilakova Trenčín University A. Dubček in Trenčín, Trenčín, Slovakia

D. Kostal Brno University of Technology, Brno, Czech Republic

N. Kovalova Technical University of Liberec, Liberec, Czech Republic

R. Kovář Technical University of Liberec, Liberec, Czech Republic

M. Kral Aerospace Research and Test Establishment, Prague, Czech Republic

Czech Technical University in Prague, Prague, Czech Republic

L. Kratky Czech Technical University in Prague, Prague, Czech Republic

J. Krepela Wikov MGI a.s., Hronov, Czech Republic

P. Kribala Czech Technical University in Prague, Prague, Czech Republic

J. Kricka Czech Technical University in Prague, Prague, Czech Republic

B. Križan Faculty of Engineering, University of Rijeka, Rijeka, Croatia

E. Kronerova University of West Bohemia, Plzeň, Czech Republic

I. Křupka Brno University of Technology, Brno, Czech Republic

M. Kubík Brno University of Technology, Brno, Czech Republic

P. Kulhavý Technical University of Liberec, Liberec, Czech Republic

J. Kuzel Aerospace Research and Test Establishment, Prague, Czech Republic

P. Lepsik Faculty of Mechanical Engineering, Technical University of Liberec, Liberec, Czech Republic

F. Lopot Faculty of Mechanical Engineering, Czech Technical University of Prague, Prague, Czech Republic

F. Lopot Charles University, Prague, Czech Republic

A. Lufinka Technical University of Liberec, Liberec, Czech Republic

R. Machalek Brno University of Technology, Brno, Czech Republic

K. Macuchova Czech Technical University in Prague, Prague, Czech Republic

R. Madaj University of Zilina, Žilina, Slovak Republic

P. Maly Czech Technical University in Prague, Prague, Czech Republic

R. Martonka Technical University of Liberec, Liberec, Czech Republic

Z. Matuszak Maritime University of Szczecin, Szczecin, Poland

M. Mazac Technical University of Liberec, Liberec, Czech Republic

I. Mazinova University of West Bohemia, Bohemia, Czech Republic

I. Mazůrek Brno University of Technology, Brno, Czech Republic

O. Milacek Czech Technical University in Prague, Prague, Czech Republic

M. Moravič Technical University in Košice, Košice, Slovak Republic

P. Mossoczy Czech Technical University in Prague, Prague, Czech Republic

J. Mrazek Czech Technical University in Prague, Prague, Czech Republic

M. Muller Czech University of Life Sciences Prague, Prague, Czech Republic

K. Nakashima Kyushu University, Fukuoka, Japan

D. Nečas Brno University of Technology, Brno, Czech Republic

M. Nemecek VŠB-Technical University of Ostrava, Ostrava, Czech Republic

S. Nemcova Czech Technical University in Prague, Prague, Czech Republic

Z. Nemec Brno University of Technology, Brno, Czech Republic

J. Nevrly Brno University of Technology, Brno, Czech Republic

L. Nohal Brno University of Technology, Brno, Czech Republic

Ondřej Novák Textile Faculty, Technical University of Liberec, Liberec, Czech Republic

P. Oganesyan Don State Technical University, Rostov on Don, Russia

South Center of Russian Academy, Rostov on Don, Russia

M. Omasta Brno University of Technology, Brno, Czech Republic

M. Ondra Brno University of Technology, Brno, Czech Republic

L. Pesik Faculty of Mechanical Engineering, Technical University of Liberec, Liberec, Czech Republic

K. Petr Faculty of Mechanical Engineering, Czech Technical University of Prague, Prague, Czech Republic

J. Petrik Technical University of Liberec, Liberec, Czech Republic

R. Petrovic Faculty for Strategic and Operational Management Belgrade, University Union Nikola Tesla of Belgrade, Belgrade, Serbia

M. Petrů Faculty of Mechanical Engineering, Technical University of Liberec, Liberec, Czech Republic

S. Poljak University of Zilina, Žilina, Slovak republic

V. Rada Faculty of Mechanical Engineering, Czech Technical University, Prague, Czech Republic

N. Radek Kielce University of Technology, Kielce, Poland

M. Radosavljevic Faculty for Strategic and Operational Management Belgrade, University Union Nikola Tesla of Belgrade, Belgrade, Serbia

S. Sadek Czech Technical University in Prague, Prague, Czech Republic

M. Sasiadek University of Zielona Góra, Zielona Góra, Poland

R. Sásik University of Zilina, Žilina, Slovak Republic

Y. Sawae Kyushu University, Fukuoka, Japan

D. Semrad Brno University of Technology, Brno, Czech Republic

L. Sevcik Technical University in Liberec, Liberec, Czech Republic

S. Shevtsov South Center of Russian Academy, Rostov on Don, Russia

Southern Federal University, Rostov on Don, Russia

R. Sigalingging Czech University of Life Sciences Prague, Prague, Czech Republic

Z. Šika Faculty of Mechanical Engineering, Czech Technical University, Prague, Czech Republic

A. Skarolek Faculty of Mechanical Engineering, Technical University of Liberec, Liberec, Czech Republic

D. Skaroupka Brno University of Technology, Brno, Czech Republic

R. Slanina VSB-Technical University of Ostrava, Ostrava, Czech Republic

D. Smejkal Brno University of Technology, Brno, Czech Republic

P. Sniehotta VSB-Technical University of Ostrava, Ostrava, Czech Republic

J. Sojka Czech Technical University in Prague, Prague, Czech Republic

P. Spacek CHEMCOMEX Praha, a.s., Prague, Czech Republic

M. Spaniel Czech Technical University in Prague, Prague, Czech Republic

P. Šperka Brno University of Technology, Brno, Czech Republic

P. Spišák University of Zilina, Žilina, Slovak Republic

P. Srb Technical University of Liberec, Liberec, Czech Republic

F. Stary Czech Technical University in Prague, Prague, Czech Republic

M. Suransky Brno University of Technology, Brno, Czech Republic

P. Svatoš Faculty of Mechanical Engineering, Czech Technical University, Prague, Czech Republic

M. Svoboda Technical University of Liberec, Liberec, Czech Republic

P. Syrovatka Czech Technical University, Prague, Czech Republic

N. Todic Faculty for Mechanical and Civil Engineering, University of Kragujevac, Kraljevo, Serbia

M. Trochta VŠB – Technical University of Ostrava, Ostrava, Czech Republic

F. Uhlir Brno University of Technology, Brno, Czech Republic

R. Uhlir Czech Technical University in Prague, Prague, Czech Republic

M. Valášek Faculty of Mechanical Engineering, Czech Technical University, Prague, Czech Republic

D. Vejrych Technical University of Liberec, Liberec, Czech Republic

M. Vrbka Brno University of Technology, Brno, Czech Republic

A. Wahab Almustaqbal University College, Hillah, Iraq

J.-K. Wu National Kaohsiung Marine University, Kaohsiung, Taiwan

S. Yarimitsu Kyushu University, Fukuoka, Japan

J. Zapotocny Trenčín University A. Dubček in Trenčín, Trenčín, Slovakia

I. Zhilyaev South Center of Russian Academy, Rostov on Don, Russia

Southern Federal University, Rostov on Don, Russia

J. Zicha Faculty of Mechanical Engineering, Czech Technical University, Prague, Czech Republic

M. Zlebek Brno University of Technology, Brno, Czech Republic

V. Zoul Czech Technical University in Prague, Prague, Czech Republic

Part I
Machine Design

L. Barton, M. Burian, M. Dub, V. Dynybyl, M. Gregor, P. Grznár, M. Haluška, J. Havlik, J. Hendrych, T. Jirout, D. Jiroutova, R. Kohar, O. Kohl, J. Kolar, L. Kratky, B. Krizan, M. Kubík, R. Machalek, K. Macuchova, R. Madaj, I. Mazurek, J. Mrazek, M. Nemecek, M. Ondra, L. Pesik, S. Sadek, R. Sásik, D. Semrad, L. Sevcik, A. Skarolek, D. Skaroupka, R. Slanina, P. Sniehotta, P. Spacek, F. Stary, M. Suransky, M. Trochta, F. Uhlir, J. Zicha, and M. Zlebek

A New Stand for Fatigue Testing of Pillars

L. Barton

Abstract A new stand for fatigue testing was designed. This stand allows testing six pillars in the same time. This contribution deals with design of stand and with influence between pillars.

Keywords Testing • Fatigue • Stand • Tram pillars

1 Introduction

Tram and trolley bus pillars are under cyclic stress. This has a negative influence on their working life time. Pillars are made from square tubes which are welded together. The demands placed on pillars put emphasis on their weight. The cyclic load is not always exactly predicted. Fatigue cracks appear due to the efforts to make pillars as light as possible.

The most crucial place for appearance of cracks is at welding sites. Figure 1 shows a pillar from the rear of a trolley bus. The pillar crack occurred due to fatigue failure. The crack appeared at a welding site and it went through the primary material of the square tube. This failure happened on a 6 year old trolley bus which had clocked up over 300 000 km.

2 The Calculation for the Stand

Fatigue tests take a lot of time and several samples are needed for relevant results. The new design of the testing stand is designed for testing six samples at one time, which saves a lot of time. The stand is designed for testing weldments made from two parts. The first part is the pillar which is attached to the wall with pitch of attachment "L". In the middle of the pillar is the welded arm of the pillar with length "L_R". The weldment is drawn in Fig. 2.

L. Barton (✉)
University of West Bohemia, Pilsen, Czech Republic
e-mail: bartonlk@rti.zcu.cz

V. Dynybyl et al. (eds.), *The Latest Methods of Construction Design*,
DOI 10.1007/978-3-319-22762-7_1

Fig. 1 The failure of a window pillar of a trolley bus [1]

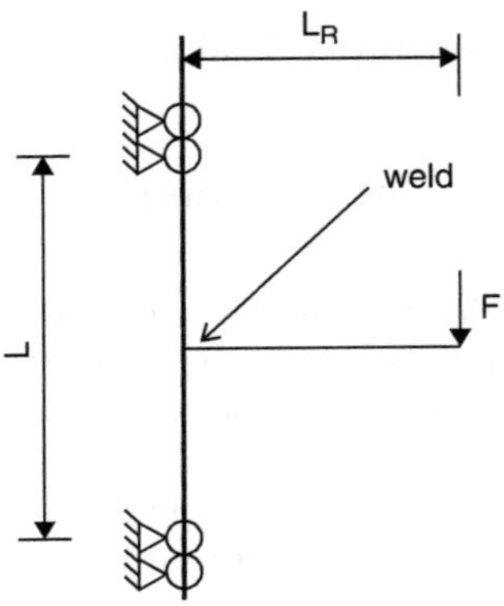

Fig. 2 Load of weldment. "L" is the pitch of attachment and it is 1000 mm long. "L_R" is length of arm and it is 500 mm

Deformation loading is considered for testing all six samples. The two hydraulic cylinders have to provide force for deformation on all six samples at one time. The deformation should be 3 mm even if the biggest considered pillars are tested.

Three kinds of deformation are considered during dimensioning. The total deformation is composed of the bend deformation of the arm, displacement from angle of deformation of the pillar and displacement from the stress deformation of the pillar. The arm is not considered as a fixed abutment but it is welded to the pillar with a certain stiffness.

The stress at the welding site is calculated from the assigned total deformation. Table 1 shows the dimensions of the square tubes in the columns on the left. Deformations are from 2.3 to 3.3 mm and stress in mega Pascals.

3 Design

An I-beam was chosen as a girder. The I-beam is 200 mm high. The fixing jigs are easily fixed on the top of the I-beam. The assembly is shown in Fig. 3. The fixing jigs are universal for all the considered sizes of the pillars.

Table 1 Stress [MPA] in the place of welding and its dependency on size of square tube and size of total deformation

Square tube			Total deformation – w [mm]										
w	h	t	2.3	2.4	2.5	2.6	2.7	2.8	2.9	3	3.1	3.2	3.3
40	40	5	93	97	101	105	109	113	117	121	125	129	133
40	60	5	139	145	151	157	163	169	175	181	187	193	199
60	60	3	139	145	151	157	163	169	175	181	187	193	199
60	60	5	139	145	151	157	163	169	175	181	187	193	199
60	100	3	230	240	250	260	270	280	290	300	310	320	330
60	100	5	230	240	250	260	270	280	290	301	311	321	331
60	120	3	276	288	300	312	324	336	348	360	372	384	396
60	120	5	276	288	300	312	324	336	348	360	372	384	396

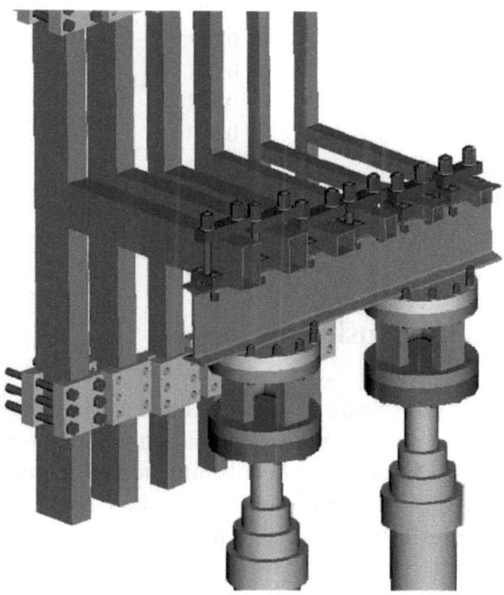

Fig. 3 The design of the stand

The girder was solved as a beam with two supports which is loaded by six forces. Mitcalc software was used for the solution. The results in Fig. 4 are the bend stress analysis and deformation analysis of the beam.

The deformation deviation between samples was calculated for total deformation of 3 mm. The deformation deviation was recalculated to stress deviation in the welding site and it was recalculated to relative deviation of stress. These values are entered into Table 2.

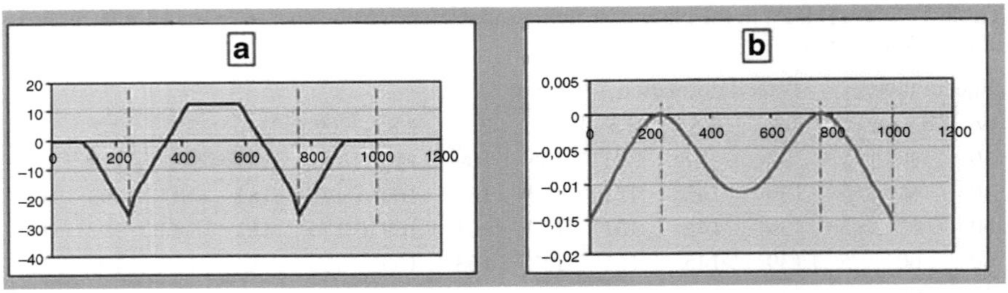

Fig. 4 The result from analysis of beam. (**a**) – Bend stress (**b**) – deformation

Table 2 The results of deviation

			Deformation deviation [mm]	Relative deviation of stress [%]
40	40	5	0.0006	0.00
40	60	5	0.0015	0.04
60	60	3	0.0013	0.04
60	60	5	0.002	0.06
60	100	3	0.004	0.13
60	100	5	0.006	0.2
60	120	3	0.006	0.2
60	120	5	0.011	0.37

4 Conclusion

From our results we can see that the relative deviation of stress is less than 0.5 %. In the real case the relative deviation will be even less. The deformation of the 200 mm I-beam has no significant influence on the deviation between testing samples. The new design can be used for testing up to six samples at one time. None of the samples influence any of the other samples.

References

1. BARTOŇ, Lukáš a Petr HELLER. CENTRUM KOMPETENCE DRÁŽNÍCH VOZIDEL, projekt č. TE01020038. *Výsledky sledování únavového poškozování skříní trolejbusů: Technická zpráva RTI-VZ-2013-5*. Plzeň, 2013

HCR Gears in the Industrial Gearbox

M. Burián, M. Trochta, and J. Havlik

Abstract Gearing with a prolonged contact ratio – so called HCR gearing is used more and more in these days, mainly for the gearboxes in the automobile industry, and for other transport machines. HCR gearing mainly influences the noise reduction in the gearing system. Relative addendum of the HCR gearing usually changes the height of the gear h_a^*, however, this value must be higher than $h_a^* > 1$.

Keywords HCR • Standard • Gearing • Industrial • Gearbox

1 Comparison of a Standard and HCR Gearing

In the Figs. 1 and 2, you can compare a standard and HCR gearing.

Advantages of HCR gearing with contact ratio coefficient $\varepsilon_\alpha \geq 2$

- reduction of gear load because they occupy two and more teeth
- bigger carrying capacity in comparison with standard gearing
- noise reduction due to larger coefficient of gear profile
- there are not spontaneous changes of stiffness as in the usage of $\varepsilon_\alpha < 2$

Disadvantages of HCR gearing

- application and production, special tools for gear production
- smaller possibility of corrections, larger amount of teeth
- higher values of slidings courses [1, 2]

2 Selection of HCR Gearing in the Industrial Gearbox MTC 42 A – 210

We had a discussion with TOS Znojmo company's designers if there should be a financial investment made for the HCR gearing production of the existing front industrial gearbox MTC 42 A – 210, see Fig. 3 [3].

M. Burián (✉) • M. Trochta • J. Havlik
VŠB – Technical University of Ostrava, Ostrava, Czech Republic
e-mail: miroslav.burian@vsb.cz; miroslav.trochta@vsb.cz; jiri.havlik@vsb.cz

V. Dynybyl et al. (eds.), *The Latest Methods of Construction Design*,
DOI 10.1007/978-3-319-22762-7_2

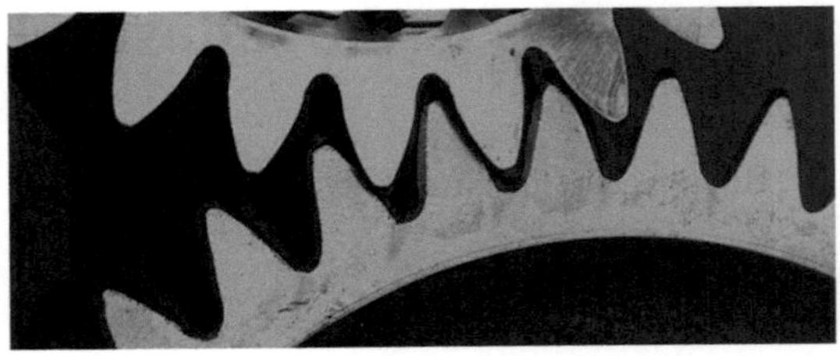

Fig. 1 Gearing with HCR profile

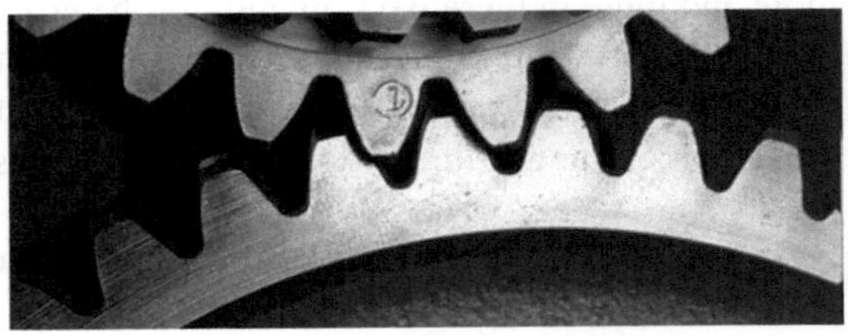

Fig. 2 Gearing with a standard gearing sprocket

Fig. 3 Front industrial gearbox MTC 42 A – 210

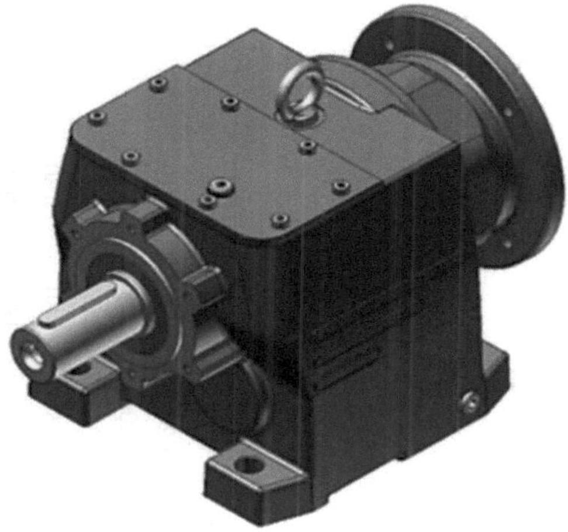

Fig. 4 Gearing production in TOS Znojmo

Table 1 Parameters of the front industrial gearbox MTC 42 A – 210

Number of teeth		Module [mm]	Angle of sprocket's inclination	Wheel's width [mm]		Cog wheel's material		Axial distance [mm]
z_1	z_2	$m_{N1.2}$	$\beta_{1.2}$	1	2	1	2	$a_{W1.2}$
23	91	1	20°	24	18	16 220	14 220	61

Company TOS Znojmo produces all cog wheels itself, see Fig. 4. Therefore if to invest funds to HCR gearing production.

Industrial gearbox is the most produced article in the company. There are no tools for non-standard HCR gearing production. The original size of gearbox, material for cog wheels, and the same gear ratio must remain. The original module, teeth number, and wheel's depth were kept.

3 Strength Calculation for Bend and Contact

I made a strength calculation of gearing for the appropriate bend and contact. Only first gearing is introduced here z_1, z_2. Values for calculation are in the Table 1.

It is impossible to determine which parameters are the best for the calculation while choosing parameters for HCR gearing. It depends on the concrete usage of industrial gearbox, peripheral speed, carrying capacity of wheels and technological possibilities of gearing production.

By the calculation of HCR gearing for gearing 1,2, I changed the relative addendum by both, the teeth of the pinion, and the cog wheel from value 1,02 to value 1,20. It is possible to change relative head clearance c_p^*, relative radius ρ_{fP}^* and mesh angle α. Alteration possibilities are in Table 2. First column is for relative

Table 2 Gear geometry of gearing

		Producer	Calculation of mesh angle α = 20°			Calculation of mesh angle α ≠ 20°			
		I.	II.	III.	IV.	V.	VI.	VII.	VIII.
Input value of the gearing 1,2	h^*_{a1}	1.00	1.05	1.10	1.15	1.20	1.10	1.10	1.20
	h^*_{a2}	1.00	1.05	1.10	1.15	1.20	1.10	1.10	1.20
	ρ^*_{fP1}	0.38	0.38	0.38	0.38	0.35	0.30	0.30	0.30
	ρ^*_{fP2}	0.38	0.38	0.38	0.38	0.35	0.30	0.30	0.30
	α	20°	20°	20°	20°	16°	18.5°	16.5°	16.5°
Calculated parameters of the gearing 1,2	ε_α	1.47	1.54	1.60	1.66	1.88	1.65	1.73	1.86
	ε_β	1.95	1.95	1.95	1.95	1.95	1.95	1.95	1.90
	ε_γ	3.43	3.49	3.56	3.62	3.84	3.61	3.69	3.82
	x_1	0.331	0.355	0.379	0.405	0.549	0.415	0.470	0.533
	x_2	0.017	−0.007	−0.031	−0.057	−0.198	−0.066	−0.119	−0.181

Table 3 Gear strength verification of gearings 1 and 2

		Producer	Calculated values						
		I.	II.	III.	IV.	V.	VI.	VII.	VIII.
Strength calculation of gearing $z_{1,2}$ *for bending*	S_{F1}	3.14	3.17	3.20	3.18	3.20	3.14	3.12	3.17
	$\sigma_{F\ MAX1}$	446	442	437	441	438	446	449	441
	$\sigma_{FP\ MAX1}$	1400	1400	1400	1400	1400	1400	1400	1400
	$\sigma_{F\ MAX2}$	502	481	477	473	512	492	511	502
Strength calculation of gearing $z_{1,2}$ *for contact*	S_{H1}	1.39	1.42	1.44	1.47	1.44	1.43	1.40	1.39
	$\sigma_{H\ MAX1}$	1.296	1.269	1.244	1.220	1.247	1.259	1.284	1.296
	S_{H2}	1.39	1.42	1.44	1.47	1.44	1.43	1.40	1.39
	$\sigma_{H\ MAX2}$	1 296	1 269	1 244	1 220	1 247	1 259	1 284	1 296

addendum tooth $h_a^* = 1$ of standard gearing profile – value of created industrial gearbox MTC 42A – 210. Calculation was done by using software "Geometrie".

By the calculation, I found out the timing coefficient of the gear profile ε_α, the timing coefficient gear profile ε_β and the total coefficient of time for the contact ratio $\varepsilon_\gamma = \varepsilon_\alpha + \varepsilon_\beta$. In Table 2 are the correction values of pinion x_1 and the wheel x_2 [4].

From the table, it is visible that the contact ratio of mesh angle $\alpha = 20^\circ$ has increased the gear profile ε_α in comparison with the original value in the first column (standard gearing). Duration of the mesh profile has increased $\varepsilon_\gamma = 3,62$. The highest value of the contact ratio $\varepsilon_\gamma = 3,82$ is for the last column VIII. In addition to the relative teethes addendum h_a^*, I also had to change the mesh angle α and relative radius of the gear heel ρ_{fP}^*. Requested axial distance $a_{w1,2} = 61$ mm showed in Table 1 must be kept. Calculated axial distance doesn't match the requested one. The difference is evened up by corrections of stabilization of the specific slips.

3.1 Simplified Strength Verification

I made a simplified strength verification of gearing based on ČSN 01 4586 – part 4 with the help of software CSNw. Results of strength verification are in Table 3 [5].

Indexes 1 are for the pinion and indexes 2 for the wheel. Particular quantities (parameters) are:

S_F – safety factor versus formation of fatigue failure in the tooth heel,
$\sigma_{F\ MAX}$ – bending stress of tooth heel in the dangerous section,
$\sigma_{FP\ MAX}$ – allowed stress during the bending = 1400 MPa,
S_{H1} – safety factor versus formation of fatigue failure on the side tooth,
$\sigma_{H\ MAX}$ – touch stress (Herz pressure) in the rolling point,
$\sigma_{HP\ MAX}$ – allowed stress during the contact = 2400 MPa.

Conditions for strength verification according to the formula (1), (2), (3) and (4):

$$\sigma_{FP\ \mathrm{MAX1}} \geq \sigma_{F\ \mathrm{MAX1}} \quad (1)$$

$$\sigma_{FP\ \mathrm{MAX2}} \geq \sigma_{F\ \mathrm{MAX2}} \quad (2)$$

$$\sigma_{HP\ \mathrm{MAX2}} \geq \sigma_{F\ \mathrm{MAX1}} \quad (3)$$

$$\sigma_{HP\ \mathrm{MAX2}} \geq \sigma_{F\ \mathrm{MAX2}} \quad (4)$$

Particular safeties, bending stress and stress in the contact haven't changed much. All conditions were fulfilled according to the formulas (1), (2), (3) and (4). Allowed stress is always higher than applied stress. There is no sense to use nonstandard HCR hearing for small modules. The Contact timing ratio of teeth profile hasn't changed much compared to original value of the standard gearing. Carrying capacity of cog wheels hasn't changed much in comparison with standard cog wheels of standard tooth profile. It is too expensive to use these teeth for this industrial gearbox. It is necessary to purchase a special tool for the production of the gearing, which the TOS company doesn't have available. Existing gearing is satisfactory.

4 Gearing from Modules 1 up to 9 mm HCR

Further, I devoted my time to HCR gearing usage during module changes. How big stress in the bending or contact will then influence (with the change of the module) imaginary axis distance? Apart from module and tooth number, the values are the same as in Table 1. I changed only relative addendum. The other parameters are the same as in the standard gearing.

4.1 Comparison of the Standard and HCR Gearing

In Fig. 5 is the comparison of bending stress in perilous (hazardous) place of pinion tooth of both standard and HCR gearing.

In the Fig. 5 is indicated the stress increase and decrease. And that at all the time by the module 2,75 mm.

Because during the calculation of bending stress in the perilous profile of the tooth heel in the pinion $\sigma_{F\ MAX1}$ and in the wheel $\sigma_{F\ MAX2}$, the coefficient slope sprocket is changing Y_β. Based on the Fig. 6 this coefficient is dependent on slope sprocket $\beta_{1,2} = 20°$ and on the contact ration of the profile ε_β.

Up to the module 2.5 mm, the contact coefficient ratio of the profile ε_β is bigger than "1". Therefore the coefficient of sprocket's angle Y_β is constant up to this value, see Fig. 6. From the module 2,75 mm, the contact ratio of the profile ε_β is smaller then "1". Therefore there is a change of the coefficient slope sprocket Y_β, which influences bending stress in the dangerous profile of tooth heel in the pinion $\sigma_{F\ \mathrm{MAX1}}$ and on the wheel $\sigma_{F\ \mathrm{MAX2}}$.

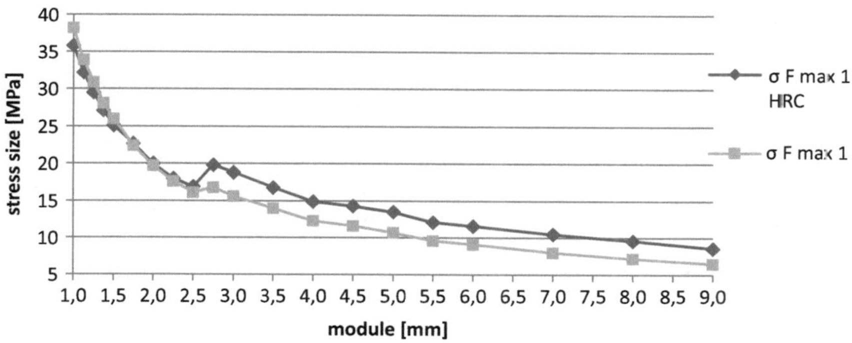

Fig. 5 Bending stress in the dangerous profile of tooth heel in the pinion

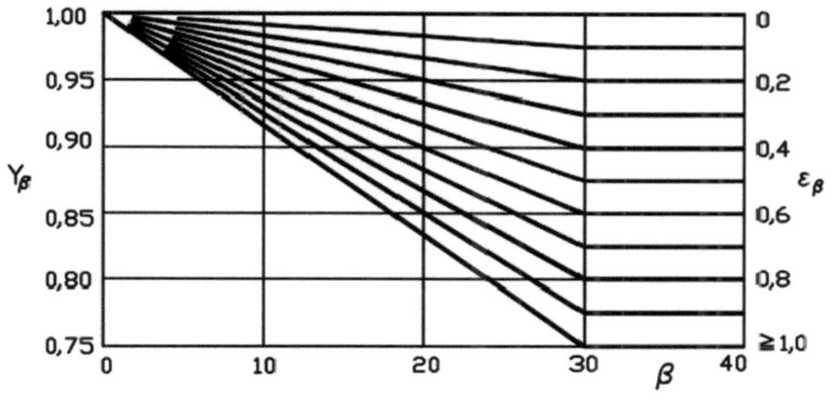

Fig. 6 Coefficient of sprocket's angle Y_β

5 Conclusion

Usage of HCR gearing is beneficial. Larger safety factor versus creation of fatigue failure in the tooth heel in the pinion and on the wheel. And to decrease of safety factor versus creation of fatigue failure of the side teeth on the wheel and in the pinion. With increasing module, there is a decrease of bending stress in the perilous profile of the tooth heel in the wheel and on the pinion. The disadvantage for the the gear production is the production of a non-standard tool. There is no sense to use non-standard HCR gearing for seller modules. It is possible that the company TOS Znojmo will not produce gearing with a small module. Depending on concrete usage, larger modules may be used.

Acknowledgments **Contribution** has been done in connection with project "Posouzení vlastností ozubených převodů z hlediska jejich geometrie a přesnosti výroby", REG. NO. SP2014/24

References

1. M. Němček, *Vybrané problémy geometrie čelních ozubených kol*. Stříbrná technická řada (Montanex, Ostrava, 2003), p. 143 s. ISBN 80-722-5111-2
2. V. Moravec, *Konstrukce strojů a zařízení II: čelní ozubená kola, teorie, výpočet, konstukce, výroba, kontrola* (Montanex, Ostrava, 2001), p. 291 s. ISBN 80-722-5051-5
3. Čelní převodovka MTC 42 A – 210, *Čelní převodovky* [online] (2014), Available via: http://www.tos-znojmo.cz/produkce/mtc/cz/index.htm. Cited 19 Mar 2014
4. GEOMETRIE. *Kontrola geometrie ozubených kol, verze 4.0* © M. NĚMČEK 2013
5. CSNw. *Pevnostní výpočet čelích ozubených kol, verze 3.0*. © M. NĚMČEK 2011

Design of the Vibration Separatory Apparatus

M. Dub, S. Sadek, J. Hendrych, and P. Spacek

Abstract Stabilization/solidification is used for liquid waste processing. For the purpose of liquid and solid phase separation the experimental vibration separatory apparatus is designed. It can be used with common cement mixer and easily transported by means of the car trailer. The device is equipped with vibration motor and it is designed as hand-operated.

Keywords Designing • Pasty material • Liquid waste • Vibration • Separation • Stabilization • Solidification

1 Introduction

Liquid waste is processed by means of stabilization/solidification. The waste is mixed with binders e.g. in mixing drum and homogenized batch is then exposed to the vibrations. Due to this process the separation of excess liquid from paste of hydrating binders is performed. Treated material doesn't release liquid phase no longer. Collected excess liquid is returned to the process of stabilization/solidification.

The aim is to design the experimental device for separation of liquid and solid phase out of the pasty material of density 1.5–2.5 kg.dm^{-3}. This device should be mobile, it means that it can be transported by car with common carriage. The assembly consists of mixing and the separatory apparatus. One batch of pasty material should be approximately 25 kg.

M. Dub (✉)
Faculty of Mechanical Engineering, Czech Technical University of Prague, Prague, Czech Republic
e-mail: martin.dub@fs.cvut.cz

S. Sadek
Czech Technical University in Prague, Prague, Czech Republic
e-mail: stepan.sadek@fs.cvut.cz

J. Hendrych
Institute of Chemical Technology, Prague, Czech Republic
e-mail: jiri.hendrych@vscht.cz

P. Spacek
CHEMCOMEX Praha, a.s., Prague, Czech Republic
e-mail: spacek@chemcomex.cz

V. Dynybyl et al. (eds.), *The Latest Methods of Construction Design*,
DOI 10.1007/978-3-319-22762-7_3

Fig. 1 The assembly of cement mixer and separation apparatus

Table 1 Parameters of vibration motor [1]

Frequency [Hz]	Revolutions [rpm]	Moment [kg.mm]	Centrifugal force [N]	Power [W]	Weight [kg]
50	3000	22	2200	120	6

2 Apparatus Design

The separatory apparatus is designed as tipping pan which is elastically imbedded into welded frame. Tipping pan is welded from stainless steel sheet because of corrosive properties of the pasty material. Due to the economical aspect common cement mixer will be used for mixing operation and that's why the separatory apparatus is designed with respect to the possibility of assemblage of both devices (Fig. 1).

For the efficient separation of liquid phase the vibration motor with elliptical oscillation is used. With respect to the weight of designed construction and the weight of the batch the corresponding power of the vibration motor is chosen. Parameters of chosen vibration motor are shown in the following table (Table 1). The vibration motor is connected to the stiffening frame of the pan.

The pan is imbedded into welded frame of separatory apparatus by means of pin and spring housing which is moreover bolted over rubber plate. At the end of one pin there is put on the operational arm for tipping of the pan. For the fixation of the pan in horizontal position the special turning holder is designed. This holder is equipped by buffer rubber springs to reduce the transmission of the vibration to the base frame. Above described parts of the separatory apparatus are shown in Fig. 2.

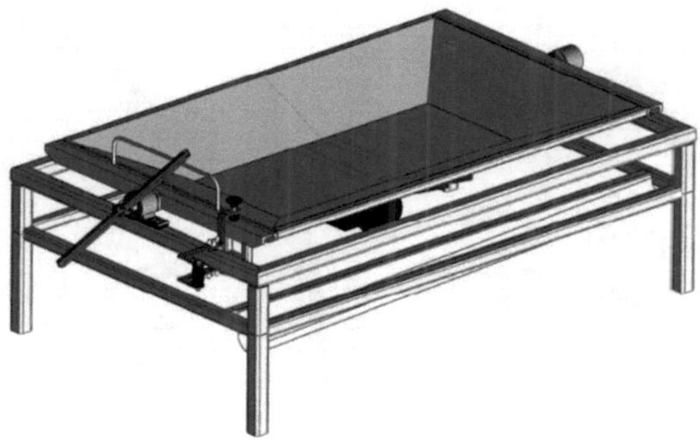

Fig. 2 The vibration separatory apparatus – 3D model

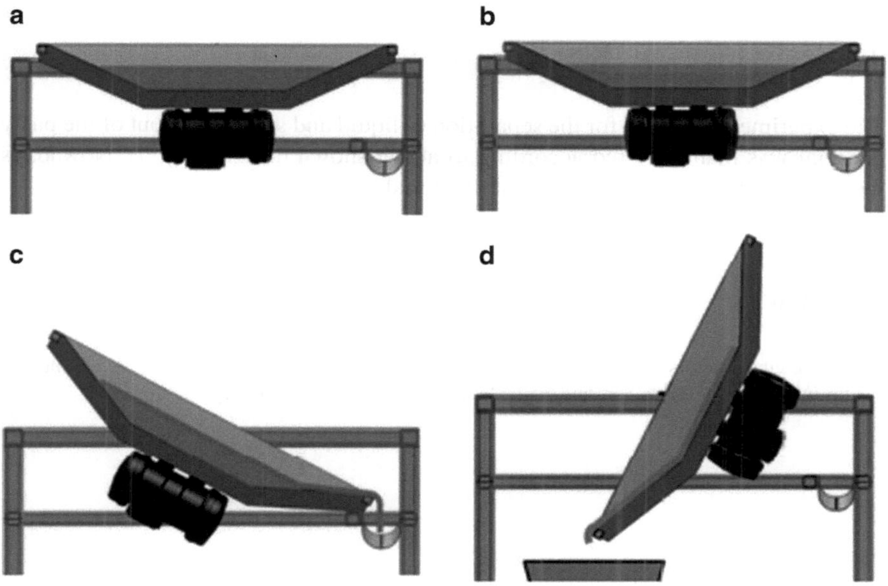

Fig. 3 The cycle of operation of separatory apparatus

3 Cycle of Operation

The cycle of operation of the separatory apparatus will be described in this chapter and it is shown in Fig. 3. Before filling the pan by batch, it has to be fixed in horizontal plane by means of turning holder.

Next the pasty material is poured from the mixer to the pan (Fig. 3a) and the vibration motor is turned on. During the vibration process the liquid is separated above the paste (Fig. 3b). The pan can be tipped to one side and the excess liquid is strained into outlet gutter (Fig. 3c). In last phase the pan is tipped to the second side and the paste without excess liquid is moved out of the pan (Fig. 3d).

Fig. 4 The vibration separatory apparatus – manufactured device

4 Realization and Results

The experimental device for the separation of liquid and solid phase out of the pasty material was manufactured according to above shown design (Fig. 4). Nowadays the practical tests with different pasty materials are carried out.

References

1. Electric vibro-motors – VV. *Webac Vibro s.r.o.* [online]. 2010 [vid. 2014-06-03]. Available via: http://www.webac-vibro.com/index.php?menu=elektricke-vibromotory-vv

Designing of Mixing Equipment, Reactors and Bioreactors

T. Jirout and D. Jiroutova

Abstract The present work deals with general description of the most common processes in the agitated batch (blending and particle suspension) and with hydrodynamic parameters and geometrical configuration of the mixing equipment (shape of vessel, baffle and impeller and their mutual arrangement) that influence the process. The optimization of the construction and arrangement of the mixing equipment were performed to ensure the specified process with minimum energy consumption.

Keywords Axial impeller • Hydrofoil impeller • Multi-stage impeller • Blending • Particle suspension

1 Introduction

A mixing of liquid media is the one of the most common operations in the chemical, biochemical, food and consumer industry. The intensification of heat and mass transfer and preparation of mixtures with required properties, e.g. suspensions and emulsions, are mixing purpose. The mixing equipment with high-speed rotary impeller, which creates forced flow in the mixing vessel, is the most frequently encountered in industry.

The homogenization of the agitated batch and ensuring the particle suspension are the most frequently encountered requirements in terms of the mixing purpose. These operations are affected by flow in the agitated batch. The primary liquid flow influencing mainly suspension of the solid phase is caused by the pumping effect of the mechanical impeller [1]. Due to the momentum transfer between the liquid flowing from the rotor area of the impeller and surrounding environment arises in the remaining volume the induce flow that affects the homogeneity of the agitated

T. Jirout (✉)
Faculty of Mechanical Engineering, Department of Process Engineering, Czech Technical University in Prague, Prague, Czech Republic
e-mail: Tomas.Jirout@fs.cvut.cz

D. Jiroutova
Klokner Institute, Czech Technical University in Prague, Prague, Czech Republic
e-mail: Dita.Jiroutova@klok.cvut.cz

V. Dynybyl et al. (eds.), *The Latest Methods of Construction Design*,
DOI 10.1007/978-3-319-22762-7_4

batch. The geometrical parameters of the mixing system, especially the shape of the impeller blade affect character of the flow in the agitated batch.

For the designing of mixing equipment, it is necessary to determine the impeller speed needed to ensure the required process taking place in agitated batch and power consumption of impeller needed for drive and gear box dimensioning. These design parameters are dependent not only on the geometrical configuration of the mixing equipment (vessel shape, baffles, type of impeller and their mutual geometric arrangement) but also on the physic-chemical properties of the agitated batch.

2 Flow in Agitated Batch

2.1 Character of the Flow in the Agitated Batch

Flow and circulation of the agitated liquid in the vessel is the primary effect of the mechanical impeller. The geometrical parameters of the high-speed rotary impellers affect axial, radial or tangential character of the flow in the agitated batch – see Fig. 1.

Mixing equipment with cylindrical vessel and centrally placed high-speed rotary impeller, which creates a forced flow in the vessel, are the most frequently encountered in the industry. In case it is necessary to suppress a tangential flow in the agitated batch and prevent the creation of axial vortex, the cylindrical vessels are

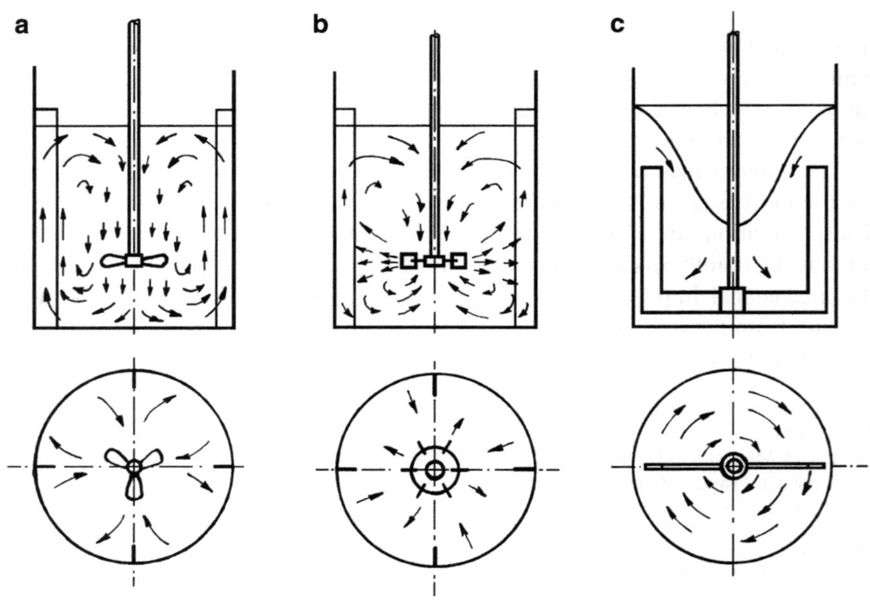

Fig. 1 Flow in agitated batch with rotary impellers [2]: (**a**) – axial-flow pattern (baffled vessel), (**b**) – radial-flow pattern (baffled vessel), (**c**) – tangential-flow pattern (unbaffled vessel)

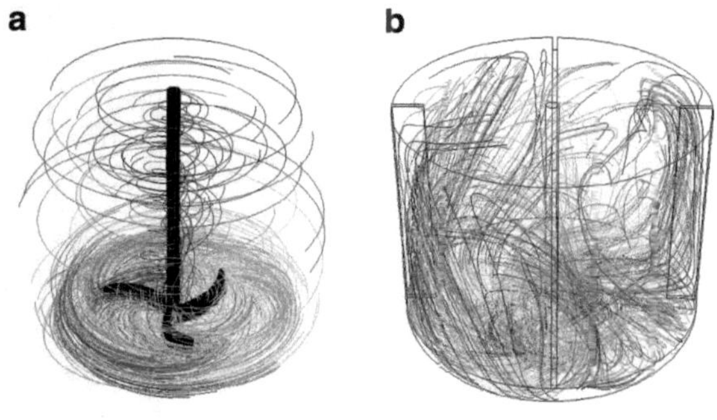

Fig. 2 Flow pattern in the cylindrical unbaffled vessel (**a**) and with two radial baffles (**b**) equipped with curved blade impeller

Fig. 3 Flow pattern in the square vessel equipped with axial impeller

equipped by differently shaped baffles (radial, tubular, special . . .). This adjustment of mixing equipment strengthens the axial flow and intensity of circulation in the agitated batch. This results in a significant improvement of the homogenization of miscible liquids, and heterogeneous mixtures in the agitated batch. Flow in the cylindrical vessel without baffles and with two radial baffles equipped with pitched blade radial impeller is shown in Fig. 2. It is not possible used vessels equipped with baffles in many applications, e.g. due to cleaning or surface finish (e.g. enamelled apparatuses). The axial flow can be enhanced by suitable impeller configuration in the case of viscous flow of agitated high viscosity liquid. Tangential flow in vessel without baffles under turbulent regime of flow can be partially suppressed by an eccentrically placed impeller in a vessel or using of impeller with special two-stage shaped blades. Square or rectangular vessel are sometimes used due to building plan or space saving. Sharp corners of vessel disrupt tangential flow and practically perform the function of baffles (Fig. 3).

Primary flow of the liquid Q_p in the agitated batch is caused by pumping effect of the high-speed impeller. Induced flow Q_i arises in the remaining volume of agitated batch due to momentum transfer between the liquid flowing from the rotor area of impeller and the surrounding environment. Primary and induced flow is shown in Fig. 4 in the velocity field of agitated batch with axial impeller.

Fig. 4 Flow in agitated batch with axial impeller

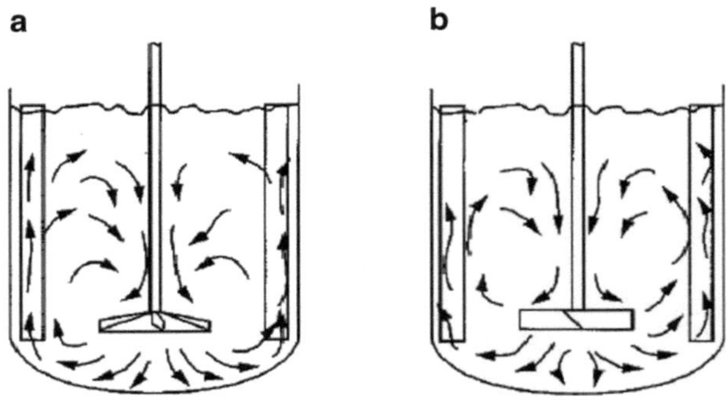

Fig. 5 Flow pattern in agitated batch equipped with axial hydrofoil impeller (**a**) and standard pitched blade impeller (**b**) [5]

2.2 *Hydrofoil Impellers*

The geometrical parameters of the mixing system and the shape of the impeller blade have essential effect on the velocity distribution in the agitated batch and thus affect processes of mixing [3, 4]. Currently the leading manufacturers of mixing equipment (e.g. Chemineer, Lightnin, Ekato and Czech company Techmix) uses its own developed new type of flow-optimizing impeller called *hydrofoil impellers*. The aim is design impeller with shaped blades that makes the most streamline of flow in agitated batch at achieving the highest hydraulic efficiency, i.e. the highest utilization of energy input to impeller in the form of its power consumption to initiate pumping efficiency and turbulences outside the rotor area of impeller. Comparison of flow in vessel equipped with an axial hydrofoil impeller and standard pitched blade axial impeller is shown in Fig. 5.

Design modifications of blade shape are consist in the adjustment of blade area and their bending or folding. Examples of development types of axial hydrofoil impellers are shown in Fig. 6. Basic no optimized type of impeller with pitched

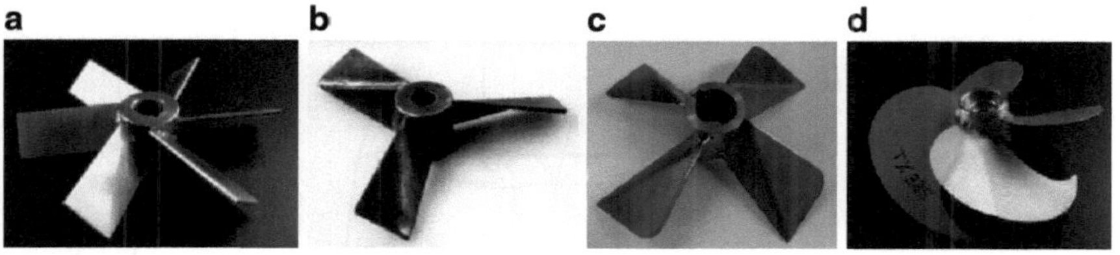

Fig. 6 Design modifications of axial impellers – from standard pitched blade to hydrofoil impeller

rectangular blades is shown in Fig. 6a. The diagonal folding of the blades is the simplest method of blade shaping (Fig. 6b). Folding of the blades results in uniform axial outflow of liquid from the impeller with almost piston velocity profile. Increasing of the blade area and adjustment of its shape its next step of design modifications. This adjustment will not only directing the flow, but also improve the process characteristics. Hydrofoil impeller proposed by author is shown in Fig. 6c [6] and hydrofoil impeller proposed by Techmix s.r.o. is shown in Fig. 6d. Shaping of blades is carried out either on the basis of working experiences or based on experimental results of process taking place in agitated batch (e.g. blending, particle suspension and dispersion) and velocity field in agitated batch (e.g. using LDA, PIV) or using CFD simulations.

The shape of the velocity profile at the outflow of the hydrofoil impeller, the streamlined flow of the batch within cylindrical vessel equipped by radial baffles and their high hydrodynamic efficiency are main qualities that predestine these impellers for enhancement homogenization [7–9] and particle suspension processes especially with regard to minimizing energy requirements. In case of the suspensions mixing in the vessel equipped by baffles where the density of particles is higher than liquid, these impellers should be placed above bottom of the vessel and pumping should be downwards the bottom [9–11]. Conversely, if the particles of the suspension are lighter than liquid, these impellers should be placed near liquid level and pumping should be upwards the level [12].

2.3 Multi-stage Impellers

To ensure homogeneity of the suspension mainly in slender vessel or for mixing of Non-Newtonian batch it is necessary to extend an intensive flow to the whole mixing batch, which can be achieved using *multi-stage axial impellers*. Generally, multi-stage impellers are commonly installed in slender vessels where individual impellers create independent circulation zones in the agitated batch. However, it is often necessary to ensure circulation in the whole mixing batch. It could be provided by the installation of multi stage impellers into standard agitated vessels with height of the liquid level approximately equal to the diameter of vessel.

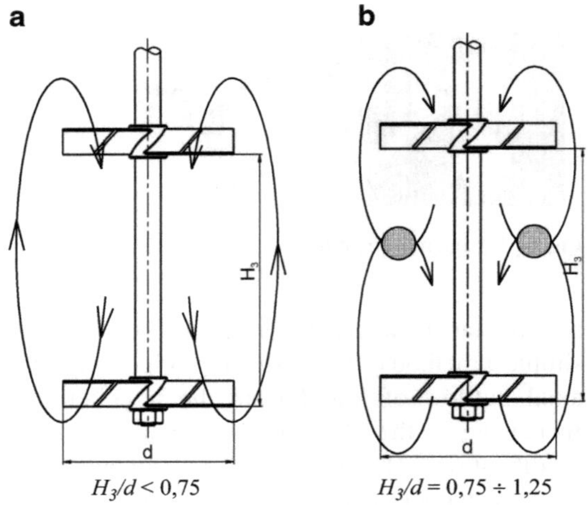

Fig. 7 Flow in vessel equipped with axial multi-stage impeller [13]. (**a**) $H_3/d < 0.75$, (**b**) $H_3/d = 0.75 \div 1.25$

The distance between impellers in a multi-stage arrangement has a strong effect on flow patterns in the mixing batch. In case of a relatively close distance between impellers i.e. at $H_3/d = 0.5 \div 0.75$, the suspension pushed out by the upper impeller is sucked by the impeller at the bottom. The multi-stage impeller creates only one circulation loop in the agitated batch (see Fig. 7a). However, for their distance higher than $H_3/d = 1.25$, the lower and upper impellers work independently that means two separate circulation loops are formed (see Fig. 7b).

The multi-stage impellers installed in the cylindrical vessel equipped by baffles are the most suitable impellers to ensure the required flow and circulation in the entire volume of the stirred batch during mixing of the concentrated, fain-grained [14] or heterogeneous suspensions [15] considering the required degree of the suspension homogeneity and energy demands of the process.

3 Special Aspects of Mixing Equipment Design

It is important during the mixing equipment design to choose mixing equipment configuration not only to minimize operating expenses, but it is necessary take into consideration choice of shape and material of functional equipment parts (impeller, shaft and vessel) with respect to their strength characteristics, corrosion resistance and mechanical wear. Mixing equipment are often exposed to aggressive environments in the chemical, food and pharmaceutical industry. It is necessary to protect mixing equipment against corrosion and wear by using of suitable surface finish, e.g. enamelling, in the aggressive environments. This equipment needs geometry modification of equipment functional parts (need to remove all sharp transitions and edges by using gradual rounding-off in case of enamelling – Fig. 8) due to machine production. These interventions to the equipment design significantly affect flow

Fig. 8 Enamelled multi-stage impeller manufactured by Tenez a.s.

character and flow intensity in the agitated batch and thus also process in it. The rounded impellers have 2 ÷ 3× higher energy requirements in comparison to no rounded impellers during blending and particle suspension. This fact must be allowed for design, e.g. enamelled mixing equipment. It is not possible used process characteristics of impellers without rounded edges to calculate the frequency and power input of enamelled impellers with rounded edges.

Environments affect the mixing equipment without adjustment against corrosion and wear and caused shaped wear or areal wear of blades. Shape wear of impeller blade results in changes of geometrical shape of the impeller. This affect decreasing of pumping efficiency and thus to change other process characteristics [16] and it can lead to deterioration of product quality. The areal wear of impeller blade results in a reduction of the blade thickness and thus their strength. It can also lead to destruction of equipment in extreme case. And this results in shut down of the mixing equipment and after that additional cost of maintenance and impeller replacement.

References

1. I. Fořt, Flow and turbulence in vessels with axial impellers, in *Chapter 14: Mixing, Theory and Practice*, ed. by V.W. Uhl, J.B. Gerry, vol. III (Academic Press, New York, 1986)
2. F. Rieger, V. Novák, T. Jirout, in *Hydromechanické procesy* II, 1 (vyd. Praha: Vydavatelství ČVUT, 2005), p. 167 (in Czech)
3. T. Jirout, *Míchání suspenzí*, Doktorská práce (Ph.D.), Praha: ČVUT v Praze, Fakulta strojní, 2005, p. 113(in Czech)
4. T. Jirout, F. Rieger, S. Jembere, Čerpací účinky axiálních míchadel, in *Zborník príspevkov 6. Medzinárodnej vedeckej konferencie Strojné inžinierstvo* (Vydavateľstvo STU, Bratislava, 2002), pp. 1–9. in Czech
5. E.L. Paul, V.A. Atiemo-Obeng, S.M. Kresta, *Handbook of Industrial Mixing: Science and Practice* (Wiley-Interscience, Canada, 2003). 1377 p
6. T. Jirout, F. Rieger, *Axiální míchadlo s velkoplošnými lomenými lopatkami*. Užitný vzor Úřad průmyslového vlastnictví, 23644, 5 Apr 2012 (Czech patent)
7. I. Fořt, T. Jirout, F. Rieger, R. Allner, R. Sperling, Study of the blending efficiency of pitched blade impellers. Acta Polytech. **41**(6), 7–13 (2001)
8. I. Fořt, T. Jirout, A study on blending characteristics of axial flow impellers. Chem. Process Eng. – Inżynieria Chemiczna i Procesowa **32**(4), 311–319 (2011)
9. F. Rieger, T. Jirout, C. Kuncewicz, D. Ceres, Homogenization and suspension production with a four-blade impeller of a new construction. Przem. Chem. **92**(4), 508–511 (2013) (in Polish)
10. T. Jirout, F. Rieger, Impeller design for mixing of suspensions. Chem. Eng. Res. Des. **89**(7A), 1144–1151 (2011)
11. F. Rieger, T. Jirout, D. Ceres, P. Seichter, Effect of impeller shape on solid particle suspension. Chem. Process Eng. – Inżynieria Chemiczna i Procesowa **34**(1), 139–152 (2013)
12. O. Svačina, T. Jirout, The effect of the stirred vessel geometry arrangement on the drawdown of floating particles, in *Procesní a zpracovatelská technika 2012 [CD-ROM]* (České vysoké učení technické v Praze/Fakulta strojní, Praha, 2012), pp. 1–16
13. T. Jirout, Pumping capacity of pitched blade multi-stage impellers. Chem. Process Eng. – Inżynieria Chemiczna i Procesowa **35**(1), 47–53 (2014)
14. J. Moravec, T. Jirout, F. Rieger, L. Krátký, Mixing system for highly concentrated fine-grained suspensions. Pol. J. Chem. Technol. **11**(4), 52–56 (2009)
15. T. Jirout, F. Rieger, *Zařízení pro míchání heterogenních suspenzí*. Užitný vzor Úřad průmyslového vlastnictví, 26644, 17 Mar 2014 (Czech patent)
16. I. Fořt, T. Jirout, The relation between the rate of erosion wear of pitched blade impeller and its process characteristics. Chem. Eng. Res. Des. **89**(10), 1929–1937 (2011)

Modern Trends in the Drive Wheelsets of Rail Vehicles

J. Kolar

Abstract The paper describes trends in individual wheelset drive and drives of independently rotating wheels of modern low-floor trams and rail vehicles. It documents the way of constructive design of individual wheelset drive (without gear box) and wheelset drive with gear boxes.

Keywords Rail vehicle • Modern AC drives • Design of mechanical parts of traction drive • Low-floor trams • Train electric units

1 Introduction

From 90 years of the twentieth century, the drive wheelset of rail vehicles uses modern AC drives. The basic powertrain consists of an asynchronous motor (ASM) or a synchronous motor with permanent magnets (PSMS). The driving torque of the electric motor is transmitted across the components (mechanical parts) of drive on the wheelset or on drive wheels portal axles of low-floor trams [1, 2].

2 Design of Traction Drives for Low-Floor Trams

Use of a low floor in throughout the length of the passenger compartment in low-floor trams required to locate drive of wheelset or drive of independently rotating wheels on the outside of the tram-wheels. The design (dimensions) of mechanical parts of independently rotating wheel drive (IRW) or wheelset drive (CW) is dependents on the cross-section which is in bogie for installation of the drive available. The lower and lateral parts cross-section is limited by kinematic gauge for bogie vehicle. The upper parts are limited by height of the floor above bogie. In the 25 years of development of low-floor trams a large number of design variants of the mechanical drive low-floor trams were developed. Some remained at

J. Kolar (✉)
Czech Technical University in Prague, Prague, Czech Republic
e-mail: josef.kolar@fs.cvut.cz

V. Dynybyl et al. (eds.), *The Latest Methods of Construction Design*,
DOI 10.1007/978-3-319-22762-7_5

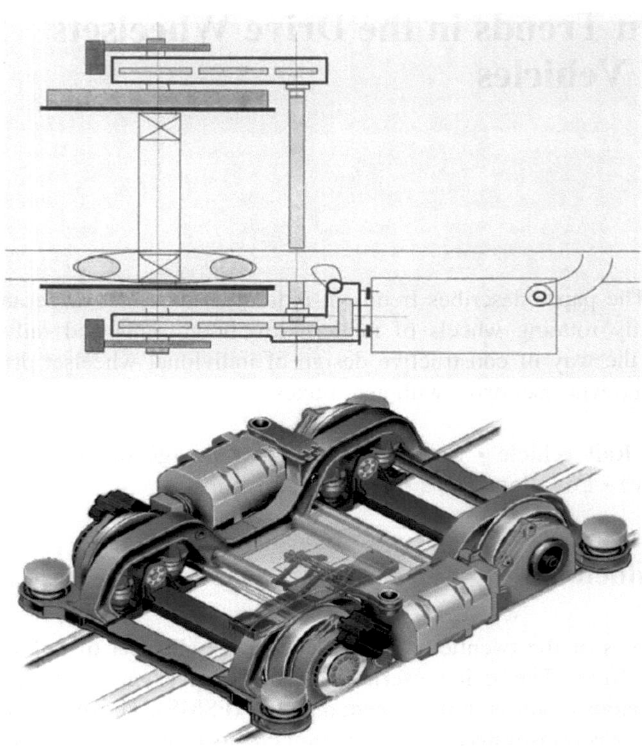

Fig. 1 Concept of traction drive in not pivoting bogie low–floor trams Sirio 7C4 [*Ansaldo Breda*]

the prototype stage, others have been successfully implemented. The successful designs can be divided into four basic groups.

2.1 Drives with Transverse Mechanical Coupling of Independently Rotating Tram Wheels

This drive concept has been motivated by a desire to maintain (during driving vehicle) sinusoidal movement of the wheelset. The concept of traction drive, Fig. 1, has this structural design. Longitudinally oriented traction motor (ASM with power of about 100 kW) is fixed to the bogie frame and the torque is transmitted through shaft coupling to the transmission gearbox. That consists of bevel gears and self-locking differential. Transmission gearbox is positioned in front of one of the axle gearboxes. Output shaft of the transmission gear box drives opposite axle gear box. Three wheeled axle gear boxes represent concept to the partly sprung traction drive. This drive concept is quite complex, with efficiency of mechanical part of traction drive of about 85 %.

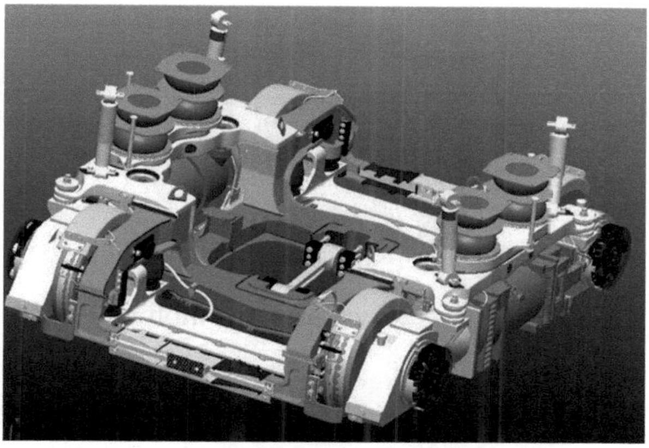

Fig. 2 Concept of traction drive in partly pivoting bogie SF 30 TF low–floor trams Avenio [*Siemens*]

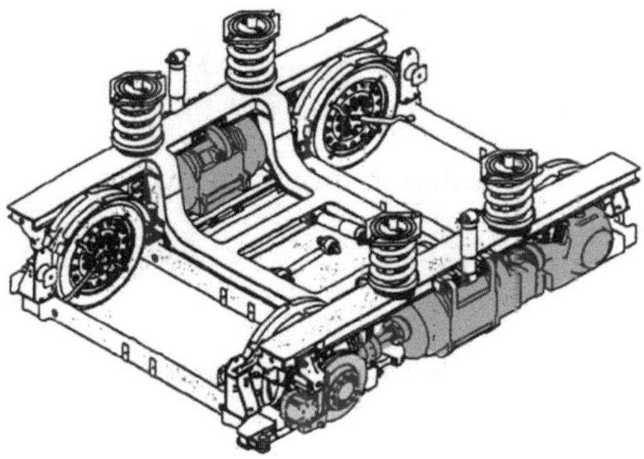

Fig. 3 Concept of traction drive in not pivoting bogie Corege low–floor trams Citadis 302 B [*Alstom*]

2.2 *Drives with Longitudinal Mechanical Coupling of Independently Rotating Tram Wheels*

This concept does not guarantee sinusoidal movement of the wheelset. The design of these traction drives is successfully resolved in two basic versions. The first variant, Fig. 2, represents a solution of fully sprung traction drive, wherein the longitudinally oriented drive motor (ASM or PMSM with power of about 110 kW), together with bevel or hypoid gear boxes constitutes an integrated unit, which is attached to the frame bogie. The hollow output shaft of the hypoid gear box is connected to the PTO driving shaft of the tram wheel. This concept has minimal unsprung mass of traction drive and efficiency of mechanical part drive is ca. 95 %.

A second variant drive is shown in Fig. 3 and represents concept to the partly sprung traction drive. Wherein is the longitudinally oriented traction motor which is mounted on the frame bogie, via two articulated joint shafts driven axle-hypoid gear box. Their output shafts drive the front and rear tram wheels on one side of a two-axle bogie. The efficiency of mechanical part of traction drive is ca. 94 %.

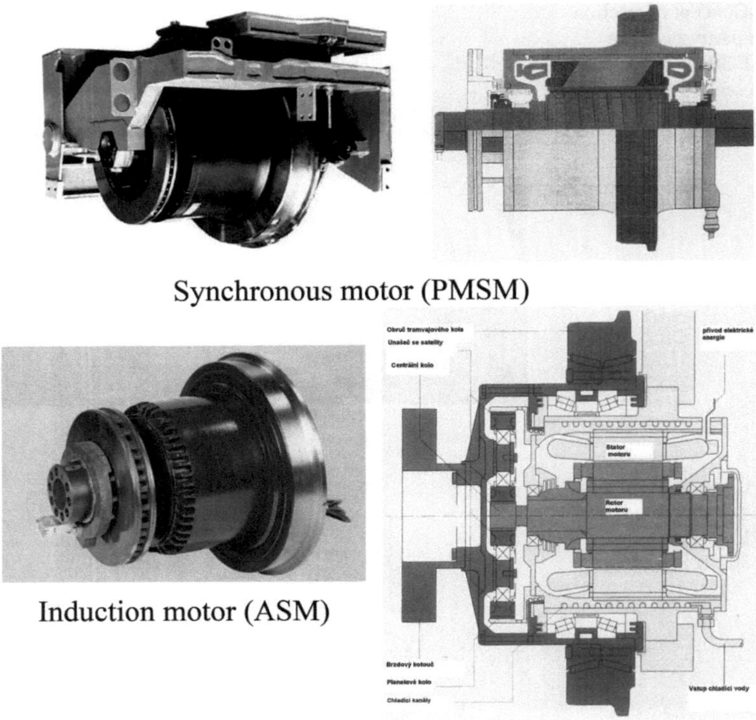

Fig. 4 Concept of traction wheel motors

2.3 *Drives Without Mechanical Coupling of Independently Rotating Tram Wheels*

This concept does not guarantee sinusoidal movement of the wheelset. This group of traction drives is represented by two concepts, see Fig. 4. The first option is a completely unsprung traction drive, represented by a wheel-traction motor with power of ca. 45 kW. In application of synchronous permanent magnet motor (PMSM) is implemented direct drive. The efficiency of mechanical part of traction drive is ca. 99 %. In application of induction motor (ASM) is used planetary gear. The efficiency of mechanical part transmission is ca. 98 %.

The modern version of a fully sprung direct traction drive is application of the synchronous motor with hollow shaft of the rotor, which is mounted on the bogie frame. In the hollow rotor shaft is mounted PTO shaft driving a tram wheel. This type of drive is used on tram Skoda 15 T For City, see Fig. 5. The efficiency of mechanical part of traction drive is ca. 99 %.

The second conceptual solution of traction drive is represented by integrated driving unit, consisting of the traction motor and wheel gearbox with hollow output

Fig. 5 Concept of drive wheel of tram Skoda 15 T

Fig. 6 Concept of integrated driving unit wheel

shafts. The output drive torque is transmitted through coupling type "Alstom" on tram wheel. The efficiency of mechanical part of traction drive is ca. 95 % (for three helical gears) or 97 % (for two helical gears). Example of this design, completely sprung drive, shown in Fig. 6. On the output shaft of gear box is mounted brake disc. Brake unit, which is not shown in Fig. 6, is mounted on the gear box.

2.4 *Drives of the Tram Wheelset with Smaller Diameter Wheel ($d_K \leq 570$ mm)*

The development engineers of low-floor trams agreed on compromise of height floor above the traction bogie to value of about 450 mm in recent years. This allows the use of a wheelset with smaller diameter wheels, which exhibit sinusoidal motion during drive. In bogie FLEXX Urban 1000 is wheelset driven by integrated drive

Fig. 7 Concept of traction drive in not pivoting bogie Flexx Urban 1000 [*Bombardier*]

Fig. 8 Concept of traction drive in partly pivoting bogie Flexx Urban 3000 [*Bombardier*]

unit with power ca. 120 kW, consisting of the traction motor and two-section gear box (with one spur gearing and one bevel gearing) or hypoid gear. From the hollow output shaft the drive torque is transmitted through special joint shaft on wheelset. The disc brake is positioned on the opposite end of the wheelset, Fig. 7. The efficiency of mechanical part of traction drive is ca. 94 %.

The FLEXX Urban 3000 bogie is used for FLEXITY 2 trams. The axle gear box (two-section) with vertical sprung support creates partly sprung mass of traction drive, see Fig. 8. The efficiency of mechanical part of traction drive is ca. 94 %.

The interesting design of the drive axle in the Ixege bogie it documented in Fig. 9.

Two traction motors (PSMS) are mounted on the bogie frame. The drive torque is transmitted by the short joint shaft (coupling) on pinion bevel gear. Bevel axle gear box is mounted on the wheelset and thus is an unsprung mass traction drive. Axle gear box is held in position by horizontal sprung support. The efficiency of mechanical part of traction drive is ca. 95 %.

For low-floor trams are dominated by the concept of traction drives with longitudinally oriented traction motors. These designs have better efficiency of mechanical part traction drive, smaller number of traction motors and simpler procedure to install larger vehicle traction power. With respect to the dynamic effects and the noise, it is preferable to apply fully sprung drives.

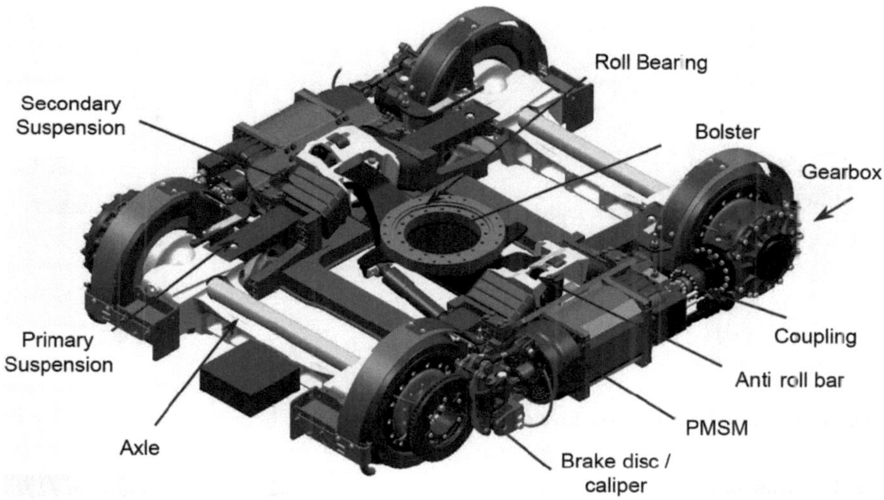

Fig. 9 Concept of traction drive in pivoting Ixege bogie [*Alstom*]

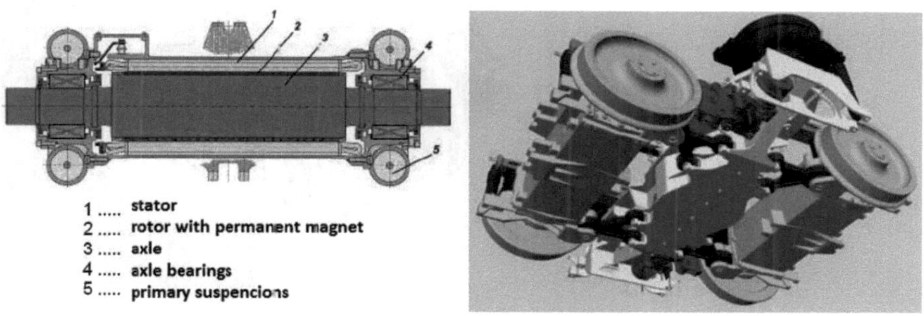

Fig. 10 Concept of traction drive in bogie Syntegra [*Siemens*]

3 Design of Traction Drives for Electric Train Unit

Electric train units use to drive wheelset traction drive with transverse axis traction motor. Design of this drive wheelset can be solved in three variants.

3.1 Direct Traction Drive of Wheelset

The first example direct of drive synchronous motor (PMSM) is represented by unsprung traction drive in bogie Syntegra for Metro unit, see Fig. 10. The efficiency of mechanical part of traction drive is 100 %, theoretically.

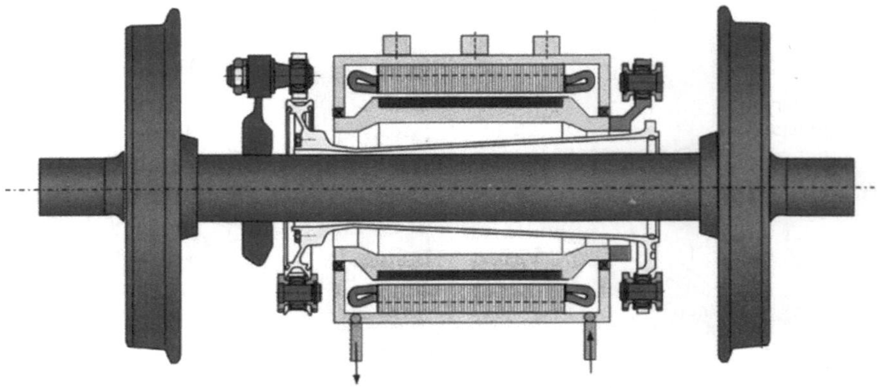

Fig. 11 Study of direct traction drive of wheelset [*Siemens*]

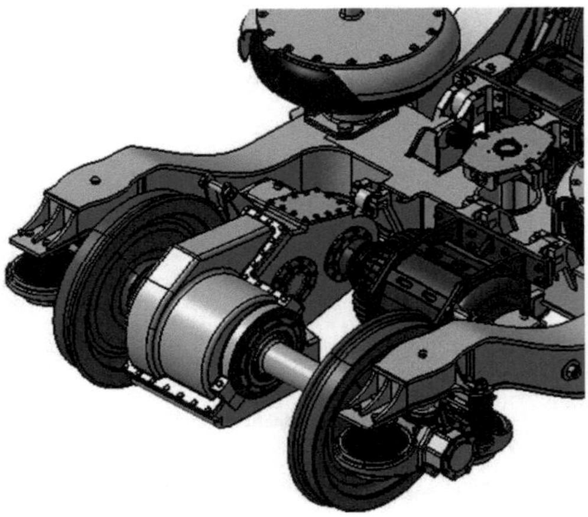

Fig. 12 Detail of axle gear box of traction drive AGV

The second example of direct drive is represented by synchronous motor with the hollow shaft rotor and with special hollow joint shaft encircles axle wheelset, see Fig. 11. The efficiency of mechanical part of traction drive is ca. 99 %.

3.2 Traction Drive of Wheelset with Axle Gear

High-speed rail units use partly sprung traction drive. Traction motor (ASM or PMSM) is sprung affixed to the bogie frame and axle gearboxes, (two-wheel or three-wheel), is affixed on the wheelset. The reaction moment of axle gearbox is absorbed via a sprung support (vertical, diagonal or horizontal) is affixed to the bogie frame [3, 4]. Between driving shaft of traction motor and pinion axle gear is inserted flexible claw coupling or short joint shaft, see Fig. 12. The efficiency of mechanical part of traction drive ca. 97 % or 95 %.

3.3 Full Sprung Traction Drive of Wheelset

Regional electrical units often use to drive wheelset full sprung traction drive, by design as integrated traction unit (motor with gearbox with constant transmission ratio), which is affixed to the bogie frame. Gear box is ordinarily designed with a lay shaft and with hollow output shaft. Driving moment is transmitted by this hollow joint shaft in wheelset. This concept has a minimal unsprung mass, and efficiency of mechanical part of traction drive is ca. 95 %.

Acknowledgments The paper was created with the financial support of the Technology Agency of the Czech Republic, project No TE01020038 "Competence Centre of Railway Vehicles".

References

1. J. Kolář, Koncepce nízkopodlažních tramvají a jejich pohonů. Nová železniční technika **15**(4), 21–33 (2007). ISSN 1210–3942
2. J. Kolář, J. Čapek, Optimal design of low-floor tram, in *From Horse-Drawn Railway to High-Speed Transportation Systems* (CTU in Prague, Praha, 2007), pp. 27–30. ISBN 978-80-01-03699-0
3. J. Kolář, Vehicles for high-speed transport systems, in *From Horse-Drawn Railway to High-Speed Transportation Systems* (CTU in Prague, Praha, 2007), pp. 113–116. ISBN 978-80-01-03699-0
4. J. Kolář, Dynamics of wheelset with axle gearbox, in *Zbornik prednášok II.-XX. Medzinárodná konferencia-Súčasné problémy v kolajových vozidlách*, vol. 2 (Vedecko technická spoločnost pri Žilinskej univerzitě, Žilina, 2011), pp. 139–148. ISBN 978-80-89276-31-8

Scale-Up of Equipment for Thermal-Expansionary Pretreatment of Lignocelluloses

L. Kratky

Abstract Thermal-expansionary treatment is a novel environmentally friendly technology that has a very high potential to be installed in various industrial biofuel technologies as a preliminary step of biomass processing. Several industrial batch apparatuses have been worldwide installed yet but there is still problem how to do it in continuous mode. Suitable process parameters (process temperature, residence time, biomass concentration) are able to be easily determined in laboratory scale. Based on these process data, this paper is dedicated to develop the methodology, how to transfer suitable process parameters to industrial scale, to present industrial basic equipment design and to solve the problematic of continual batch decompression into storage vessel.

Keywords Biofuel • Lignocelluloses • Scale-up • Thermal-expansionary treatment • Wheat straw

1 Introduction

Currently operating biofuel plants are based on the treatment of lignocellulosic biomass, which is included in such materials as agriculture and forestry wastes, municipal solid waste, waste paper, wood and herbaceous energy crops. However, the inherent properties caused by composite structure make them resistant against enzymatic attack. Hence, pretreatment of biomass is an essential step to increase cellulose, and hemicellulose accessibility, and also hydrolysis effectivity.

Biofuel plants are currently endeavoured to implement a new physical pre-treatment technology called thermal-expansionary pretreatment (TEP). Scheme of the laboratory batch TEP unit is shown in Fig. 1A and its principle is clearly demonstrated in Fig. 1B. Biomass is firstly stuffed into a hydrolyser *(1)*,

L. Kratky (✉)
Czech Technical University in Prague, Prague, Czech Republic
e-mail: Lukas.Kratky@fs.cvut.cz

V. Dynybyl et al. (eds.), *The Latest Methods of Construction Design*,
DOI 10.1007/978-3-319-22762-7_6

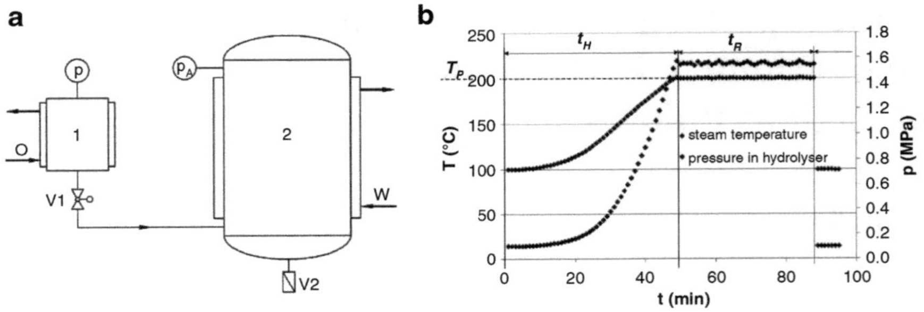

Fig. 1 The lab thermal-expansionary pretreatment. (**a**) Lab TEP in batch mode. (**b**) Temperature and pressure history. *1* hydrolyser; *2* storage vessel; *V1* ball valve; *V2* dump valve; *O* oil heating; *W* cooling

heated and boiled in hot water that is kept by pressure in liquid state, under a predetermined process temperature T_P and a predetermined process time t_R. Biomass boiling ensures water penetration into its porous structure and dissolution of cell walls. Both these changes primarily cause decrease in a strength yield of cell walls. Secondly, when a process time expires, ball valve *(V1)* is rapidly opened and a batch is immediately decompressed into storage vessel *(2)* in a few seconds, where atmospheric pressure is maintained inside. Liquid water changes phase to vapour because of rapid expansion and associated volumetric change causes intensive disruption of substrate and cell walls. Krátký [1] observed that thermal-expansionary pretreatment is able to increase methane production e.g. of wheat straw up to 50 % towards to untreated. Commonly used industrial pretreatment technologies (milling, dilute acid, alkaline) are able to increase biogas yield only by 5–25 % [2]. Moreover, the major advantages of TEP are no addition of chemicals, no sludge is generated, no formation of inhibitors and low-cost reactor construction because of low-corrosion potential. On the other hand, the major disadvantage is high heat demand caused by high process temperature.

Based on these facts, TEP has a very high potential to be installed in various industrial biofuel technologies as a preliminary step of biomass processing. Suitable process parameters (process temperature, residence time, biomass concentration) are able to be easily determined in laboratory scale [3]. Nevertheless, there is no information how to design it in industrial scale and in continuous mode especially. Therefore, this paper is dedicated to develop a methodology, how to transfer suitable process parameters to industrial scale, to present industrial basic equipment design and to solve the problematic of batch decompression into storage vessel, all in continual mode of operation.

2 Industrial Thermal-Expansionary Pretreatment Technology

The general flow sheet of an industrial thermal-expansionary pretreatment in continuous mode was designed, see Fig. 2. Pre-milled biomass is initially fed into a balancing storage vessel ***B-110***, wherefrom it is continuously dosed to a heated extruder ***R-111*** with process or waste water together. An output feedstock containing up to 10 % wt. of biomass is prepared and pre-heated there. Then an extruder ***R-111*** carries this feedstock to subsequent processing in a hydrolyser ***B-112***. Several serially connected hydrolysers are supposed to be installed in the technology to reach a required residence time at a given process temperature. Then a feedstock is fed to an extruder ***R-113***, it is passed through its expansion head and expands into an expansion cyclone vessel ***B-114*** in seconds. Solid hydrolyzate and vapour are formed there and they are ready for subsequent processing.

As for the process parameters of thermal-expansionary pretreatment, its effectiveness generally depends on the composition and concentration of the lignocellulosic biomass in aqueous suspension, on batch acidity, on process temperature and also on residence time. These process parameters are easily determinable in lab-unit working in batch mode [3]. They are directly transferable and applicable to industrial scale except of process time. In the terms of scale-up, it applies, that process time in batch mode is equal to residence time in continual mode. The typical values of process temperature and residence time range between 170–200 °C (i.e. 0.80–1.55 MPa) and 5–20 min [4]. Krátký [1] recommended process temperature of 185 °C, residence time of 40 min. and wheat straw concentration of 5 % wt. TS in feedstock. Maximum increase in methane yield by 50 % was observed for these process parameters. Generally known, biomass concentration,

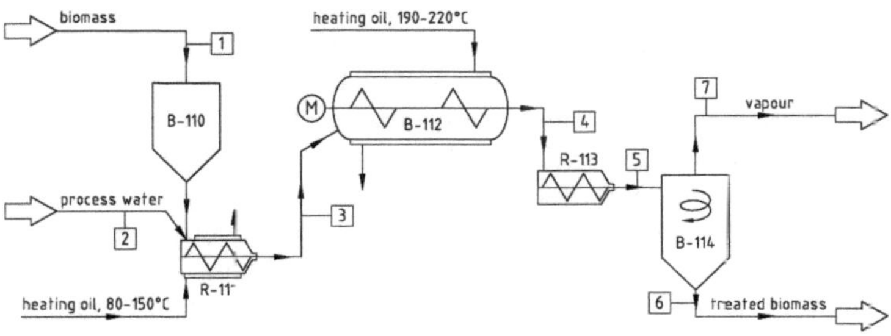

LIST OF APPARATUSSES	
Apparatus No.	Characterization
B-110	storage vessel
R-111	extruder
B-112	hydrolyser
R-113	extruder with expansion head
B-114	expansion cyclone vessel

	PROCESS STREAMS						
	Milled biomass 1	Process water 2	Biomass to TEP 3	Biomass in TEP 4	Biomass after TEP 5	Biomass to process 6	Vapour 7
moisture (%hm)	>40	100	>80	>80	>80	88–92	100
dry matter (%hm)	<60	0	<20	<20	<20	8–12	0
particle size (mm)	<10	–	<10	<10	<10	<10	–
temperature (°C)	–	>60	80–90	185	–	30–40	100

Fig. 2 Flow sheet of industrial continual thermal-expansionary pretreatment

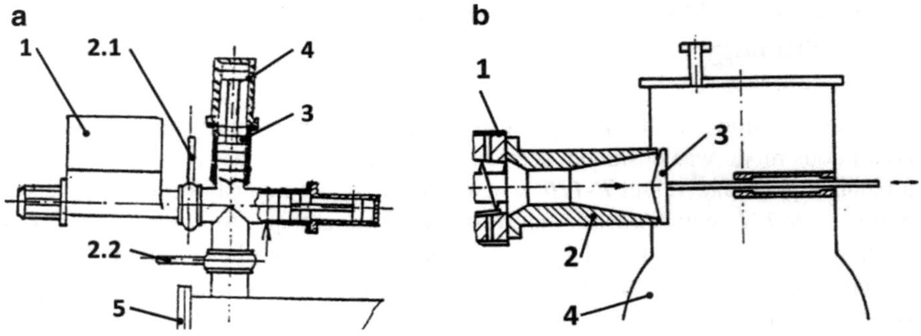

Fig. 3 Basic design of biomass filling into pressurized area. (**a**) Filling unit [6]. *1* storage vessel; *2* slide valve; *3* filling cylinder; *4* drive with piston; *5* hydrolyser. (**b**) Filling screw unit [7]. *1* extruder; *2* filling head; *3* adjustable control piston; *4* hydrolyser

referred to as the liquid-to-solid ratio, is recommend to be lower than 10 [4] and batch acidity must be kept in a range of 4–7 [5] to avoid saccharides' degradation.

The fundamental part of thermal-expansionary technology is pressurized hydrolyser, where biomass is treated under recommended process temperature of 185 °C, i.e. equivalent pressure of 1.11 MPa. Therefore, there are problems of feedstock filling into pressurized vessel, hydrolyser design and rapid continual batch expansion into expansion vessel.

2.1 *Filling of Biomass into Hydrolyser*

There are generally two possibilities, see Fig. 3, how to fill biomass suspension into pressurized apparatus. The first system is composed of two slide valves, filling chamber and piston pump, see Fig. 3A. As for its mechanism, slide valve 2.1 is firstly opened, slide valve 2.2 closed and a feedstock is filled into a chamber by screw conveyer. Secondly, valve 2.1 is closed, valve 2.2 is opened a piston pump feeds raw material into hydrolyser. This system could be also used for batch decompression. Nevertheless, its disadvantage is discontinuous mode of operation. The continual one is plotted in Fig. 3B. There, a biomass is pressed by screw through the gap between a filling head and an adjustable piston. Control piston is used as a segment to set an overpressure in filling head that is equal to working pressure in hydrolyser at least. The maximum overpressure up to 2.8 MPa can be reached in the system [7]. The main advantage of a screw extruder installation in the technology is its usage as an anti-puncture-shock pressure chamber from pressurized hydrolyser.

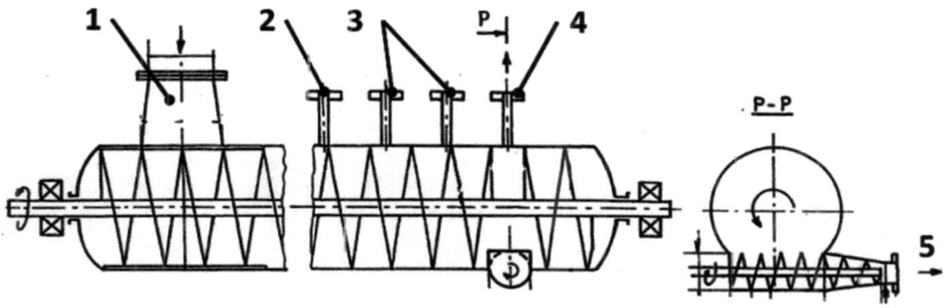

Fig. 4 Basic design of continual hydrolyser [8]. *1* biomass input; *2* thermometer; *3* pressure gauge; *4* inerts output; *5* biomass output

2.2 *Hydrolyser and Continual Batch Expansion*

The basic design of continual hydrolyser is shown in Fig. 4. Hydrolyser is a horizontally placed cylindrical vessel equipped with a mixing system, with necks for input and output of feedstock and with control necks for temperature and pressure measurements. The process volume of apparatus depends both on an amount of processed feedstock, on the time to reach a required process temperature and also on residence time. There is also possibility to divide overall process volume to several parts and design a hydrolyser as serially connected hydrolysers with lower process volume. A low-speed mixing system is frequently installed in hydrolyser. It is usually formed by paddle or screwed mixers with hollow shaft and tubes. Based on the knowledge of processing amount of biomass, screw geometry, heating and residence time, rotational speed of mixer is able to be calculated. Its value typically ranges in units of revolutions per hour to per minute. Batch tempering is primarily ensured by saturated steam or heating oil that flow in double jacket of hydrolyser or secondly in hollow mixers. The usage of heating oil has the advantage of its non-pressure loading on a main cylinder of hydrolyser. Nevertheless, its main disadvantage is a potential risk of feedstock contamination.

The output of treated feedstock out of a hydrolyser is typically provided by two systems. The first one was actually mentioned above, see Fig. 3A. The second one is shown in Fig. 5. The end of an extruder is equipped with an expansion head that provides continual batch expansion into storage vessel. Fibred lignocellulosic biomass usually plugs a gap in head and creates overpressure against working pressure in a hydrolyser. Controlling revolutions of extruder and stem positions, the continual rapid batch expansion is able to be easily achieved.

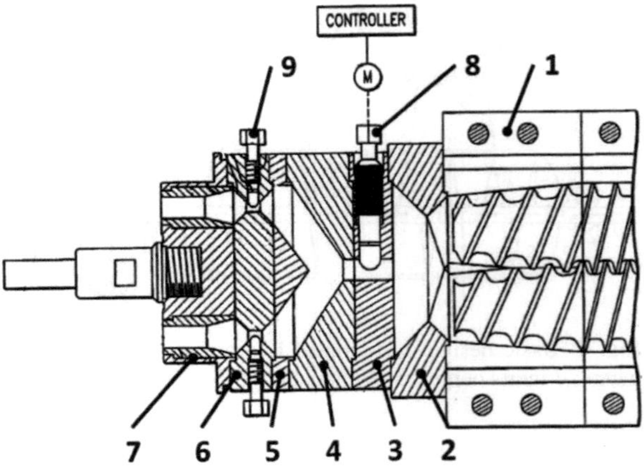

Fig. 5 Expansion head for extrusion-cooked aquatic feeds [9]. *1* twin screw extruder; *2* inlet block; *3* throttle block; *4* secondary block; *5* ring block; *6* valve block; *7* insert flange block; *8* throttle stem; *9* valve stem

3 Conclusion

Thermal-expansionary pretreatment is a novel, environmentally friendly and very efficient preliminary processing of lignocellulosic wastes in biofuel technologies. There is still problem how to do it industrially in continuous mode. Therefore based on recent patents and experimental layout, a general flow sheet of thermal-expansionary pretreatment in continuous mode was designed and crucial equipments parts were generally described.

Acknowledgments This work was supported by the Grant Agency of the Czech Technical University in Prague, grant *No. SGS14/060* and by Thomas Bata Foundation.

References

1. L. Krátký, Technologie a zařízení pro předúpravu surovin při výrobě biopaliv, Ph.D. Thesis, Czech Technical University in Prague, Faculty of Mechanical Engineering, Department of Process Engineering (January 2014), p. 174
2. J.M. Taherzadeh, K. Karimi, Pretreatment of lignocellulosic wastes to improve ethanol and biogas production: A review. Int. J. Mol. Sci. **9**, 1621–1651 (2008). ISSN 1422–0067
3. L. Krátký, T. Jirout, J. Nalezenec, Lab-scale technology for biogas production from lignocellulose wastes. Acta Polytech. **52**(3), 54–59 (2012). ISSN 1210–2709
4. S.E. Jacobsen, C.E. Wyman, Xylose monomer and oligomers yields for uncatalyzed hydrolysis of sugarcane bagasse hemicellulose at varying solids concentrations. Ind. Eng. Chem. Res. **41**, 1454–1461 (2002). ISSN 1520–5045
5. A. Hendriks, G. Zeeman, Pretreatments to enhance the digestibility of lignocellulosic biomass. Bioresour Technol. **100**, 10–18 (2009). ISSN 0960–8524
6. F. Machek, Z. Kratochvíl, *Method and apparatus for complex production of bioethanol from renewable sources.* Inventors: F. Machek, Z. Kratochvíl, Registered 26.09.1996, Int. Cl. C13K1/02, Patent No. 283872, 30 Apr 1998

7. Z. Kratochvíl, *Method of processing of lignocellulose materials by continuous pressure hydrolysis and equipment*. Inventors: Z. Kratochvíl, I. Roušar, F. Machek, Z. Kratochvíl jr., Registered 08.02.1995, Int. Cl. C07D307/50, Patent No. 281504, 08 Feb 1995
8. Z. Kratochvíl, F. Bouška, *Complex sugars yield hydrolysis of lignocellulosic materials for biotechnological production*. Inventors: Z. Kratochvíl, F. Bouška. Registered. 21.11.2000, Int. Cl. C07H 1/08, Patent No. 2000–4328, 17 Jul 2002
9. B.W. Hauck, *Pressure-controlled die apparatus for the production of extrusion-cooked aquatic feeds*. Inventor: B.W.Hauck, Registered 18.07.1996, Int. Cl. A23P1/00, Patent No. US005700510A, 23 Dec 1997

Should We Teach Polymeric Machine Elements as Well?

B. Križan

Abstract The growing importance and use of polymeric (plastic) parts in many industrial product categories isn't generally accompanied by adequate education at the university level. In response, an overview of relevant literature on polymeric machine elements is presented. Most important differences in characteristics of the polymeric and metallic materials are also highlighted, and the contents of a Croatian textbook on polymeric machine elements are used as a reference.

Keywords Machine element • Polymer • Teaching • Textbook

1 Introduction

Polymeric materials have been produced for more than 100 years. The world production in the year 1900 was 20,000 tons, in 1950 1.7 million tons, in 1990 more than 100 million tons, in 2010 already 265 million tons, and is anticipated to be 300 million tons in 2015 [1]. Constructional thermoplastics including acrylonitrile butadiene styrene (ABS), polyamide (PA), polycarbonate (PC), poly (methyl methacrylate) (PMMA), poly (butylene terephthalate) (PBT), polyoxymethylene (POM), polytetrafluoro-ethylene (PTFE), as well as elastomers (rubber), have adequate properties for use as machine elements. In 2010 the share of constructional thermoplastics used in domains such as building, construction, car industry, electrical wiring, electronics, and home appliances, was 21.5 million tons in the world, of which 2.9 million tons were used in Europe [2]. It is obvious that there is growing application and importance of polymeric parts in many industrial product categories. However, it seems that university curricula don't reflect the quick growing production and usage of products containing polymeric machine elements.

B. Križan (✉)
Faculty of Engineering, University of Rijeka, Rijeka, Croatia
e-mail: krizan@riteh.hr

V. Dynybyl et al. (eds.), *The Latest Methods of Construction Design*,
DOI 10.1007/978-3-319-22762-7_7

2 Properties and Suitability of Polymeric Materials for Machine Elements

Taking into consideration that most of engineering curricula are still dealing mostly with metallic materials and their technology, the topics related to polymeric materials are indubitably welcome. It is very important to emphasize that the mechanical properties of polymers are completely different than those of metals. Following can be pointed out [e.g. 3–7]:

(a) Polymers can be extremely extensible and flexible with large elastic deformation. The modulus of elasticity for polymers $E < 7500$ N/mm^2, and for metals is in the range $E \approx 30{,}000\ldots210{,}000$ N/mm^2.
(b) Polymers have a tensile strength about 50…100 N/mm^2, i.e. ten times less than steel.
(c) Polymers are roughly five times less dense than metals, but have nearly equivalent strength-to-weight ratio.
(d) Mechanical properties change greatly with variations in temperature. The polymeric materials become brittle at low temperatures.
(e) Polymers creep even at room temperature and polymer machine element under load may, with time, acquire a permanent set.
(f) Water percentage in polymers interferes with their mechanical properties.

Apart from these differences in mechanical properties and some other disadvantages, the following desirable characteristics promote the increasing use of polymeric machine elements:

(a) Polymers are easy to shape.
(b) Polymers can have very good electrical and thermal insulation properties.
(c) Most of the polymers have good chemical resistance.
(d) Sliding without lubrication is possible.
(e) Some of the thermoplastics can be transparent.
(f) Polymers can be coloured throughout.
(g) Possibility to create composites.

The manufacturers are continuously improving the properties of polymeric materials by inventing new materials, mixing different kinds of polymers and applying different additives (e.g. fillers or reinforcements). All this significantly improves the suitability of polymers for use as machine elements.

3 Contents of Standard Textbooks on Machine Elements

Standard textbooks on machine elements typically deal with polymeric parts in following chapters:

(a) Rubber, i.e. elastomeric springs, which are mostly used as supports for compression and shear loads.
(b) Seals for shafts and bearings.
(c) Flexible couplings which can contain elastic parts made of elastomers.
(d) Belt drives, because belts are made mostly of elastomers with polyamid or polyester interior tension chords.
(e) Friction drives that can contain rubber wheels.

Apart from these five cases, some of the most used and well known textbooks in German and English language are dealing with polymers generally and polymeric machine elements to a rather small extent. Here is an overview:

- Niemann, G. (1975, 1983, 1983): *Maschinenelemente – Band I & II & III* [8–10]; design of polymeric parts (1 page), lightweight design (1), polymeric materials (8), polymeric materials for sliding bearings (<1), polymeric gears (12).
- Niemann, G.; Winter, H.; Höhn, B.-R. (2001): *Maschinenelemente – Band 1* [11]; design of polymeric parts (1), lightweight design (1), polymeric materials (6), polymeric sliding bearings (7).
- Shigley, J.E.; Mischke, C.R. (1989): *Mechanical Engineering Design* [12]; polymeric materials (1).
- Steinhilper, W.; Röper, R. (1993, 1994): *Maschinen- und Konstruktionselemente 2 & 3* [13, 14]; welded joints (<1), polymeric materials for sliding bearings (3).
- Hamrock, B.J.; Jacobson, B.; Schmid, S.R. (1999): *Fundamentals of Machine Elements* [3]; polymeric materials (2) + composites (2), snap joints (3).
- Decker, K.-H. (2004): *Decker – Maschinenelemente* [15]; welded joints (2), polymeric sliding bearings (6), design and calculation of polymeric gears (7).
- Wittel, H.; Muhs, D.; Jannasch, D.; Vossiek, J. (2009): *Roloff/Matek – Maschinenelemente* [16]; riveted joints (<1), inserted nuts (<1).

Although the mentioned textbooks have been published over a time span of 34 years, only the books of Niemann [8–11] and Decker [15] pay any substantial attention to polymeric machine elements.

One of the most important and comprehensive books on polymer machine elements has been written in German language by G. Erhard and E. Strickle under the title *Maschinenelemente aus thermoplastischen Kunststoffen* (*Thermoplastic Machine Elements*) and published in two parts in 1974 and 1985 [17, 18]. Practically all machine elements have been elaborated on 460 pages.

The book *Ozubljenja i zupčanici* (*Toothings and Gears*) written by E. Oberšmit [19] has been published in Croatia in 1982. There are 14 pages dealing with the calculation and design of polymeric gears.

The book *Plastzahnräder* (*Plastic Gears*) edited in 1985 by W. Krause [20] gives on 159 pages comprehensive explanations of the calculation, design, control and manufacture of polymeric gears.

4 Contents of Books Dealing with the Manufacturing Process and Design of Polymeric Parts

The books quoted in this chapter are concerned primarily with the manufacturing process for the production of polymeric parts and with the design of polymeric parts generally, not machine elements specifically. However, some of the polymeric machine elements are considered in following books:

- Wimmer, D. (1991): *Kunststoffgerecht konstruieren* [21]; welded joints (18 pages), bonded joints (23), threaded joints (24), press-fit joints (9), snap joints (17), hinges (5), sliding bearings (23), gears (40).
- Erhard, G. (1993): *Konstruieren mit Kunststoffen* [4]; threaded joints (16), snap joints (20), springs (9), sliding bearings (18), gears (32), rollers (15).
- Tres, P.A. (1994): *Designing Plastic Parts for Assembly* [22]; welded joints (48), bonded joints (4), riveted joints (6), press-fit joints (26), snap joints (36), hinges (35).
- Malloy, R.A. (1994): *Plastic Part Design for Injection Molding* [5]; welded joints (32), bonded joints (15), threaded joints (24), press-fit joints (7), snap joints (21).
- Michaeli, W.; Brinkmann, T.; Lessenich-Henkys, V. (Ed.) (1995); *Kunststoff-Bauteile werkstoffgerecht konstruieren* [23]: welded joints (6), threaded joints (13), press-fit joints (6), snap joints (14), hinges (6), gears (13).
- Ehrenstein, G.W. (2002): *Mit Kunststoffen konstruieren* [6]; welded joints (5), bonded joints (9), threaded joints (24), press-fit joints (7), snap joints (11), hinges (2), rolling bearings (2), sliding bearings (14), gears (10), rollers (6).
- Ehrenstein, G.W. (Ed.) (2004): *Handbuch Kunststoff-Verbindungstechnik* [24]; welding and welded joints (190), bonding and bonded joints (88), riveted joints (21), threaded joints (53), press-fit joints (14), snap joints (36), hinges (14).

Although each book covers machine elements to a different extent, these books are very useful help for the design and calculation of some of the polymeric machine elements.

5 Polymeric Machine Elements in University Teaching

To the author's knowledge, with the exception of the Erhard & Strickle's books *Thermoplastic Machine Elements* [17, 18] and Krause's *Plastic Gears* [20], there are no textbooks dealing systematically with such machine elements. This fact, as well as the fact of growing application of polymeric parts in a wide range of industrial products, point to a need of introducing these topics in university textbooks and curricula. That's why the textbook *Polymeric Machine Elements* [7] (266 pages, in Croatian language, written by the author of this paper and co-author R. Basan) has been published at the Faculty of Engineering – University of Rijeka in Croatia. The content of this book has been chosen by studying previously mentioned books and with the idea that it should be appropriate to the students of undergraduate studies of mechanical engineering. After a short introduction, the book chapters are as follows:

- Properties and types of polymers (15 % of the book content): a short overview of polymers, additives and reinforcements; influential factors for the selection of adequate material; mechanical properties; influences on mechanical properties; friction.
- Joints (45 %): threaded joints, snap joints, bonded joints, welded joints, riveted joints, shaft-hub joints, hinges and springs. Special attention has been paid to threaded and snap joints.
- Bearings (13 %): sliding bearings and (less important) rolling bearings.
- Gears (13 %): the basics of the calculation are as for metallic gears; specific is the control of the tooth deflections.
- Belt drives (3 %): as this kind of transmission is normally presented in standard textbooks on machine elements, only basic explanation are given.
- Wire rope sheaves (2 %): short description.
- Rollers (5 %): although rollers are usually not described in machine elements textbooks, their wide application merited their inclusion.
- Clutches and couplings (4 %): as polymers are widely used in flexible couplings which are always elaborated in standard machine elements textbooks, only some specific polymer clutches and couplings are described.

Lectures on polymeric threaded, snap, and welded joints have already been included in the course "Machine Elements I" at the Faculty of Engineering in Rijeka since 2010. The experience of the author of this paper is that students also sometimes need information about polymeric machine elements when solving problems in other creative study projects and that they usually can't find it in the available literature.

6 Conclusion

Today's designer has the option to choose from a great variety of different polymeric materials because polymers can often compete with other materials such as wood, sheet metal, cast or forged metals, glass, or ceramics. The designer must correlate the end-use performance requirements of the product with the properties and cost of the individual polymeric materials in an attempt to obtain the best material for the application. The designer must also be aware of all advantages as well as limitations of the polymeric machine elements and polymeric products generally in order to perform an optimal job. It is of the outmost importance to make the students of mechanical engineering aware of the differences between the metallic and polymeric materials and the possibilities that afford the use of polymers. One of the important prerequisites for this is introduction of polymeric machine elements in university teaching.

References

1. G. Barić, Kretanja na svjetskom tržištu plastike. Polimeri **30**(1), 38–40 (2009). ISSN 0351-1871
2. G. Barić, *Proizvodnja i prerada plastike i gume u svijetu i Europi*. Presentation at the Faculty of Mechanical Engineering and Naval Architecture in Zagreb, Croatia (2011)
3. B.J. Hamrock, B. Jacobson, S.R. Schmid, *Fundamentals of machine elements* (McGraw-Hill, Boston, 1999). ISBN 0-256-19069-0
4. G. Erhard, *Konstruieren mit Kunststoffen* (Carl Hanser Verlag, München/Wien, 1993). ISBN 3-446-17397-8
5. R.A. Malloy, *Plastic part design for injection molding* (Carl Hanser Verlag, München/Vienna/New York, 1994). ISBN 3-446-15956-8
6. G.W. Ehrenstein, *Mit Kunststoffen konstruieren* (Carl Hanser Verlag, München/Wien, 2002). ISBN 3-446-21295-7
7. B. Križan, R. Basan, *Polimerni konstrukcijski elementi* (Zigo – Rijeka & Tehnički fakultet Sveučilišta u Rijeci, Rijeka, 2009). ISBN 978-953-7142-43-8
8. G. Niemann, *Maschinenelemente – Band I* (Springer-Verlag, Berlin/Heidelberg/New York, 1975). ISBN 3-540-06809-0
9. G. Niemann, H. Winter, *Maschinenelemente – Band II* (Springer-Verlag, Berlin/Heidelberg/New York/Tokyo, 1983). ISBN 3-540-11149-2
10. G. Niemann, H. Winter, *Maschinenelemente – Band III* (Springer-Verlag, Berlin/Heidelberg/New York/Tokyo, 1983). ISBN 3-540-10317-1
11. G. Niemann, H. Winter, B.-R. Höhn, *Maschinenelemente – Band 1* (Springer-Verlag, Berlin/Heidelberg, 2001). ISBN 3-540-65816-5
12. J.E. Shigley, C.R. Mischke, *Mechanical engineering design* (McGraw-Hill, New York, 1989). ISBN 0-07-056899-5
13. W. Steinhilper, R. Röper, *Maschinen- und Konstruktionselemente 2* (Springer-Verlag, Berlin, 1993). ISBN 3-540-55863-2
14. W. Steinhilper, R. Röper, *Maschinen- und Konstruktionselemente 3* (Springer-Verlag, Berlin, 1994). ISBN 3-540-57429-8
15. K.H. Decker, *Decker – Maschinenelemente* (Carl Hanser Verlag, München/Wien, 2004). ISBN 3-446-21525-5
16. H. Wittel, D. Muhs, D. Jannasch, J. Vossiek, *Roloff/Matek – Maschinenelemente* (Vieweg +Teubner, Wiesbaden, 2009). ISBN 978-3-8348-0689-5
17. G. Erhard, E. Strickle, *Maschinenelemente aus thermoplastischen Kunststoffen – Grundlagen und Verbindungselemente* (VDI-Verlag, Düsseldorf, 1974). ISBN 3-18-40 0290-X
18. G. Erhard, E. Strickle, *Maschinenelemente aus thermoplastischen Kunststoffen – Lager und Antriebselemente* (VDI-Verlag, Düsseldorf, 1985). ISBN 3-18-400712-X
19. E. Oberšmit, *Ozubljenja i zupčanici* (SNL, Zagreb, 1982)
20. W. Krause (ed.), *Plastzahnräder* (VEB Verlag Technik, Berlin, 1985)
21. D. Wimmer, *Kunststoffgerecht konstruieren* (Hoppenstedt Technik Tabellen Verlag, Darmstadt, 1991). ISBN 3-87807-158-2
22. P.A. Tres, *Designing plastic parts for assembly* (Carl Hanser Verlag, München/Vienna/New York, 1994). ISBN 3-446-17594-6
23. W. Michaeli, T. Brinkmann, V. Lessenich-Henkys (eds.), *Kunststoff-Bauteile werkstoffgerecht konstruieren* (Carl Hanser Verlag, München/Wien, 1995). ISBN 3-446-17535-0
24. G.W. Ehrenstein (ed.), *Handbuch Kunststoff-Verbindungstechnik* (Carl Hanser Verlag, München, 2004). ISBN 3-446-22340-1

Design of Semi-active Magnetorheological Valve

M. Kubík and I. Mazůrek

Abstract This paper presents a methodology of design of a semi-active magnetorheological (MR) valve. The MR valve has been used successfully for a long period of time in many technical applications. When the valve is used as a semi-active element, problems occurred. The application of this valve is limited, mainly due to the slow response time and its low dynamic range. The methodology consists of flow analysis of a non-Newtonian fluid and FEM analysis of a magnetic circuit. A parallel-plate model was used together with Bingham fluid to describe the flow in the valve. A static and transient model of the magnetic circuit was solved by the FEM program Maxwell. The semi-active MR valve design was based on the presented methodology. In our research the magnetic circuit was made from ferrite that significantly reduces the response time of the magnetic field. The valve was designed only to operate only until the velocity at the breaking point of the F-v curve. Therefore, now there is a large dynamic range in this area. It is reasonable to expect that the designed modifications of the MR valve allow us to use this technology in semi-active suspension systems.

Keywords Magnetorheological valve • MR valve • Semi-active valve • MR fluid

1 Introduction

Today, modern mechatronic systems cannot do without having the possibility of controlling the damping elements [1]. Active or semi-active damping elements may be present in the system. The semi-active element in a mechatronic system is what controls the damping force. In contrast to the semi-active elements, the active element is able to supply energy to system. However, this introduces the risk of destabilising the system and requires a lot of power to operate. Therefore, semi-active elements appear to be more suitable than active ones. A semi-active element is often known as a fail-safe system. This is because semi-active elements can often operate as a passive system when power failure occurs. The MR valve is a progressive semi-active element in a mechatronic system.

M. Kubík (✉) • I. Mazůrek
Brno University of Technology, Brno, Czech Republic
e-mail: y115760@stud.fme.vutbr.cz; mazurek@fme.vutbr.cz

V. Dynybyl et al. (eds.), *The Latest Methods of Construction Design*,
DOI 10.1007/978-3-319-22762-7_8

MR valves or dampers are devices which can resize the dissipated energy depending on the current in the coil. An MR valve is usually constructed from coil, which creates a magnetic field, magnetic circuit and MR fluid. The MR fluid is colloidal suspension formed by micron-sized magnetic particles (most often powdered iron) and carrying fluid [2]. Upon the application of an external magnetic field, the MR fluids are able to change their behavior from a fluid state to a semi-solid or plastic state, and back, in a couple of milliseconds [7]. This phenomenon is known as the MR effect. This field-dependent behavior is often represented as a Bingham plastic having a variable yield stress [5]. Thus, it is possible to modify the characteristics of damper. Many authors were involved with the designing of MR dampers; [2], [3] or [4]. These dampers (valves) can be found some seats in automotives [3], in the control of seismic vibrations in buildings [2] or in guns [4]. When these valves are used for fast semi-active control, they are limited by slow time response damping forces on the control signal. For instance, a large seismic MR damper has a time response of under a tenth of a second (0.06 s [2]). This is a very slow response for semi-active control. This delay damping force on the control signal is only to a small extent due to the MR fluid [7]. The MR valve is influenced greatly by its design and material.

The main goal of this paper is to introduce a methodology design for a fast semi-active MR valve. This methodology has not been published yet. On based this methodology has been designed MR valve. This MR valve was designed by 4th year students of Brno University of Technology faculty of Mechanical engineering department Mechanical engineering design in project-oriented teaching.

2 Materials and Methods

2.1 Rheological Model

MR fluids in the presence of a magnetic field exhibit Non-Newtonian behavior [5]. This behavior is often represented as a Bingham plastic with a variable yield stress [5]. The Bingham model is beneficial in the damper design phase due to its simplicity. However, this model does not include all the characteristic of an MR fluid, such as shear thinning [2]. In the absence of a magnetic field we expect Newtonian behavior from the MR fluid [5]. The Shear stress in the Bingham model is given by the equation:

$$\tau = \tau_y(H) + \eta\dot{\gamma},\ \tau \geq \tau_y \tag{1}$$

where H is the magnetic flux intensity [A/m], $\dot{\gamma}$ is the shear rate [1/s], η is the dynamic viscosity in off-state (H = 0 A/m) [Pa.s], τ is the shear stress in the MR fluid [Pa] and τ_y is the yield stress of the MR fluid [Pa].

The MR fluid in a valve is usually flowing across an annular gap. In the design phase, it is more advantageous to use a simplified model. A parallel plate model would sufficiently approximate an annular gap. The error that occurs due to this simplification is less than 0.5 % for the ratio valve geometry h/R <0.2 [2]. The mean velocity of flowing MR fluid in the presence of a magnetic field is given by Eq. (2). This equation includes the plug effect of the MR fluid [6]. The validity of this relation is limited to a laminar flow.

$$v_s = \frac{-\Delta p.h^2}{12\eta.l}\left(1 - 3\frac{y_0}{h} + 4.\left(\frac{y_0}{h}\right)^3\right) \tag{2}$$

Mean velocity is described by pressure difference Δp[Pa], length of gap l [m], size of gap h [m] and size of plug y_0[m]. The size of the plug is determined by the yield stress and pressure difference [6]. Therefore, plug size is especially function of magnetic field.

In the MR valve, MR fluid can flow through two different types of gaps. The first one is a magnetic gap in which it is possible to change the yield stress of the MR fluid. The second one is a simple connection inlet and outlet of the MR valve through geometric constrictions. This type of MR valve is often called bypass. These gaps can be connected in various ways to obtain the required F-v characteristics of the MR valve. The magnetic gap and bypass are frequently connected in parallel. In parallel, the bypass size effects the slope of F-v characteristics. The magnetic gap and MR fluid properties have effect the breakpoint of the F-v curve. The dynamic range of an MR valve is defined as force in on-state divided by force in off-state in at a specific velocity. The dynamic range reaches its maximum point below the breakpoint. After that it is decreasing nonlinearly. The size of the bypass changes the dynamic range and it also moves breakpoint to the higher velocity on the F-v graph.

2.2 *Magnetic Model*

The magnetic circuits in MR valves are usually made from low carbon steel [2, 3]. However, low carbon steel also has some disadvantages. For instance, this steel is magnetically and electrically conductive. This causes the creation of eddy currents in the material when it the polarity of the current is changed rapidly. Xinchungan [8] published that eddy currents in steel have a significant effect on the response time of an MR valve. Therefore, for a high speed response, steel is unsuitable. For high speed response, material which is magnetically conductive and electrically non-conductive is desirable. This material prevents the creation of eddy currents. Ferrite materials meet these requirements; however their disadvantage is that they have a lower magnetic saturation limit than steel.

Low carbon steel and ferrite materials are nonlinear magnetic materials. Non-linearity is described by the magnetic B-H curve. However, the analytical solution is complicated. Therefore, numerical solution is more suitable and the finite-element program Ansys Maxwell was used for the design of the magnetic circuit. The magnetic circuit was solved as a 2D axisymmetric model. Magnetostatic and transient analysis was performed. An adaptive mesh with maximal error 1 % was used.

In the MR valve analysis, MR fluid MRF-132DG from the LORD company was used. This fluid has a dynamic viscosity $\eta = 0.112 \pm 0.02\,\text{Pa.s}$ (40 °C) in the off-state [5].

3 Results

The main goal of students was designed MR valve on the presented methodology. The MR valve should achieve a force of 1200 N in on-state at a velocity of 0.15 m/s. Minimal dynamic range of 3.2 was required.

Based on the presented methodology, a rheological and magnetic model of an MR valve was created. The magnetic circuit was designed from ferrite material N95 (1, 5, 8), and the outer tube of the valve (4) was designed from steel AISI 1018 (ČSN 11 523). Steel was also included in the magnetic circuit. This tube was cut longitudinally. The cut surface was electrically insulated. This method reduces the influence of eddy currents on the time response. Other components of the MR valve were designed from aluminum (6, 7). Magnetic (D) and bypass (E) gaps are connected in parallel in the MR valve. MR valve was created from iterative process from these models.

The design of the MR valve has a magnetic gap of size 0.6 mm and length 32 mm. The diameter of the inner cylinder is 30 mm (1). The bypass gap has a diameter of 1.5 mm and length 33 mm (3). The magnetic circuit is composed of three coils. The coils are wound in a way such that everyone adjacent coil has a different winding direction (Fig. 1).

Position A and B shows the direction of the magnetic field (winding direction). Every coil includes 150 turns of a 0.6 mm wire (2).

The magnetostatic solution was carried out in Maxwell. Current 1 A in the coil creates a magnetic flux intensity in the magnetic gap of 78 kA/m. This magnetic flux intensity creates a yield stress 24 kPa in the MR fluid 132-DG. Using the transient magnetic solution in Maxwell, the time response of the magnetic field in the valve was calculated. The results show that the time response is less than 2 ms.

Based on the rheological solution, the F-v characteristic of the designed MR valve was determined. A parallel plate model was used (0.6/15 = 0.04). If the piston velocity is 0.15 m/s than the velocity in the bypass gap is 50.2 m/s. The Reynolds number is 2028. Therefore, the flow is laminar (Fig. 2).

The maximal dynamic range is 14 in the designed MR valve. The presented dynamic range applies to the breaking point.

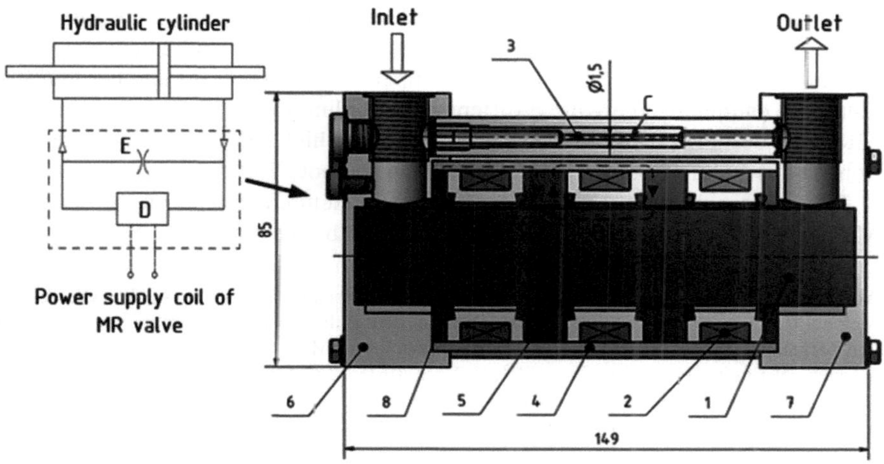

Fig. 1 MR valve

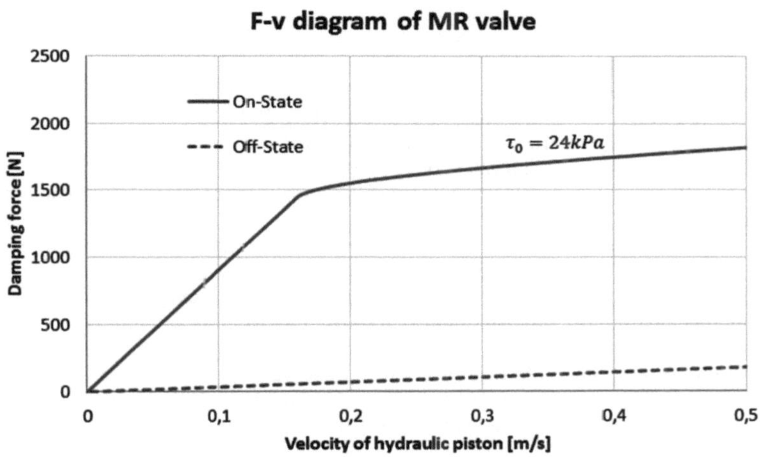

Fig. 2 F-v characteristic of MR valve

4 Conclusion and Discussion

This paper deals with the methodology of designing an MR damper. Based on this methodology, students from the Brno University of Technology designed an MR valve. The magnetic circuit in the valve was designed from ferrite material which decreased the response time of the MR valve to less than 2 ms. This MR valve is appropriate to operate under breakpoint on F-v characteristic. This results in a maximal dynamic range. The bypass gap has a major effect on the dynamic range.

The MR valve designed by the students should have a minimal force 1200 N in the on-state at a velocity of 0.15 m/s in the piston. This requirement was met. However, this applies to when it is at a temperature of 40 °C. At another temperature, the damping force will be different. According to the FEM simulation, the MR valve designed by the students can be used for high speed semi-active regulation. The switch from on-state to off-state can be about 500 times per second.

Future research in this area leads to the manufacture and tests on the designed MR valve. The proposed methods will be verified by the manufactured MR valve.

Acknowledgments This work is an output of research and scientific activities of NETME Centre, regional R&D centre built with the financial support from the Operational Programme Research and Development for Innovations within the project NETME Centre (New Technologies for Mechanical Engineering), Reg. No. CZ.1.05/2.1.00/01.0002 and, in the follow-up sustainability stage, supported through NETME CENTRE PLUS (LO1202) by financial means from the Ministry of Education, Youth and Sports under the "National Sustainability Programme I".

References

1. M. Unsal, C.D. Crane, C. Niezrecki, Vibration control of parallel platforms based on magnetorheological damping, pp. 1–6 (2006)
2. G. Yang, B.F. Spencer, J.D. Carlson, M.K. Sain, Large-scale MR fluid dampers: Modeling and dynamic performance considerations. Eng. Struct. **24**(3), 309–323 (2002)
3. S.B. Choi, M.H. Lee, B.K. Lee, Vibration control of MR seat damper for commercial vehicles. J. Intell. Mater. Syst. Struct. **11**(12), 936–944 (2002)
4. Z.C. Li, J. Wang, A gun recoil system employing a magnetorheological fluid damper. Smart Mater. Struct.. 2012, roč. 21, č. 10
5. M.R. Jolly, J.W. Bender, J.D. Carlson, Properties and applications of commercial magnetorheological fluids. J. Intell. Mater. Syst. Struct. **10**(1), 5–13 (1999)
6. I. Mazůrek, J. Roupec, M. Klapka, Z. Strecker, Load and rheometric unit for the test of magnetorheological fluid. Meccanica **48**(3), 631–641 (2013)
7. F.D. Goncalves, M. Ahmadian, J.D. Carlson, Investigating the magnetorheological effect at high flow velocities. Smart Mater. Struct. **15**(1), 75–85 (2006)
8. G. Xinchun, G. Pengfei, O. Jingping, Study of the response time of MR dampers, in *Proceedings of the SPIE 7493*, (Weihai, China, 2009). DOI: 10.1117/12.840217

Support Tools in the Development of Bearings for Wind Turbines

R. Madaj and R. Kohar

Abstract The aim of this paper was to show the way to build a model roller bearings in the program execution Adams for dynamic analysis and obtain information about different parts of the bearing during the simulation. Bearing model was designed to be as close as possible to the actual deposit, which allows us to get an idea of traffic load, the dynamic conditions of the bearing parts and interactions between them. And in the second part describes the preparation of a prototype designed for durability tests.

Keywords Vacuum casting • Vacum chamber • Bearing cage

1 Introduction

This article details the creation of a virtual bearing model allowing the analysis of individual bearing components under dynamic behavior. Two models of tapered roller bearings will be employed: one with steel cage and the other with plastic cage, both of which are pictured in Fig. 1 along with the assembled bearing model.

2 Dynamic Simulations

Dynamic simulations of the tapered roller bearing were performed in the MSC. Adams system. A precise geometrical model of the bearing was necessary in order to perform the said simulations. The 3D model has been created based on available drawing documentation and incorporates various methods with regards to the overall model complexity. Model design was performed in Pro/Engineer Wildfire 5 (Fig. 1), which, when compared to the MSC. Adams environment, allows simpler model creation and subsequently easier bearing geometry modifications. The bearing model assembly was transformed from Pro/Engineer into Adams environment using the Parasolid file format and was further processed based on analysis

R. Madaj (✉) • R. Kohar
University of Zilina, Žilina, Slovak Republic
e-mail: rudolf.madaj@fstroj.uniza.sk; robert.kohar@fstroj.uniza.sk

V. Dynybyl et al. (eds.), *The Latest Methods of Construction Design*,
DOI 10.1007/978-3-319-22762-7_9

Fig. 1 Tapered roller bearing model in Pro/Engineer

Table 1 Material properties of individual bearing parts

	Density [kg.mm^{-3}]	Young modulus [MPa]	Poisson constant [−]
Inner ring	$7.85 \cdot 10^{-6}$	$2.02 \cdot 10^{5}$	0.29
Outer ring	$7.85 \cdot 10^{-6}$	$2.02 \cdot 10^{5}$	0.29
Roller	$7.85 \cdot 10^{-6}$	$2.02 \cdot 10^{5}$	0.29
Steel cage	$7.85 \cdot 10^{-6}$	$2.02 \cdot 10^{5}$	0.29
Plastic cage	$1.1 \cdot 10^{-6}$	3000	0.42

requirements. The first step included material definition for individual bearing components. The bearing consisted of inner and outer ring, cage and rollers. Table 1 lists values assigned to individual parts.

In the next step we defined the contacts between individual bearing components. Contact type "solid to solid" was chosen for the afore mentioned operation, defining two objects coming into contact. This was due to the geometrical complexity of the model and inability to determine all bearing parts that come into contact. This contact type requires the definition of the following parameters: Stiffness, Exponent, Max Damping and Penetration Distance. Contact pairs were formed between inner ring and rollers, outer ring and rollers and between ring and rollers. Parameter values for contact between cage and rollers have been defined as per [1]. Parameter values for contacts of rollers with outer and inner ring have been defined based on the Hertz theory of contact pressure. Table 2 lists values of parameters for individual contacts. We also considered a friction model based on Coulomb friction force calculation. Values of static and dynamic friction coefficient have been set according to [2] and values of transmission velocity according to [1].

Next we defined the geometric and kinematic constraint conditions and load force. Axial load force of the outer ring was associated with "Fixed joint" constraint, which resulted in removal of all degrees of freedom. Inner ring was associated with "Cylindrical Joint" constraint condition, which allowed rotation and translation along the x axis. "Rotational Joint Motion" of type "Velocity" has been assigned to the "Cylindrical Joint" constraint, allowing rotational movement.

Table 2 Contact parameters used in the tapered roller bearing model

Parameter	Contact pairs			
	Roller – outer ring	Roller – inner ring	Roller – cage	
			Steel	Plastic
Stiffness [N.mm^{-1}]	9,870,000	9,870,000	100,000	80,000
Exponent [–]	1.5	1.5	1.5	1.8
Max. Damping [N.s.mm^{-1}]	10,000	10,000	100	
Penetration Distance [mm]	0.15	0.15	0.1	
Static coefficient [–]	0.12	0.12	0.12	0.1
Dynamic coefficient [–]	0.12	0.12	0.12	0.1
Stiction transition velocity [mm. s^{-1}]	100	100	100	
Friction transition velocity [mm. s^{-1}]	1000	1000	1000	

Fig. 2 CAD model of segment and exported model in STL format

This movement has been defined via the STEP function [1] and corresponds to bearing rotational speed n = 15.5 min-1. Loading force was defined via gravitation acceleration "Gravity" and "Axial Force" of magnitude 518,000 N in x axis direction, influencing the inner ring. Figure 2 left shows a model with axial load force with defined geometrical and kinematic constraint conditions.

Constraint conditions for dynamic analysis with radial load force have been defined as following: "Revolute Joint" has been assigned to the inner ring and allowed inner ring rotation around the x axis. "Translational Joint" has been assigned to the outer ring, allowing movement in direction of the y axis. "Rotational Joint Motion" of type "Velocity" has been assigned the aforementioned constraint, allowing rotational movement. This movement has been defined in a similar fashion as described above using the STEP function. Load force was defined via gravitation acceleration "Gravity" and "Radial Force"of magnitude 4,500,000 N has been applied in y axis direction, influencing the outer ring. Figure 2 right shows the model under radial load force with defined geometric and kinematic constraint conditions.

3 Design Verification by the Durability

Another reported beneficial activity is the prototype production of the bearing parts namely plastic cage. For the size of the cage, which must ensure the optimum setup and keeping the roller cage segments need to be divided into several segments, which are combined in the bearing assembly into a single unit. The relative position of the individual components is ensured by positive engagement and after the installation of a cage connected by elastic rings to secure the desired position. Such a solution allows the production of large plastic cage using a standard injection molding tool, which keeps the cost of mass production of which is orders of magnitude lower compared to the standard use of the cage. Also, inertial forces are based on the weight of the cage in the order below.

Since it is a preparation kit for the development, where the cost of producing a serial instrument for injection were enormously high, comes to the use of technologies for rapid prototyping. The cage itself will be designed in CAD system. When designing into account all aspects that affect accuracy, tolerances and geometrical dimensions. Subsequently, the model is exported to the transmission format, usually it is the format. STL, which is supported for transfer between different software platforms with no memory effect. Basically it is a representation of the outer shell CAD models created using triangles, which also can be called triangulation surface. The size and quantity of the triangles depends on the accuracy of details and size of the generated data. Among these there is a direct correlation, the higher the resolution the larger the file. The disadvantage of large files is that their treatment is lengthy and direct correspondingly increase errors when exporting. The number of elements, such as the triangle tells you can also edit the generated but of course it is effective only in the direction of the zoom elements, thereby losing the detail section. Therefore, choosing the correct settings export very important.

After export of the model is it loaded into the 3D printer, where is prepared for the implementation of the data to the model. The basis of preparation is to review and implement the correct orientation of the model because of the conditions and specifications of the press. Every kind and type of such a device has its own specific requirements and implementation of the model for different types of printers can turn out differently (Fig. 3).

With limited mechanical properties and two printed parts is required to use the technology working with forms for a limited amount of pieces. This copy technology uses a model made on a 3D printer, for example, to the effect that takes flexible form. Such a tool cavity is then filled under vacuum with polyurethane, the material properties are identical with the material to be used in mass production. The form of such an instrument is able to produce only a limited amount of pieces. The rule is one of 15–20 parts, but the number of significantly affected by the used polyurethane aggressiveness. The preparation for this kind of prototype includes a number of corrections dimensions. According to the properties required to be compiled dimensional chain and proceed according to him, since production preparation. The final dimensions of influence especially polyurethane shrinkage during

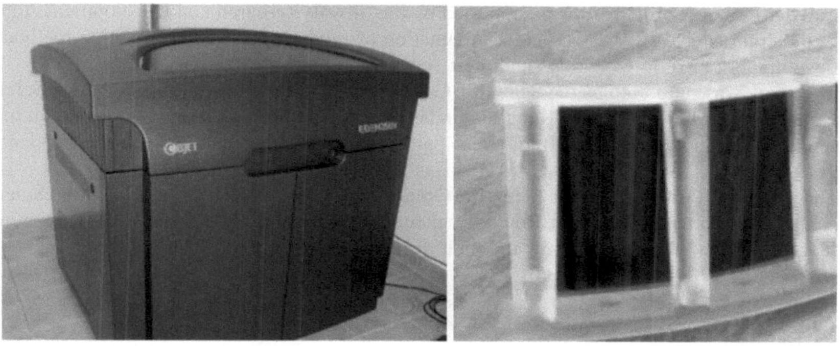

Fig. 3 EDEN 350 V, and printed segment

Fig. 4 Durability test

solidification, the casting mould dilatation but also a chemical reaction, the dimensional accuracy of prints for the production of moulds. Only after mastering and testing of these corrections can be made part of large bearing, in the required quality, durability and geometric tolerances. Than the durability test is correct and denounced the actual situation at the specified load.

Segments produced using the processes described with the use of modern technologies were applied in durability tests carried out in the framework of testing bearings that take place in the premises of Faculty of Mechanical Engineering, University of Zilina (Fig. 4).

Description given in this article only describes one iteration of each of the described area. It goes without saying that the above iteration quantity contributes to a better result and the total number of such cycles is dependent on the skill and experience of the investigators tasks.

As noted earlier in this article, in principle it is possible to use these resources in any developmental role, however, to optimize the methodology for a particular type of task for which the process will be used.

Acknowledgments This work was supported by the Slovak Research and Development Agency under the contract no. APVV-0419-11 – Adaptation of modern computer-simulation methods to the development of rolling bearings and their verification in real conditions.

This work was supported by the Scientific Grant Agency of the Ministry of Education, Science, Research and Sport of the Slovak Republic under the contract no. V-1/0396/14 – Research of the influence of construction and technological parameters of rolling bearings on the service life thereof.

This work was supported by the Scientific Grant Agency of the Ministry of Education, Science, Research and Sport of the Slovak Republic under the contract no. V-1/0844/13 – Research on guidance elements of rolling bearing and their design.

References

1. Ľ. Kučera, M. Lukáč, L. Jurák, F. Brumerčík, Hydromechanical automatic transmission. Komunikacie **11**(2), 33–35 (2009)
2. I. Gajdáč, S. Hrček, M. Mikita. Skúšobné zariadenie pre skúšky nápravových skríň a ložísk, in: ProIN : dvojmesačník Stredoeurópskeho technologického inštitútu (CEIT). – ISSN 1335–5961. – Roč. 13, č. 5 (2012), s. 15–16

Note to an Undercut Limit of Involute Gearing

M. Nemecek

Abstract An undercut of an involute tooth flank is a routine matter. Moreover there is no need to be concerned with an undercut if there is no problem with teeth strength. A problem occurs if undercutting strikes not only a root but an involute flank too. It even does not matter until an operational part of an involute is not cut off. In this case the size of decreasing of a transverse ratio coefficient ε_α is necessary to specify. This contribution deals with two approaches to the determination of start of undercut involute and performs their evaluation with possible implications. At the conclusion the contribution is highlighting to one unpublished fact.

Keywords Involute profile • Undercutting • Start of involute

1 Undercut Limit

An undercut limit is simply determined for the case when the end point of a linear edge of a tool L "meshes" with an involute just in its beginning on a base circle – see Fig. 1. For this situation it is easy to derive a number of teeth at the undercut limit and needed minimum addendum shifting coefficient (for achieving of the undercut limit).

$$z_{mez} = \frac{2 \cdot \cos\beta}{\sin^2\alpha_t} \cdot \left(h_{a0}^* - c_{a0}^* - x\right) \tag{1}$$

$$x_{mez} = h_{a0}^* - c_{a0}^* - z \cdot \frac{\sin^2\alpha_t}{2 \cdot \cos\beta} \tag{2}$$

2 Start of an Undercut Involute (the Usual Way)

Once the end point of a linear edge of a tool L "meshes" under point N – undercutting takes place. Thereby a truncation of an involute above a base circle occurs. A guideline how to determine a diameter of a start of an undercut involute is

M. Nemecek (✉)
VŠB-Technical University of Ostrava, Ostrava, Czech Republic
e-mail: milos.nemcek@vsb.cz

V. Dynybyl et al. (eds.), *The Latest Methods of Construction Design*,
DOI 10.1007/978-3-319-22762-7_10

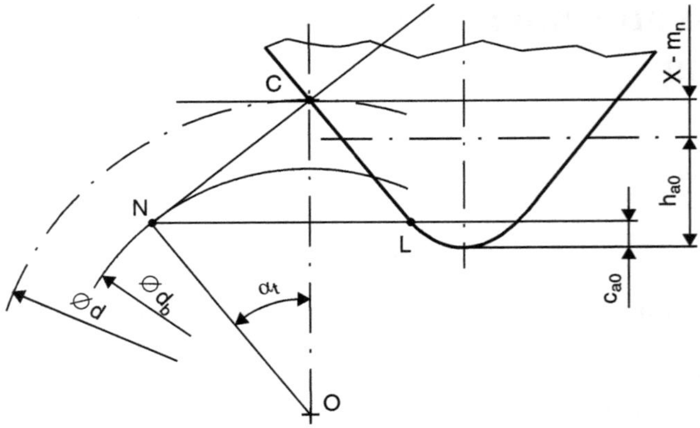

Fig. 1 Undercut limit

presented in [1] and [2]. An influence of a rounded tool crest is not considered. The system of three transcendental equations is solved. This must be carried out numerically. See Fig. 2 end Eqs. (3), (4) and (5).

$$\frac{d}{d_b} - \frac{y}{0,5 \cdot d_b} - \frac{d_L}{d_b} \cdot \cos\left(\alpha_t + \beta + inv\ \delta\right) = 0 \tag{3}$$

$$\frac{d_L}{d_b} - \frac{\beta}{\sin\left(\beta + inv\ \delta\right)} = 0 \tag{4}$$

$$\tan\left(\arccos\frac{d_b}{d_L}\right) - \arccos\frac{d_b}{d_L} - inv\ \delta = 0 \tag{5}$$

3 Exact Start of an Undercut Involute

Although one might expect that a rounded tool crest, which turns aside from a linear edge of a tool, will not cause further undercutting – so contrary is the case. This rounding (which is creating a tooth fillet) cuts off in addition a part of involute indeed, which has already been undercut by a linear edge of a tool. Two curves are seen on the Fig. 3. Curve k_L is manufactured by an end point of a linear edge of a tool. The second one labelled as k_p is a fillet curve which is manufactured by a rounded tool crest (an envelope of its positions). It is seen that an additional undercutting really takes place. This truncation is otherwise very small but when accurate calculations are making, it is necessary to take it into account.

Finding of a point of intersection between a curve k_p and an involute is rather difficult. The most commonly used technique is finding such a diameter (bigger than d_b) for which thicknesses on the involute s_e and on the fillet curve s_p have the same size, see Fig. 4. Only one solution exists above a base diameter. It is important to realize that a rounded tool crest will change into ellipse in transverse section in

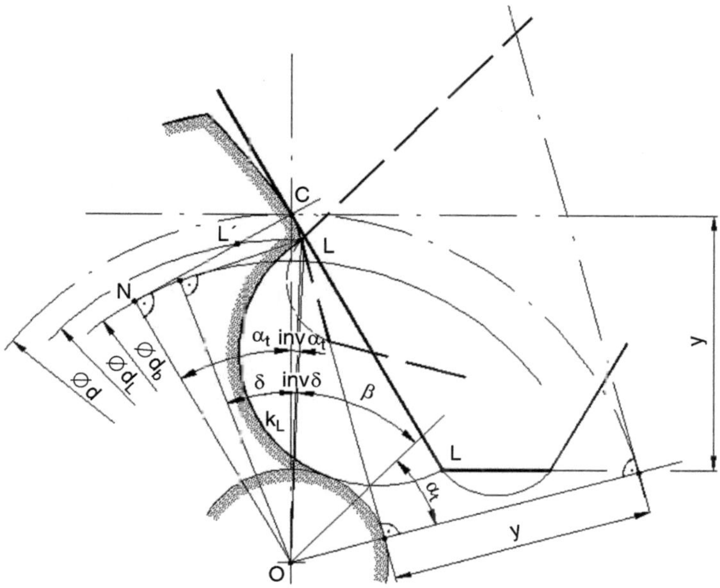

Fig. 2 Start of an undercut involute (the usual way)

Fig. 3 Two curves

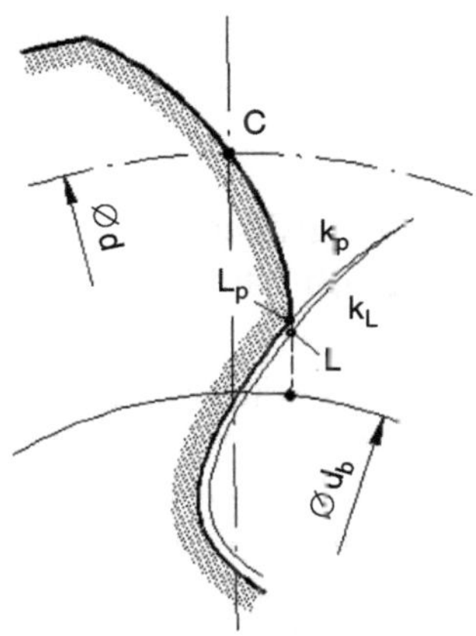

which the solving is made. A fillet curve k_p is an envelope of this ellipse discrete positions.

Except this method there is another method for finding of a point of intersection between a curve k_p and an involute. This method is published among others in the standard ISO 6336, part 2, Annex A [3]. Analysis there is made in polar coordinates. But potential user must be warn of quite a number of typing errors and mistakes in equations and figures.

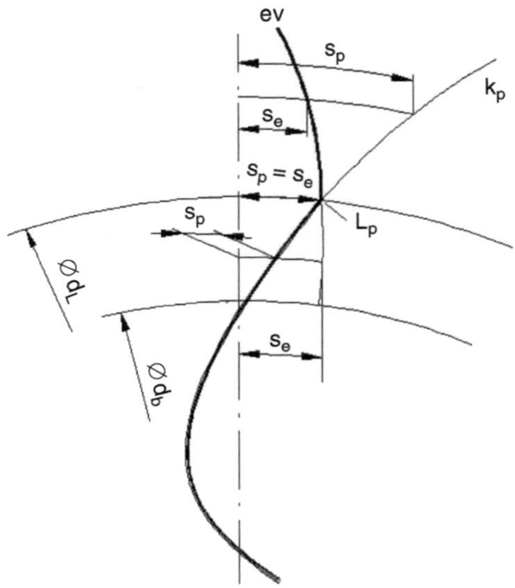

Fig. 4 Comparing of thicknesses

Table 1 Two examples

No. of teeth	Normal module	Helix angle	Shifting coefficient	Diameter		
					Usual way	Exact
z	m_n	β	x	d_b	d_L	d_L
10	1	0	−0.2	9,3969262	9,4993911	9,5018880
20	1	20	−0.6	19,84612599	19,8546048	19,8546368

4 Magnitude of the Difference of Undercutting Between Both Methods

Accuracy of the calculation for both methods will be shown in two examples. The standard basic profile will be used (Table 1).

Differences are practically insignificant. It is possible to use the first simpler method for routine tasks. Its sufficiency is verified in practice. However be aware that the slightest change of a diameter evokes a substantially greater change in a position on a meshing line. For high accurate calculations (contact stresses) it is necessary to use the second method. This method is exactly accurate. Its accuracy was proved by CAD drawing, see Fig. 5. The diameter d_L obtained by this method is 19,854637 mm.

5 Conclusion

It would seem that the difference between both calculation methods is insignificant. And it is true in the majority of cases (Figs. 6 and 7). But for some FEM calculations it is necessary to know the absolutely precise tooth form. Also for precise geometry

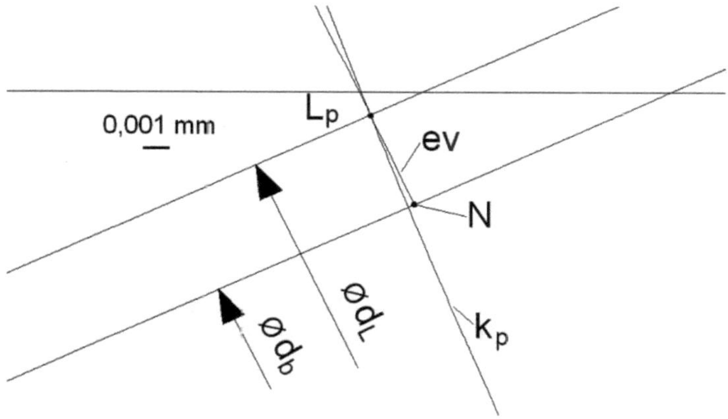

Fig. 5 CAD verification

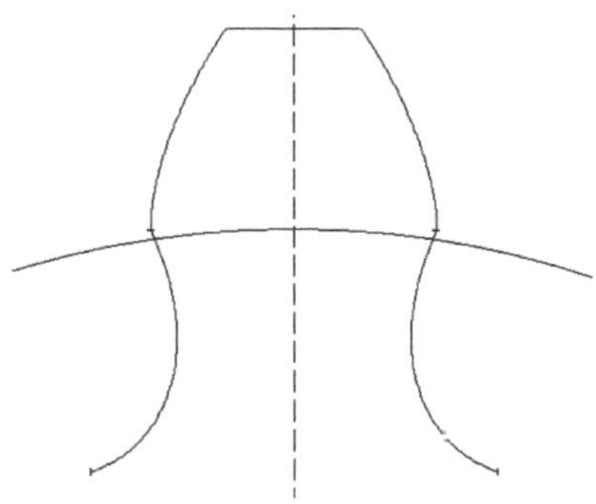

Fig. 6 z = 10, β = 0, x = −0.2

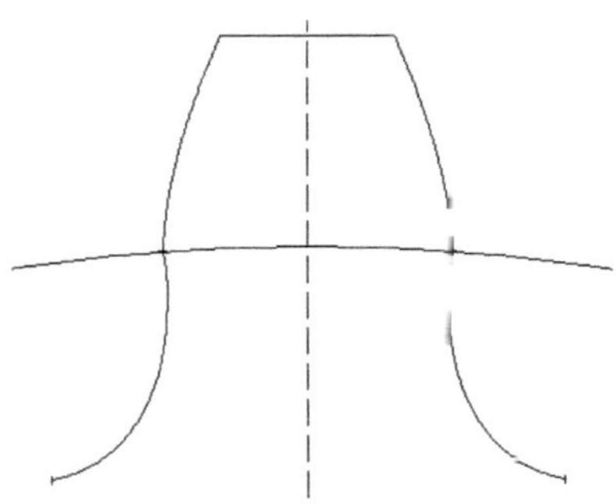

Fig. 7 z = 20, β = 20, x = −0.6

calculations it needs to be concern with it. Very interesting thing is that well known Eqs. (1) and (2) are thereby inaccurate ones because they are derived using the first not very precise method. As long as a designer must know the undercut limit exactly, he must calculate with an additional undercutting (although relatively small one) of an involute and thus with slightly different results.

Acknowledgments This paper was elaborated with the support of Project TE01020020 – Josef Božek Competence Centre for Automotive Industry (Czech Republic). Authors gratefully acknowledge it.

References

1. M. Němček, *Vybrané problémy geometrie čelních ozubených kol* (Montanex a.s, Ostrava, 2003). ISBN 80-7225-111-2
2. M. Šejvl, *Theorie a výpočty ozubených kol (díl 1)* (SNTL, Praha, 1957)
3. ISO 6336, *Calculation of Load Capacity of Spur and Helical Gears* (International Standard Organization, Geneva, 2006)

Assessment of Industrial Design Projects with Use of Online Evaluation System

M. Ondra, R. Machalek, D. Skaroupka, D. Semrad, and F. Uhlir

Abstract During study of Industrial design, students create multiple projects, which are always evaluated in terms of aesthetic value and are assessed afterwards with a grade. Since perception of such designs is a subjective, objectivity can't be achieved by single teacher. The paper analyses current methods of assessing de-sign projects and examines their application in online evaluation system, which is used during education. Method for assessing projects is based on multiple evaluations by teachers. Relationship between ordinary and online assessment are studied and described.

Keywords Industrial design • Evaluation • Online assessment • Design project

1 Introduction

Article deals with types of evaluation of students design projects on Industrial design department FME BUT. The subject of design project evaluation is largely a visual nature of the proposal, which is subjectively perceived by expert evaluator. Common method of evaluation at Department of industrial design, that committee consisting of 3–5 teachers evaluate on base of oral defence of the project. For smaller projects only one teacher evaluates and this brings subjectivity into assessment. In other universities a method of portfolio evaluation is used. Students work on their projects and all their projects are reviewed afterwards. This may take into account differences between previous and actual works.

The importance of on- line evaluation systems will increase with the amount of student works that are submitted in digital form. The aim of any evaluation should be objectivity and impartiality. For this reason, the authors decided to design and verify on- line evaluation system in practical tuition. This system should extend the assessment with independent method designed either for verification, correction, or in some cases, a complete replacement of traditional method.

M. Ondra (✉) • R. Machalek • D. Skaroupka • D. Semrad • F. Uhlir
Brno University of Technology, Brno, Czech Republic
e-mail: y101172@fme.vutbr.cz; ymacha33@stud.fme.vutbr.cz; skaroupka@fme.vutbr.cz; y116022@stud.fme.vutbr.cz; filip.uhlir@teapot.cz

V. Dynybyl et al. (eds.), *The Latest Methods of Construction Design*,
DOI 10.1007/978-3-319-22762-7_11

The aim of this article is to verify proper function and usefulness of the pro-posed on- line evaluation system using statistical methods and compare it with the existing evaluation methods.

2 Problem Statement

In terms of interaction between evaluators we find two types of evaluation categories: influenced (group) / independent (online)

2.1 Influenced Evaluation

Group of experts (teachers) discusses proposals in an open debate, then each of them evaluates individually. Every one of them may be influenced by the views of his colleagues. The second method is evaluating exclusive individual student works, which are not discussed before assessment but are evaluated by 3 experts individually in sequence. In this method, the evaluator is influenced by assessment of previous expert. This means that only the first one evaluates independently.

In the online evaluation system each teacher uses scale of. Evaluator sees all works and uses slider or bracket for giving points to individual projects (Fig. 1).

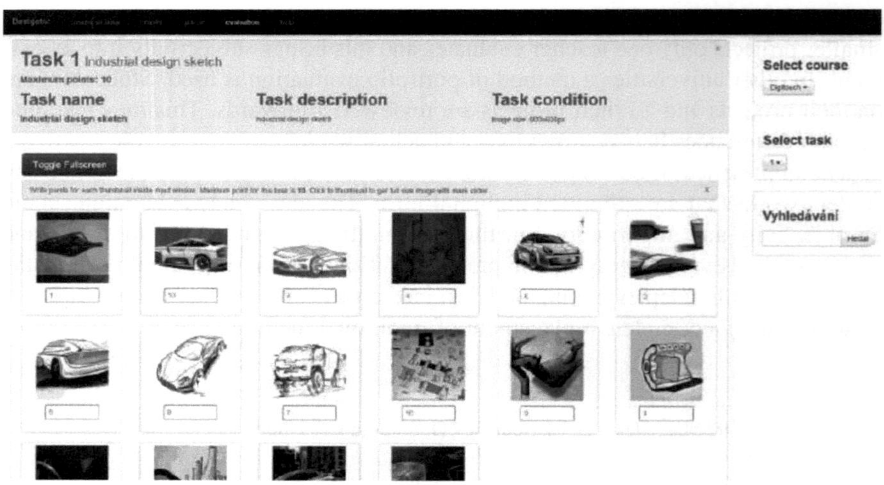

Fig. 1 GUI of Desigator system

3 Methods

To verify the accuracy of the evaluation procedures, we used the following two methods: W Test and Pearson's correlation coefficient.

Firstly it is necessary to check if the representations of individual evaluation stages from the total number of marks from all evaluators correspond to the nor-mal (Gaussian) distribution. For this purpose we used the W test (Shapiro – Wilk test for Normality). W test provides a statistic for evaluation of supposed normality of complete samples [1] and enables exploring the distribution of a sample and their test for non – normality [3]. If we assume that the results of student evaluations respond to the quality of their work, we expect that the rating scales will be distributed normally.

The second step was to verify the degree of independence of individual ratings using Pearson's correlation coefficient, which describes the relationship between the evaluators and the resulting assessment. Pearson's Correlation Coefficient "is a measure of the strength of the association between the two variables" [4]. We assume that the evaluation is credible if the inter-evaluator agreement is not too high or too low.

4 Experimental Results

Figure 2 describes basic results from distribution of marks student projects received from on-paper and online evaluation. The main result are the scales of assessment teachers use. For on-paper method teachers seem to have C as average result, whereas online evaluators are close for average result at 50 % of given points (mark E-D).

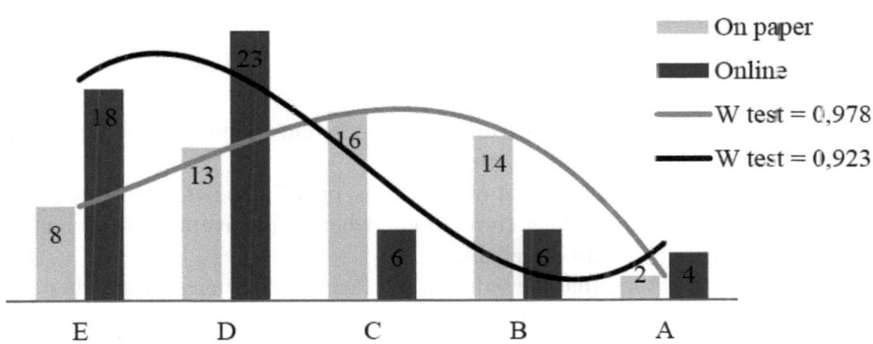

Fig. 2 Graph of received marks distribution on-paper vs. online evaluation

Table 1 Results of W test

Task no.	A	B	C	D	E	F	Mean	Cases	Scale	Limit
1	0,919	0,929	0,891	0,927	0,931	x	0,948	21	20	0,908
2	0,932	0,943	0,897	0,858	0,91	x	0,931	22	20	0,911
3	0,89	0,887	0,922	0,866	0,945	x	0,936	22	10	0,911
4	0,942	0,951	0,882	0,968	0,853	x	0,973	21	30	0,908
5	0,975	0,912	0,95	0,942	0,931	0,945	0,889	26	30	0,92
6	0,963	0,936	0,961	0,964	0,945	x	0,979	26	40	0,92

Table 2 Results of Pearson's correlation coefficient

Task no.	A	B	C	D	E	F
1	0,742	0,659	0,612	0,652	0,523	x
2	0,453	0,658	0,623	0,279	0,436	x
3	0,791	0,363	0,414	0,734	0,742	x
4	0,609	0,599	0,467	0,676	0,466	x
5	0,669	0,871	0,807	0,8	0,833	0,781
6	0,69	0,759	0,673	0,668	0,785	x
Mean	0,659	0,651	0,599	0,635	0,631	0,781

4.1 W Test

Table 1 contains the coefficients obtained by analysing online evaluation with W-test method. Individual values for the task are listed in rows. In the column "Mean" normality coefficients are given for the analysed tasks. Column "Limit" shows the limit value W test for $p < 0.05$ [1]. In the event that the value of "Mean" is larger than the belonging value "Limit", we consider the distribution as normal.

From the results in Table 1 we see that, only the task no. 5 is not normally distributed, with mean value 0.889 which is less than limit of 0.920 [2].

4.2 Pearson's Correlation Coefficient

Accordingly to interpretation of Cohen [5] coefficients about 0.1 are significant from 0.5 values, the correlation is strong.

As can be seen from Table 2 out of 31 ratings, only 7 reviews may be classified differently than with the "strong correlation", whereas the average coefficient for all evaluators again strongly correlates.

From Table 3, we see that the value of the Pearson coefficient are higher than 0.8. It can be deduced that due to the values of the coefficients used in psychological research [5] it is an extremely high value.

Table 3 Result of exclusive individual task evaluation – on-paper method

Evaluator	Avrg. (30)	St. Dev.	W test	Pearson
A	22,8	3,4	0,963	0,885
B	22,2	3,4	0,978	0,906
C	21,4	3,3	0,972	0,936
Mean	22,1	3,1	0,978	x

Table 4 Result of exclusive individual task evaluation – online

Evaluator	Avrg. (30)	St. Dev.	W test	Pearson
D	19	6,3	0,939	0,663
E	20,6	4,7	0,947	0,606
F	20,4	5,8	0,958	0,628
Mean	20	3,5	0,923	x

5 Conclusion and Discussion

With W test and Pearson methods were found coefficients that verify the functionality of described online evaluation system Desigator. These results will be further verified in subsequent research through additional tests and other assessments of student design projects. The authors suppose that due to the nature of the evaluation process, their validity will be confirmed.

The online evaluation system was designed (and included in tuition) for assessment of students projects. A particular student project was evaluated experimentally and the results were processed by statistical methods that allow judging of the accuracy (bias and objectivity) of the entire evaluation process. Evaluation of the same task on-paper (influenced evaluation) was subjected to same analysis.

By comparing statistical data from both types of evaluation that:

1. Online evaluation has normal distribution.
2. Online evaluation has greater independence of evaluators than used on-paper method (comparison Tables 3 and 4).

Further research will focus on the verification of these results and development and modification of parameters of the online evaluation system.

Acknowledgments Funding for this research was provided by specific research FSI – J – 13–2102.

References

1. S.S. Shapiro, M.B. Wilk, An analysis of variance test for normality (complete samples). Biometrika **52**(3/4), 591–611 (1965), ISSN: 0006–3444
2. P. Royston, A remark on algorithm AS 181: the W-test for normality. J. R. Stat. Soc. Ser. C Appl. Stat. **44**, 547–551 (1995)

3. *Normality tests* [online]. StatsDirect Limited. [citováno 15.5.2014]. Dostupné z: http://www.statsdirect.com/help/default.htm#parametric_methods/normality.htm
4. *Data analysis* [online] University of the West of England. [citováno 15.5.2014]. Dostupné z: http://learntech.uwe.ac.uk/da/Default.aspx?pageid=1442
5. *Effect Size* [online]. Emory University. [citováno 15.5.2014]. Dostupné z: http://www.psychology.emory.edu/clinical/bliwise/Tutorials/SCATTER/scatterplots/effect.htm

Vibration Isolation Pneumatic System with a Throttle Valve

L. Pesik, A. Skarolek, and O. Kohl

Abstract The paper shows a possibility of the tuning mechanical system by means of two pneumatic springs in a differential configuration connected with a throttle valve. The springs are inserted into the lead mechanism and connected to its parts, and to its supporting platform. The vibrations, transferred from the kinematic excitation of the base, are intended to be minimized. The vibration isolation by means of pneumatic springs is available in many technical systems, e.g. in supports of heavy machinery as well as in systems characterized by the human interaction, such as driver seats, ambulance couchettes, etc. The pneumatic springs provide the option of adaption of the stiffness, and herewith the adaption of the natural frequency of the system according to the exciting frequency. In cases of application of the object vibration isolation, they can change the load characteristics in a relative large range. In the studied case of the differential spring configuration, the springs are connected with an air pipe to the throttle valve. The air being exchanged during the motion period comes through the valve, the cross-section of which determines the time delay of the pneumatic sub-system thus creating a hysteresis of load characteristic of the spring support. This brings an additional, controllable damping to such a system that is profitable in most vibration isolation cases.

Keywords Pneumatic spring • Vibration isolation system

1 Introduction

There are many technical problems that are connected with a necessity of the minimization of the vibrations. Some of them are present during the run of the machines when intensive dynamic forces are transmitted into their foundations. In these cases, the elastic supports are used for an effective dynamic isolation. Here, the basic principle is

L. Pesik (✉) • A. Skarolek
Faculty of Mechanical Engineering, Technical University of Liberec, Studentská 2, 461 17 Liberec, Czech Republic
e-mail: lubomir.pesik@tul.cz; antonin.skarolek@gmail.com

O. Kohl
Technical University of Liberec, Liberec, Czech Republic
e-mail: ondrej.kohl@tul.cz

V. Dynybyl et al. (eds.), *The Latest Methods of Construction Design*,
DOI 10.1007/978-3-319-22762-7_12

grounded on the sufficient difference between the force excitation frequency and resonant frequency of the dynamic system. For processing machines, the excitation frequency is usually constant and relates to the operation speed. The vibration isolation system supporting these objects is created by springs with unchangeable load characteristics. On the other hand side, especially in the automotive industry, there are many types of equipment which have an excitation through the movement of the foundation. The typical instance of this set is a driver seat. With vibration isolation of these objects, there are further problems that come from variable excitation frequency. It is clear that the condition of the sufficient difference between the force excitation frequency and resonant frequency of the dynamic systems cannot be ensured at all times. In these cases, it is necessary to use special supports with the possibility of the stiffness changing, and to tune the natural frequency of the system appropriately.

2 Principle of the Resilient Pneumatic Support

The resilient support is created by a lead mechanism, in which two pneumatic springs are inserted, see (Fig. 1). Here they have their ratios of transmission. One of the springs has a positive ratio and the other one a negative one. This means that the first of the pneumatic springs brings the support upwards and the second one downwards. The force effect of the second spring on the support has a similar effect of its loading. This configuration gives possibility to adapt the stiffness as required by the changing of the air pressure in the second spring. This change can be executed in the still stand or during the movement of the supported object.

In the studied case, the pneumatic springs are connected not only by the lead mechanism but also by pneumatic tubing with a throttle valve. This configuration is called differential. The effective cross-section of the valve can be adjusted standstill or can be controlled during the motion. In this paper, we deal only with the former situation, i.e. the selecting of the optimum throttle valve cross-section to achieve the most promising amplitude-frequency characteristic of the mechanism. Considering the two boundary cases of the throttle valve setup, the closed valve and the fully open valve with the infinite cross-section, the valve would have following effects. Fully closed valve does not permit any air exchange between the springs; the mechanism is stiff with no additional energy dissipation. The pressures inside the springs change in time with 180° phase shift. For a very large cross-section of

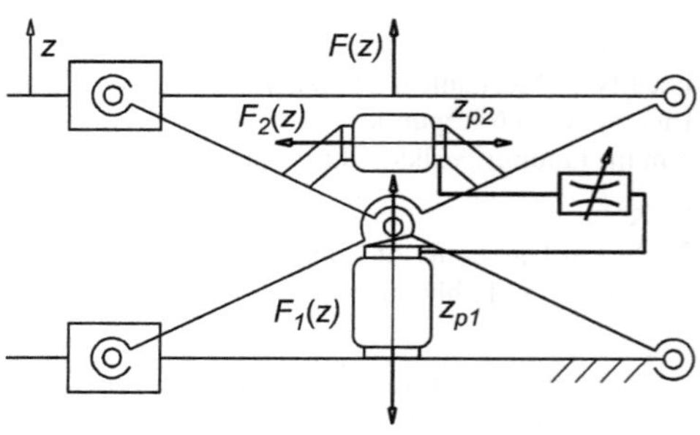

Fig. 1 Principled scheme of the differential pneumatic support with throttle valve

the valve, the stiffness of the mechanism is significantly reduced, the pressure inside the springs is changing in phase as they are now equal all the time, but again there is virtually no additional energy dissipation. The valve setup between the two extremes will allow for maximizing energy dissipation by creating a certain phase shift between spring pressures and thus spring forces [1–3].

3 State Equation of the Pneumatic Mechanical System

Equation of motion of the mechanism platform with reduced mass *m* (including the part of the mechanism, seat and the passenger) is simple

$$m\frac{d^2}{dt^2}(z(t)+u(t))+b\frac{d^2}{dt^2}(z(t)-u(t))=-mg+F, \tag{1}$$

Equation of motion of the mechanism platform with reduced mass *m* (including the part of the mechanism, seat and the passenger) is simple

$$m\frac{d^2}{dt^2}(z(t)+u(t))+b\frac{d^2}{dt^2}(z(t)-u(t))=-mg+F, \tag{2}$$

where $z(t)$ is the absolute displacement of the platform, $u(t)$ is the displacement of the base under the kinematic excitation, b is the construction damping of the mechanism, which can be observed experimentally, mg the static load of the mechanism [4, 5].

The function F is the equivalent force from the springs

$$F=i_{p1}(z)\cdot F_1(z_{p1},p_{p1})+i_{p2}(z)\cdot F_2(z_{p2},p_{p2}), \tag{3}$$

determined by means of transmission functions $i_{p1}(z)$ and $i_{p2}(z)$

$$i_{p1}(z)=\frac{z_{p1}(z)}{z},\qquad i_{p2}(z)=\frac{z_{p2}(z)}{z}. \tag{4}$$

Spring displacements z_{p1}, z_{p2} are also determined by these ratios

$$z_{p1}=i_{p1}(z)\cdot z,\qquad z_{p2}=i_{p2}(z)\cdot z. \tag{5}$$

Air pressures inside the springs obey the state equation of ideal gas

$$p_{p1}=\frac{m_{a1}r\ T}{V_1(z_{p1})},\quad p_{p2}=\frac{m_{a2}r\ T}{V_2(z_{p2})}, \tag{6}$$

where m_{a1} and m_{a2} are masses of the air enclosed inside the springs, r and T is specific gas constant and temperature respectively [6–8]

The air exchange between the springs is described by isentropic air flow through the throttle valve. In the next two equations, the rate of air exchange depends on pressures p_A and p_B; the pressure p_A signs the higher pressure of p_{p1}, p_{p2} at given time, p_B is the other one. Which of the two pressures is higher determines the sign of the flow rate. The rate of air mass is then

$$\frac{\mathrm{d}m_{a1(a2)}}{dt} = A_v\, c\, p_A \sqrt{\frac{2}{rT}\frac{\kappa}{\kappa-1}\left(\left(\frac{p_B}{p_A}\right)^{\frac{2}{\kappa}} - \left(\frac{p_B}{p_A}\right)^{\frac{\kappa+1}{\kappa}}\right)}, \tag{7}$$

for subcritical flow conditions, where $p_B/p_A \geq \beta^*$ or

$$\frac{dm_{a1(a2)}}{dt} = A_v\, c\, p_A \sqrt{\frac{2}{rT}\frac{\kappa}{\kappa+1}\left(\frac{p_B}{p_A}\right)^{\frac{2}{\kappa-1}}}, \tag{8}$$

otherwise. Critical pressure ratio $\beta*$ is

$$\beta^* = \left(\frac{2}{\kappa+1}\right)^{\frac{2}{\kappa-1}}, \tag{9}$$

where κ is specific heat ratio for the air. In the Eqs. (7) and (8) are c discharge coefficient and A_v cross-section of the throttle valve.

Differential Eq. (6) supplemented by differential equation for air mass inside the springs fully describe the presented pneumatic-mechanical system. We have considered closed pneumatic system, so the air masses are bound by condition [9]

$$m_{a1} + m_{a2} = m_a\,. \tag{10}$$

4 Simulation Results

On the Fig. 2, there are depicted three load characteristics of the mechanism under kinematic excitation of 0.6 Hz (graph A) and 1.1 Hz (graph B). The selected cases are 1, 1.4 and 2 mm of the throttle valve nozzle [10].

5 Conclusion

This solution presents a pneumatic spring system with a differential configuration and the throttling of the air flow between the springs. The springs act one against the other. The cross-section of the throttling element permits the changing of the amplitude-frequency characteristic with regards to the position and magnitude of the peak resonant amplitude. This system can be used with benefits to the vibration

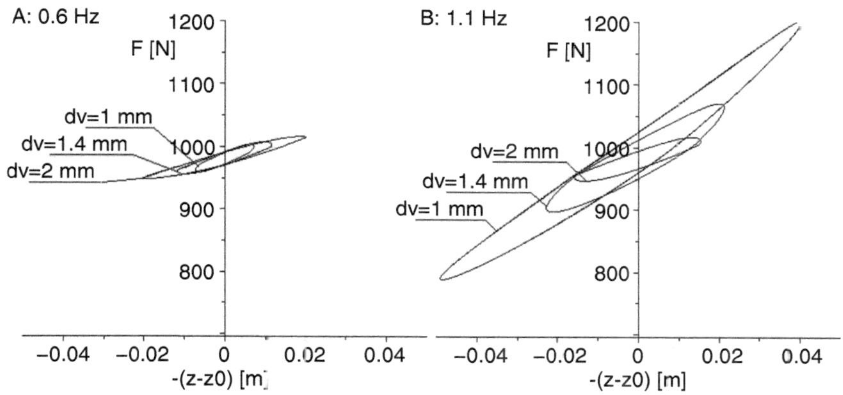

Fig. 2 Mechanism dynamic load characteristic, three valve setup at two excitation frequencies A: 0.6 Hz, B: 1.1 Hz

isolation of the objects of the systems with kinematic excitation, e.g. driver seats, ambulance couchettes, etc. Based on the frequency spectrum of excitation, it is possible to choose the optimum cross-section of the throttling element and achieve efficient vibration isolation in a relatively broad range of low excitation frequencies.

References

1. C.M. Harris, *Shock and Vibration Handbook*, 5th edn. (McGraw-Hill, NewYork, 2005)
2. R. Uhlir, P. Mossoczy, P. Maly, Experimental testing of agricultural trailer, in *Book of Proceedings of 54th International Conference of Machine Departments* (Technical University of Liberec, Liberec, 2013), pp. 353–358. ISBN 978-80-7372-986-8
3. D.J. Whitehouse, *Handbook of Surface and Nanometrology*, 2nd edn. (CRC Press, Boca Raton, 2011). ISBN 978-1-4200-8201-2
4. Y. Girfhar, A. Xu, B. B. Dey, M. Meghjani, F. Shkurti, I. Rekleitis, G. Dudek, MARE: Marine Autonomous Robotic Explorer, in *Mobile Robotics Lab* [online] (McGill University, 2011), [vid. 2011_10_10]. Available via: http://www.cim.mcgill.ca/~mrl/pubs/anqixu/iros2011_boat.pdf
5. V. Zeman, Z. Hlaváč, *Kmitání mechanických soustav* (Západočeská univerzita v Plzni, Plzeň, 2010). ISBN 978-80-7043-337-9
6. Z. Bradský, R. Vrzala, *Mechanika III. Dynamika* (Skripta VŠST, Liberec, 1986)
7. H. Dresig, F. Holzweißig, *Maschinendynamik*, 8th edn. (Springer, Berlin, 2008)
8. L. Pešík, A. Skarolek, Heat transfer effects on vibration isolation differential pneumatic system, in *Transactions of the Universities of Košice: Research Reports from the Universities of Košice* (Technická univerzita v Košiciach, Košice, 2012), pp. 109–114. ISBN 1335-2334
9. L. Pešík, A. Skarolek, Tuning of vibration isolation differential pneumatic system by means of trottle valve, in *Transactions of the Universities of Košice: Research Reports from the Universities of Košice* (Technická univerzita v Košiciach, Košice, 2011), pp. 191–196. ISBN 1335-2334
10. M. Mansouri, Q. Ilham. H. Snoussi, C. Richard, Adaptive quantized target tracking in wireless sensors networks. Wirel. Netw. **17**(7), 1625–1639. ISSN: 1022-0038. 2011

Development of the Assembly Set for the Logistic Transport Solution

R. Sásik, M. Haluška, R. Madaj, M. Gregor, and P. Grznár

Abstract Reconfigurability has become as a new engineering technology which has a significantly impact on the changes in design. Reconfigurable manufacturing systems should be designed so that they can effectively respond to changes in the product requirements. Currently reconfigurable products will be a great benefit for customers and also a big challenge for designers of the manufacturing systems.

Keywords Assembly set • Logistic transport solution • Prototype • Rapid prototyping

1 Introduction

The aim of today's enterprises is to produce a different variants of the products that ensure their profitability. Due to the fact the enterprises must to focus on the customer requirements and low cost production. An acceptable alternative is the production of modular product components that can be easily and quickly mounted on the basis of the actual requirements of customer (personalized product). However the customers will much more prefer a product that can be easily reconfigured. Each configuration is represented by the specific product. But it is necessary to take into account the modular architecture of the product, which allows us to propose a reconfigurable product. Modularity of the product is a sufficient but not necessary condition for creating reconfigurable product. Reconfigurable product must be convertible from one shape to another or scalable.

According to the Koren [1] the basic characteristics of reconfigurable product can include:

- **Modularity** – Product modularity enables the change of product shape to fit the various customers and utilizations.
- **Integrability** – Integrability refers to defined module interfaces (mechanical, electrical, and information) for ease of product reconfiguration and customization.

R. Sásik (✉) • M. Haluška • R. Madaj • M. Gregor • P. Grznár
University of Zilina, Žilina, Slovak Republic
e-mail: robert.sasik@fstroj.uniza.sk; michal.haluska@fstroj.uniza.sk;
rudolf.madaj@fstroj.uniza.sk; milan.gregor@fstroj.uniza.sk; patrik.grznar@fstroj.uniza.sk

V. Dynybyl et al. (eds.), *The Latest Methods of Construction Design*,
DOI 10.1007/978-3-319-22762-7_13

- **Customization** – Customization provides customers with just the product functions they need, and not features that they will not use.
- **Convertibility** – Convertibility enables products to be changed to fulfil various utilizations of the same product.
- **Scalability** – Scalability allows a product to be scaled up or down to fit to the user's body or applications.
- **Diagnosibility** – Diagnosibility means that a product is designed with embedded diagnostic functions, for easy service and maintenance.

Due to the rapidly changes customer requirements, manufacturers must have manufacturing system that will offer different levels of response to changing the customer demand but also reckon with new practices in the field of design [2].

2 Assembly Set of the Prototype

In laboratory ZIMS was designed and subsequently produced the first mounting assembly prototype of the logistic transport solution. The mission of the prototype is to integrate the advanced technologies and workplaces in the laboratory. It should be also noted that the product will be used to developing of different variants, which include production of product with different features. From the point of view, the product must be modular and should suggest also other modules, which will be easy integrable and implementable. Replacing of the components can affect the added value, which can product created.

Basic components of the assembly set include front axle, rear axle, shell, I. part of the logical box, II. part of the logical box and screws. In the mounting assemblies is also possible to incorporate chassis that is navigated through bluetooth chip and also chassis which is navigated through infrared communication interface.

2.1 Rapid Prototyping of the Designed Model

Assembly set was designed in Creo Paremetrics 2.0 and SketchUp 2013. Prepared digital parts of assembly for 3d printing should be as a solid object with closed surfaces. The functional model has scale 1:5. Digital model was exported to STL format for 3d printing. There was used 3d printer Stratasys Dimension SST 768 for rapid prototyping. The parameters of the 3d printer are:

- printed method FDM
- used material ABS+
- Thickness of the printed layer 0.25 mm
- System Dimensions: $914 \times 686 \times 1041$ mm – max. dimensions of the printed part

After printing was carried out surface finishing of each component which included cementing, sanding, polishing of outer surfaces and spattering with airbrush.

Assembly set is represented by the reconfigurable product which possesses of essential characteristics of the reconfigurability as modularity, integrability of new functions and convertibility. It is also possible to easy change and integrate individual components.

The next step of the development of the designed model is a manufacturing of the miniseries used vacuum casting. We have modified the model to scale 1:10 for vacuum casting. Master model was printed with 3d printer Formiga P100. This 3d printer uses laser sinthering method. It has smoother printed layer 0.1 mm. We had to prepatre printed 3d model by smoothing and spraying of surfacer.

In addition to developing of the transport solution model, we develop the model of the electric vehicle EDISON. It is static model nowadays, but we want to develop its own power drive.

3 Functional Analysis

There was also performed functional significance analysis of the individual system components from which is evidently clear that the most important link in the functional set is chassis that consists of logical components.

Simultaneously was also identified satisfaction fulfilment rate for each function which shows the direction of the further improving. Among the critical functions of the product can be assign control of movement direction, protection against external environment, reusability and quality of image recognition.

4 Figures

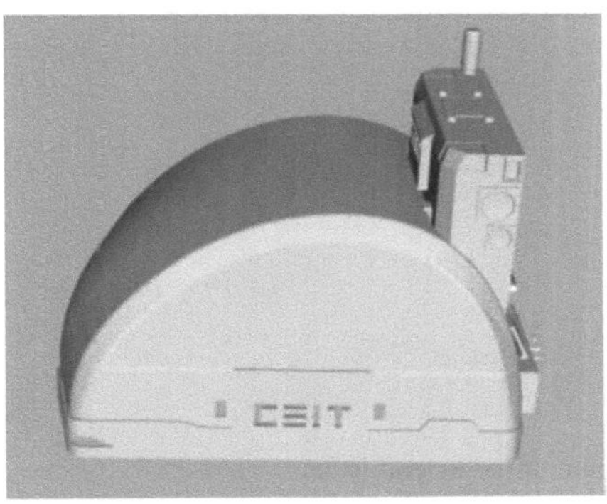

Fig. 1 Digital model of the assembly set

Fig. 2 Prototype of the assembly set

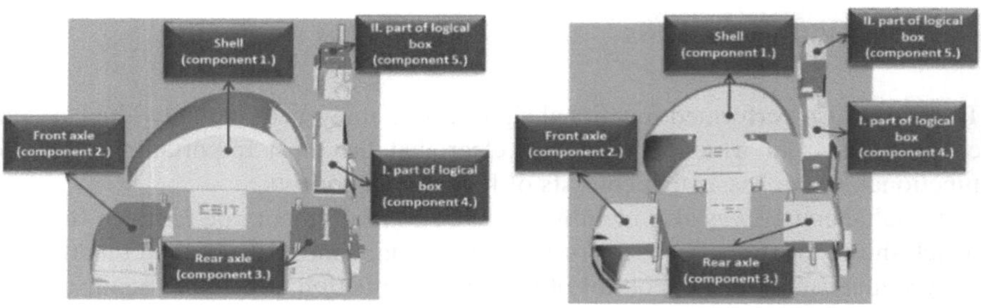

Fig. 3 Components of the assembly set

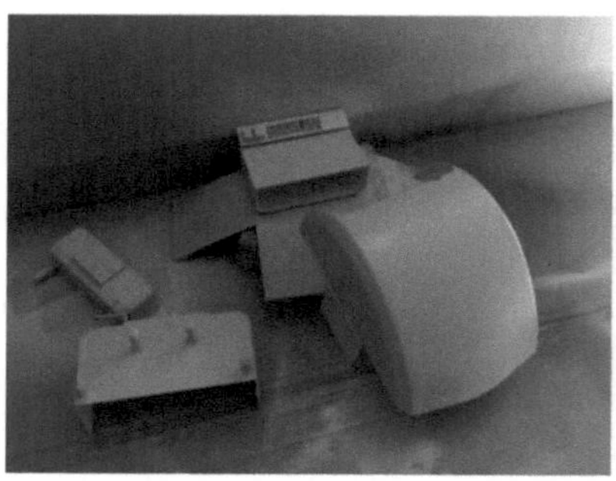

Fig. 4 Components scheme of the assembly set

Fig. 5 Surface finishing of the prototype

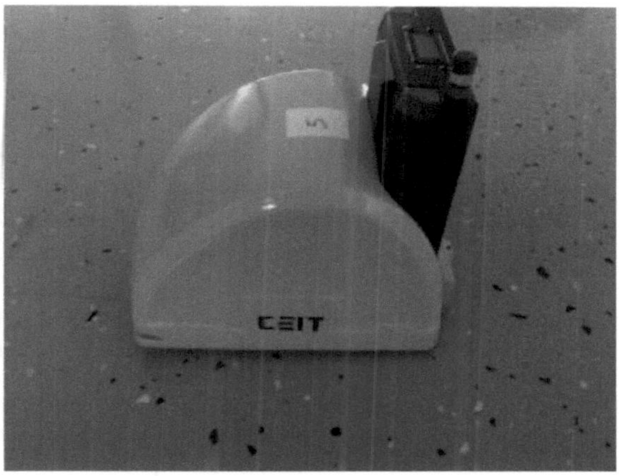

Fig. 6 Functional model produced by vacuum casting

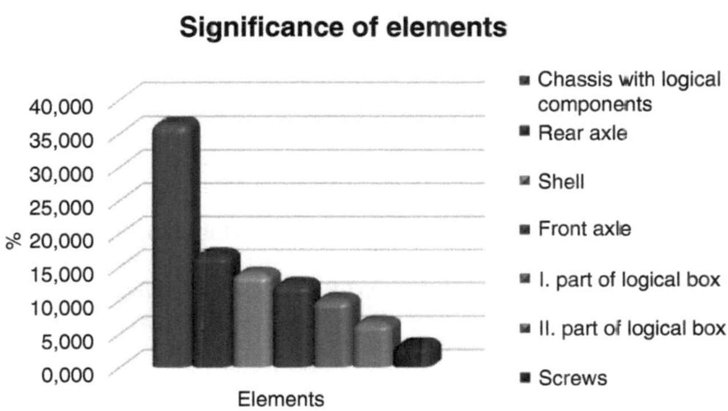

Fig. 7 Graphic representation of the significance of individual elements

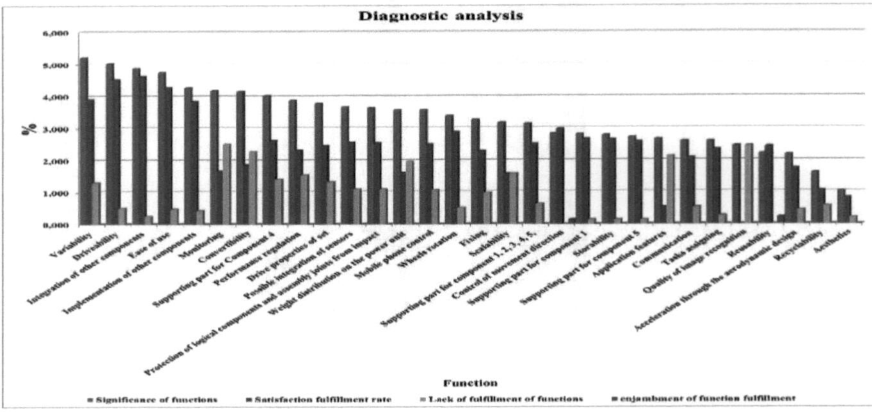

Fig. 8 Diagnostic analysis of functions

5 Conclusion

From the functional analysis is evident that product has still deficiencies in control but also in the level of possible reconfiguration. The product will be used not only for the integration of technologies in laboratory ZIMS but also for the direct development of the own logical platform and product variants.

Acknowledgments This work was supported by the Slovak Research and Development Agency under the contract no. APVV-0419-11 – Adaptation of modern computer – simulation methods to the development of rolling bearings and their verification in real conditions.

This work was supported by the Scientific Grant Agency of the Ministry of Education, Science, Research and Sport of the Slovak Republic under the contract no. V-1/0396/14 – Research of the influence of construction and technological parameters of rolling bearings on the service life thereof.

This work was supported by the Scientific Grant Agency of the Ministry of Education, Science, Research and Sport of the Slovak Republic under the contract no. V-1/0844/13 – Research on guidance elements of rolling bearing and their design.

References

1. Y. Koren, *The Global Manufacturing Revolution* (Willey, Hoboken, 2010). 333 p. ISBN 978-0-470-58377-7
2. B. Mičieta, H. Tureková, *Inovačný manažment: Podpora vzniku inovácií* (Georg, Žilina, 2010)

The Design of the Device for the Manufacture of Planar Filter

L. Sevcik

Abstract The contribution describes a device for folding flat filter masks civil protection with nano fiber layer. This layer is in a protective case and requires special treatment. The aim of the research was the development of a new type provided that the production of the filter will be large series. Activity in the folding machine must be in controlled mode, without human intervention. The main problem is the mechanical rigidity of the needles that are control by pneumatic system. These needles are essential for equal foundation fabric. In this paper will be made the strength check by FEM. Was checked deformation and stress of the needles. They were designed structural measures to ensure the function of the mechanism.

Keywords Masks • Filters • Layer of nanofibers • Design • Assembly

1 Introduction

Connection with the filter masks was implemented through the surface masks – flat surface with seal [1] – seating area filter in the face of the mask must be sealed in the area, namely in front of the filter. The filter should have a greater rigidity than the abutment face masks that could be moulded. Thrust mechanisms must be spaced around the circumference.

The advantage is its simplicity of shape of filter – no need to screw surfaces [2].

The disadvantage of this connection is the size of the area and a small contact pressure, the possibility of penetration filtered pollutants.

A common feature for composing the filter material is to get as much surface pleated fabric into the smallest possible volume. The offered design uses supporting form with open ends.

Design of the particle filter is based on the mechanical properties of textile filter material and the possible design dimensions in the mask, Fig. 1a. The goal was to filter area should be 1.5 dm^2, material "Spanbond" in two layers of polypropylene and between them is a layer of nanofibers [3]. This textile sandwich structure has a

L. Sevcik (✉)
Technical University in Liberec, Liberec, Czech Republic
e-mail: ladislav.sevcik@tul.cz

V. Dynybyl et al. (eds.), *The Latest Methods of Construction Design*,
DOI 10.1007/978-3-319-22762-7_14

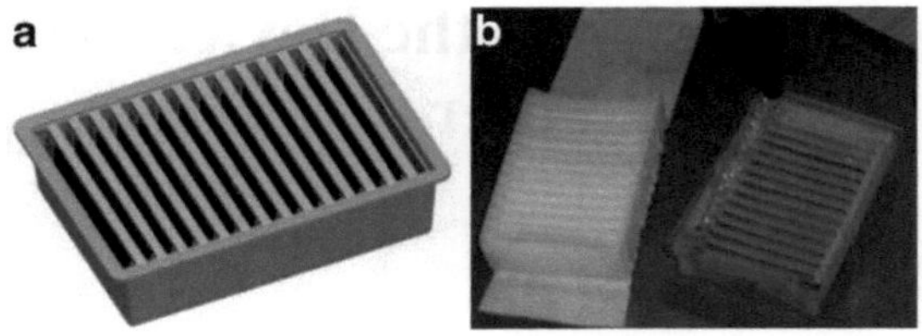

Fig. 1 Designed filter (**a**) and assembly of filter with hot glue (**b**)

relatively low flexural rigidity and low buckling and so it was decided to positively support the filter fabric layer parts cover. The following image is composed of outdoor filter cover and support work, which fits into the inner support part, in which there is pre-stacked filter fabric. These parts fit together and cover both sides of the filter layer. Parts can be glued. Fabrics in inner comprise a pair of interlocking fingers alternately, so that one finger holds the previous tooth of inner part of the filter the filter material and the second finger stubs to the next tooth.

Filter is composed of two plastic parts. In one of them is stacked filter fabric with nanofibers. At the bottom of this section there are the cover ribs preventing damage to the filter layer. The second plastic part comes from the upper side and protects the filter layer from the outside. On the upper part the glue is applied. Both these parts including the filter layer are compressed and combined with glue, Fig. 1b. The filter element is supported by ribs and sealing facets around the circumference of the filter.

2 Design of Automatic Assembly Machine

For the automatic assembly of the filter was designed functional sample device, consisting from the machine frame, secure mounting space, mounting filter device, device for applying adhesive, control panel and the electrical equipment. Designed equipment was based on an experience with manual assembly process. Many tests of conducted composing and assembling filters were made. It is important, that the filter folding mechanism worked smoothly, without harming of thin nanofiber layer, from which is derived the effectiveness of the filter. On the basis of these tests was assembled machine. The distribution of aggregates and process equipment operations was determined according to Table 1.

Some operations must be checked subjectively, especially attaching the filter material and its possible racking when folding. It is also necessary check quality adhesive tape applied and quality of the trimming filter layer.

Frame of the device was assembled from AL profile; top board was made from aluminum plate, the bottom board from chipboard. The doors are protected by a safety switch and the input operator was use an electronic barrier. Device in 3D computer space was drawn with CAD model system Pro/Engineer see Fig. 2a. In this system was created also technical drawings. According to this documentation was made functional model of automated assembly machinery filters. The device Fig. 2b has been tested.

Table 1 Time diagram with the activities of individual units

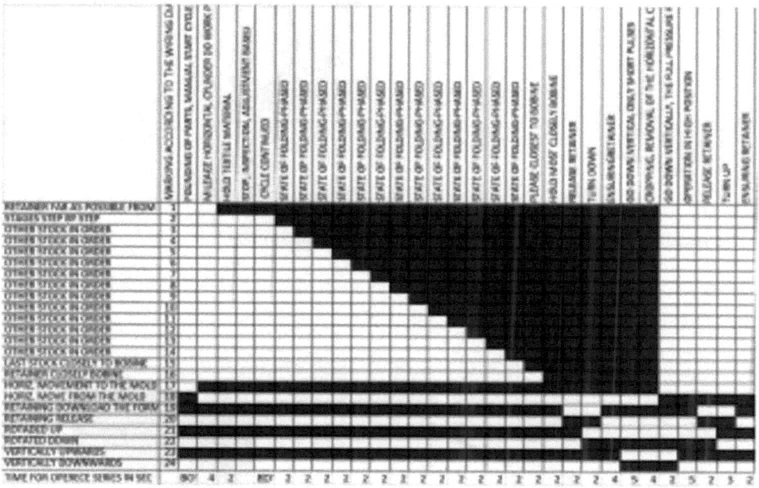

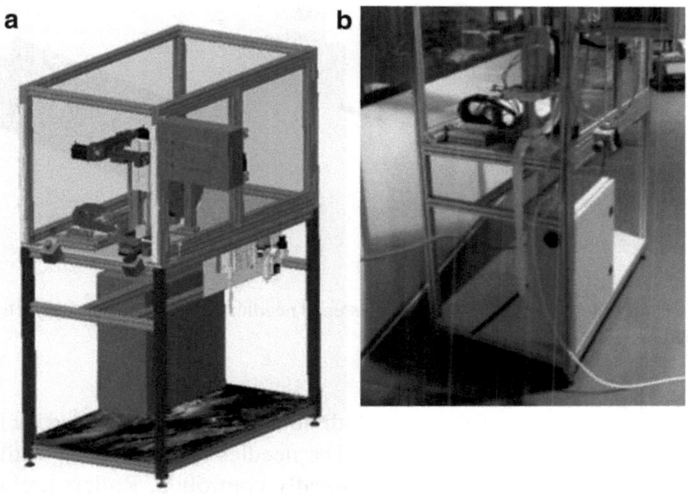

Fig. 2 CAD model (**a**), and equipment manufactured equipment (**b**)

3 Strength Calculations of Principal Part of the Machine

In Fig. 3a is model of the main part of the folding mechanism of the machine. The mechanism is composed from two parts. The moving parts located on the pneumatic cylinder with linear guide and static part, attached to the machine frame. The moving part of the filter must pull out before own press-fit. The movable part comprises foldable needles that create folds acting on the filter material. Pleats are

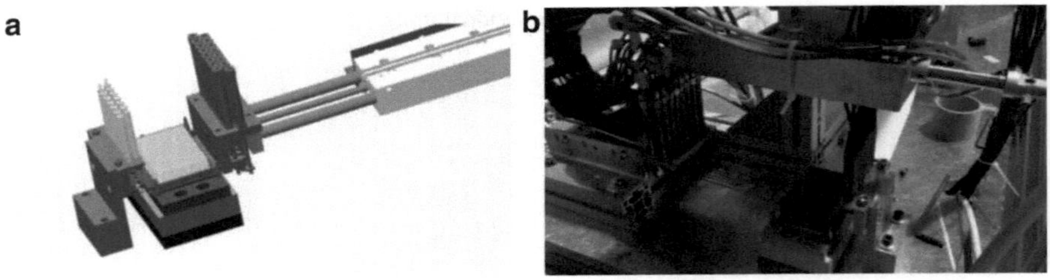

Fig. 3 The mechanism of folding model (**a**) real mechanism (**b**)

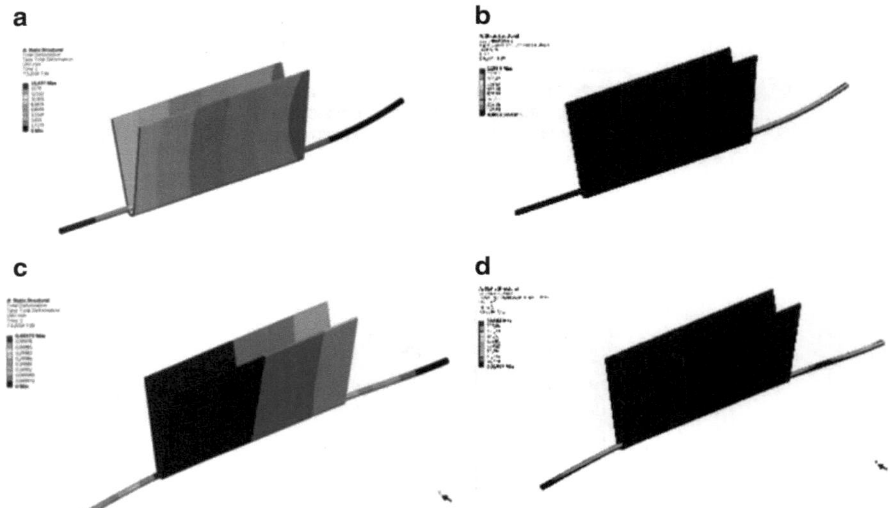

Fig. 4 (**a**) Deflection (**b**) tension of one side fastened needle, (**c**) deflection and (**d**) tension of both side fastened needle

created in the inner ridge of the filter. Individual needles are guided by rack. This determines the position of the needles. The needles are associated with a single-acting pneumatic cylinder – programmatically controlled. Rollers are alternately switched on as scheduled in the Table 1. The cylinders work in pairs with the fixed-side roller. Rollers on the fixed side help to the formation of the pleat.

Figure 4a shows the strain one side imposed the needle during the formation of the pleat; in Fig. 4b is shown tension in needle. Stress and strain are too large and so was elected thrust needle from the other side. This reduced the deflection of the needle 50 times see Fig. 4c and tension in the needle six times, see Fig. 4d.

4 Conclusion

On the basis of calculations was made design of needle mechanism. Compliance of the needle causing unequal refill pleated material. The results of simulations and tests needed to design the fixed part with single-acting cylinders. Involvement of solid cylinders in pairs with movable cylinders helps refill the same both sides of the filter.

Acknowledgments Thank you to the project of Ministry of the Interior of the Czech Republic. This paper was supported by the project VG20122014078.

References

1. D. Kimmer, *Research and Development of Filter Materials Based on Nanowires and their Application to Filter Protective Face Masks*, Czech Republic. (MPO, 2012)
2. J. Reischel, D. Knivsland, US Patent Des. 378,610, Full Face Respirator, 1997
3. P. Lenfeld, *Applied Research on a New Generation of Protective Masks with Nano to Increase the Protection of the Design Technological and Material Aspects*, Czech Republic. (MV, 2013)

The Structural Design of Industrial Vacuum Cleaner for Dental Laboratories

R. Slanina and P. Sniehotta

Abstract Developed industrial vacuum cleaner, is designed for use in dental laboratories. Primarily is the vacuum cleaner designed for suction of fine dust generated during the manufacture of dental prostheses. In the medical field in general, emphasis is placed on quality workmanship, durability and cleanliness of the environment. This represented a challenge in the design of the device with respect to the required performance and price. Compared with similar vacuum cleaners on the market, with the same power, there is a much better ratio of the basket volume, to the overall dimensions of the vacuum cleaner. The hallmark of the vacuum cleaner is the ability to connect two separate hoses to the vacuum cleaner. Unique about the vacuum cleaner is a triple filtration system and noise, which is lower than competitors. This article describes the development of the vacuum cleaner.

Keywords Vacuum cleaner • Dental laboratories

1 Introduction

Development of industrial vacuum cleaners Applebox DL2013 for dental laboratories, was only one part of the development of equipment for newly established training center YESdent in Jesenik. Part of the development was, apart from vacuum cleaner Applebox, especially the development of laboratory table MIG and suction module EM1. Strong emphasis was laid on design for all devices.

For the vacuum cleaner was essential, besides the required suction power and volume of basket in the design of vacuum cleaner was a crucial point of the resulting noise level. Production of dentures is a challenging work for accuracy. Long-term high noise, reduces concentric and can affect the overall quality of work. For this reason, apart from design of the vacuum cleaner, the noise reduction was a major problem (Fig. 1).

The tables were designed from polished stainless steel and artificial stone. To this was adapted the design of the vacuum cleaner. For hygiene reasons, simpler

R. Slanina (✉) • P. Sniehotta
VSB-Technical University of Ostrava, Ostrava, Czech Republic
e-mail: radim.slanina@vsb.cz; patrik.sniehotta@vsb.cz

V. Dynybyl et al. (eds.), *The Latest Methods of Construction Design*,
DOI 10.1007/978-3-319-22762-7_15

Fig. 1 Training center YESdent – Jesenik

Fig. 2 Table visualization

production and price, a galvanized sheet was chosen, painted with white comaxite color. This contrasts well with the silver color of stainless steel (Fig. 2).

2 Design

The basis on which it was necessary to build, were the dimensions of the box in the table (H×W×D 550 × 250 × 530 mm), where the vacuum cleaner should be placed. The vacuum cleaner should contain a basket with a minimum volume of 15 L, which the space for other components was very small. A specific feature of the vacuum cleaner Applebox is the ability to connect two suction hoses to one

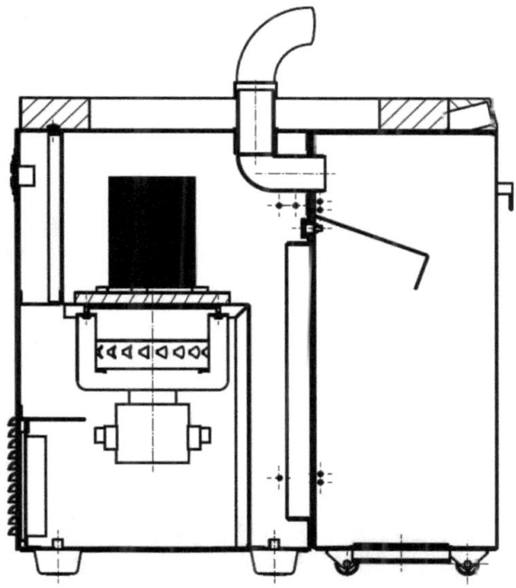

Fig. 3 Cut through the vacuum cleaner

vacuum cleaner. This is very important because the work tables in laboratories are for two laboratory technicians.

The body of the vacuum cleaner and basket was made of 1.5 mm galvanized steel sheet. Galvanized sheet was chosen because of better weldability compared to a stainless steel sheet. Thicker sheet reduced the risk of resonance and increase rigidity vacuum cleaner. Through it came to noise reduction. The disadvantage, however, is the higher weight (Fig. 3).

Basket of the vacuum cleaner was welded to ensure the tightness. The rest of the vacuum cleaner was riveted whit blind rivets, respectively screwed together on parts that should be releasable. A problematic point was the mount of engine bracket, to the cover sheet of the vacuum cleaner. In terms of design it was undesirable to have visible connections on the sides. For this reason, the engine bracket was riveted to cover only the top and bottom. The sides were glued together whit MS polymer (Silyl Modified Polymers terminated) adhesive. This increased rigidity. At the same time using PU adhesive a flexible connection was created, which reduces the transmission of vibrations on a the vacuum cleaner cover and thus contributes to reducing the resulting noise level.

The space in the vacuum cleaner can be divided on the suction and exhaust section. The suction motor, positioned in the exhaust section should have at least 1300 W input power and flow of 3 m^3/min. It was selected motor Hevo SMO.116 127-00, which is also designed for wet suction and has a separate cooling. For reducing noise and vibrations from the motor, it was important to "separate" the motor from rest of the vacuum cleaner, to reduce vibration and noise. Therefore, the engine was placed on the silentblocks and between the motor contact surface and the body of the vacuum cleaner, were inserted rubber rings (Fig. 4).

Fig. 4 Sample of the silentblocks

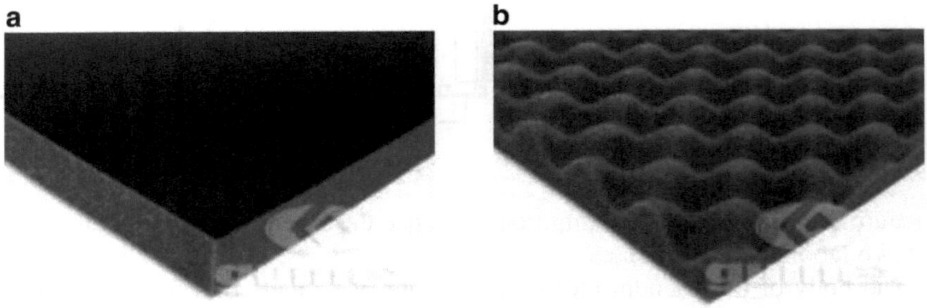

Fig. 5 Noise reduction foam whit PU film (**a**) and wavelet (**b**) – Gumex

Noise reduction has little effect on the function of the vacuum cleaner, but it is essential for the quality work environment. The aim was to achieve a noise level below 60 dB at maximum operation. For this reason, all the sides of the vacuum cleaner body were provided with acoustic foam Gumex type 501 with a thickness of 20 mm. To avoid capturing dust into the foam, a version with black polyurethane film was chosen (Fig. 5).

At the bottom, in the exhaust section and front of the output filter, for better noise absorption, it was used the noise reduction foam Gumex type 505 (in the form of wavelets) with a thickness of 50 mm.

For vacuum cleaner is an important part the filtration of the sucked material. Choosing the right type of filter has a large influence on the resulting noise level. In this case it was a very fine dust which arises in the production of dentures. The dust should not escape from the cleaner space, so we chose a three-stage filtration and quality sealing of the connection between the basket and the body of the cleaner.

The first filter is placed already in the basket vacuum cleaner. We had to make a frame filter (class F5), according to our dimensional requirements, manufactured at the company EWAC (Fig. 6).

Fig. 6 Frame filter for basket – EWAC

Fig. 7 Motor filter – Hiflofiltro (**a**) and output HEPA filter – Electrolux (**b**)

The second stage consists a cylindrical filter Hiflofiltro HFA 3501. It is in the suction section and protects the motor against the ingress of dust, which the basket filter does not capture.

The third filter is located in the exhaust section, just before the exhausting the air into the room. For absolute filtration has been used Electrolux EFH12N filter, filtration class H13 (HEPA) (Fig. 7).

Noise level of the vacuum cleaner was then measured in an anechoic chamber, using the methodology according to DIN EN ISO 3745 on a hemispherical surface. According to these measurements, at maximum operation the noise level reaches values of the time-averaged sound pressure Lp = 51.139 dB, with the index Vl = 3552 dB and values of sound power Lw = 54,139 dB with standard deviation σtot = 3.587 dB.

A separate part is the solution of the vacuum cleaner electronics. The vacuum cleaner has an on/off button and a dial to regulate the power of the vacuum cleaner. Solution of electronics has been designed so, that the control system worked (for safety reasons) on the 24 V voltage. Solution of electronics was assign to company MARPOS. The vacuum can have a conventional arrangement controls (the controls

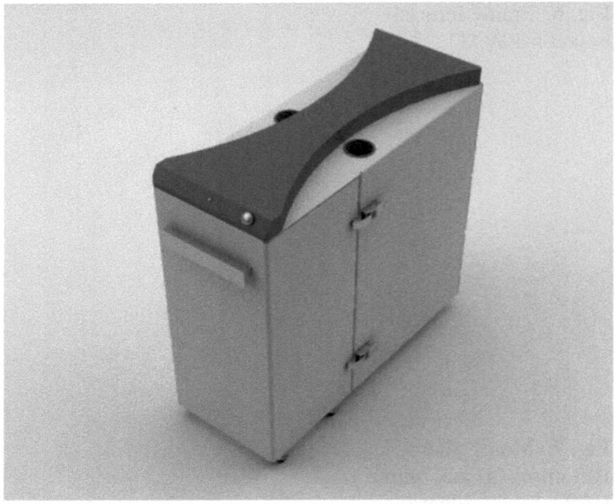

Fig. 8 The final vacuum cleaner Applebox DL2013

are located in the upper part made of artificial stone), or the second option is to lead the controls out of the vacuum cleaner. This allows to integrate the controls into the laboratory table (Fig. 8).

3 Conclusion

The final design solution is functional and several pieces of the vacuum cleaner have been already sold. Despite the development of the vacuum cleaner continues. The main goals for the future is to simplify emptying the basket, which is full very heavy. Another point of development is to expand options in the electrical part. Specifically preparation electronics for engine whit higher performance and system for starting the vacuum cleaner when you turn on another device.

Acknowledgments The research work reported here was made possible by Ministry for Education, Youth and Sports, Czech Republic, project – specific university research Nr. SP2014/24.

Reading of Operating Values of End Effector During Goods Manipulation

F. Stary, V. Dynybyl, and J. Mrazek

Abstract The article describes methodology of reading of operating values of end effector during goods manipulation. Among read values will be gripping force of end effector jaws, amount of grip distance and weight of manipulated goods. In case of right forming of mathematical model of the end effector, the software should be able to determine actual position of manipulated goods with consideration of acceleration influence.

Keywords End effector • Grip force • Weight • Strain gauge

1 Introduction

This article follow-up to articles [1–3], that describe development of wrapping machine with integrated manipulator. In mentioned articles were described concept designs of wrapping machine and manipulator. This article is focused on manipulator, which is based on concept variant mentioned in [1].

2 Final Design of Manipulator

In article [1] was designed concept variant shown in the Fig. 1. This concept variant was modified to final variant shown in the Fig. 2. Main changes are in mean of drive, where drive with toothed rack was substituted with toothed belt. Another change is configuration of vertical axis of manipulator as telescope. With this change is lowered demand on space above machine. Further was changed method of guiding of axes from linear guides to combination of linear guide and eccentric rollers.

F. Stary (✉) • J. Mrazek
Czech Technical University in Prague, Prague, Czech Republic
e-mail: frantisek.stary@fs.cvut.cz; jiri.mrazek@fs.cvut.cz

V. Dynybyl
Faculty of Mechanical Engineering, Czech Technical University of Prague, Prague, Czech Republic
e-mail: vojtech.dynybyl@fs.cvut.cz

V. Dynybyl et al. (eds.), *The Latest Methods of Construction Design*,
DOI 10.1007/978-3-319-22762-7_16

Fig. 1 Concept design of manipulator

Fig. 2 Final design of manipulator

3 Methodology of Reading Operating Values

Among read values will be gripping force of end effector jaws, amount of grip distance and weight of manipulated goods. In case of right forming of mathematical model of end effector, the software should be able to determine actual position of manipulated goods with consideration of acceleration influence, alternatively evaluate instability (shaking) of end effector.

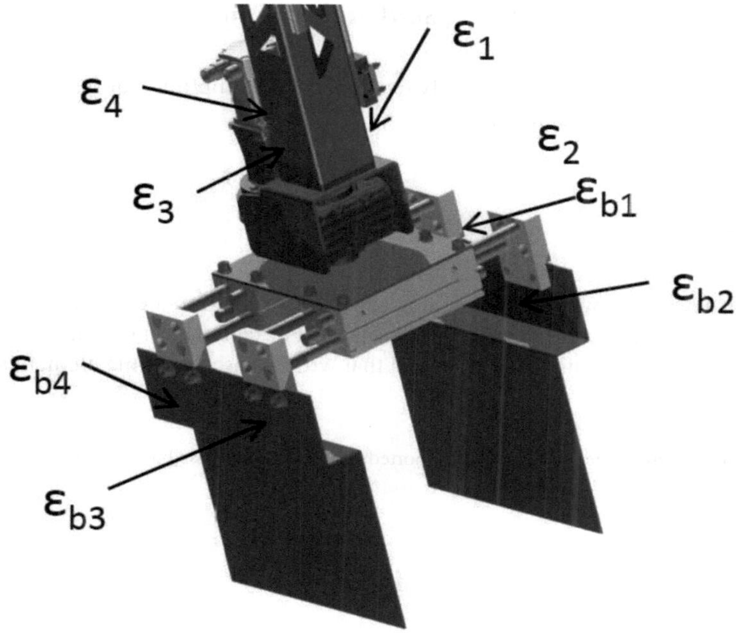

Fig. 3 Detail of end effector with marked measuring points

3.1 Weight Sensing

The tensile tension in profile of vertical axis of manipulator will be used for determination of weight of manipulated goods. Actual weight will be calculated from formula (1) where E means elastic modulus, S is sectional area of vertical beam of manipulator and g is acceleration of gravity. Tensile tension is evaluated from two (four) opposite strain gauges to eliminate flexion strain (2).

$$m = \frac{E \cdot \varepsilon \cdot S}{g} \tag{1}$$

$$\varepsilon = \frac{\varepsilon_1 + \varepsilon_2}{2} \quad or \ \varepsilon = \frac{\varepsilon_1 + \varepsilon_2 + \varepsilon_3 + \varepsilon_4}{4} \tag{2}$$

3.2 Sensing of Gripping Force

To evaluate the gripping force of jaws will be used flexion strain of jaws (ε_{b1}, ε_{b2}, ε_{b3}, ε_{b4} in Fig. 3) near their attachment to pneumatic motor flange. Evaluation of gripping force is not easy and it necessary to create mathematic model, which will be considering gripping of goods in whole length of jaws, as well as on the right side, in the middle and on the right side of jaws. This model will be created with help of finite element method calculations and experimentally verified.

3.3 *Sensing of Instability (Shaking) of End Effector*

Shaking of end effector can be easily evaluated from amplitude of flexion strain on the vertical beam of manipulator.

4 Conclusion

In the article is shown, how to evaluate weight of manipulated goods. Further is shown how to measure gripping force of jaws. For calculation of gripping force is necessary to create mathematical model that will be based on finite element method calculations.

Acknowledgments The research work reported here was made possible by project FR-TI3/281.

References

1. F. Stary, J. Mrazek, Design of the wrapping machine with integrated manipulator, in *53rd International Conference of Machine Design Departments* (Vysoké učení technické v Brně, Brno, 2012), pp. 275–280. ISBN 978-80-214-4533-8
2. F. Stary, V. Dynybyl, J. Mrazek, Wrapping machine with integrated manipulator, in *Experimental Stress Analysis 51* (Trans Tech Publications, Uetikon-Zurich, 2014), pp. 223–226. ISSN 1660-9336. ISBN 978-3-03785-977-3
3. F. Stary, J. Mrázek, J. Kamenický, P. Kříbala, Different approaches to calculation of rotating ring deformations of wrapping machine, in *Book of Proceedings – 54th IC of Machine Design Department* (Technická univerzita Liberec, Liberec, 2013), pp. 445–450. ISBN 978-80-7372-986-8

Building Platform for Identification of Important Parameters for 3D Printing of Huge Parts

M. Suransky and M. Zlebek

Abstract 3D printing of big and very big parts is used mostly in mechanical engineering, automotive industry and industrial design. 3D printing can be used to manufacture prototypes of complicated shapes, which could not be manufactured using different technologies and it is not needed to manufacture expansive tools. To be able to print enormous parts with FDM technology many obstacles must be overcome and several unknown parameters have to be revealed. Primary focus of research and development in this paper is description of building process of experimental device, which will be used to identify important parameters of printing process of huge parts. Some and most important of these parameters are supposed to be influence of scanning speed as well as speed of filament movement and influence of melting temperature. Very important issue to observe and take care of is thermal stress inside the printed material and of course final strength of the part. We assume that we will be able to show experimentally that all of these parameters are important and influence the final part quality.

Keywords Fused deposition modelling • Experimental device • Parameters • Influence

1 Introduction

3D printing, called by many also "the future of manufacturing" has several extraordinary advantages. That is why technicians have very big expectations from the technology on the one side and very severe demands on the other. 3D printing has already brought revolutionary changes in manufacturing of prototypes, but what stays a real challenge is to develop the 3D printing technology to be able to manufacture huge parts instead of small to medium parts. Nowadays the biggest advantages of 3D printing are manufacturing topologically optimised designs, manufacturing porous parts or parts with complicated geometrical shapes. All you need to have is the proper CAD model.

M. Suransky (✉) • M. Zlebek
Brno University of Technology, Brno, Czech Republic
e-mail: y131124@stud.fme.vutbr.cz; y133952@stud.fme.vutbr.cz

V. Dynybyl et al. (eds.), *The Latest Methods of Construction Design*,
DOI 10.1007/978-3-319-22762-7_17

3D printing can be divided in three main categories. FDM – fused deposition modelling, SLS – selective laser sintering and LOM what stands for laminated object manufacturing. FDM process is so far restricted to build prototypes of small or medium sizes. Several tries to build parts of nominal dimension of 1 m were successfully done, but further research in the field is still important. Mainly because the production time is very high. To print a former mentioned 1 m part takes up to 24 h. If the parameters of the process are well defined, this production time could be shortened. Of course producing parts bigger, what means 2 m or even several meters of nominal dimension, will require changing the overall view on the FDM printing machine. As for example in [1] a printing device was used with 5 mm wide nozzle. Nozzle this big in diameter caused significant thermal wrapping.

The main goal of this paper is to introduce building process of experimental FDM printing machine, which will be used to identify important parameters of printing process of huge parts. As stated in [2] some and most important parameters influencing the manufacturing process are scanning speed as well as speed of filament movement. Of course as shown in [3, 4] also environmental temperature is a crucial parameter. Very important issue to observe and take care of is thermal stress inside the printed material and of course final strength of the part.

2 Experimental Device

The device is designed in the way, that the most important printing process parameters, what means influence of scanning speed as well as speed of filament movement, influence of melting temperature and base plate heating can be experimentally observed.

2.1 Design Concept

3D printer is mechatronic system and therefore, mechanical construction as well as electronic and software solution must be taken into consideration. The developed printer must follow these requirements (Table 1):

Table 1 Requirements on printer

Printing volume	Printing material	Nozzle diameter
X direction, 850 mm	ABS, ∅ 3 mm	1 mm
Y direction, 380 mm	PLA, ∅ 3 mm	
Z direction, 100 mm		

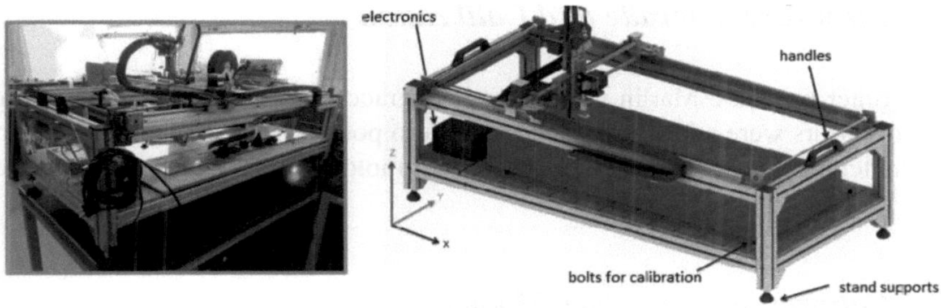

Fig. 1 Mechanical design of testing device, draft on the *right*, real on the *left*

Mechanical construction was designed from Alutec profile 45 × 45 mm. Overall dimensions of the frame are: height – 300 mm, width – 1500 mm, depth – 550 mm (Fig. 1).

2.2 *Nozzle Travels*

Nozzle have to be movable in all directions. Therefore a design similar to milling machines was used, where the drill is moved in all directions.

Travel in the X direction is actuated with stepper motor with a pulley which is rigidly connected to travel cart on both sides of the printer. Carts are mounted on two interposer bars. These bars have diameter of 12 mm and length of 520 mm. Travel in the Y direction is actuated with stepper motor with a pulley which is rigidly connected to travel cart. The cart moves along two vertically placed interposer bars, which are fixed on moving plate of X direction. Travel in the Z direction is actuated with stepper motor which turns through a pulley with the levelling screw. Levelling screw is through part connected to the plate, which is through carts connected to prismatic piloting.

2.3 *Hardware*

To control the printer an Arduino Mega 2560 was used with a special module RAMPS 1.4. This device allows to control up to six stepper motors with maximal current of 2A (if cooled).

2.4 *Software Control*

Communication interface starts with CAD software, that can save data in .stl format. Next step is to translate the .stl files into a G-code, that is standardly used for NC machines. This is done in freewares Pronterface or Repetier-Host, that can easily communicate with Arduino unit.

2.5 Firmware Upgrade and Calibration

As a printer firmware Marlin v1 with little upgrades was used. Only three end-of-motion sensors were added for detection of zero position in each axis. Furthermore printing temperatures were defined and stepper motors calibrations were performed.

2.6 Calibration of Stepper Motors

Movement of stepper motor is measured in steps. What is meant by calibration is to define the ratio between number of steps and one unit of desire. In our case the ratio between number of steps and 1 mm of extruded material is defined for extruder motor and number of steps and 1 mm of movement for axis motors. Formulas for calculations of ratio between number of steps and 1 mm of extruded material are Eq. (1).

$$R = \frac{k \cdot \left(\frac{1}{16}\right)^{-1} \cdot i}{\pi \cdot D} \tag{1}$$

$$R = \frac{k \cdot \left(\frac{1}{16}\right)^{-1} \cdot i}{b \cdot z} \tag{2}$$

Where k – is number of steps per revolution, I – is gear ration of printing head, D – diameter of extruding wheel, b – distance between belt teeth, z – number of teeth of pulley, (1/16) – defines number of microsteps per step.

When calibrating motor for vertical axis (Z axis) instead of two parameters namely distance between belt teeth (b) and number of teeth of pulley (z) only one parameter pitch of the threaded bolt was used.

2.7 Calibration of Base Plate

It is very important that the base plate is as planar as possible. Therefore glass board is placed onto MDF board and four bolts are mounted, one in each corner to be able to set the base plate in the plane.

3 Testing of Printing Process

After all the motors were calibrated, first tests of printing process could be performed. Parameters as for example height of first layer and height of other layers, temperature of the nozzle, speed of the nozzle movement and of course material adherence on the base plate need to be considered.

Table 2 Adhesion of base plate properties

Base plate type, recommendation	Base plate type, recommendation
Sugar water, yes	Corundum glass (2 bar), no
Stratasys base plate, yes	Hair lacquer, no
Corundum glass (5 bar), with heating	Adhesive tape for painting, no
ABS melted in acetone, with heating	Kapton adhesive tape, with heating
Corundum glass (3,5 bar), no	Pure glass, no

3.1 Testing of Base Plate Adhesion Properties

Because so far this device does not have heated base plate it was decided to use PLA material instead of ABS, because PLA has better adhesive properties on non-heated one. Several different types of the base plate were tried out, in order to find the best adhesion properties. For example glass plate, glass plate covered with hair lacquer or with acetone with melted ABS. Next base plate was from corundum glass of different roughness but the best was glass covered with simple sugar water, which has shown the best adhesion properties. Outcome of adhesion tests is shown in table below (Table 2).

3.2 Testing of Other Parameters

After the best base material and cover was chosen (glass cover with sugar water) the testing of first layer height and the best temperature of the nozzle for the first layer printing was tested. In Fig. 2. there can be seen how extremely the quality of printed part (simple rectangle) is improved only by tuning these two parameters. On the upper left the part is shown, which was printed with first layer height too low and nozzle temperature too high. That is why the part is so flat and pushed. Otherwise on the bottom right, there is part with first layer heights and nozzle temperature well set. The best first layer height is about 0.6 mm and first layer nozzle temperature is 220 °C, what is a bit higher than 210 °C for other layers. A bit higher temperature increases an adhesion on the base plate.

Another very important issues were the speed the nozzle movement and filament movement speed. The maximal speed of the filament movement is 10 mm in a second, because if it was higher the material would not melt enough and would stack in the nozzle. The highest speeds of the nozzle movement were found out experimentally are were set to be 80 mm per second. If the speed was higher the printing head started to vibrate too much.

Fig. 2 Same parts printed with different first layer heights and nozzle temperature

3.3 Testing the Building Accuracy

To test the building accuracy two parts, the cylinder and the wall were chosen and were scanned with 3D scanner ATOS Triple scan. Technical data of ATOS Triple scan are shown below. Data from 3D scanner were compared to the CAD model using a GOM software, which was used for the scanning as well. CAD model was imported in the GOM software and fitted with the best fit function on the scanned object and deviations were evaluated. Some deviations are up to half millimeter. This deviations are caused because of big nozzle diameter. If the diameter prints 1 mm thick lines it is impossible to print 2.5 mm thick wall (Fig. 3 and Table 3).

4 Conclusion

This paper deals with construction and testing of the device for 3D printing of big parts. Based on this vision, students from the Brno University of Technology designed the printing device and tested properties of the device. These properties are described by parameters as for example height of first layer and height of other layers, temperature of the nozzle, speed of the nozzle movement and of course material adherence on the base plate. All of these parameters were tested and the best possible values in terms of the best part quality are stated. It was found out that the first layer needs to be a little lower than other successive layers because of improvement of adhesion properties of the first layer. Adhesive properties of the first layer and thereby the overall part quality strongly depends on the base plate adhesion property. Therefore it is very important parameter of the printing process and very consistent study is done to find out how base plate glass with different roughnesses and with different cover substance affects the adhesive properties. As cover substance was, for example used hair lacquer, acetone with melted ABS and

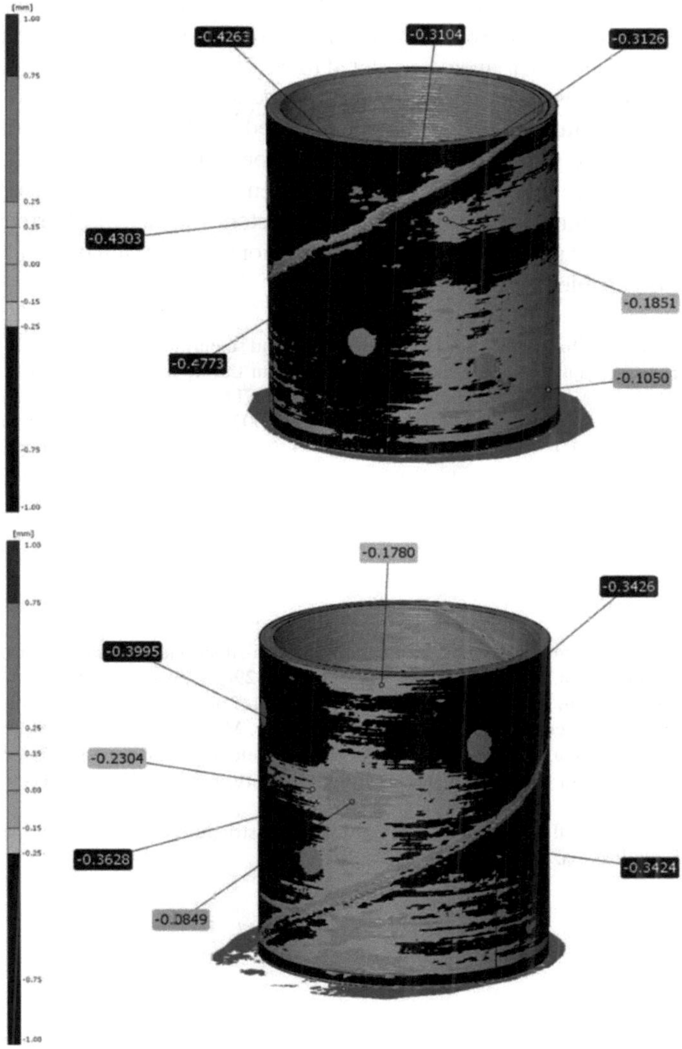

Fig. 3 Comparison of scanned and CAD model data of the cylinder

Table 3 ATOS triple scan parameters

ATOS parameter, value	ATOS parameter, value
Camera pixels (px), 2×8.10^{6}	Lamp, LED
Measuring area (mm^2), 38×29–2000×1500	Optics (mm), SO $170 \times 130 \times 130$
Point spacing (mm), 0.01–0.61	Point distance (mm), 0,055
Working distance (mm), 490–2000	Recomm. ref. points, 0.8 mm

simple sugar water, which turned out to be the best cover substance. It is cheap and assures the best adhesion properties. Furthermore optimal nozzle temperature for printing the first and other layers was stated to be around 210 °C. For the first layer a little more to increase the adhesivity and for the successive layers a little less is sufficient. A maximum filament movement speeds and speed of the nozzle in all axis directions were stated. A consisted calibration of the stepper motors was done in order to improve printing accuracy. The demonstrative comparison of parts printed with well-tuned and badly tuned parameters was shown. Furthermore the comparison of CAD models and real part dimensions is done. It has been shown that the printer is capable of printing accurately.

Acknowledgments This work is an output of research and scientific activities of NETME Centre, regional R&D centre built with the financial support from the Operational Programme Research and Development for Innovations within the project NETME Centre (New Technologies for Mechanical Engineering), Reg. No. CZ.1.05/2.1.00/01.0002 and, in the follow-up sustainability stage, supported through NETME CENTRE PLUS (LO1202) by financial means from the Ministry of Education, Youth and Sports under the "National Sustainability Programme I".

References

1. S. Keating, N. Oxman, Compound fabrication: a multi-functional robotic plaform for digital design and fabrication. Robot. Comput. Integr. Manuf. **29**(6), 439–448 (2013)
2. I.T. Ozbolat, H. Chen, Y. Yu, Development of 'multi-arm bioprinter' for hybrid biofabrication of tissue engineering constructs. Robot. Comput. Integr. Manuf. **30**(3), 295–304 (2014)
3. J.-W. Choi, F. Medina, C. Kim, D. Espalin et al., Development of a mobile fused deposition modeling system with enhanced manufacturing flexibility. J. Mater. Process. Technol. **211**(3), 424–432 (2011)
4. I. Gajdos, J. Slota, Influence of printing conditions on structure in FDM prototypes. Tehnicki Vjesnik-Technical Gazette **20**(2), 231–236 (2013)

On Cardan Joint: Some Special Applications

J. Zicha and K. Macuchova

Abstract Cardan joint (Gerolamo Cardano 1501–1576) is one of the scientific results of the renaissance scientist (mathematics, astronomy, astrology, philosophy). At present time is this one applied as mechanical joint which allows the transport of the torsion moment between intersecting axes. It is known that this transport is dependent on the angle between both axes. In this contribution it will be described some other applications for example in the metrology and experiments with redundant kinematics system.

Keywords Cardan joint • Angle measurements • Mechatronics

1 Introduction

Besides all uncountable and good known applications in design of cars it is possible to use the Cardan joint also for next specified uses.

Henry Augustus Rowland (1848–1901) which is known with his work about diffraction gratings applied the Cardan joint in the design of the grinding machine as a connection element between precise screw and table with the grating. By means of this solution he compensated the influence of the first harmonic of the screw pitch. The sophisticated design of pivots enables compensation also of skew between the axis of the screw and the axis of the linear guidance. This arrangement was used in the design of microphotometer at the Astronomical Institute of Czech Academy of Sciences in Ondřejov [1], [2].

Some unusual application is placing of the base of the transportable measurement station realized by ENKI Company in Třeboň. The base of the stand is connected with the frame of the device by means of Cardan joint. It might appear that we can use the inclinable double row ball bearing but in the time of manipulation with the stand we need some blocking of the rotation about the longitudinal axis of the stand.

There exists a connection between the Cardan joint and mounting of the astronomical instruments. For example: the mountings "alt-alt" (altitude-altitude) and

J. Zicha (✉) • K. Macuchova
Czech Technical University in Prague, Prague, Czech Republic
e-mail: josef.zicha@fs.cvut.cz; karolina.vosatkova@fs.cvut.cz

V. Dynybyl et al. (eds.), *The Latest Methods of Construction Design*,
DOI 10.1007/978-3-319-22762-7_18

alt-azimuthal which are largely known. The kinematical principle of some alt-azimuthal mounting is used in the design of experimental device for the study of optical properties of rastered glazing.

The basic idea of the Cardan joint is also used in the design of the platform of measurement devices which is connected with the airship. Movement of this airship must be compensated so the axes of the measurement devices are kept vertical. The damping effect is realized by means of the prestressed Teflon bearing shell. The material combination of metal-Teflon is working without stick-slip effect.

The latest application of the Cardan joint is the metrological Cardan joint. The design allows us the measurement of angular displacement in all three axes. In this case is not the Cardan joint used as coupling but as an instrument which is able to describe spherical rotation around three axes.

2 Three Axes Measurement Device

This special type of the Cardan joint is used in the field of the research of redundant mechanical systems in mechatronics. One partial question is real feedback of the platform controlled by the redundant mechanism. Basic requirement is to measure the rotational movement (spherical movement) in three degrees of freedom. To fulfill the requirements it is possible only by means of mechanical separation of these three together perpendicular axes. Boundary conditions are practically zero passive resistance by the rotation, the limit by the angular range bigger than $\pm$ 90° and the installation of the three incremental encoders by the Renishaw Company.

Now we can show you the chain of the first idea up to realized device. The basic idea is not too complicated for imagination. The second step is a lot of work because we have to accept different mechanical parts and sensors. The description is in the Fig. 1 – development documentation.

After the discussion with the producer of the mechanical system – BMD Company in Teplice – the production documentation was prepared. This documentation is directly connected with the CNC production center and after an agreement the parts were produced and assembled [3].

The device is a measurement instrument which is often in contact with human sweat in fingerprints and therefore we used anti-corrosion materials for all important parts. For nice appearance of the whole device, non-functional surfaces were sanded.

This device is the last one in a row of other different prototypes. The other will be mentioned during the oral presentation of this paper.

The main part 1 is the axial cross for the first two axes. This cross is not symmetrical because we need a little more space for the installation of the Renishaw sensors 5. The third axe is represented with the detail 6 and sensor 17. Minimal passive resistance is realized by means of two skew ball bearings 7002 2 and 7004 7. These pairs of bearings are slightly prestressed. According to this picture we get the guarantee about the possible angular motion which is quite $\pm$ 150°. In the body of the parts are made holes for placing bunched cables. For this short information it is not necessary to describe the design further (Figs. 2 and 3).

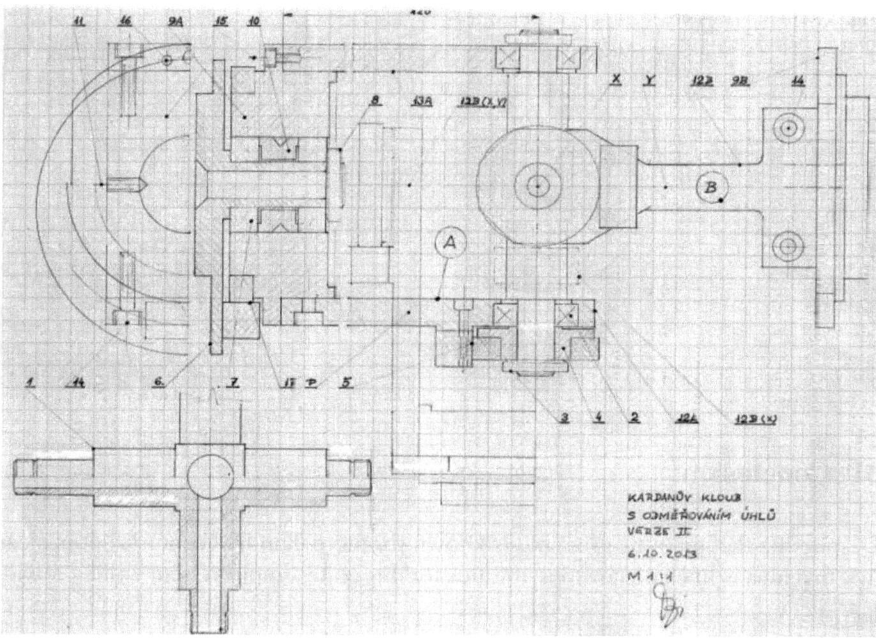

Fig. 1 Development documentation with description

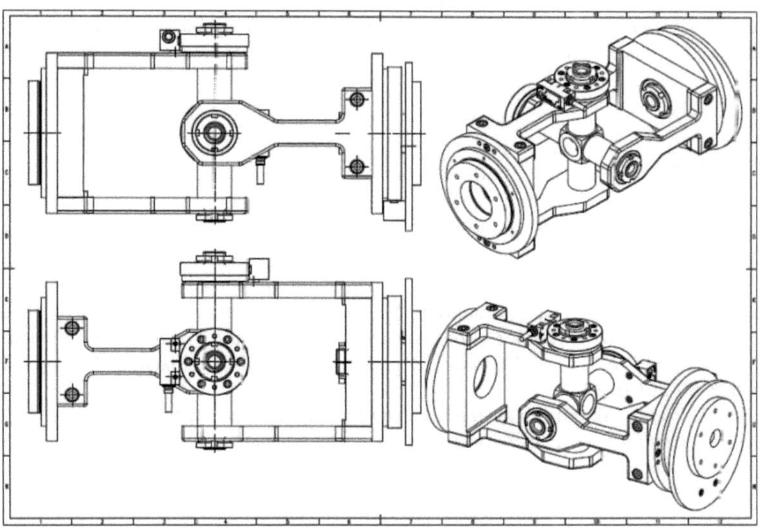

Fig. 2 Production documentation

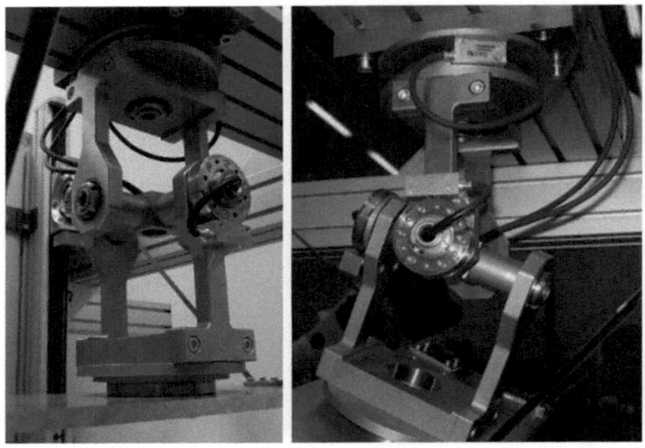

Fig. 3 Three axes Cardan joint in laboratory

3 Conclusion

Despite the fact that this principle is known for more than five centuries, it is clear that this idea is live and new possibilities for this quite simple system will be found.

Acknowledgments The authors appreciate the kind support of the presented work by the grant GAP101/111627 "Tilting Mechanisms Based on Fiber Parallel Kinematic Structure with Antibacklash Control" of the Czech Science Foundation and by the grant TA01020698 "Technologie a systém určující fyzikální a prostorové charakteristiky pro ochranu a tvorbu životního prostředí a pro zvýšení potenciálu energetických zdrojů" of the Technology Agency of the Czech Republic and by the grant 2B06023 "Vývoj metody stanovení toků energie a látek ve vybraných ekosystémech, návrh a ověření principů hodnocení hospodářských zásahů pro zajištění podmínek autoregulace a rozvoje biodiversity" of the Ministry of Education, Youth and Sports.

References

1. J. Zicha, J. Čáp, B. Šourek, V. Jirka, J. Červený et al., Simulátor slunečního svitu pro testování optických rastrů, in *Jemná mechanika a optika* (2007), roč. 52, č. 6, s. 194–198. ISSN 0447-6441
2. P. Svatoš, Z. Šika, J. Zicha, M. Valášek, V. Rada, Optimization and design of fibre-driven spherical mechanism, in *29th Conference with International Participation Computational Mechanics 2013 – Extended Abstracts* (Západočeská Universita, Plzeň, 2013), pp. 123–124. ISBN 978-80-261-0282-3
3. V. Jirka, J. Zicha, Aktuální stav realizace transportabilní terénní měřicí stanice, in *Sborník odborného semináře Nové metody a postupy v oblasti přístrojové techniky, automatického řízení a informatiky 2011* (Ústav přístrojové a řídicí techniky FS ČVUT, Praha, 2011), s. 33–38. ISBN 978-80-01-05041-5

Part II
Tribology

M. Bosansky, J. Broncek, M. Dzimko, P. Fabian, J. Fryza, R. Galas, M. Hartl, S. Hornak, B. Kopilakova, D. Kostal, I. Křupka, K. Nakashima, D. Necas, M. Omasta, N. Radek, Y. Sawae, D. Smejkal, P. Sperka, M. Vrbka, S. Yarimitsu, and J. Zapotocny

Experimental Analysis of Tribological Properties of Heat Treated Graphitic Cast Irons

J. Broncek, P. Fabian, M. Dzimko, and N. Radek

Abstract The results of research investigations of tribological properties of heat-treated cast iron with globular graphite are presented. The samples of investigated materials, including the reference material steel C45 were heat treated to the same hardness value. Tests were carried out on the experimental device with a linear reciprocal movement with variable speed using the Ball on Plate method. The device simulates real changed conditions in different environments of the tested contact pairs, such as sliding bearing, hinges, brackets and the like. The results of measured tribological properties of materials show that for practical applications and for specific environments and dynamic conditions the replacement of the original material C45 by isothermally tempered cast iron may be realistic.

Keywords Experimental device • Tribology • Steel C45 • Friction • Wear

1 Introduction

Isothermally heat treated cast iron with globular graphite ADI (Austempered Ductile Iron) belongs to new perspective construction materials, determined mainly for dynamically stressed parts [1, 2].

ADI is made of the cast iron with globular graphite by means of isothermal heat treatment. Microstructure of ADI contains globular graphite, which is distributed in a matrix. The matrix consists of two phases – bainitic ferrite and stabilized austenite. The combination of these two phases in the structure of the matrix enables to reach good mechanical properties of this material [2].

Within an experimental research there are followed tribological material properties in more detail at the Department of construction and parts. The aim of the

J. Broncek (✉) • P. Fabian • M. Dzimko
University of Zilina, Žilina, Slovak Republic
e-mail: jozef.broncek@fstroj.uniza.sk; peter.fabian@fstroj.uniza.sk

N. Radek
Kielce University of Technology, Kielce, Poland
e-mail: norrad@tu.kielce.pl

V. Dynybyl et al. (eds.), *The Latest Methods of Construction Design*,
DOI 10.1007/978-3-319-22762-7_19

research is to complete or gain another knowledge necessary for an application of this material under such conditions where mechanical energy losses and material loss caused by wearing are occurred. These losses are occurred where friction – immanent part of component motion is appeared.

2 Experimental Device

A new experimental device a linear microtribometer is being used for evaluation of tribological properties at the Department of construction and engine parts.

Mechanism of an experimental device changes a rotary motion of an electromotor to a reversible straight line motion, whose characteristics has a sinus curve The speed of a feed is possible to change arbitrarily by increasing of engine speed, which is interconnected with a frequency convertor.

The testing is based on the method Ball – on – Plate, which is a standard method of tribological material properties testing. The main part of a microtribometer is a tribological knot which consists of a ball (fixed on the end of a roller) and a flat sample in a shape of a plate. The motion of a ball is not uniform, linear reversible. The moving ball is pressed to the sample, fixed in a bowl, by its head weight and a plumb. The bowl is fixed on a freely placed plate supported in three points and it is fixed to the tensometric detector. The value of the frictional force F_T is determined by tensometric measurement of the plate position. The values of a friction coefficient are counted from known values F_T and F_N.

Gained information from the measuring system is evaluated by PC software LabView. The test is possible to realize in different environments – atmospheric, water bath, in a solution, etc. Figure 1 shows the scheme of the linear microtribometer [3].

Fig. 1 The scheme of a building structure of the linear microtribometer

3 The Preparation of Experimental Samples

Within the solution of later described experiment part there were compared two materials – the reference material steel STN 050 (C45) and cast iron with globular graphite (isotermically heat treated to the cast iron ADI). Valuated material has been chosen to reach the same mechanical properties for both compared materials (the same hardness, toughness, etc.) [4].

The second tested body which we used, was a ball made of a standard bearing material 14 209 with a diameter of 3 mm. The production of the cast iron with a globular graphite was performed in a casting laboratory at the Department of technological engineering. The cast iron was produced by remelting of raw iron, steel, ferroalloys and LEGUR. The chemical composition of cast iron has been chosen with the consideration of its following processing and supposed cast thickness 30–50 mm. The heating to the temperature 910 °C with the holding time 20 min. was performed in a resistance furnace ELOP 1200 to reach hardness 60HRC. Isothermal heat treatment was performed in a chamber furnace KPO 7/5 in a salt bath AS 140 to the temperature 210 °C with the holding time 90 min. By means of this heat treatment the reached hardness was 55 HRC in whole material.

Metallographic evaluation of the structure of material was performed by the light microscope AxioImager.A 1 m (Fig. 2).

From the metallographic evaluation of the sample from ADI cast iron is seen that the structure of a matrix ADI is created by a bainitic structure, consisting of soft needles of bottom bainite and balls of graphite and stabilized austenite. The character of a microstructure and mechanical properties ADI is strongly dependent on the temperature and the time of a bainitic transformation. ADI cast iron with the microstructure created by upper bainite has lower hardness and tensile strength in comparison with ADI with the microstructure created by bottom bainite. The lower temperature of isothermal transformation is, the softer bainitic microstructure and

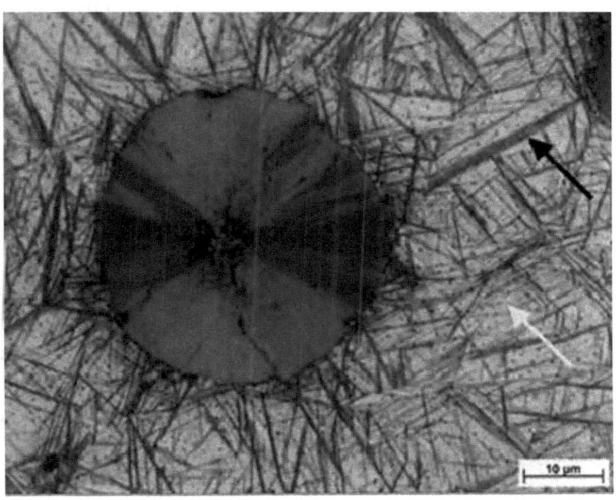

Fig. 2 Microstructure of material ADI, stabilized austenite (*white arrow*), bottom bainite (*black arrow*)

Fig. 3 Microstructure of steel 12 050 after heat treatment

higher hardness and tensile strength are. On the other hand it is necessary to notice that increasing of hardness and tensile strength has a negative influence on fatigue properties of ADI cast iron. Gained hardness HV 10 (STN EN 1043 – 1) at a testing sample is 55 HRC.

Reference material steel STN 12 050 is a suitable material for heat treatment, exactly for surface hardening and heat treatment. This material is usually used for dynamically stressed parts.

The reference sample was made of common used steel STN 12 050. Hardening was realized at the hardening temperature of 830 °C with the holding time of 20 min in a resistant furnace MLM with following cooling into the water. After measuring of hardness (61 HRC) tempering at the temperature 310 °C and the holding time 40 min. was followed. By means of this heat treatment there was reached the hardness value of the reference sample 55 HRC in its whole volume. From the metallographic evaluation of the reference sample of steel STN 12 050 after heat treatment is seen that the matrix is created by tempered martensite (Fig. 3).

4 Testing Method

The testing method has been chosen to find out the dependency of a friction coefficient for different values of loading at the chosen speed v. The test was realized in three environments under the atmospheric conditions, in the water and in the salt bath. The time of individual experimental tests was 60 min at the constant speed $v = 0{,}02\ \mathrm{m.s^{-1}}$ and for the loading $F_N = 5$ N. The total trajectory length of a tribological food was 80 m for each test (Fig. 4).

Fig. 4 Tribological track during the test

5 Results

Figures 5 and 6 show measured values of the friction coefficient for compared materials in three different environments atmospheric conditions, in the water and in one – mole solution NaCl, for loading $F_N = 5$ N.

5.1 Heat Treated Steel 12050 (C45)

From the course of the values of the friction coefficient under the atmospheric conditions (a blue curve) (Fig. 5) is seen that the strain of the friction coefficient increase was gradual, but the next course was nonuniform and unstable. The friction coefficient was within the values 0,75–0,87 and an average value was stable at $f \sim 0{,}8$ (–).

From the course of friction in the water (a green curve) is seen that the strain was gradual, smooth and stable. The friction coefficient was within the values 0,25–0,45 and an average value was $f \sim 0{,}35$ (–).

From the course of analysis of the friction coefficient in the solution (a red curve) is seen that friction in an introductory part of the testing (running in) skipped from the zero value to the value of 0,42 then it decreased after that the value of the friction coefficient became stable. The friction coefficient was within 0,3–0,35. An average value was stable at the value $f \sim 0{,}35$ (–).

From the comparison of gained values of the friction coefficient found out on steel 10 050 in different testing environments is seen that in atmospheric conditions there was reached the highest value of the friction coefficient for given material pair.

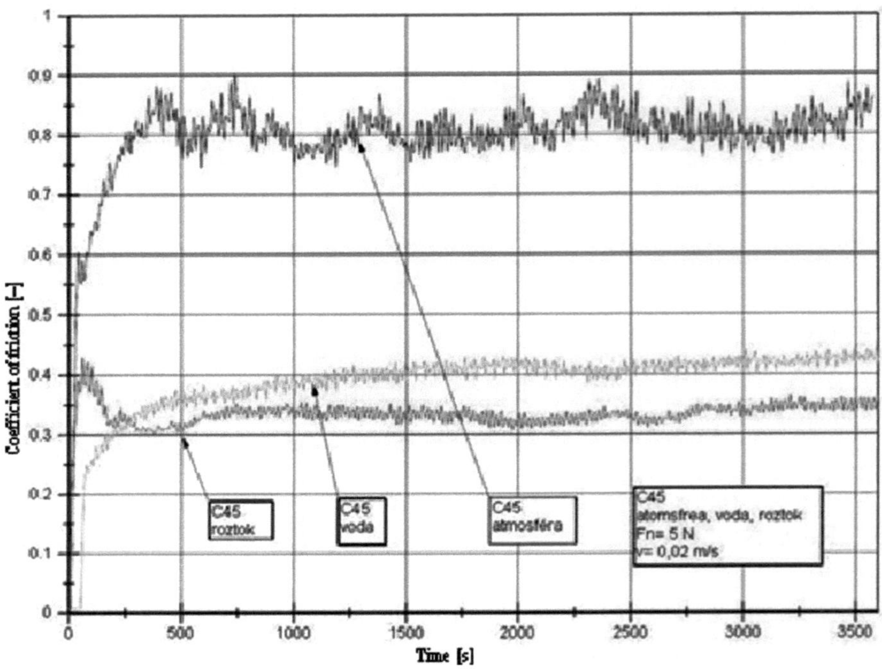

Fig. 5 Courses of the friction coefficient of steel c45 dependent on time in different environments

Friction in the water bath and in the solution was much smaller (approximately 2×) than under the atmospheric conditions. The course was smooth and there wasn't any adhesion between surfaces.

It is possible to explain it by the fact that for given conditions water and a solution work as a lubricant. From the comparison of the courses is evidently seen that the smallest friction is gained in the solution of the salt bath.

5.2 *Isothermally Heat Treated Cast Iron ADI*

From the course of the friction coefficient under the atmospheric conditions (blue curve) (Fig. 6) is seen that the friction coefficient was at the beginning of the test distributed within the values 0,8–1. After stabilization of conditions in the half of the testing its average value was 0,9. After all we can say that the value of the coefficient of friction reached high values connected with adhesion of material ADI for these conditions.

From the course of friction in the water (a green curve) we can see the increase and a gradual stabilization of the curve to the value of $f \sim 0, 55$ (–).

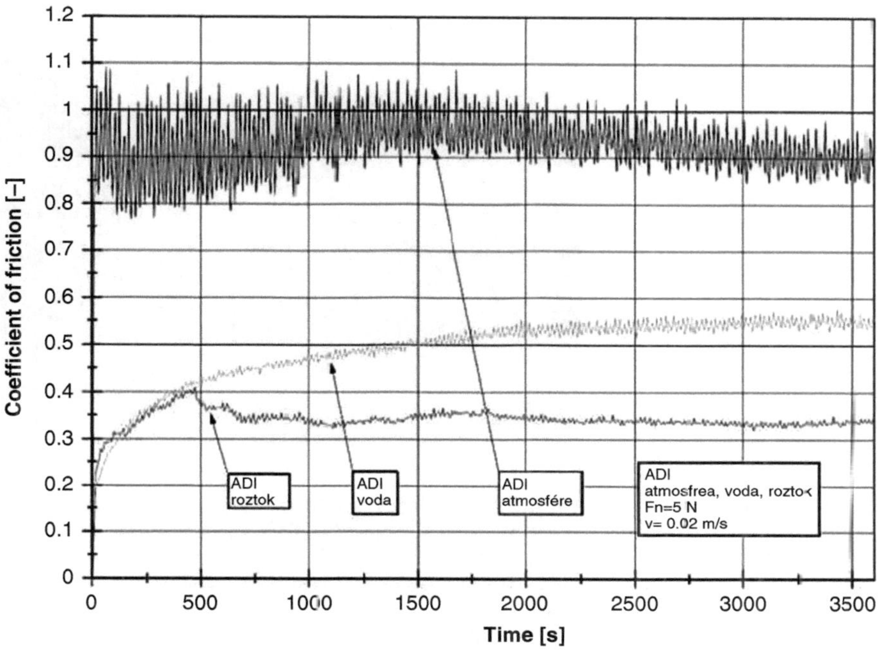

Fig. 6 Courses of the friction coefficient of steel c45 dependent on time in different environments

In the solution the curve of the friction coefficient characterizes in the first phase a gradual increase to the value (0,4) then the decrease of the value which is followed by stabilizing of the testing value to the value 035 and its holding time to the end of the test.

From the comparison of reached values of the friction coefficient found out for isothermally heat treated cast iron ADI in different testing environments is seen that in atmospheric conditions there was reached the highest value of the friction coefficient 0,9 for given material pair. The friction in the water bath was 1,6× smaller and in the solution 3× smaller than under the atmospheric conditions.

5.3 *Microstructure of the Friction Path on a Sample*

After testing the surface of the samples in the place of tribological mark was examined by the light microscope.

In the Figs. 7 and 8 are shown microscopic pictures of the surface of steel (C45) and material ADI after the test under the atmospheric conditions with the projection of the area in the place of a tribological track.

From the visual observing it is seen that both valuated samples are remarkably worn. Wearing is connected with the degradation of the surface.

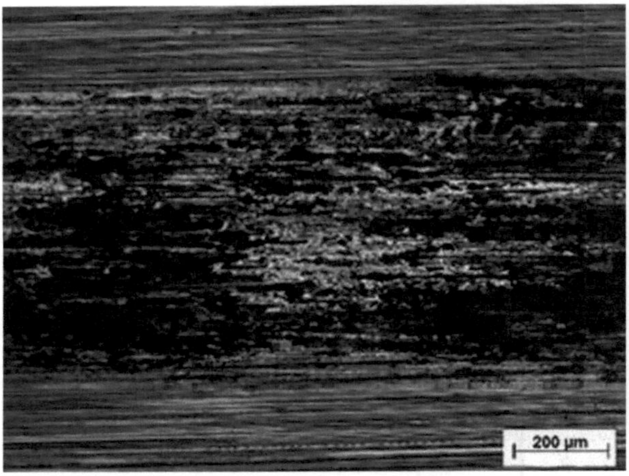

Fig. 7 Tribological track on the surface of steel (C45) in atmospheric environment

Fig. 8 Tribological track on the surface of ADI in the atmospheric environment

On the base of analyzed results it is possible to say that in the contact places between bodies there are strong adhesive connections. The friction was performing between plastically tough surface of the sample material and equally plastically tough ball material.

6 Conclusion

On the base of gained experiments results cast iron ADI has a tendency to seize for dynamically changing conditions of the work in a friction knot under the atmospheric conditions. It is necessary to take into an account a wide range of found out tribological parameters to improve mechanical parameters in setting and to decrease friction losses by an application of isothermal heat treatment cast iron ADI.

Considered tribological properties are values of contact pressure, sliding speed, kind of lubrication, influence of temperature, polluting and contamination of the surface by other elements and cyclic and dynamic loading as well.

Acknowledgments The research work reported here was made possible by the grant project VEGA V-1/0582/12 and International project APVV SK-PL 0034-12.

References

1. K. Hanzlíková, Vliv doby izotermické transformace na mikrostrukturu a únavové vlastnosti ADI, in *Vedecké spisy VUT v Brně*, Edice Ph.D. thesis, sv. 378 (1996). ISSN 1213-4198
2. M. A. Jescas-Gonzalez, *Modeling the Microstructure and Mechanical Properties of Austempered Ductile Irons*, A dissertation submitted for the degree of Doctor of Philosophy, University of Cambridge (2001)
3. J. Bronček, M. Dzimko, B. Hadzima, J. Takeichi, Experimental investigation of Aluminium alloys 2024 –T 3 form in terms of trobocorrosion characteristics. Acta Metall. Slovaca **20**(1), 97–104 (2014). doi:10.12776/ams.v20i1.273. ISSN 1335-1532
4. J. Bronček, M. Dzimko, M. Kovalíček, Y. Takeichi, Friction characteristic of CrN/DLC2 layer deposited on AISI 304 in contact with AL2O3 ball examined under atmospheric and vacuum conditions, in *Modern Methods of Construction Design*, ed. by L. Ševčík et al. (Springer, 2014) pp. 597–602. doi:10.1007/978-3-319-05203-8_79

The Experimental Determination of the Grease Amount to Effective Wear Reduction in the Wheel-Rail Contact

J. Fryza and M. Omasta

Abstract The wear mechanisms within the wheel flange and rail gauge contact affect safety and operating costs of railway transport. This contact occurs when a rail vehicle reaches a curved track. A low coefficient of friction (COF) is required. This COF and wear rate are influenced by operating conditions (load, velocity and temperature), lubrication and geometry (i.e. the track curvature radius). The damage to the contact surfaces can be decreased by using on-board lubrication systems. Grease is applied to the first wheelset of a driving car by a nozzle. The rate of wear reduction depends on the grease amount and its application intervals introduced by the on-board lubrication system into the contact. In this paper, the full-scale laboratory apparatus has been used to study friction, wear and lubricant film distribution under pure sliding conditions in the wheel flange contact. The grease amount and its appropriate application intervals were found.

Keywords Wheel flange lubrication • Wheel-rail contact • Wear • Seizure • Grease

1 Introduction

Railway transportation is an energy-effective way to transport cargoes and passengers over medium distances. Demands for rail transportation have steadily increased since the early nineteenth century. These brought not only the increase of transport speed, quantity of goods and persons, but also requirements for reliability, safety and environmental friendliness of this service. All of these aspects are determined by a contact with a typical size of 1.5 cm^2 which connects the wheel with the rail.

Experimental investigations are focused primarily on two types of wheel-rail contacts. These types of contacts depend on their position. The first type is the wheel tread-rail head contact occurring in straight sections or very high radius curves of a track. The tread contact is very important for transmission of traction and braking forces (performance and safety issues, respectively). In this field,

J. Fryza (✉) • M. Omasta
Brno University of Technology, Brno, Czech Republic
e-mail: fryza@fme.vutbr.cz; omasta@fme.vutbr.cz

V. Dynybyl et al. (eds.), *The Latest Methods of Construction Design*,
DOI 10.1007/978-3-319-22762-7_20

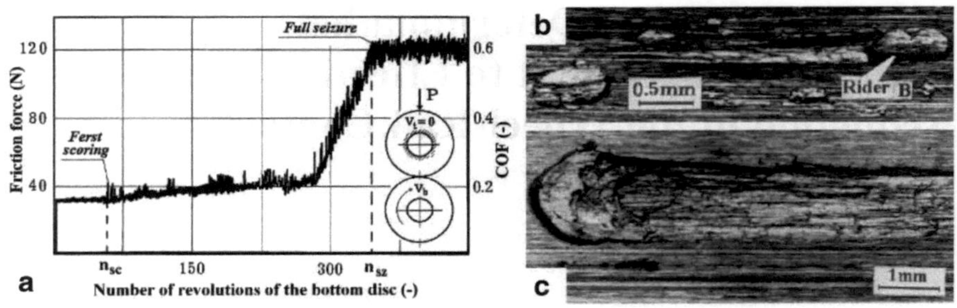

Fig. 1 Seizure development for constant load and sliding speed (**a**) progress of COF and friction force; (**b**) appearance of scoring on bottom disc surface; (**c**) appearance of welding protrusion (galling) on bottom disc surface (Adapted from [4])

research is focused on the influence of high friction modifiers (friction enhancers and sand) and contaminants (leaves, solid particles or water) on adhesion, noise and wear or fatigue. On the other hand, low COF is required in the second type of contact. It is the wheel flange-rail gauge corner contact. The flange contact prevents a rail vehicle against derailment in curved sections of a track. The effects of low friction modifiers (lubricants – especially greases) and operating conditions on wear and fatigue are examined. This paper deals with the second mentioned type of contact.

The maps of wear regimes and transitions based on laboratory tests and field measurements were produced by Lewis and Olofsson [1]. Wear rate of the flange contact can grow from the severe to the catastrophic wear regime unlike the tread contact which belongs to the mild regime of wear. The basic mechanisms of mild wear are oxidation and abrasion which lead to smooth surfaces with small oxide particles. The contact pressure, sliding velocity and temperature are increased due to change of the contact position from the tread on to the flange. The flange contact position results in abrasive and adhesive wear mechanisms (severe wear regime). The transition into the catastrophic regime is probably caused by reaching the contact temperature about 200 °C when a drop in yield strength of carbon-manganese steels occurs. The loss of material volume per sliding distance is up to 10 times higher in the catastrophic regime in comparison with the mild regime.

The problem of railway materials is their tendency to seizure that can occur even at low loads [2, 3]. The seizure process as mode of wear was studied on a twin disc machine under pure sliding conditions by Markov [4]. Seizure is composed mainly of two stages of adhesive wear – scoring and galling. Figure 1a shows the dependence of COF and friction force on number of bottom disc revolutions. The COF value remains almost constant before reaching the scoring threshold (the first peak in the diagram). Scoring is exhibited by local plastic deformations of both surfaces with transfer of materials between surfaces due to the formation of hard riders (ridges) and grooves, as shown in Fig. 1b. Scoring frequency and contact temperature gradually grow during disc revolutions. These increases lead to galling. It is formation of welding protrusions on one surface only (see Fig. 1c) which is associated with a sudden increase in COF and a significant damage to opposite surface.

As outlined above, the wear rate of the flange contact is strongly influenced by the contact temperature. Numerical estimates [5] have indicated that maximum temperatures in the flange contacts are in the range of 450–850 °C for common operating conditions. Therefore, it is necessary to use greases with extreme pressure and solid lubricants additives such as graphite or molybdenum disulphide. Another requirement is an organic lubricant degradability, especially in biologically sensitive areas, because flange lubrication is a loss lubrication system.

A lubricating mixture [6] is produced in the flange contact during field experiments. The mixture is composed of grease, wear metal particles and mineral particles from a track surrounding environment. Concentration and size of ingredients in this mixture (solid particles, grease soaps, etc.) affect their distribution within the contact and thus their effect on lubrication. Smaller particles enter the contact more easily. Large particles or excessive concentration of particles lead to their accumulation at the inlet to the contact [6, 7]. Conversely, isolated particles give a high local contact pressure when passing through the contact [8]. Therefore it is advisable to also take into consideration the distribution of the mixture within the contact.

The purpose of this work is to obtain the amount of the grease and the appropriate intervals of its application to achieve an effective wear reduction in the flange contact. It is expected that the excessive amount of grease is squeezed out of the contact and does not help to the lubrication process.

2 Materials and Methods

In the present paper, a full-scale laboratory apparatus is used to study friction, wear and lubricant distribution in the flange contact. Contact parts of the apparatus are the UIC-ORE (the standard European railway profile) wheel with its 800 mm diameter and the BK7 glass specimen with the contact radius 13.1 mm or the UIC900A steel specimen. The steel specimens are cut out of a rail with the UIC 60 profile and the hardness value HB262. The wheel made of the R7T carbon-manganese steel with the hardness value HB280 is driven through the planetary gear by the AC motor. The glass or steel specimen is attached to the linear guide which moves in the normal direction to the flange root surface. The specimens are pushed on to the flange by the load lever mechanism.

All tests are carried out under pure sliding conditions at ambient temperature 25 °C ± 2. Friction force is measured by a strain gauge force transducer which is attached to the steel specimen. Wear of the steel specimens is inspected by a 3D optical profiler. The glass specimens are used together with a microscope and a high-speed CMOS camera (the optical method) to obtain the lubrication film distribution and size of the flooded area of the contact by grease. The contact is lubricated with the environmentally adapted grease containing rapeseed base oil, Al-based thickener and graphite. The grease amount, duration and interval of its application are driven by an on-board lubrication system. The minimum amount of grease is limited to 0.2 ml by a pump of lubrication system. The grease is applied by nozzle for 3 s on to the flange root surface during the wheel rotation.

Experiments with the steel specimens are performed for a fully starved and fluid lubricated contact at the constant sliding velocity 0.33 m/s and the maximum Hertzian pressure of 0.33 and 0.48 GPa, respectively. Before testing, the wheel flange is smoothed out by sandpapers and both contact surfaces are cleaned by acetone. Thereafter 0.4 ml of grease are applied and cleaned by acetone again to ensure fully starved conditions. For fluid conditions of the contact, running-in without grease is performed to reach the first signs of wear and then the grease is sprayed on to the flange.

Tests with the glass specimens are carried out under fluid lubricated conditions for the sliding velocity of 0.11, 0.22 and 0.55 m/s at the pressure just 0.26 GPa due to limited properties of the BK7 material. The no running-in phase is included. The contact surfaces are acetone cleaned, lubricated by 0.2, 0.4 or 0.6 ml of grease and the tests are finished before starving of the contact. The presented values of the flooded area width for each of the test conditions correspond to the median value of flooded sizes. These sizes are obtained from eight equally spaced positions around the circumference of the wheel during two wheel revolutions.

3 Results and Discussion

Initially, the amount of grease was established via the optical method to effective use of lubricant in terms of its distribution and squeezing out of the contact. Subsequently, the application intervals for this amount of grease were found to prevent the seizure development.

The values of the flooded area width are almost equal for 0.4 and 0.6 ml of grease under different sliding velocities, as shown in Fig. 2a. This means that these amounts were excessive. It led to squeezing the grease out of the contact and accumulation of the lubricant near to the inlet of the flange contact (see Fig. 2b). Figure 2b also shows that the mixture of grease and wear particles were distributed

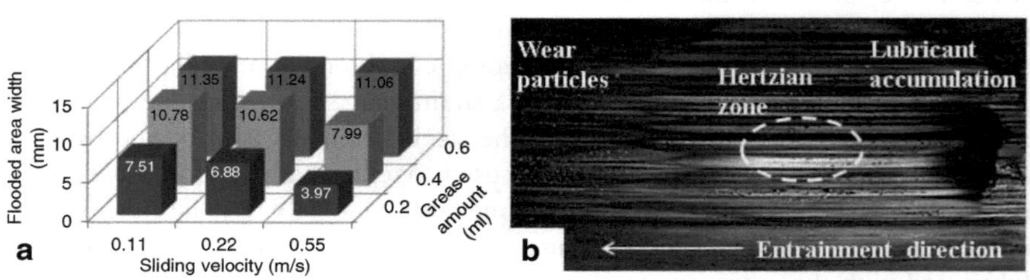

Fig. 2 Width of flooded area of flange contact by grease (**a**) and lubricant distribution within flange contact for 0.6 ml of grease at sliding velocity 0.11 m/s (**b**)

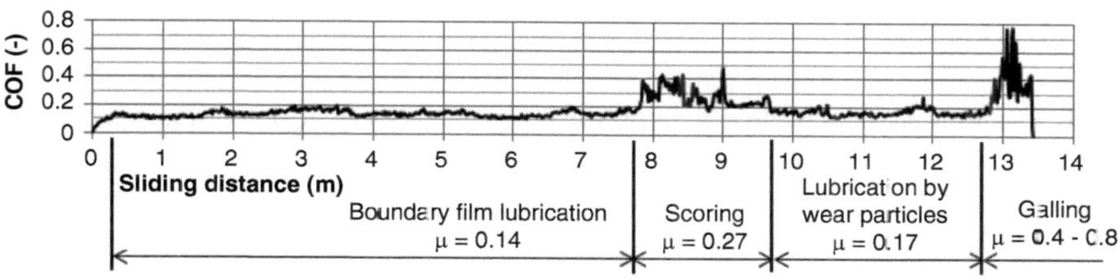

Fig. 3 Development of COF and wear for fully starved conditions at sliding velocity 0.33 m/s and pressure 0.33 GPa

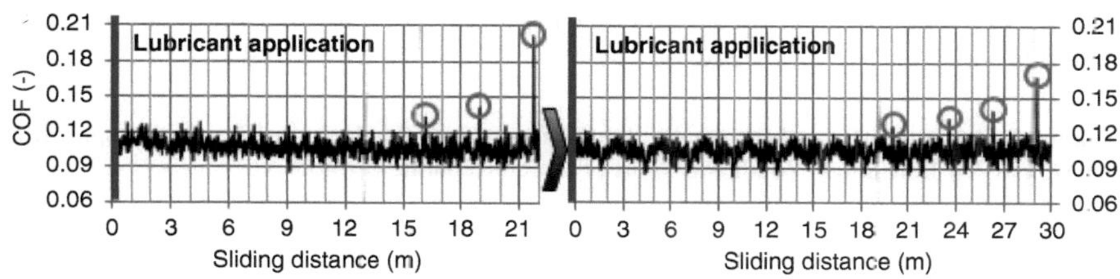

Fig. 4 Application intervals to prevent flange contact from development of seizure for 0.2 ml of grease, sliding velocity 0.33 m/s and pressure 0.48 GPa

mainly through negative asperities on the flange surface. Several wear and graphite particles were stuck on the glass specimen surface behind the outlet of the contact.

Figure 3 represents the development of COF and wear under fully starved conditions. There is an interesting part of the curve with lower median value of COF which follows the scoring section of the curve. The reason for this decrease in COF is not clear, but it could be caused by the time limited lubrication of the contact by the particles of wear before their led to the galling. The observed wear of the steel specimens and the flange surface were similar as is given in Fig 1b, c.

The application intervals of the 0.2 ml grease are defined by the sliding distance from the point of lubricant application to the seizure threshold (the local COF peak highlighted by circles in Fig. 4). Based on 10 tests, the seizure threshold appeared randomly from 16 to 35 m of sliding in the same test conditions with the median value of 20 m. It is well illustrated in Fig. 4 that the seizure development is prevented and a new sliding interval without seizure occurs after reapplication of grease. For experiments with the contact pressure of 0.33 and 0.42 GPa the application intervals were 60 and 37 m of sliding, respectively. Therefore, it may be possible that the application intervals between 0.33 and 0.48 GPa have a linear dependence on the contact pressure.

4 Conclusions

This research paper has clearly shown that repeated application of suitable amount of lubricant is a way to effective wear reduction in flange contacts. It was found that the application of 0.2 ml of the grease every 20 m of sliding prevents the seizure development of wheel-rail materials under pure sliding conditions. Accumulation of an excessive lubricant ahead of the contact inlet and its squeezing out were confirmed by the optical observation of the lubricant distribution within the flange contact. This work has only been able to touch on the very limited part of this parametrically extensive issue. It is obvious that many further studies are needed for better understanding of tribological processes within wheel flange contacts and before we will be able to satisfactorily predict or control behaviour of these contacts.

Acknowledgments The authors would like to express their thanks to the financial support from the Ministry of Education, Youth and Sport of Czech Republic (Project No. FSI-S-14-2336).

References

1. R. Lewis, U. Olofsson, Mapping rail wear regimes and transitions. Wear **257**(7–8), 721–729 (2004). doi:10.1016/j.wear.2004.03.019
2. J. Sundh, U. Olofsson, K. Sundvall, Seizure and wear rate testing of wheel-rail contacts under lubricated conditions using pin-on-disc methodology. Wear **265**(9–10), 1425–1430 (2008). doi:10.1016/j.wear.2008.03.025
3. J. Sundh, U. Olofsson, Seizure mechanisms of wheel-rail contacts under lubricated conditions using a transient ball-on-disc test method. Tribol. Int. **41**(9–10), 867–874 (2008). doi:10.1016/j.triboint.2007.12.011
4. D. Markov, U. Olofsson, Laboratory tests for seizure of rail and wheel steels. Wear **208**(1–2), 91–104 (1997). doi:10.1016/S0043-1648(96)07400-5
5. M. Spiryagin et al., Numerical calculation of temperature in the wheel-rail flange contact and implications for lubricant choice. Wear **268**(1–2), 287–293 (2010). doi:10.1016/j.wear.2009.08.014
6. S. Descartes et al., Wheel flange/rail gauge corner contact lubrication: tribological investigations. Wear **271**(1–2), 54–61 (2011). doi:10.1016/j.wear.2010.10.019
7. H.D. Huang et al., An investigation on tribological properties of graphite nanosheets as oil additive: tribological investigations. Wear **261**(2), 140–144 (2006). doi:10.1016/j.wear.2005.09.010
8. P.O. Larsson et al., Pressure fluctuations as grease soaps pass through an EHL contact: tribological investigations. Tribol. Int. **33**(3–4), 211–216 (2000). doi:10.1016/S0301-679X(00)00033-5

The Effect of Friction Modifier on the Wheel-Rail Contact

R. Galas and M. Omasta

Abstract One of the key factors affecting the rolling-sliding contact is traction, which is expressed by the adhesion coefficient. It is a major characteristic for safety, efficiency, reliability and maintenance costs of rail transportation. The adhesion is influenced by many factors such as contaminants, environmental and operating conditions. The friction modifier (FM) is applied on the top of rail in order to control the adhesion coefficient in required range (0.3–0.35). This range is often called an intermediate level of friction. This study has two aims. The first aim is to introduce tribological aspects of FM. The second aim is to investigate the influence of the FM on the adhesion coefficient. The experiments were carried out on the ball-on-disc apparatus.

Keywords Wheel-rail contact • Friction modifier • Adhesion • Adhesion coefficient • Traction • Friction • Wear

1 Introduction

Railway transportation is one of the most used ways of transportation. This transportation is characterized by many advantages, in comparison with other modes of transportation, such as safety, energy effectiveness and environmental favourableness. The key factor of railway transportation is the wheel-rail interface, where two contact zones are distinguished, Fig. 1 [1]. A wheel tread-rail top contact enables to transfer the traction forces from wheel to rail. This physical phenomenon is called adhesion which is expressed by an adhesion coefficient. If a vehicle goes through a curve the wheel flange-rail gauge corner contact is occurred. There is a problem if opposing requirements on magnitude of friction in these contacts. High friction is required within the wheel tread-rail top contact which is necessary for traction and braking of vehicle whereas low friction is required within the wheel flange-rail gauge corner contact concerning wear and damage. It should be noted that the wheel-rail interface is an open system and it means that the contacts are influenced by many factors such as natural contaminants, water, oil and wear

R. Galas (✉) • M. Omasta
Brno University of Technology, Brno, Czech Republic
e-mail: galas@fme.vutbr.cz; omasta@fme.vutbr.cz

V. Dynybyl et al. (eds.), *The Latest Methods of Construction Design*,
DOI 10.1007/978-3-319-22762-7_21

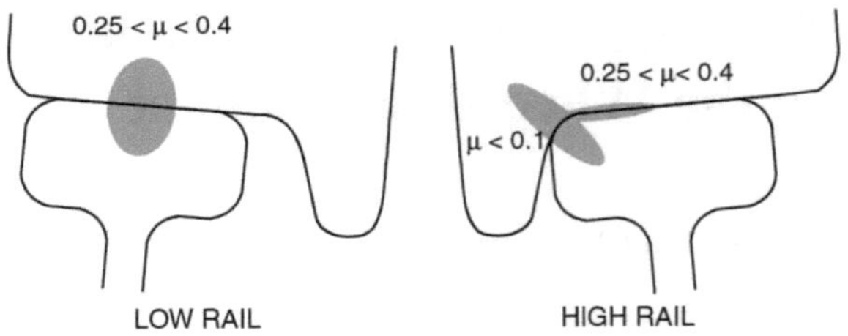

Fig. 1 The ideal adhesion coefficients within the wheel-rail contact [1]

debris. In addition, the operating conditions such as vehicles speed, wheel slip, contact pressure and many other factors also have a significant effect on tribological aspects of the wheel-rail interface [2, 3]. For this reason, the friction in contact is modified by application of oil, sand and FM. This approach is often called the friction management.

1.1 Friction Management

As mentioned above in the text, there are two basic approaches to modify friction level within the wheel-rail contact. The first approach is used in order to decrease in friction in the wheel flange-rail gauge corner contact when a vehicle goes through a curve. In this case, the lubrication is sprayed on the wheel flange. It immediately ensures decrease in friction level. In other words, wear of the contact bodies is reduced [4]. The second approach is used in order to increase in the adhesion coefficient in the wheel tread-rail top contact. If this contact is contaminated by for example: water, snow, dust, oil, leaves it causes that the adhesion coefficient is reduced from typical dry level (0.2–0.4) to level (0.025–0.2) [5], Table 1. This level can be termed as insufficient range because the adhesion coefficient which is essential for safety braking and traction should be in the range of 0.09–0.2 [6]. For this reason, sand is applied on the top of rail. Although sand application enables to reach the desired adhesion coefficient this process is associated with the side effects such as wear, damage to track circuits, electric isolation (a problem with train detection) and air pollution [7]. These side effects have a significant impact on maintenance cost, reliability and also on safety of railway transportation. However sanding is still the most used method to improve adhesion. Although the adhesion problem is already known several decades and many scientific articles have been focused on the adhesion problem this problem is still persisting. Recently the field of the friction management has been carried out extensively. The result of this research is a new type of an adhesion enhancer, which is called the friction modifier (FM).

Table 1 Parameters of the experiments

Parameters	Situation 1	Situation 2	Situation 3
Hertz max. pressure	0.7 GPa	0.7 GPa	0.7 GPa
Velocity	200 mm/s	200 mm/s	200 mm/s
SRR	10 %	5 %	10 %
Lubrication	FM	FM	Oil
Amount of lubrication	<2 ml	<2 ml	<2 ml
Material of the ball	AISI 52 100	AISI 52 100	AISI 52 100
Material of the disc	AISI 52 100	12 050	AISI 52 100
Hardness of the ball	(800–920) Hv	(800–920) Hv	(800–920) Hv
Hardness of the disc	(720–780) Hv	214 Hv	(720–780) Hv
Roughness of the ball	Ra 0.1	Ra 0.4	Ra 0.1
Roughness of the disc	Ra 0.1	Ra 0.4	Ra 0.1

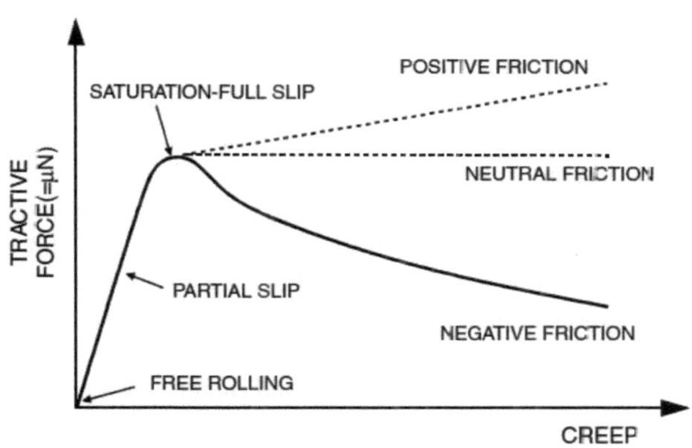

Fig. 2 The traction curve [8]

1.2 *Friction Modifiers*

The FM offers the opportunity to control friction properties in the wheel-rail interface. A function principle of the FM can be explained by the traction curve, Fig. 2. This is a fundamental characteristic for assessment of friction properties in the wheel-rail contact. This curve can be divided into two sections, an effective section and an ineffective section. The transition point between these sections is called saturation point.

In dry contact or contact with sand, the friction is decreased if the creep exceeds saturation point. This friction is termed as negative friction. In this case, if the wheel-rail contact operates in the vicinity of saturation point the stick-slip (roll-slip) is occurred. This results in the corrugation of the rail and wheel which causes the unpleasant noise and vibration [8]. These side effects can be eliminated by change of friction behind saturation point from negative friction to positive friction. Positive friction can be achieved by applying the suitable FM. The presence of

FM in the wheel-rail contact also causes that the maximum value of adhesion coefficient is reduced. It ensures decreasing in wear rate, noise and crack generation [9, 10]. Nowadays, the producers offer both liquid and solid FM.

The liquid FM can be consisted of base composition (water or oil), a solid lubricant (graphite, molybdenum disulphide), a binding agent, parts for modification of friction (calcium carbonate, magnesium carbonate) and a wetting agent. The composition of FM is modified according to desired friction properties and interfacial layer which is formed between wheel and rail. This layer is often called "the third layer" and the rheological behaviour of this layer plays an important role. The third layer is composed of contaminates and oxidized wear debris [11]. The FM can be designed for optimal interaction with this third body.

2 Experimental

The experiments were carried out using Mini Traction Machine (MTM). The main parts of this apparatus were the steel ball (diameter 19.05 mm) and the steel disc diameter (46 mm). Both of these parts were propelled independently. It enabled to create a rolling and sliding contact. In addition, MTM enabled to simulate different temperature within the pot.

During the experiments, the liquid FM and synthetic oil were used. Experiments were carried out under ambient temperature and the others parameters of experiments are shown in Table 1. There are two types of materials which were used in experiments. First material was the steel AISI 52 100, which is usually used to bearings. This steel is one of the basic materials, which is used in manufacturing of the samples for MTM. The second material was the steel 12 050 which is equal to the steel used in the manufacture of wheel and rail.

3 Results

The two types of timed tests were carried out. The first test showed how the adhesion coefficient is changed after application of FM and oil. In this test, the FM and oil were applied only once whereas in the second test, which illustrated durability of the FM for two materials of the disc, the FM and oil were applied several times. In both tests, FM and oil were applied after initial (200–400) seconds to ensure that the adhesion coefficients for dry contact have been reached.

In the first test, when the oil is applied the adhesion coefficient is reduced from 0.65 to 0.06, Fig. 3a. Subsequently, the adhesion coefficient remains stable. In contrast, the first test with FM has shown that adhesion is increased after the FM application. In the second test, when the adhesion coefficient reached value about 0.45, the FM has been applied again where the durability of the FM was observed, Fig. 3b. It is obvious, for both materials that the recovery time is different as well as

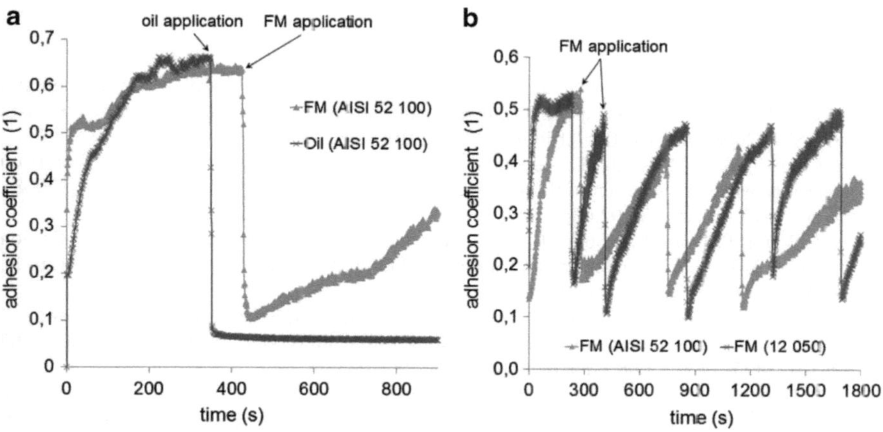

Fig. 3 The effect of oil and the FM on the adhesion coefficient (**a**) and the durability test of the FM (**b**)

the minimum value of the adhesion coefficient. On the other hand, is evident that the material AISI 52 100 exhibits shorter recovery time than the material 12 050.

4 Conclusion

The wheel-rail interface is a key factor of railway transportation. Therefore, friction properties within contact zones should be optimized concerning wear, traction, safety and so on. The conventional methods such as sanding are associated with many side effects. For this reason, the new type of adhesion enhancer, which is termed as FM, was developed. The FM is able to modify the friction according to composition of interfacial layer which is formed between wheel and rail. There are two aims for the FM application: achievement of intermediate level of friction (0.3–0.35) and positive friction after reaching saturation point.

The function of the FM was tested by using MTM apparatus. The adhesion was improved by the application of the FM. On the other hand, the intermediate level of friction was not achieved. It can be caused by high sensitivity of adhesion on FM amount and forming of the oxide layer. Another research is necessary.

Acknowledgments The authors would like to express their thanks to the financial support from the Ministry of Education, Youth and Sport of Czech Republic (Project No. FSI-S-14-2336).

References

1. J. Sundh, U. Olofsson, M. Ishida, T. Nakahara, Seizure mechanisms of wheel-rail contacts under lubricated conditions using a transient ball-on-disc test method. Tribol. Int. **41**(9–10), 867–874 (2008). doi:10.1016/j.triboint.2007.12.011
2. H. Chen, T. Ban, M. Ishida, T. Nakahara, Experimental investigation of influential factors on adhesion between wheel and rail under wet conditions. Wear **265**(9–10), 1504–1511 (2008). doi:10.1016/j.wear.2008.02.034
3. W.J. Wang, P. Shen, J.H. Song, J. Guo, Q.Y. Liu, X.S. Jin, Experimental study on adhesion behavior of wheel/rail under dry and water conditions. Wear **271**(9–10), 2699–2705 (2011). doi:10.1016/j.wear.2011.01.070
4. J. Sundh, U. Olofsson, K. Sundvall, J. Guo, Q.Y. Liu, X.S. Jin, Seizure and wear rate testing of wheel-rail contacts under lubricated conditions using pin-on-disc methodology. Wear **265** (9–10), 1425–1430 (2008). doi:10.1016/j.wear.2008.03.025
5. E.A. Gallardo-Hernandez, R. Lewis, K. Sundvall, J. Guo, Q.Y. Liu, X.S. Jin, Twin disc assessment of wheel/rail adhesion. Wear **265**(9–10), 1309–1316 (2008). doi:10.1016/j.wear.2008.03.020
6. P. Clayton, R. Lewis, K. Sundvall, J. Guo, Q.Y. Liu, X.S. Jin, Tribological aspects of wheel-rail contact: a review of recent experimental research. Wear **191**(1–2), 170–183 (1996). doi:10.1016/0043-1648(95)06651-9
7. O. Arias-Cuevas, Z. Li, R. Lewis, J. Guo, Q.Y. Liu, X.S. Jin, Investigating the lubricity and electrical insulation caused by sanding in dry wheel-rail contacts: a review of recent experimental research. Tribol. Lett. **37**(3), 623–635 (2010). doi:10.1007/s11249-009-9560-1
8. D.T. Eadie, J. Kalousek, K.C. Chiddick, J. Guo, Q.Y. Liu, X.S. Jin, The role of high positive friction (HPF) modifier in the control of short pitch corrugations and related phenomena: a review of recent experimental research. Wear **253**(1–2), 185–192 (2002). doi:10.1016/S0043-1648(02)00098-4
9. D.T. Eadie, M. Santoro, K.C. Chiddick, J. Guo, Q.Y. Liu, X.S. Jin, Top-of-rail friction control for curve noise mitigation and corrugation rate reduction: a review of recent experimental research. J. Sound Vib. **293**(3–5), 747–757 (2006). doi:10.1016/j.jsv.2005.12.007
10. D.T. Eadie, D. Elvidge, K. Oldknow, R. Stock, P. Pointner, J. Kalousek, P. Klauser, The effects of top of rail friction modifier on wear and rolling contact fatigue: full-scale rail-wheel test rig evaluation, analysis and modelling. Wear **265**(9–10), 1222–1230 (2008). doi:10.1016/j.wear.2008.02.029
11. K. Hou, J. Kalousek, E. Magel, Rheological model of solid layer in rolling contact. Wear **211**, 134–140 (1997). doi:10.1016/S0043-1648(97)00097-5

Measurement Methodology of Low-Shear Viscosity of Reference Liquid Under High-Pressure

S. Hornak

Abstract Accurate value of low shear viscosity under high pressure is important for predicting film thickness in elastohydrodynamic lubrication. This can be obtained theoretically from numerical simulations or by experiments on viscometers. Calibration of the viscometer has significant importance for precision. For this purpose, reference liquids such as squalane are used. Viscosity of these liquids can be easily compared with results published by other authors. In this work, methodology of measurement with the falling-body type viscometer with optical detection of falling-body speed is described.

Keywords Falling body viscometer • Viscosity • Squalane • Calibration

1 Introduction

Accurate value of low shear viscosity under high pressure is important for predicting film thickness and traction in elastohydrodynamic lubrication. Sometimes even small change in chemical structure leads to significant difference in the viscosity at high pressure. There are still many differences between theory and real behaviour of pressure viscosity correlation of liquids [1].

For recently built viscometer, in laboratories of Institute of Machine and Industrial Design of Brno University of Technology, verification and calibration measurement must be made. For this purpose, reference liquids such as squalane are used. Variation of viscosity with pressure of these liquids is well known and can be easily compared with results published by other authors [2, 3]. In this paper the measurement methodology is made.

S. Hornak (✉)
Brno University of Technology, Brno, Czech Republic
e-mail: y165927@stud.fme.vutbr.cz

V. Dynybyl et al. (eds.), *The Latest Methods of Construction Design*,
DOI 10.1007/978-3-319-22762-7_22

2 Falling Body Viscometer

The most simple and reliable device for measuring viscosity under high pressure is falling body type viscometer [4]. Pioneer of this technique was P.W. Birdgman [5]. Experimental method is based on calculating viscosity from times of passage of sinker by two points when velocity is in the steady-state [1].

$$\mu = Ct\frac{\rho_S - \rho}{\rho_R} \tag{1}$$

where μ is low shear viscosity, C is calibration factor, t is fall time, ρ_S is estimated density of sample, ρ is average density of sinker, and ρ_R is reference density chosen to be an approximation to $\rho_S - \rho$.

Sinker falls in chamber filled with the liquid sample. If the fall is not vertical the speed of body is increased. Centration is necessary either mechanically or hydrodynamically. There are more types of sinkers for different purposes [4]. In our case we have been used the solid cylinder sinker. Detection of the sinker position is provided optically by couple of lamps and photodiodes.

Pressure vessel consists of four cylinders pressed to one. Three of them is made of steel and the fourth smaller one is from tungsten carbide. Estimated maximum pressure for this vessel is up to 3 GPa, but for safeness, maximum is 1.5 GPa. Pressure is generated by hydraulic hand pump and increased to high pressure by intensifier with different diameters. Cross section of the viscometer is on Fig. 1.

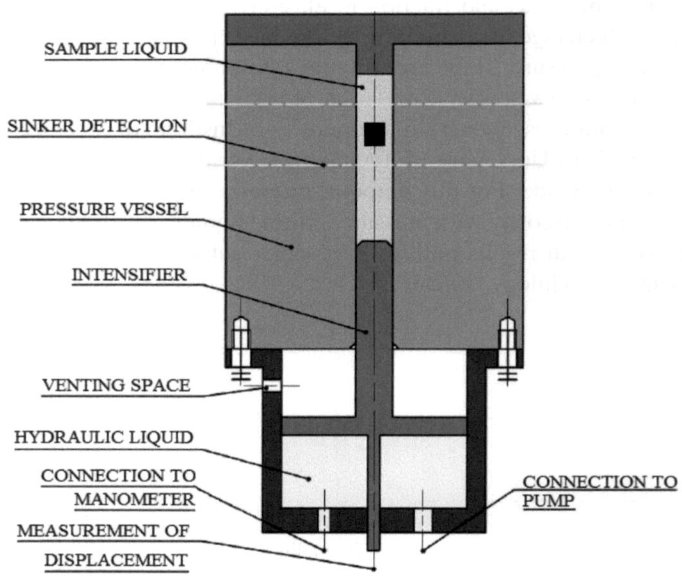

Fig. 1 Cross section of viscometer [8]

3 Reference Liquid

ISO standard for measuring viscosity of Newtonian liquid is water at 20 °C. This is impractical for viscometer calibrating. Selected reference liquid must have higher viscosity and have to be accurately characterized by many researchers. Other criterions are availability and relatively inexpensiveness.

These requirements fulfil squalane (SQL). It is low-molecular-weight (422.81 g/mol) branched alkane, 2, 6, 10, 15, 19, 23-hexamethyltetracosane. Because of its low-molecular weight inlet zone in EHL should be Newtonian. Calculations of film thickness can be compared with measurement [2, 3].

Experimental data can be compared with work of Comunas et al. [3]. In three laboratories four falling-body viscometers were used for pressures to 350 MPa. Some of their results can be seen in Table 1. Squalane as a reference liquid for EHL is recommended also by Bair [2]. Measurement of pressure viscosity dependence up to 1.2 MPa is in Table 2. Other results can be found in work of Harris [6] and Pensado [7].

Table 1 Experimental viscosities of SQL at high pressure [3]

T/K	p/MPa	μ/mPa·s	T/K	p/MPa	μ/mPa·s
University of Pau					
303.15	20	34.78	323.15	20	15.77
303.15	40	51.64	323.15	40	22.36
303.15	60	74.96	323.15	60	31.17
303.15	80	107,09	323.15	80	42.71
303.15	100	146.47	323.15	100	57.75
303.15	120	210.53	323.15	120	77.38
303.15	140	305.40	323.15	140	102.35
303.15	160	422.8	323.15	160	175.94
University of Santiago de Compostela ViscLPT1					
313.15	10	16.8	343.15	10	7.3
313.15	15	18.9	343.15	15	8.0
313.15	25	23.8	343.15	25	9.4
313.15	50	39.2	343.15	50	13.6
313.15	75	61.2	343.15	75	19.2
313.15	100	92.3	343.15	100	26.5
313.15	125	136.4	343.15	125	36.2
313.15	150	199.0	343.15	150	49.1
University of Santiago de Compostela ViscLPT2					
313.15	75	59.6	343.15	75	18.7
313.15	100	91.3	343.15	100	27.2
313.15	125	134.8	343.15	125	38.1
313.15	150	194.1	343.15	150	52.0
313.15	175	275.3	343.15	175	69.9
313.15	200	386.2	343.15	200	92.8
313.15	225	537.7	343.15	225	122.1
313.15	250	744.7	343.15	250	159.6

Table 2 The viscosity of squalane in mPa s [2]

p/MPa 1	293,15 K	313.15 K	338.15 K	373.15 K
0.1	35.9	15.7	6.73	2.95
50	102.9	41.0	15.7	6.08
100	260	94.2	32.0	11.04
200	1429	405	108	30.9
300	5670	1390	319	74.0
400	20,400	4230	866	168
500	67,600	11,920	2200	
600	215,000	32,400	5240	
700	681,000	86,000	11,950	
800	2,130,000	225,000	25,900	
900	7,070,000	584,000	54,600	
1000	24,400,000	1,605,000	115,000	
1100		4,500,000	250,000	
1200		13,160,000	525,000	

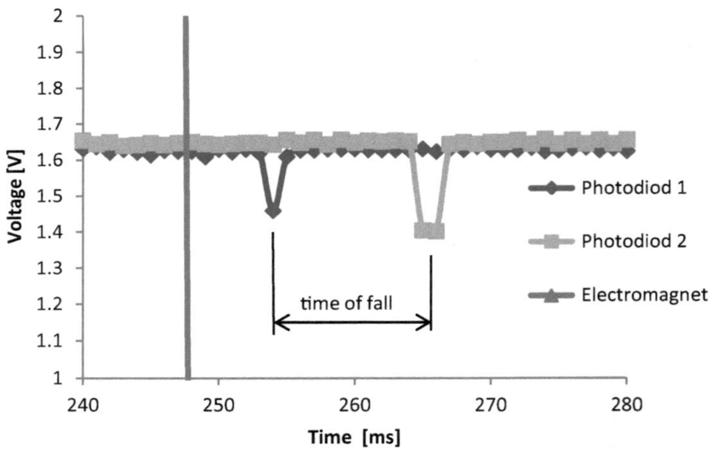

Fig. 2 Time of sinker fall [8]

4 Measurement Methodology

Sample of the reference liquid is injected into viscometer chamber by syringe. Then the chamber is sealed with the intensifier. Sinker is forced to bottom of the chamber by electromagnet. Then viscometer inverted and sinker is released from the electromagnet. Time of fall is the difference from times registered by photodiodes, as can be seen on Fig. 2.

Viscometer is turned back until sinker reached bottom. Then is again forced by electromagnet and, turned upside down. Liquid is then pressurised by the pump to higher pressure. Displacement of the intensifier is measured for calculation of liquid density. Measurement is repeated for required times.

5 Conclusion

Methodology of measurement of pressure viscosity correlation of the reference liquid was described. From the comparison of results by other authors and from this laboratory, the calibration factor of viscometer can be adjusted. This factor is essential for further measurements with the viscometer.

Acknowledgments This work is an output of research and scientific activities of NETME Centre, regional R&D centre built with the financial support from the Operational Programme Research and Development for Innovations within the project NETME Centre (New Technologies for Mechanical Engineering), Reg. No. CZ.1.05/2.1.00/01.0002 and, in the follow-up sustainability stage, supported through NETME CENTRE PLUS (LO1202) by financial means from the Ministry of Education, Youth and Sports under the "National Sustainability Programme I".

References

1. S. Bair, A routine high-pressure viscometer for accurate measurements to 1 GPa. Tribol. Trans. **47**(3), 356–360 (2004)
2. S. Bair, Reference liquids for quantitative elastohydrodynamics: selection and rheological characterization. Tribol. Lett. **22**(2), 197–206 (2006)
3. M.J.P. Comunas, X. Paredes, F.M. Gacino, J. Fernandez et al., Viscosity measurements for squalane at high pressures to 350 MPa from T = (293.15 to 363.15) K. J. Chem. Thermodyn. **69**, 201–208 (2014)
4. L. Kulisiewicz, A. Delgado, High-pressure rheological measurement methods: a review. Appl. Rheol. **20**(1), 1–15 (2010)
5. P.W. Birdgman, The effect of pressure on viscosity of forty-three pure liquids. Proc. Am. Acad. Arts. Sci. **61**(1926), 57–99 (1926)
6. K.R. Harris, Temperature and pressure dependence of the viscosities of 2-ethylhexyl benzoate, bis(2-ethylhexyl) phthalate, 2,6,10,15,19,23-hexamethyltetracosane (squalane), and diisodecyl phthalate. J. Chem. Eng. Data **54**(9), 2729–2738 (2009)
7. A.S. Pensado, M.J.P. Comunas, L. Lugo, J. Fernandez, High-pressure characterization of dynamic viscosity and derived properties for squalane and two pentaerythritol ester lubricants: pentaerythritol tetra-2-ethylhexanoate and pentaerythritol tetranonanoate. Ind. Eng. Chem. Res. **45**(7), 2394–2404 (2006)
8. L. Skalicky, Diploma thesis, Brno University of Technology, 2013. Supervisor Ivan Křupka (2013). Available via: http://dl.uk.fme.vutbr.cz/zobraz_soubor.php?id=1894

Modification of Tooth Flank as the Possibility of Increasing Resistance of Gearing to Pitting

B. Kopilakova, M. Bosansky, and J. Zapotocny

Abstract In this article are shown the influence of the tooth flank to reduce damage to the working tooth flanks in terms of the pitting. The introduction of the article describes a fundamental difference between the involute and noninvolute gearing and the influence of some parameters, especially of the slip ratio to damage of gearing. The paper also shows the basic results of the experiment, which was executed on two types of gearing (HCR gearing and C-C gearing) on the Niemann's stend in term of the damage to pitting [1].

Keywords Pitting • Tooth flank • Fatigue wear • C-C gearing • HCR gearing

1 Introduction

The influence of time-variable load in transfers commute to the mechanical fatigue abuse. As is well known, during the engagement of the tooth under the influence of surface and subsurface tension is formed stress, which results to the pitting. Touching of the surface of the working portion of the tooth leads to removing of the metal, to splitting of the particles of surface and to the formation of the holes, i.e. to the formation of pitting. The emergence of pitting are also influence the factors such as: the geometrical parameters of gearing, the ratio of traction determining the loading tooth by the external forces, the resolved shear ratios on the tooth faces, the frictional forces, the finish of tooth faces, selected material and hardness of the working tooth faces of the pinion and the gearing, the properties of the lubricating oil and operating conditions [2].

B. Kopilakova (✉) • J. Zapotocny
Trenčín University A. Dubček in Trenčín, Trenčín, Slovakia
e-mail: beata.kopilakova@tnuni.sk; jan.zapotocny@tnuni.sk

M. Bosansky
Slovak University of Technology in Bratislava, Bratislava, Slovakia
e-mail: miroslav.bosansky@stuba.sk

V. Dynybyl et al. (eds.), *The Latest Methods of Construction Design*,
DOI 10.1007/978-3-319-22762-7_23

2 Characteristic of the Individual Types of the Gearing

Touching of two tooth is the point of touch during the rotation moved. The geometric point of all the items image that progressively come to the gear is the trajectory – mating line. According to the shape of the mating curve and its position in term of the rotating centre of the profile, is also defined the type of the gearing. The gearing is clearly defined by the shape of the line of contact, however the shape can be generally arbitrary.

General planar gearing can be understood as the gearing with the axis of rotation O_1, O_2 of the both together engaging wheels. It is defined by the line of contact “S”, composed by two circular curves that centres lies on the any straight line crossing the point C. The centres of the circular curves can be in the headquarter or out of it and the ratio of the circular curves could be arbitrarily sized [2]. In general the gearing can be divided on the bases of the process of the line of contact, whereupon the gearing is distinguished by the following tooth flank:

1. In case that the curvatures of the line of contact “S” are not identical with the curvatures of the rolling circle, both together engaging wheels and the centres of the arc S_{kh} and S_{kd} that form the line of engagement and lie on the centre joins of the rotating of the wheels, that is the case of cycloid gearing (Fig. 1a).
2. In case that the angle of the mating line and the radius of the both parts of curvatures, which are representing the line of contact and having infinitely large radius. The curvatures of the circle are degenerating to the line, what defines the involute gearing (Fig. 1b).
3. In case that the line of contact “S” is composed of two circular curvatures, which centres are not on the headquarter of both wheels then it is the special case of general planar gearing – convex-concave (C-C) gearing (Fig. 1c). Its properties are depended on the geometrical parameters of the line of contact: the angle of the slope of the line of contact α_C, radius of the circular curves of the line of contact r_{kd}, r_{kh} defined by the angle shot and by the endpoints of arcs of the mating lines defined by the angle α_A, α_E or by the coordinates of relevant points $A[X_A, Y_A]$, respectively $E[X_E, Y_E]$.

C-C gearing could be generally understood as any gearing, which the tooth flank forms a curve with the convex - concave portion, where in the line of contact may be composed of two non-symmetrical (Fig. 2a) or symmetrical arcs (Fig. 2b) [3]. If the line of contact is defined by one or more of general curve, it is a case of the virtually identical case which is shown to Fig. 1c. In practice, we can meet with a combined line of contact, for example when in the vicinity of pint C the gearing is defined by the linear line of contact and in the other points by circular curves [3]. From this it is clear that the curve of the tooth flank and may take the form of a general curve (involute, cycloid, epicycloid, hypocycloid, etc.).

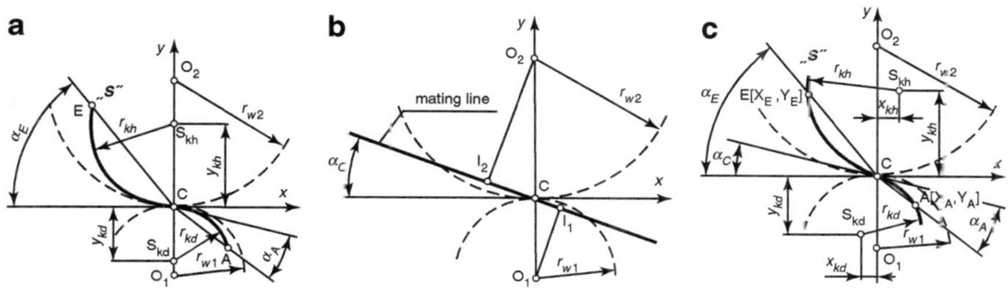

Fig. 1 Characteristic of the mating line definig the: (**a**) cykloid gearing [2]; (**b**) involute gearing [2]; (**c**) planar C-C Gearing [3]

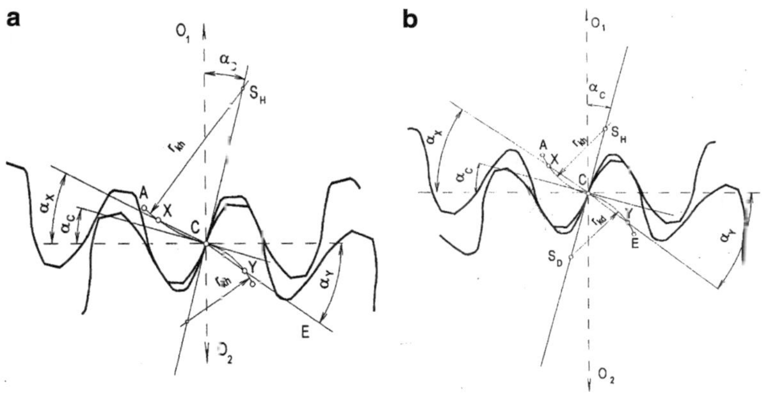

Fig. 2 C-C Gearing [3]: (**a**) non-symmetrical; (**b**) symmetrical

3 Evaluation of the Pitting in the Gearing

Geometric and qualitative parameters of gearing can strongly influence their failures. In case of pitting, the failures are mainly measuring deviations. Since the goal of this presentation is to compare involute HCR gearing and non-involute C-C gearing, there are developments of deviation rates for a sprocket and a wheel shown in the Fig. 3. In the Fig. 4, the development of distribution of strain of teeth along line of contact is described.

In the Fig. 5a, pitting on involute HCR gearing is shown and in the Fig. 5b, it's the pitting on non-involute C-C gearing. It is obvious that in case of involute HCR gearing, the pitting developed under rolling circle and in case of non-involute C-C gearing it developed around or above the circle. The origin of pitting in non-involute C-C gearing strongly influences its geometric parameters and the size of contact pressure. If we change some of the parameters α_C, r_{kh}, r_{kd}, z or the combination of them, we can get a change of the curve of the tooth flank [3].

The change cannot be random, but it is limited to making such tooth flank that protuberant part of tooth and convex part of anti-tooth are in mesh in the gear and

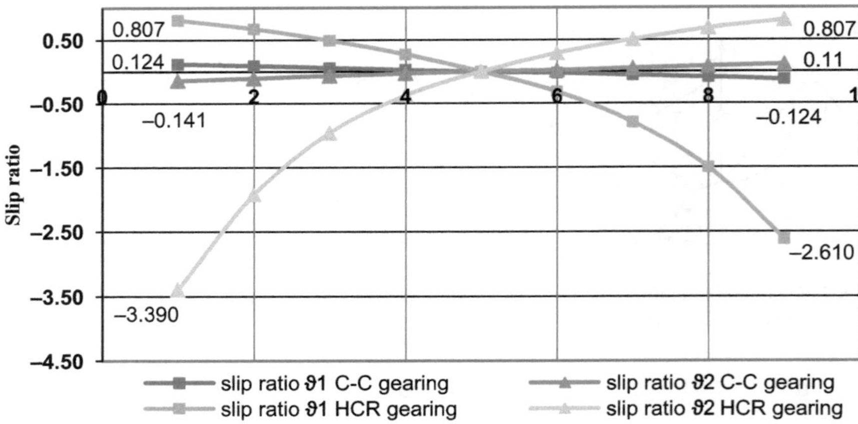

Fig. 3 Processes of the slip ratio of involute and C-C gearing

Fig. 4 Allocation F_n along mesh in: (**a**) HCR gearing [2], (**b**) C-C gearing

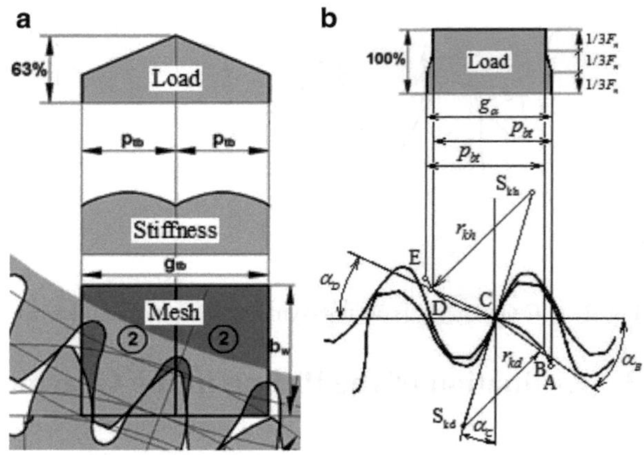

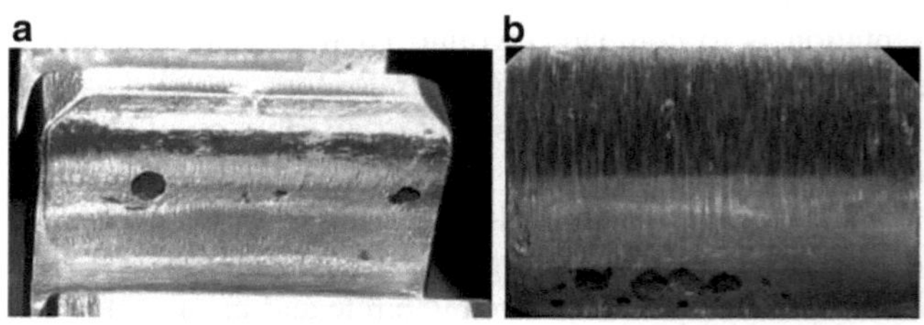

Fig. 5 Pitting: (**a**) non-standard gearing HCR[10], $\varepsilon_\alpha = 2.003$; (**b**) C-C gearing, $\varepsilon_\alpha = 1.116$

vice versa, so convexo-concave stroke happens. The change of geometric parameters of tooth flank in involute gearing is reached by adjusting.

Mesh conditions in C-C gearing and in involute gearing are different. That is why we have to verify the theoretical thinking about contact solidity of C-C gearing which we gained through MKP method through an experiment and find out if the usual procedures of involute gearing are applicable on C-C gearing, too. One of these parameters is fatigue limit in contact $\sigma_{H\,\mathrm{lim}}$. Niemann stend, standard testing machine with enclosed output in accordance with norm DIN 51354 according to methods FZG, was used for experimental finding out of values of fatigue limit $\sigma_{H\,\mathrm{lim}}$ for C-C gearing as well as for the standard involute gearing in accordance with norm DIN 3990 T5 [4].

During the experiment C-C gearing was tested under given strain based on analysis of suitable geometrical parameters at ÚDTK SjF STU in Bratislava and made of non-hardened carbon steel C60.

The basic principal of testing gears on load machines DIN 51354 consists of gradual loading of gear and finding out the number of finished cycles until damaging the testing sample. Eight different tests were made at three different load levels A÷C with gradual decline of load (Table 1). Three experiments were made at first two levels. Two experiments were made at the third level near to high-cycle fatigue. The load at each level was set by lever mechanism [2]. The experiment was interrupted after given number of cycles so the area damaged by pitting could be evaluated.

There are many ways of evaluating in which the condition of damaged tooth flanks by pitting. In the past the damaged area have been deducted through the squared network of the graph paper, what was too laborious and time-consuming. Later, the situation was simplified by using digital cameras and their interface with PC. The disadvantage of this method is the necessity of a lot of configuration and compliance of lighting conditions during the taking the photography for the best possible quality of the image [5].

In consideration of the weaknesses of the used methods so far, we have proposed and used a new method macrophotography [6], the principle of which is shown in Fig. 6. The digital microscope connected via USB to a personal computer was obtained a digital image of each tooth flanks of the testing gearing. The image was recorded directly to a PC. To evaluate the percentage of damage of each tooth flanks was used the Pitting – check software that allows to process digital images. The program includes utilities that enable simple system of check sampling areas affected by pitting. Also the bevelled edges are considered etc. Then the content of the damaged surface is percentually shown and compared with the value limit state. This new method streamlines the procedure of evaluating of the damaged state during the experiment. On the base of the extent of damage was determined following step of the experiment or its termination.

Table 1 Loading levels

Levels i	Load torque $M_{k(i)}$
A	$M_{k(A)} = 265\ \mathrm{N \cdot m}$
B	$M_{k(B)} = 185\ \mathrm{N \cdot m}$
C	$M_{k(C)} = 125\ \mathrm{N \cdot m}$

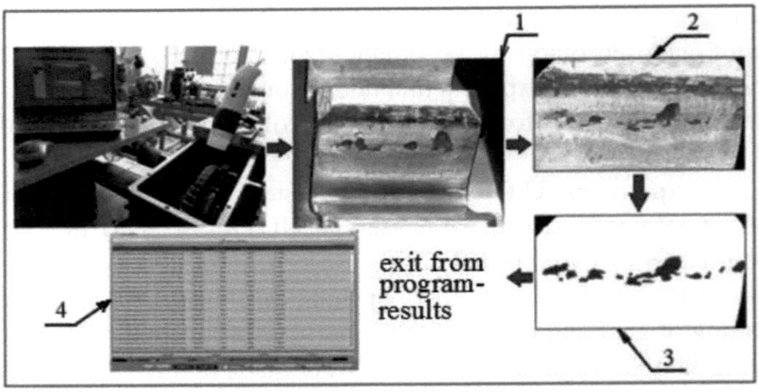

Fig 6 The principle of evaluation pitting makrofoto graphical method (*1*) created a digital image, (*2*) processed a digital image, (*3*) with program marked damage area, (*4*) evaluation of damage (%)

4 Conclusion

The experiment was done at the STU in Bratislava on Niemann's stende under the same conditions and load for HCR and C-C gearing. The results that have been obtained are the following: in the tests of the C-C gearings the samples were tested in the range (265 ÷ 125) Nm, what corresponding to the period of the marked examination to failure (437 ÷ 1483) hours and to the range of probable number of cycles to failure, approximately $(18 \div 65).10^6$ cycles. The description of the pitting formation shows that in C-C gearing, this type of fatigue wear much less aggressive than in HCR profiles, where the initialization of the crack occurs at greater depth. At the first load rate on the surface of A (Table 1) in the C-C gearing, where $M_k = 265$ Nm and the marginal state of the contact stress occurred at 19.10^6 cycles. The percentage of damage to tooth flank was evaluated by software Pitting – check and showed 4.04 %. On the reduced load rate on the surface of B, where $M_k = 185$ Nm, this state occurs at 50.10^6 cycles with damage of 3.83 % on one of the tooth flank. In the third stage the lowest load for $M_k = 125$ Nm, the damage was 0.4 % at 64.10^6 cycles when the experiment was terminated. From this it is clear that the extreme case (4 %) the damage by the pitting to such surface would be created only after extremely large number of cycles burdensome. The result is that in the case of C-C gearing was achieved a higher contact fatigue.

Acknowledgments The research work reported here was made possible by grant projects VEGA 1/0277/12, 1/1035/12.

References

1. M. Černý, J. Filipek, P. Mazal, Pitting damage of gears. Acta univ. agric. et silvic. Mendel. Brun. **LVIII**(2), 51–60 (2010)
2. M. Bošanský, M. Vereš, J. Rusnák, P. Tököly, Design II (STU Bratislava, 2009), 325 s., ISBN 978-80-227-3510-0
3. M. Bošanský, M. Vereš, P. Tököly, A. Vanya, Non-Standard Gearing, (STU Bratislava, 2012), 159 s., ISBN 978-80-227-3713-5
4. M. Bošanský, M. Vereš, J. Gaduš, Theory of Convex-Concave and Plane Cylindrical Gearing (STU Bratislava, 2006), 180 p., ISBN 80-227-2451-3
5. J. Medzihradský, Experimental and theoretical research on HCR gearing as well as in terms of its optimization and resilience to damage to pitting, Dissertation, STU Bratislava, 2009
6. M. Bošanský, B. Kopiláková, J. Zápotočný, Makrofotografická metodika pre hodnotenie poškodenia zubov z hľadiska poškodenia boku zuba opotrebením, vytrhávaním okolitého materiálu v tvare jamiek. Úžitkový vzor 6760

Including the Method of Lubricant Film Thickness Measurement with the Use of the Monochromatic Interferometry to the Education of Tribology

D. Kostal, I. Křupka, and M. Hartl

Abstract Measuring of distance as presented at the universities is usually limited to dimension down to microns. This orders of distance are sufficient for common engineer's application. However dimension are a lot smaller when engineers deal with the tribology research. One of the most important progresses in the field of tribology in last century is without doubts the elastohydrodynamic lubrication (EHL). This lubrication regime can be found for example in rolling bearings or gears. Common thickness of the lubricant film in the EHD contact is in order of tens or hundreds of nanometers. Optical method which enables measurements of such distances was developed by Gohar and Cameron in 1960s. This method is based on interference of monochromatic light and provides possibility of obtaining lubricant film thickness from known wavelength of used light and its intensity in the lubricated contact. Students can create central film thickness prediction with use of the analytical formulae of the Hamrock and Dowson. This prediction is then confirmed by hands-on experiment in laboratory. Contact simulator with microscope and red light is used to record sequence of the interferograms. These interferograms are then processed in Matlab® software in order to obtain film thickness.

Keywords Interferometry • Education • Tribology

1 Introduction

University students are familiar with common methods of measuring length such as ruler, Vernier scale or different types of calliper gauges. These tools provides possibility of measuring lengths in orders down to hundredths of millimetres. This scale is sufficient for common engineering applications such as controlling of dimension of ordinary engineering parts. Problem occurs when one needs to deal with dimensions or shapes from microns down. Mentioned mechanical methods are

D. Kostal (✉) • I. Křupka • M. Hartl
Brno University of Technology, Brno, Czech Republic
e-mail: kostal@fme.vutbr.cz; krupka@fme.vutbr.cz; hartl@fme.vutbr.cz

V. Dynybyl et al. (eds.), *The Latest Methods of Construction Design*,
DOI 10.1007/978-3-319-22762-7_24

not sufficient in such cases because of increasing demands for production precision of the tools. Necessity of measuring smaller length led to development of new methods. Problem with mechanical precision of measurements tools were sufficiently overcome by establishing methods which are using light beams.

One of the most method used is an interferometry. This method is suitable for measuring length from micrometres down to nanometres or less. Phenomenon used in this method is light interference which occurs when two light beams combines together. The amplitude and the brightness of a resultant beam is then different then original one. Only brightness of the light is changing if both light sources are of a same single wavelength and we call this method a monochromatic interferometry. This method was among first used and is, therefore, used for purpose of education at our institute. Another methods such as colorimetric interferometry are more precise but they lack the simplicity and cannot be explained simply.

2 Process of Education

This main part of the article should provide the process of the course as is demonstrated to the students. First part is theoretical and students are acquiring knowledge necessary to finish the course. Example of using the monochromatic interferometry method is illustrated on tribology application of measuring thin lubricant film in the elastohydrodynamically lubricated point contact. The lubricant film in this type of contact can be of values from tens of nanometres to the few microns depending on the velocity and lubricant properties.

2.1 Theory of the Monochromatic Interferometry

The interferometry is presented on the contact simulator which was originally used by Gohar and Cameron [1].

It is ball on disc contact simulator and the interference is observed between two light beams originating from single light source. Light is split into two beams on the semi-reflective chromium layer (Fig. 1). Part of the beam is reflected and second part is maintaining the direction and is fully reflected from the steel ball's surface. These two beams then interferes and they are slightly out of phase because each of them travelled different distance. Distance can be calculated from the change in the light intensity. When phase shift is equal to the half of the wavelength, destructive interference is observed and light intensity is decreased. Constructive interference is observed in case of phase shift equal to wave length or its multiplication. In this case is observed bright light.

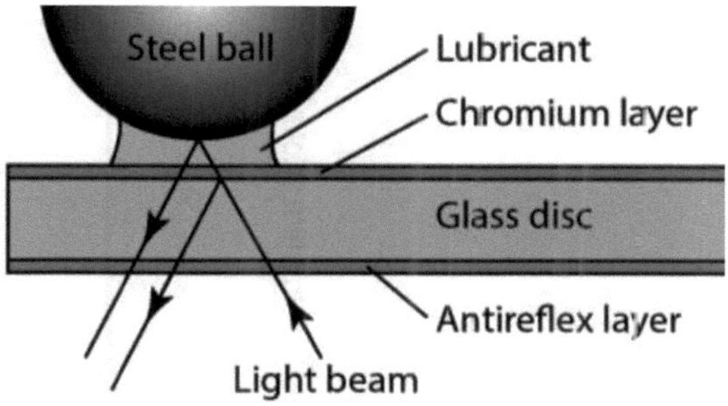

Fig. 1 Illustration of the interferometry principle

2.2 *Prediction*

First step is to calculate the lubricant central film thickness from given conditions. Given conditions are lubricant properties, rolling velocity range, contact couple material and geometrical properties, temperature and load. Calculations should be made for range of velocities originating from the zero film thickness thus zero speed. This step eliminates the necessity of calibration. Equation (1) used for calculation was presented by Hamrock and Dowson [2].

$$\frac{\mathbf{h_c}}{\mathbf{R'}} = \mathbf{2.69}\left(\frac{\mathbf{U\eta_o}}{\mathbf{E'R'}}\right)^{\mathbf{0.67}}\left(\boldsymbol{\alpha}\mathbf{E'}\right)^{\mathbf{0.53}}\left(\frac{\mathbf{W}}{\mathbf{E'R'^2}}\right)^{\mathbf{-0.067}}\left(\mathbf{1-0.61e^{-0.73k}}\right) \quad (1)$$

Result of the prediction should look like the plot in Fig. 2. Relation of the central film thickness and the rolling speed is linear when plotted in logarithmic axes.

2.3 *Experiment*

Experiment is conducted on the contact simulator where contact is created between a glass disc and a steel ball. Base mineral oil is used as a lubricant with syringe pump which provides sufficient amount of oil during the experiment to achieve fully flooded conditions in order to maintain validity of the Hamrock and Dowsons formulae. Interferograms are captured by high speed camera with rate of 4000 fps connected to the industrial microscope system. One thousand watt Xenon lamp is used as light source. This lamp provides bright white light which needs to be filtered in order to separate single wavelength. In case of red filter is wavelength equal to $\lambda = 700$ nm, but it depends on filter type.

Example of the interferogram sequence can be seen in Fig. 3. Changing colours from bright to dark and back can be seen when observing contact centre of interferograms. This is the mentioned constructive and destructive interference. Outcome of 1 s of the experiment with increasing rolling speed is the set of 4000 images of this type. Therefore, evaluation is made in Matlab® software.

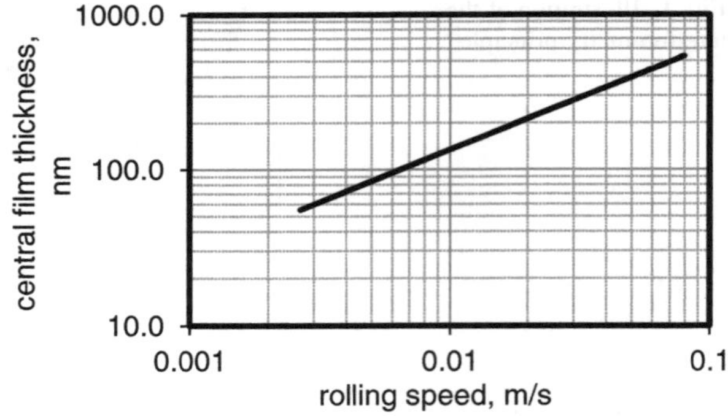

Fig. 2 Result of the prediction

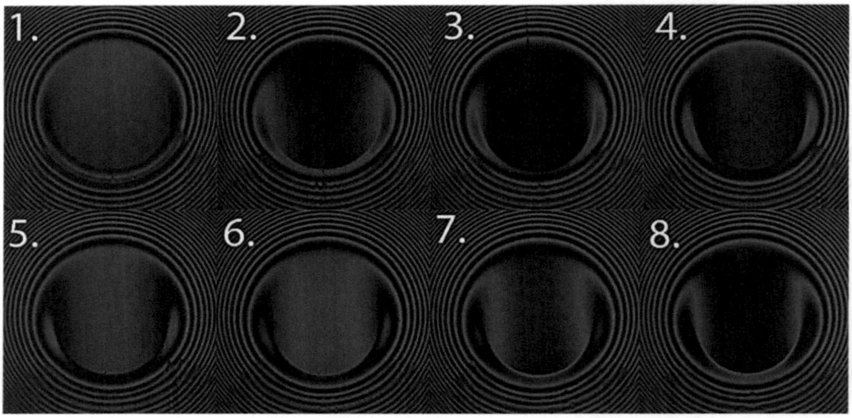

Fig. 3 Example of the data output

2.4 *Evaluation*

Evaluation consist of processing few thousand of interferograms (*.bmp file) like those presented in Fig. 3. Matlab script firstly let you choose the central part of the first interferogram which will be designated as contact centre or sample point. Small area around this point is then averaged. After this initial process a batch script is run. Script loads all images sequentially and stores one value which represents the red channel intensity as average of designated area. After this batch can be data plotted as function of the image number. Data from matlab script might look like curve in Fig. 4.

Time is related to the number of image, because high speed camera has precise digital shutter and takes desired amount of images per second with great accuracy. For example of 4000 frames per second is time between two consequent images equal to 1/4000 s.

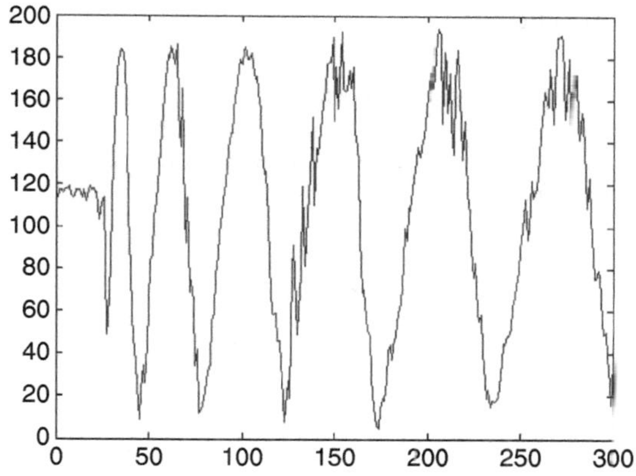

Fig. 4 Light intensity as function of image count

Film thickness can be now extracted from number of peaks in the plot. Each peak is equal to increase in the film thickness by $\lambda/4 = 175$ nm for red light. From this information and fact that data starts from the zero film thickness can be plotted same trend as in Fig. 2.

Final step is to compare prediction and experimentally obtained data.

3 Conclusion

This set of procedures should help students to understand the optical method used in many scientific and engineering application with hands-on attitude. These methods can be much more precise compared to commonly used methods. Interferometry used for example in the mapping of the surface topography can provide vertical precision down to 0.1 nm. Another benefit is that students needs to use high speed camera for capturing data so they experience work with high speed cinematography. Last but not least skill is the mapping of the elastohydrodynamic film thickness which is one of the most important aspect in the field of tribology. Whole exercise is created to cover multiple areas of interest and to be interesting for students by utilizing of the new technologies.

Acknowledgement David Kostal is Brno Ph.D. Talent Scholarship Holder – Funded by the Brno City Municipality. This work and conference contribution is funded by University project FOND VEDY FV 14-07.

References

1. A. Cameron, R. Gohar, Theoretical and experimental studies of the oil film in lubricated point contacts. Proc. R. Soc. London Ser. A. **291**, 520–536 (1966)
2. B.J. Hamrock, D. Dowson, *Ball Bearing Lubrication, the Elastohydrodynamics of Elliptical Contacts* (Willey, New York, 1981)

Frictional Properties of PVA Hydrogel

D. Nečas, M. Vrbka, S. Yarimitsu, K. Nakashima, Y. Sawae, P. Šperka, I. Křupka, and M. Hartl

Abstract Total hip replacement is one of the most successful surgical treatments of modern medicine. Typically, at present, hard-on-hard bearing surfaces are widely used for components of artificial hip joints. Hard-on-hard means that both components have high modulus of elasticity in range of hundreds of GPa. However, these materials suffer from relatively high friction and wear rate. This is connected especially with occurring lubrication regime. To approach conditions presented in natural joints, it is necessary to think about artificial cartilage. One of the anticipated materials for artificial cartilage is polyvinyl alcohol (PVA) hydrogel. PVA hydrogel has water content about 85 % and its elastic modulus is approximately $E \approx 1.2$ MPa, which is similar to natural cartilage. The main disadvantage of PVA hydrogel is its lower strength. In this study, commercial mini traction machine (MTM) was used to determine friction coefficient for various slide-to-roll ratios (SRR). Bovine serum was used as a lubricant and the tests were carried out under ambient temperature for three various speeds $u_1 = 25$ mm/s; $u_2 = 50$ mm/s; $u_3 = 100$ mm/s and two different loads $F_1 = 5.2$ N; $F_2 = 9.8$ N, respectively. As expected, friction coefficient was very low, less than 0.05 under some conditions. In future, optical method based on the principle of fluorescent microscopy will be used for studying lubricant film thickness and protein adsorption on bearing surfaces.

Keywords Biotribology • PVA hydrogel • Compliant • Friction

D. Nečas (✉) • M. Vrbka • P. Šperka • I. Křupka • M. Hartl
Brno University of Technology, Brno, Czech Republic
e-mail: necas@fme.vutbr.cz; vrbka.m@fme.vutbr.cz; sperka@fme.vutbr.cz; krupka@fme.vutbr.cz; hartl@fme.vutbr.cz

S. Yarimitsu • K. Nakashima • Y. Sawae
Kyushu University, Fukuoka, Japan
e-mail: yarimitsu.seido.681@m.kyushu-u.ac.jp; nakashima.kazuhiro.306@m.kyushu-u.ac.jp; sawa@mech.kyushu-u.ac.jp

V. Dynybyl et al. (eds.), *The Latest Methods of Construction Design*,
DOI 10.1007/978-3-319-22762-7_25

1 Introduction

Total hip replacement became one of the mostly applied orthopaedic treatments in the world of today. Initial attempts to replace human joints were made around 1850. First artificial joints were fabricated from ivory, which was then substituted by wood. Year 1962 is usually mentioned as a milestone. That year, British orthopaedist John Charnley employed polymer known as ultra-high molecular weight polyethylene (UHMWPE) as a material for acetabular cup. Other important historical point is discovery of CoCr alloy and ceramics [1].

Typically, hard-on-hard materials are used for artificial joints at present. One of the limitations of these prostheses is its service life; therefore it is necessary to perform revision operation after 15 or 20 years. Moreover, in case of younger patients (under 50 years), service life of hip replacement rapidly decreases, and in 28 % of all cases re-operation is needed in less than 10 years [2].

Because re-operations lead to deterioration of life quality, researchers are trying to study effects influencing lifetime of prosthesis. Biotribology is relatively new research area which deals with tribology aspects in relation to human body [3].

Previously, biotribologists focused on wear of artificial hip joints. Several studies were conducted using simulators to evaluate wear rate of bearing surfaces [4–7]. Later, researchers tried to determine the film thickness between the components to better understand tribological processes in replacements [8–11]. Important point, which was not fully described yet, is protein film formation in the contact; however, this explanation could bring important information for further development of implants in relation to its design and geometry.

Especially in last tens of years, more and more scientists are thinking about replacement of human cartilage. In natural joints, cartilage separates bearing surfaces and according to its softness and continuous hydration, it ensures almost ideal tribological conditions. In real joints, there is no such a soft layer between surfaces, while this absence leads to increase of friction and wear [12].

One of the anticipated materials for artificial cartilage is polyvinyl alcohol (PVA) hydrogel. PVA hydrogel has water content around 85 % and its young modulus is $E \approx 1.2$ MPa. Its main disadvantage is lower strength [13].

This study is focused on the measurement of coefficient of friction (COF) between CoCrMo ball and PVA hydrogel disc. PVA discs were prepared by freezing-thawing (FT) method which enables to prepare non-transparent samples of constant thickness. PVA samples prepared by cast-drying (CD) method were also studied. Even if these specimens are transparent, which could be advantageous for observation using optical methods, CD technique does not allow controlling the thickness of samples yet, so the results were not good.

In future, CD method is going to be improved to control the thickness of prepared layer. It would allow to study protein film formation and to measure film thickness between PVA hydrogel and CoCrMo or ceramic alloys by using optical method based on fluorescent microscopy described in previous study [14].

2 Materials and Methods

Commercial mini traction machine (MTM, PCS Instruments, UK), schematically shown in Fig. 1, was used to study traction curves under various loading and kinematic conditions. COF is measured in configuration ball-on-disc. Both components are driven independently to reach different slide-to-roll ratios (SRR). The ball is loaded against a disc; the friction force is measured by transducer [15].

CoCrMo ball of diameter of ½ inch and 2 mm thick PVA hydrogel disc were used as the test specimens. PVA sample was fixed on the steel disc for better stability during rotating. Before the test, PVA specimens have to be stored in distilled water all the time to ensure sufficient hydration of it. CoCrMo ball was cleaned by detergent solution, distilled water and isopropyl alcohol prior to testing.

COF was measured under three various speeds $u_1 = 25$; $u_2 = 50$; $u_3 = 100$ mm/s, respectively. Two different loads were applied $F_1 = 5.2$ N, $F_2 = 9.8$ N (resulting contact pressures $p_1 = 0.4$ MPa, $p_2 = 0.5$ MPa). All the measurements were conducted at room temperature T = 22 °C. Because of non-stable loading force at lower speeds (25; 50 mm/s) according to structure of PVA specimens, it was decided to not to measure the traction curves directly, but to determine the COF for different SRR (from 0 % to 200 %) as a function of time and to create the traction curves from mean values of COF at each SRR. At the highest speed (100 mm/s) it was possible to measure traction curves directly, because the load was stable.

To substitute the human synovial fluid, 25 % bovine serum (B9433, Sigma Aldrich, US) solution was used. Total amount of lubricant was 12 ml; the tests were carried out under fully flooded conditions. The lubricants were frozen at – 22 °C and were removed from the freezer 2 h before experiments to thaw naturally.

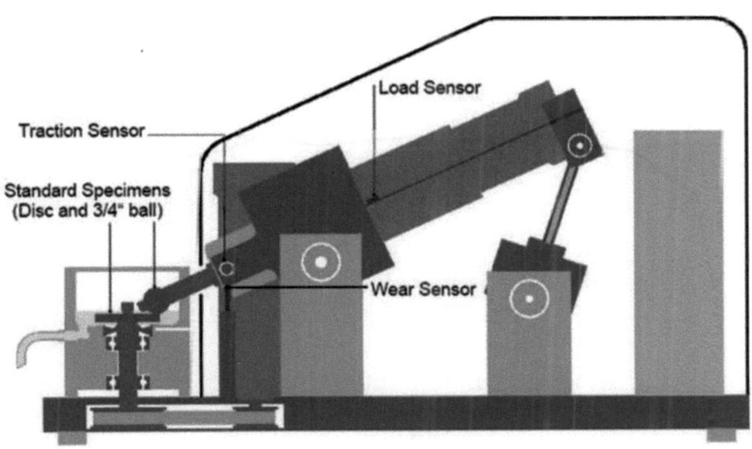

Fig. 1 Schematic of MTM 2 [15]

3 Results

As can be seen in Figs. 2 and 3, COF is strongly dependent on the entrainment speed; however load value seems to be not so important. While applied load was 5.2 N, COF decreased with increasing speed. When the entrainment speed was 100 mm/s, COF was lower than 0.05. These values are much lower than COF between hard-on-hard prosthesis components published elsewhere [8].

Behaviour is a little bit different in case applied load is 9.8 N. COF is around 0.08 at entrainment speed 25 mm/s. Increasing the entrainment speed to 50 mm/s leads to an increase of COF. However, further increase of speed leads to decrease of

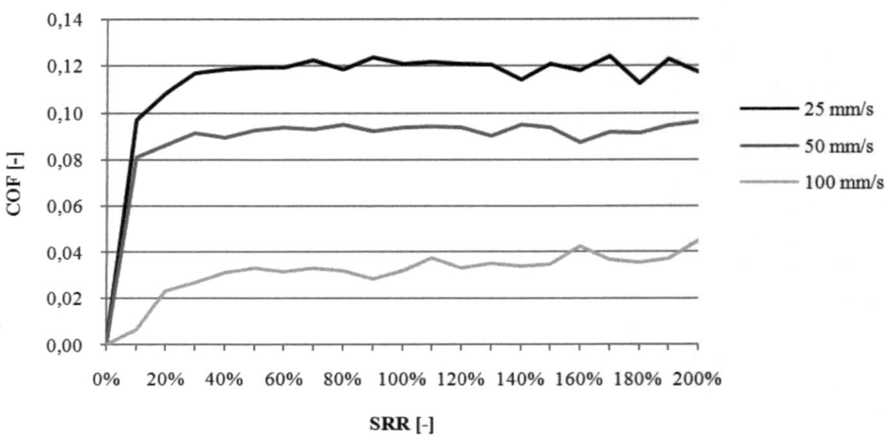

Fig. 2 Traction curves for various speeds for load $F_1 = 5.2$ N

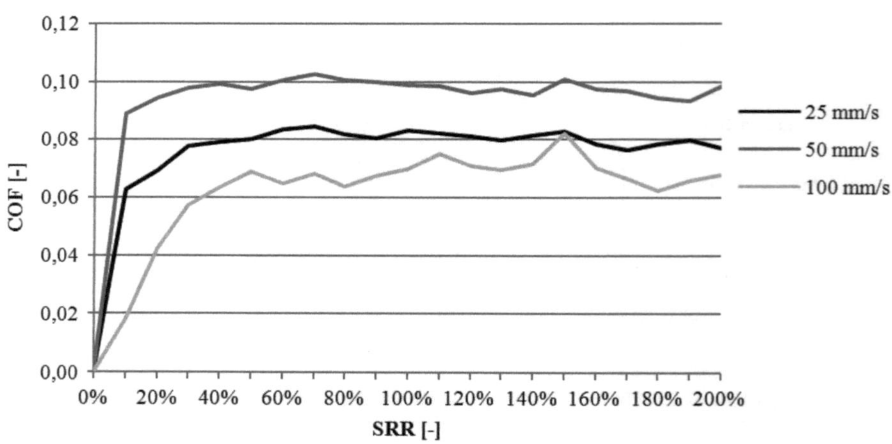

Fig. 3 Traction curves for various speeds for load $F_2 = 9.8$ N

COF up to 0.07. To explain this phenomenon, it is necessary to perform more analyses and to study protein film formation to better understand tribological processes occurring inside the contact.

4 Discussion

Several analyses were conducted to get traction curves of PVA hydrogel vs. CoCrMo contact pair. In most cases, COF seems to be remarkably lower than in hard-on-hard contacts studied by other authors [8].

To better understand the results, it is necessary to study protein film thickness and film formation between surfaces. Conventional methods used in tribology research, such as optical interferometry or electrical methods, cannot be used in this case, because PVA hydrogel is non-reflective compliant material with very low electrical conductivity. So, for this purpose, optical method based on fluorescence microscopy have to be established, as well as, preparation procedure of cast-drying PVA hydrogel needs to be improved to provide flat transparent samples of constant thickness.

While thinking about replacement of natural synovial cartilage, PVA hydrogel seems to be very suitable material; however, some further research hove to be done to improve some characteristics like strength, lifetime and so on. Some experiments on hip joint simulator based on principle of pendulum were also conducted. In this case, real conformity of both components was respected, which led to significant decrease of contact pressure. Ceramic head vs. PVA hydrogel cup contact pair was studied. It was observed that PVA layer leads to much better lubricating conditions. These results will be presented in further paper.

Acknowledgments This work is an output of research and scientific activities of NETME Centre, regional R&D centre built with the financial support from the Operational Programme Research and Development for Innovations within the project NETME Centre (New Technologies for Mechanical Engineering), Reg. No. CZ.1.05/2.1.00/01.0002 and, in the follow-up sustainability stage, supported through NETME CENTRE PLUS (LO1202) by financial means from the Ministry of Education, Youth and Sports under the "National Sustainability Programme I".

References

1. S.A. Pramanik, K. Agarwal, K.N. Rai, Chronology of total hip joint replacement and materials development. Trends Biomater. Artif. Organs **19**(1), 15–26 (2005). Available via http://www.angelfire.com/space2/trends/pdf/19-1-15-26.pdf
2. K.L. Corbett, E. Losina, A.A. Nti, J.J.Z. Prokopetz, J.N. Katz, F.P. Rannou, Population-based rates of revision of primary total hip arthroplasty: a systematic review. PLoS One **5**(10), e13520 (2010). doi:10.1371/journal.pone.0013520
3. D. Dowson, Bio-tribology. Faraday Discuss. **156**, 9 (2012). doi:10.1039/C2FD20103H. Available via http://xlink.rsc.org/?DOI=c2fd20103h

4. J.G. Lancaster, D. Dowson, G.H. Isaac, J. Fisher, The wear of ultra-high molecular weight polyethylene sliding on metallic and ceramic counterfaces representative of current femoral surfaces in joint replacement. Proc. Inst. Mech. Eng. H J. Eng. Med. **211**(1), 17–24 (1997). doi:10.1243/0954411971534647
5. D. Dowson, New joints for the millennium: wear control in total replacement hip joints. Proc. Inst. Mech. Eng. H J. Eng. Med. **215**(4), 335–358 (2001). doi:10.1243/0954411011535939
6. A. Wang, A. Essner, G. Schmidig, The effects of lubricant composition on in vitro wear testing of polymeric acetabular components. J. Biomed. Mat. Res. **68B**(1), 45–52 (2004). doi:10.1002/jbm.b.10077
7. L. Mattei, F. Di Puccio, B. Piccigallo, E. Ciulli, Lubrication and wear modelling of artificial hip joints: a review. Tribol. Int. **44**(5), 532–549 (2011). doi:10.1016/j.triboint.2010.06.010. Available via http://linkinghub.elsevier.com/retrieve/pii/S0301679X10001647
8. A. Mavraki, P.M. Cann, Friction and lubricant film thickness measurements on simulated synovial fluids. Proc. Inst. Mech. Eng. J J. Eng. Tribol. **223**(3), 325–335 (2009). doi:10.1243/13506501JET580
9. A. Mavraki, P.M. Cann, Lubricating film thickness measurements with bovine serum. Tribol. Int. **44**(5), 550–556 (2011). doi:10.1016/j.triboint.2010.07.008. Available via http://linkinghub.elsevier.com/retrieve/pii/S0301679X10001775
10. M. Vrbka, T. Návrat, I. Křupka, M. Hartl, P. Šperka, J. Gallo, Study of film formation in bovine serum lubricated contacts under rolling/sliding conditions. Proc. Inst. Mech. Eng. J J. Eng. Tribol. **227**(5), 459–475 (2013). doi:10.1177/1350650112471000
11. M. Vrbka, I. Křupka, M. Hartl, T. Návrat, J. Gallo, A. Galandáková, In situ measurements of thin films in bovine serum lubricated contacts using optical interferometry. Proc. Inst. Mech. Eng. H J. Eng. Med. **228**(2), 149–158 (2014). doi:10.1177/0954411913517498
12. S. Yarimitsu, K. Nakashima, Y. Sawae, T. Murakami, Effects of lubricant composition on adsorption behavior of proteins on rubbing surface and stability of protein boundary film. Tribol. Online **3**(4), 238–242 (2008). doi:10.2474/trol.3.238. Available via http://joi.jlc.jst.go.jp/JST.JSTAGE/trol/3.238?from=CrossRef
13. K. Nakashima, Y. Sawae, T. Murakami, Influence of protein conformation on frictional properties of poly (vinyl alcohol) hydrogel for artificial cartilage. Tribol. Lett. **26**(2), 145–151 (2007). doi:10.1007/s11249-006-9185-6
14. D. Nečas, M. Vrbka, P. Šperka, M. Druckmüller, P. Skládal, P. Štarha, I. Křupka, M. Hartl, Qualitative analysis of film thickness in rolling EHD contact by fluorescence technique, in *Modern Methods of Construction Design*, ed. by L. Ševĉik, P. Lepšík, M. Petrů, I. Mašín, R. Martonka. Lecture Notes in Mechanical Engineering (Springer International Publishing, Cham, 2014), pp. 615–622. ISBN: 978-3-319-05202-1. ISSN: 2195-4356
15. PCS Instruments: Leaders in Tribology Test Equipment. PCS Instruments [online]. Available via http://www.pcs-instruments.com/mtm/mtm.shtml#page=page-1. Cited 15 May 2014

Effect of Sanding on Adhesion Under Oil and Water Condition Between Wheel and Rail

D. Smejkal

Abstract Adhesion in contact between wheel and rail is significant process in railway systems. Adhesion is affected by many factors such as speed, load and especially contaminants like leaves, water, oil, dust and other. Contaminants can decrease adhesion to very low adhesion and can causes serious problems such as extended stopping distances, or, in traction, it may lead to time delay of train. Sand has been usually used for effective increase of adhesion, but its disadvantage is abrasive effect on wheel and rail. A systematic laboratory investigation of the effect of sanding on wheel and rail has been performed using twin disk simulation facility, which consists of disk simulation of wheel and disk simulation of rail. Discs have equal diameters, due to simulating equivalent contact zone different geometry compared to real conditions. The results show that sand increases of adhesion coefficient enough in water and oil contaminated contact. Wet leaves rapidly drop adhesion coefficient, sand in this condition enhances adhesion less because of the presence of steady layer of leaves. Sand has problem to break this layer and effectively decrease adhesion coefficient.

Keywords Wheel/rail • Adhesion • Sanding • Twin disk • Contaminants

1 Introduction

Poor adhesion in contact between a wheel and a rail has implication for acceleration and deceleration of a train. A both phase are important. The general phenomenon low adhesion during braking and acceleration is sliding wheel on track. Low adhesion in braking phase causes extend stopping distances and it could leads to problems connected with safety [1]. Acceleration in low adhesion has effect on time delay of trains. Poor adhesion respectively loss of adhesion usually occur like third layer in contact wheel and rail. Representative of third layer occurred on the line may be dew, water usually mixed with dust and debris, or leaves.

D. Smejkal (✉)
Brno University of Technology, Brno, Czech Republic
e-mail: smejkal@fme.vutbr.cz

V. Dynybyl et al. (eds.), *The Latest Methods of Construction Design*,
DOI 10.1007/978-3-319-22762-7_26

The most used method for increasing levels of adhesion is sand application between wheel and rail. Sand particles enter into contact where are crashed and effectively contribute to increase level of adhesion. In braking, sand is used to reduce distance for stop the train. In traction mode sand helps decrease sliding and thereby enables transfer more power from wheel to rail. Sanding can increase adhesion coefficient but on the other site, sand causes damage the wheel/rail surfaces [2].

Recently, many studies deal with investigate the causes and mechanism of adhesion loss in wheel/rail contact. Wang [2] investigated the adhesion behavior of wheel/rail under oil, water and sanding conditions. They observed that oil remarkably decrease adhesion in compare with water and dry condition. Sanding can better improving of adhesion under water condition. Cuevas [3] study influence of the particle size on adhesion and wear in wheel/rail contact. The results show that large particle size causes higher adhesion coefficient, but it may significantly increase degree of wear which is depend on the slip. Kumar [4], evaluates wheel/rail wear and adhesion with and without sand for a North American Locomotive. It was observed that liquid contaminants such as water, sand, diesel fuel, lubrication oil, san after oil and conversely decrease adhesion coefficient and increase creepage resp. slip between wheel and rail. Clayton [5] summarized problems with low adhesion. It was founded that a rail is covered a layer consisting water, oil and solid material. I was observed that water has impact on wheel/rail adhesion. Moisture and dew with this layer can cause very low adhesion. In case of rain the adhesion coefficient is fairly high and keep stabile value approximately 0, 2.

Various attempts have been made to remove layers of water, oil and autumn leaves, but all have been done without any significant success. Sand is still use like the most effective approach at improving adhesion coefficient, but with well know disadvantages.

2 Experimental Details

2.1 Apparatus

Twin disk experimental facility was used to simulate the wheel/rail contact in rolling sliding condition. A scheme of the machine is shown in Fig. 1. The facility was developed with goal evaluating effects of sanding on adhesion in contact between wheel and rail. The twin disk consists of upper and lower discs; machine has two electromotor which could rotate independently. A simulation of appropriate condition is reached by using 15 kW motor, transmission with reducing rotation and frequency converter. When a slip is applied, one of the electromotors works as motor and other as a generator and generated electric energy is recovered in the system.

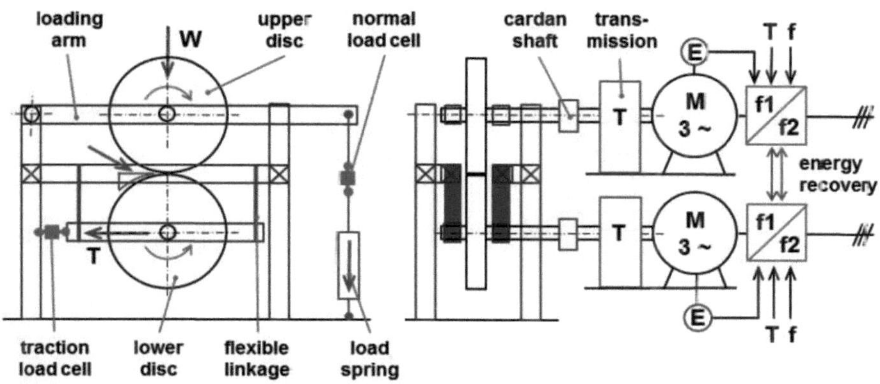

Fig. 1 The test rig

Upper disc is loaded with loading arm against lower disc, which is mounted on freelance frame and transfers traction force to traction load cell. The test rig contains a sealed chamber which covers the discs and allows the application of sand and contaminants without pollution.

The bases of measuring system are two load cells. First load cell is used to measure vertical force, which is subsequently recalculated on normal force in the contact zone. Second load cell is used to measure traction force. The shaft encoders are used to check running speed, resp. slip (Eq. 1) that is set by user.

$$s = \frac{w_{wheel} \cdot r_{wheel} - w_{rail} \cdot r_{rail}}{w_{wheel} \cdot r_{wheel} + w_{rail} \cdot r_{rail}} \cdot 200 \quad (1)$$

Where w_{wheel} and w_{rail} are the rotational speed of discs and r_{wheel} and r_{rail} are the radii of discs. The load is always set before the beginning of a test. Subsequently, the adhesion coefficient is calculated by Eq. (2).

$$\mu = \frac{W}{T} \quad (2)$$

where W is a normal force and T is traction force.

2.2 *Test Specimen*

The dimension and shape of test discs is shown in Fig. 2. An initial roughness of discs in contact area is 0,4 μm. Dimension and shape of discs are made in scale rate 1:3, considering to real rail wheel and rail profile (UIC 60). Both discs are made of C45 steel, which is an equivalent material with UIC 900A and R7T steel.

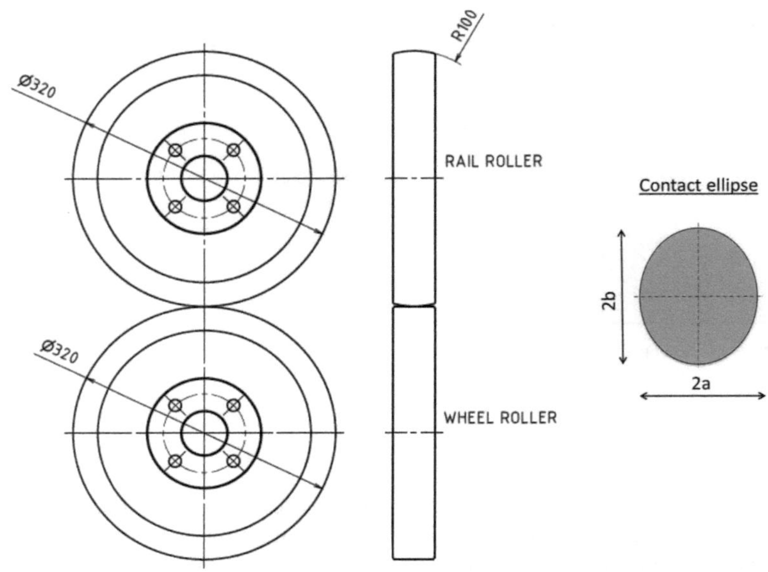

Fig. 2 Contact state of wheel and rail discs

2.3 *Procedure*

A load is applied in the wheel using lever mechanism. The Hertzian maximum contact pressure is approximately 1200 MPa, whereas tests were done with pressure 1 GPa. For tests were used only two speed because the sand causes fast damage of contact area. Slip was varying from 1 % to 10 % at steps (1, 3, 5, 10)%. Higher speed can lead to resonance which can affect results. Moreover, the temperature was set at 30°; this was stabile during tests.

The surface of the test discs was grinded by sandpaper and cleaned by acetone, before the experiments were removed contaminants such as sand, leaves, oil, and water. On one clear disc were applied leaves and on another disc drip the water (Fig. 3). During the test were applied dose of sand and there was observed change of coefficient adhesion. Measured data were recorded into PC through A/D converter.

3 Results

3.1 *Effect of Leaves and Sand on Coefficient of Adhesion*

Figure 3a shows the curves for adhesion coefficient, of wheel/rail in wet condition and when fresh tree leaves is suddenly added to the contact between wheel and rail, the change of adhesion coefficient is show in graph. The adhesion coefficient has

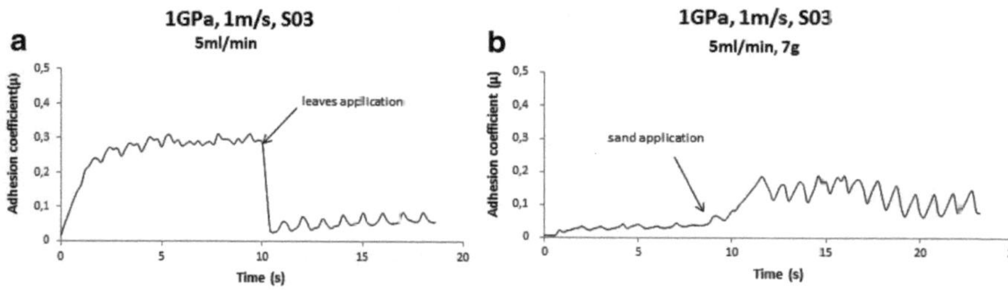

Fig. 3 Adhesion coefficient of wheel/rail when (**a**) leaves (**b**) sand is added to the contact

average value about 0.3 in wet condition, after leaves addition to the wheel/ rail contact, the adhesion decrease suddenly. There is no significant regeneration in coefficient of adhesion after leaves application so it keeps a relative low steady state.

Figure 3b shows effect of sanding on adhesion coefficient under leaves contaminated contact. When sand is added to wheel/rail contact, the adhesion coefficient increase from 0.03 to 0.2. However, after a couple of cycle the coefficient of adhesion begins slightly descend with time.

When the sand is added to the wet contact it causes increase of adhesion coefficient as show Fig. 4. However, when the sand is added to the contact it causes sudden decrease the adhesion coefficient. It is obviously causes by process of grinding of sand particle. The adhesion coefficient drops from 0.35 to 0.05 when the sand is crashed the adhesion coefficient increase up to 0.6 and then a slight reduction.

4 Discussion

The results from this study indicate that low adhesion coefficient may occur under water contaminated contact. The average adhesion coefficient of wheel/rail under wet condition is about 0.35. It is found that fresh tree leaves can cause significantly decrease of adhesion coefficient under 0.1.

Using of sand effectively increase of adhesion coefficient as shows Figs. 3b and 4. Non – effective drop in adhesion is occurred suddenly after application of sand as show Fig. 4. It is probably causes by crushing new sand particles which come through the wheel/rail interface. Increase adhesion coefficient is obvious when sand change structure from particle to powder.

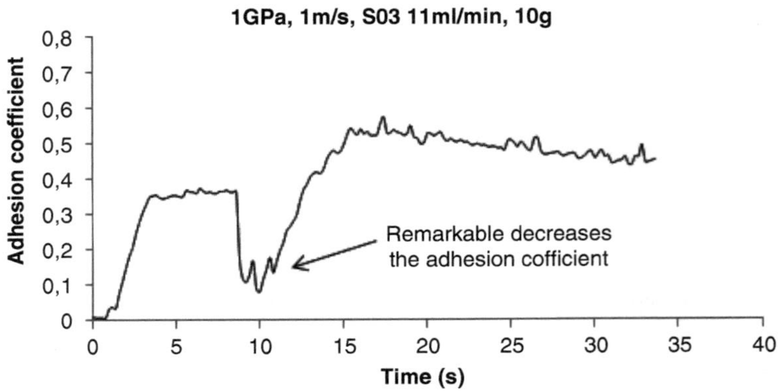

Fig. 4 Effect of sand on adhesion coefficient

5 Conclusion

This paper focuses on the effect of water leaves and sand on adhesion coefficient by using the twin – discs machine under various speed and slip condition. The experimental results show the adhesion coefficient depends on environment condition. Leaves decrease adhesion coefficient much more than presence of water in wheel/rail contact. Sand effectively improve adhesion coefficient however decreasing of adhesion occur immediately after application of sand.

Acknowledgments This work is an output of research and scientific activities of NETME Centre, regional R&D centre built with the financial support from the Operational Programme Research and Development for Innovations within the project NETME Centre (New Technologies for Mechanical Engineering), Reg. No. CZ.1.05/2.1.00/01.0002 and, in the follow-up sustainability stage, supported through NETME CENTRE PLUS (LO1202) by financial means from the Ministry of Education, Youth and Sports under the "National Sustainability Programme I".

References

1. R. Lewis, R.S. Dwyer-Joyce, Wear at the wheel/rail interface when sanding is used to increase adhesion. Proc. Inst. Mech. Eng. F J. Rail Rapid Transit **220**(1), 29–41 (2006). doi:10.1243/095440905X33260
2. W.J. Wang, H.F. Zhang, H.Y. Wang, Q.Y. Liu, M.H. Zhu, Study on the adhesion behavior of wheel/rail under oil, water and sanding conditions. Wear **271**(9–10), 2693–2698 (2011). doi:10.1016/j.wear.2010.12.019
3. H. Chen, T. Ban, M. Ishida, T. Nakahara, Experimental investigation of influential factors on adhesion between wheel and rail under wet conditions. Wear **265**(9–10), 1504–1511 (2008). doi:10.1016/j.wear.2008.02.034. Dostupné z http://linkinghub.elsevier.com/retrieve/pii/S0043164808001646
4. M. Mansouri, Q. Ilham, H. Snoussi, C. Richard, Adaptive quantized target tracking in wireless sensors networks. Wirel. Netw. **17**(7), 1625–1639 (2011). ISSN 1022-0038
5. P. Clayton, Tribological aspects of wheel-rail contact: a review of recent experimental research. Wear **191**(1–2), 170–183 (1996). doi:10.1016/0043-1648(95)06651-9

Part III
Hydraulics – Fluid Mechanisms

M. Andjelkovic, T. Ibrahim, R. Petrovic, M. Radosavljevic,
N. Todic, and A. Wahab

Experimental Research and Optimization of Characteristic Parameters of the Valve Plate of the Axial Piston Pump/Motor

R. Petrovic, M. Andjelkovic, M. Radosavljevic, and N. Todic

Abstract Modern design of the axial piston pump/motor, based on computer aided design, requires description of all processes and parameters in the pump/motor. The hydrodynamic and dynamic processes in the axial piston pump (cylinder, intake and discharge chambers, discharge valve, and high pressure pipeline) are very complex, so they demand a thorough physical and mathematical analysis. Compression losses, which are caused by sudden pressure changes in the cylinders of cylinder block, result in energy loss and increased noise. The compression losses can be decreased by identification of optimal dimensions and shapes of the port at the valve plate.

Keywords Piston pump • Geometrical flow section • Radial clearance • Diameter of cylinder • Numbers of cylinders • Dynamic viscosity • Modulus of elasticity • Flow coefficient • Dynamic processes • The driving shaft angle • Experiment

1 Introduction

The experimental research is based on the ultra speed measurement system, mathematical modeling of non-stationary high dynamic processes, and optimization method. The paper also includes the analysis of the influence of working and design parameters of the axial piston pump/motor on the position of intake and discharge ports of the valve plate. To solve the problem the special program (AKSIP) has been developed and presented in the paper.

R. Petrovic (✉) • M. Andjelkovic • M. Radosavljevic
Faculty for Strategic and Operational Management Belgrade, University Union Nikola Tesla of Belgrade, Belgrade, Serbia
e-mail: radovan4700@yahoo.com

N. Todic
Faculty for Mechanical and Civil Engineering, University of Kragujevac, Kraljevo, Serbia
e-mail: ntodic@gmail.com

V. Dynybyl et al. (eds.), *The Latest Methods of Construction Design*,
DOI 10.1007/978-3-319-22762-7_27

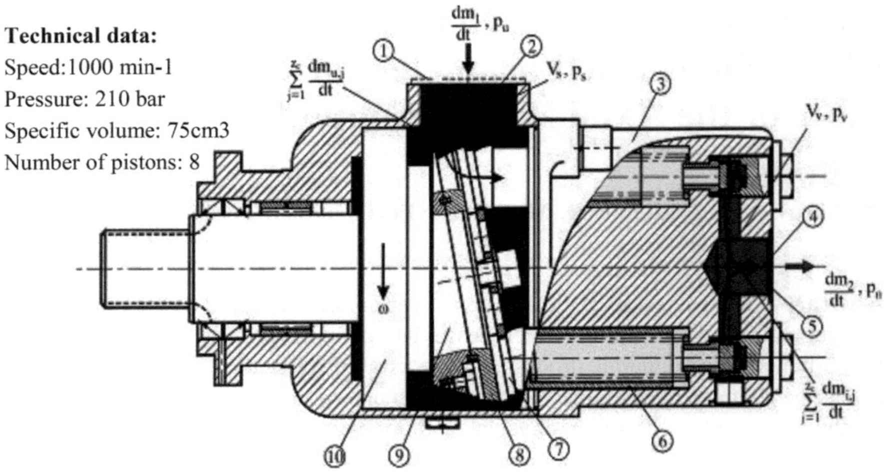

Fig. 1 Piston axial pump type PPT 3112. 750.02C/1DL, firm PPT, Trstenik, Serbia. *1*. intake pipe line connection; *2*. intake space of the pump; *3*. cylinder block; *4*. discharge space of the pump; *5*. discharge pipeline connection; *6*. piston; *7*. split panel; *8*. acute panel; *9*. the pump shaft; *10*. in bearing the pump shaft

2 Mathematical Model of a Pump Process

Mathematical model is given for each element, considering the complexly of some processes and their mutual dependence as well as the need for further mathematical modeling [1, 2]. This makes programming module and their further improvement and monitoring much easier [1]. Piston axial pump are shown in Fig. 1.

Mass flow through the opening 1, on the entrance place into the intake space of the pump of fluid, Fig. 1:

$$\frac{dm_1}{dt} = \sigma_1 \mu_1 A_1 \sqrt{2\rho_s |p_u - p_s|} \tag{1}$$

where: $\sigma_1 = 1$ for $p_u \geq p_s$, $\sigma_1 = -1$ for $p_u < p_s$.

Mass flow of fluid through the split pump organ during filling one of the pump cylinders:

$$\frac{dm_u}{dt} = \sigma_u \mu_u A_u \sqrt{2\rho_s |p_s - p_c|} \tag{2}$$

where: $\sigma_u = 1$ for $p_s \geq p_c$, $\sigma_u = -1$ for $p_s < p_c$

Mass balance of the intake space is:

$$\frac{dm_s}{dt} = \frac{dm_1}{dt} - \sum_{j=1}^{z_c} \frac{dm_{u,j}}{dt} \tag{3}$$

where: j = 1,2,. . ., zc order number of cylinder, zc the numbers of cylinders.
Differential pressure equation in the intake pump space:

$$\frac{dp_s}{d\phi} = \frac{E}{V_s\rho_s}\left(\frac{dm_1}{d\phi} - \sum_{j=1}^{z_c}\frac{dm_{u,j}}{d\phi}\right) \tag{4}$$

Differential pressure equation in the pump cylinder:

$$\frac{dp_c}{d\phi} = \frac{E}{V_c}\left[\frac{A_c v_k}{\omega} + \frac{1}{\rho_c}\left(\frac{dm_u}{d\phi} - \frac{dm_i}{d\phi}\right)\right] \tag{5}$$

Where: $V_c = V_{c\min} + V_{cx}; V_{cx} = A_c \cdot x_k$ – immediate volume of the cylinder; the change of the volume of the pump cylinder caused by piston moving $\frac{dV_c}{dt} = -A_c v_k$,
Mass balance of the discharge space is:

$$\frac{dm_v}{dt} = \sum_{j=1}^{z_c}\frac{dm_{i,j}}{dt} - \frac{dm_2}{dt} \tag{6}$$

where: j = 1,2,. . ., z_c order number of cylinder
Mass flow streaming out of the discharge space into the discharge pipe is:

$$\frac{dm_2}{dt} = \sigma_2\mu_2 A_2\sqrt{2\rho_t|p_v - p_n|} \tag{7}$$

where: $\sigma_2 = 1$ for $p_v \geq p_n$, $\sigma_2 = -1$ for $p_v < p_n$
Differential pressure equation in the discharge pump space:

$$\frac{dp_v}{d\phi} = \frac{E}{V_v\rho_v}\left(\sum_{j=1}^{z_c}\frac{dm_{i,j}}{d\phi} - \frac{dm_2}{d\phi}\right) \tag{8}$$

Mass flow through a concentric clearance between the cylinder and the piston:

$$\frac{dm_z}{dt} = \frac{\pi \cdot D_c \cdot \Delta r^3}{12 \cdot \eta \cdot x_k(\phi)} \cdot (p_c - p_s) \cdot \rho_c \tag{9}$$

2.1 Modeling the Streaming in the Intake and Discharge Pipe Line of the Pump

During mathematical modeling of a process in a pump, it is also necessary to include and consider a series of suppositions for a process modeling occurring in the intake and discharge pipe line of the pump. For the most general model the following suppositions for streaming of the operational fluid in the intake and discharge pipe line are taken and considered:

The fluid streaming is one-dimensional. The pipes are of a constant cross section. Temperature and streaming fields per cross section of the pipe are homogeneous. Velocity vector laps the direction of the axis of the pipe at any moment and in any section.

Viscosity friction between some layers of the fluid inside the pipe is neglected. The friction forces appear on the inside walls of the pipe.

The processes in the pipes are isentropic. The change of entropy caused by friction, heat and mixing of fluid parts are neglected. Forces of the field (gravitational, magnetic, etc) are neglected.

In the scope of dynamic of one-dimensional streaming, such streaming is considered as "non stationary streaming in a streaming fiber" [3, 4].

2.2 Continuity Equations

The equation of continuity of pressed fluid with functions p, w, ρ at the isentropic change of the state:

$$\frac{\partial p}{\partial t} + w\frac{\partial p}{\partial x} + a^2\rho\frac{\partial w}{\partial x} = 0 \tag{10}$$

where: $a = \sqrt{\left(\frac{\partial p}{\partial \rho}\right)_s}$, the velocity of sound in the fluid,

where: $p = p(t, x)$ and $\rho = \rho(t, x)$ are functions of time t and coordinate x.

2.3 Momentum Equations

$$w\frac{\partial \rho}{\partial t} + \rho\frac{\partial w}{\partial t} + w\frac{\partial}{\partial x}(\rho w) + \rho w\frac{\partial w}{\partial x} + \frac{\partial p}{\partial x} = -f_r\rho \tag{11}$$

3 Experimental Testing the Process in the Axial Piston Pump

The pressure changes in the cylinder, discharge chamber and discharge pipeline were measured within the experimental research work. The vibrations of the pump housing were also analysed depending on the shaft angle. All pressures and vibrations were simultaneously measured at cca 0.09o of the pump shaft (exactly 4096 times per shaft rotation). The incremental angle sensor is an optical sensor with 1024 pulses per rotation. The pulses of the angle sensor were quadrupled by the interface of the ADS 2000 system [5]. Thus, 4096 pulses per shaft rotation were achieved. Ten consecutive cycles were measured in order to repeat the consecutive cycles with the continuous performance. At the same time, the period from one angle to the next one was also measured in order to identify the uniformity of the shaft speed and to control the incremental angle sensor. All analogous signals (pressures, vibrations) were simultaneously converted into cipher form by four ultraspeed convertors working simultaneously. The total number of measured data was $(4 + 1) \times 4096 = 20{,}480$ per rotation, i.e. 204,800 for ten consecutive cycles. The number 4096 was not chosen by chance; it was chosen in order to apply Fast Fourier's Transformation (FFT) of measured signals. The testing was done in the Laboratory for Development and Research in College of Applied Engineering – Center for Power Control Hydraulics (CPCH), which is situated in Trstenik, Serbia and in cooperation with the Institute for Motors in the Faculty of Mechanical Engineering in Belgrade University. The pump was, for the purpose of this experiment processed, to ensure that the necessary measuring up converter needed for measuring the characteristic parameters of the axial piston pumps with a combined distribution of working fluid.

3.1 Testing Devices and System for Data Acquisition

The axial piston pump was tested by the devices which are made especially for this experiment done at the testboard in the Laboratory for Development and Research. Basic component of the testboard is driving electromotor whose power is 137 kW and whose number of revolutions and torque are controlled by electromotive drive. Figure 2 shows the testboard used for testing the characteristic parameters of the axial piston pump having the following components: electromotor whose power is 137 kW, speed 1450 min^{-1}, with controlled number of revolutions(1); redactor(2); axial piston pump 3112 -750.020/02, (3); angle marker(4); measuring converter of vibrations(5); measuring converter of pressure in discharge chamber(6); measuring converter of pressure in cylinder(7); system for measuring and acquisition ADS 2000 – CADEX(8) [5].

The following quantities were measured: pressure in the cylinder which depends on the angle of driving shaft; pressure in the vale chamber which depends on the angle of driving shaft; pressure in the discharge pipeline which depends on the angle of driving shaft; vibrations of the pump housing which depends on the angle of driving shaft; angle of driving shaft; pump flow; temperature of working fluid; number of revolutions of driving shaft.

Measuring converters of pressure, type P3MA, were used to measure pressure. These converters are based on measuring tapes and they are made by the company "Hottinger", Germany; their measuring range is 500 bar, class of accuracy is 0.1 and transmission range is 100 kHz. Flow was measured by measuring turbine, type RE2 25/180 l/min. Its class of accuracy is 0.4 and it is made by the company

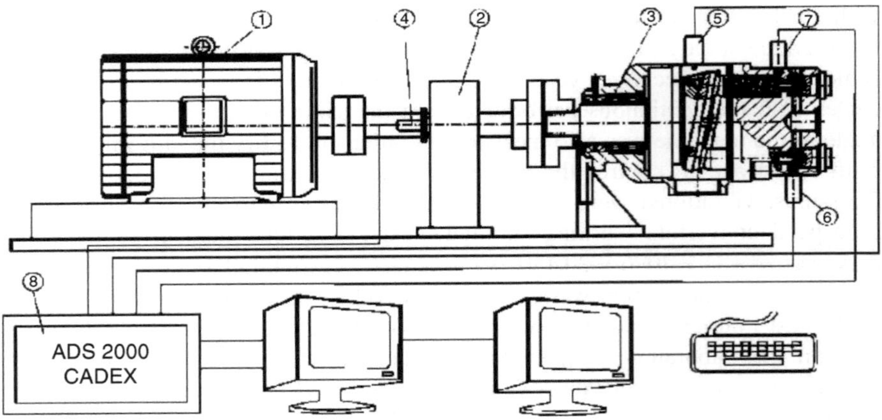

Fig. 2 Testboard used for testing the axial piston pump

"Hydrotechnik", Germany. Incremental sensor, type ROD 426E, made by the company "Heidenhain", Germany, was used to measure the rotation angle of the pump shaft. It has 1024 optical markers and its maximum number of rotation is 12,000 min^{-1}. The accelerometer whose measuring range is up to 5 $[m/s^2]$ and is made by the company "Brüel&Kjaer", Denmark, was used to measure the vibrations of the pump housing. Ultra speed measuring system ADS 2000-CADEX was used for acquisition. The system provides continual measurement and calculation of characteristic parameters of working cycle in real time. The system ADS 2000-CADEX, which is used to develop highly dynamic mechanical objects by integrated measuring and calculating technique, is based on the following components: VME-bus CPU with graphics, VME-bus ADC, VME-bus PGA and CDM Interface. The processor receives data from A/D converter in real time directly in CPU-DRAM. The processor simultaneously controls the amplifying and multiplexer cards in real time. The software was developed especially for this system in order to measure cyclic and non-cyclic processes with graphical on-line display. Statistic processing of measured data is done with graphical display. Fifty million data were measured at maximum speed of 3 MHz by 6…12 simultaneous A/D converters. Up to four VME-CPU cards with processors Motorola 68020…68060, Intel Pentium, Digital Alpha or Motorola Power PC, can be integrated into the system. VME-bus ADC contains two A/D converters with simultaneous working speed of $2 \cdot 350$ kHz at 1 bit and 1 timer. Start of each conversion is done by pulses of angle sensor or timers with hardware registration of referent mark of incremental angle sensor in order to 100 % control the proper work of angle sensor in real time. It is possible to integrate up to six A/D cards into the system. i.e. 12 A/D converters. Two VME-bus ADC modules were installed into this system. VME-bus PGA multiplexer and amplifier modul has six fast instrumental amplifiers used for direct connection between the sensor and measuring tapes tied in full bridge with DC supply of 5 V (options 12 or 15). Maximum speed of conversion is 150 kHz. Four

Fig. 3 Axial piston pump with integrated measuring sensors

VME-bus PGA modules were integrated in this system. The system has an interface for incremental angle sensors with DC supply and for multiplying the pulses of angle sensor. Up to four interfaces can be integrated into the system [6].

The applied measuring system enables simultaneous measuring at four fast analogous canals with parallel measuring of time periods from the angle mark to the mark of incremental angle sensor (Fig. 3).

3.2 *Results of Measuring of Parameters of Working Processes of the Piston Axial Pump*

In the scope of performed experimental testing was done a measuring of the pressure flow in the cylinder, discharge space and intake pipeline as well as vibration of the pump housing in dependence of the passed angle of the pump shaft.

All pressures and vibrations were measured completely parallel on each cca 0.09o of the pump shaft (exactly 4.096 times per shaft rotation).

As incremental giver of the angle an optical giver with 1024 pulses per rotation was used. Pulses of the giver of the angle were four times increased by the interface for the angle givers on the ADS 2000 system and so 4096 pulses per shaft rotation were obtained. In order we might see the repetitions of the consecutive cycles with the unchanged work regime ten consecutive cycles were measured. At the same time, a time interval from angle to angle was measured as well in order to determine an even angle speed of the shaft and work control of the incremental giver of the angle.

All the analogue signals (pressure, vibrations) were converted into cipher form by means of four ultra speed converters working simultaneously (parallel). The total number of measured data was $(4+1) \times 4096 = 20{,}480$ per rotation (cycle),

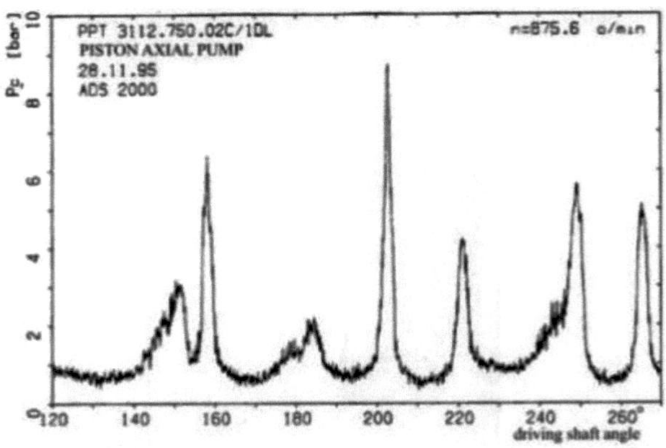

Fig. 4 The pressure history in the cylinder (p_c) and delivery chamber (p_v) for the average cycle

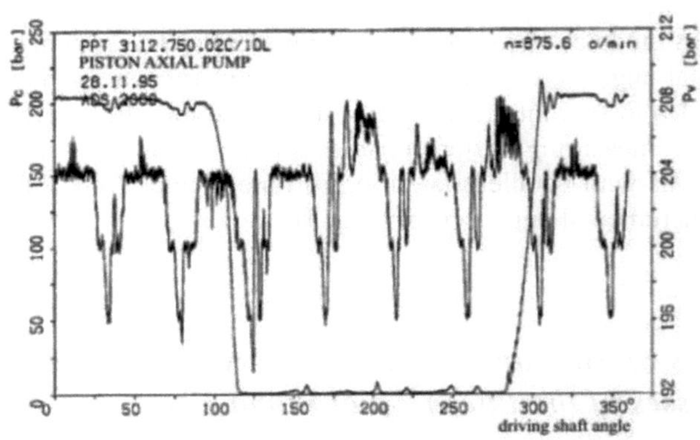

Fig. 5 The pressure history in the cylinder (p_c) in the 120°÷°270 interval for one cycle

that is, 204,800 for ten consecutive cycles. The number of samples of 4096 was not chosen by chance, but purposely with the aim of the application of the fast Furrier's transformation (FFT) of measured signals. Measures were done for seven working regimes [7].

Figures 4, 5, 6, 7, and 8, shows the measured pressure flow for individual, that is, ten consecutive cycles of the piston axial pump. Big similarity of measured pressures for the first of ten consecutive cycles (MERF) in relation to the middle of ten consecutive cycles (MERM). Figure 1 show the measured pressure flow in the cylinder (pc) for one, that is the middle of ten consecutive cycles in the function of the angle of the shaft. The diagram shows the visual pressure gradients at the pressure stage and expansion as well as the appearance of peaks during intake. Figure 1 also show the pressure flow in the discharge space (p_v) for one, that is middle of the ten consecutive cycles in the function of the angle of the shaft. The pressure pulses in the discharge space depend on the number of the cylinders what

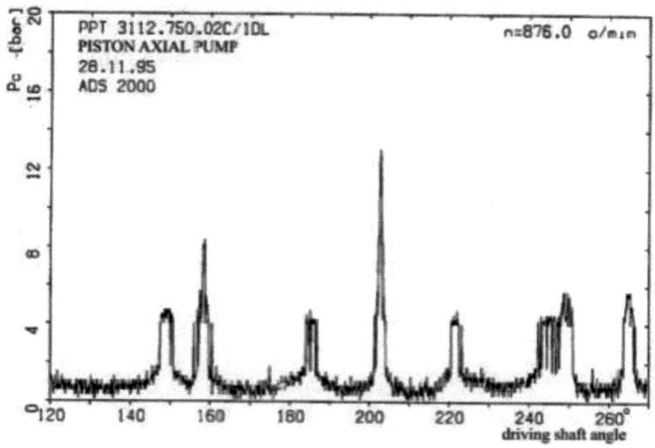

Fig. 6 The pressure history in the cylinder (p_c) in the 120°÷°270 interval for the average cycle

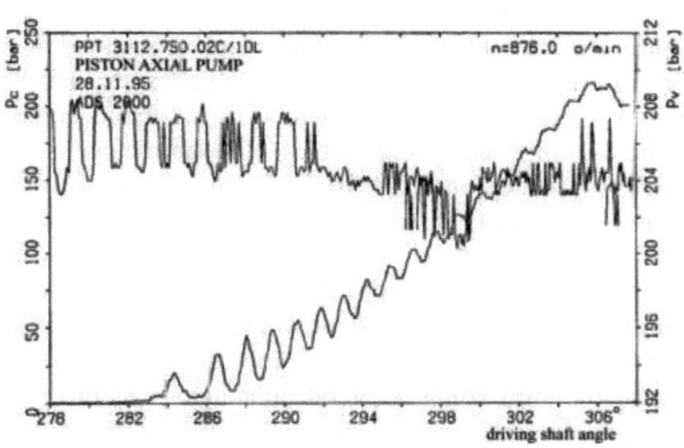

Fig. 7 The pressure history in the cylinder (p_c) and delivery chamber (p_v) in 278°÷°307 interval for one cycle

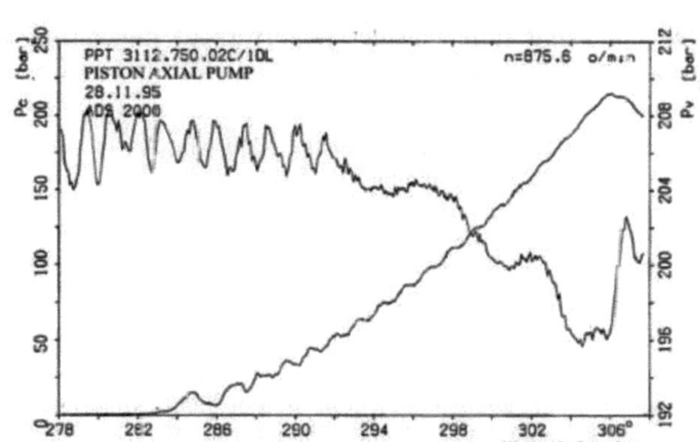

Fig. 8 The pressure history in the cylinder (p_c) and delivery chamber (p_v) in 278°÷°307 interval for the average cycle

is obvious in this case, because it deals with the pump with eight cylinders. The appearance of peaks at the intake stage for one, that is, middle of ten consecutive cycles, is shown in Figs. 5 and 6. Figures 4 and 8 present the measured pressure flow in the cylinder (p_c) for one, that is middle for ten consecutive cycles with the aim to analyze in detail the gradient growth of pressure at the stage of pressing. The same diagram, at the same interval, shows the pressure pulses in the discharge space. The Figs. 4, 5, 6, 7, and 8, stand for diagrams of measured pressure at the work regime $n = 875.6\ \text{min}^{-1}$ and $p_c = 210$ bar [8].

4 Conclusion

It is not possible to give a precise determination of parameters of hydrodynamic processes of a piston axial pump neither experimentally nor by a mere mathematical modeling only. Sufficiently exact parameters can be obtained by combining the application of measuring the pressure flow in the cylinder, mathematical modeling of a real hydrodynamic process and the method of nonlinear optimization which enables, at the same time, the determination of systemic measuring errors and unknown parameters. The computer AKSIP program gives possibilities to combine 56 influential pump parameters in order to achieve optimal solution, in regard to flow losses, flow inlet etc. Further research is possible in the construction of the piston-axial pumps with a bent cylinder block and splitting of working fluid by means of a split panel. Mathematical model would be, in that case, expanded by a dynamic cylinder block and hydrodynamic processes in clearances between the cylinder block and the split panel. General conclusions of the presented results relate to all tested working regimes of the axial piston pump. It is not possible to define precisely the parameters of hydrodynamic processes in the axial piston pump by experiments only, or by mathematical modeling only. Precise parameters can be obtained if the following methods are combined: measurement of pressure changes in the cylinder, mathematical modeling of real hydrodynamic process and nonlinear optimization. At the same time, systematic errors of measuring and unknown parameters can be defined this way.

Acknowledgement The part of this research is supported by Ministry of Education and Science, Republic of Serbia, Grant TR 32036 and Grant TR 35038.

References

1. R. Petrovic, Mathematical modeling and identification of multi cylindrical axial piston pump parameters, Ph.D. thesis, Faculty of Mechanical Engineering, Belgrade, 1999
2. M. Bergemann, Systematic research of noise of piston pumps with odd number of cylinders, *Edition Mainz* (Wissenschaftsverlag, Aachen, 1994)

3. J. Darling, Piston-cylinder dynamics in oil hydraulic piston pumps, Dissertation, Universität Bath, 1985
4. R. Petrović, R. Jankov, M. Lazović, Experimental study of hydraulic and vibration axial piston pumps, Scientific Conference "Heavy Machinery", Kuala Lumpur, 1996
5. R. Jankov, ADS 2000-CADEX, system for the development of facilities mechanical application of integrated measurement and calculation techniques, Faculty of Mechanical Engineering Belgrade, Belegrad, 2001
6. R. Petrović, Mathematical modeling and experimental research of characteristic parameters of hydrodynamic processes of an axial piston pump. Strojniški vestnik J. Mech. Eng. **55**(4), 224 (2009), UDK 621.785.4
7. R. Petrović, Mathematical modeling and experimental verification of operating parameters of vane pump with double effect. Strojniški vestnik J. Mech. Eng. **55**(1), 26 (2009), UDC 621.9.04
8. R. Petrovic, Z. Glavcic, Systematic research of characteristic parameters of the piston axial pump, in *III international workshop on computer software for design, analysis and control of fluid power systems* (Ostrava/Malenovice, 2003)

The Effects of the Temperatures on the Performance of the Gas Turbine

A. Wahab and T. Ibrahim

Abstract The parametric study of thermodynamic performance on gas power plant. The variation of operating conditions, exhaust temperature. Gas turbine inlet air cooling has been considered for the power output and efficiency of gas turbine during hot seasons. In this research, the power requirements of several inlet air cooling techniques for gas turbine power plant. The considered techniques are evaporative cooling and chiller coils system for decrease the inlet air temperature to the compressor. The results show that the inlet air temperature, compressor ratio, inlet gas turbine temperature are.

Keywords Gas turbine • Compressor ratio • Air temperature • Gas temperature

1 Introduction

The gas turbine performance is affected by the temperature of the inlet air to compressor and turbine working temperature. There are several parameters that affect its performance including the compressor compression ratio, inlet air temperature compressor. The gas turbine performance is affected by components efficiencies and turbine working temperature. The effect of temperature is very important. Overall efficiency of a gas turbine cycle depends primarily upon the pressure ratio of the compressor. In the gas turbine the processes of compression, combustion and expansion do not occur in a single component as they occurred in a reciprocating engine. It is well known that the performance can be qualified with respect to its efficiency, power output, and specific fuel consumption as well as work ratio [4]. Basically, the gas turbine power plant consist of four components, including the compressor, combustion chamber, turbine. Fresh air at ambient conditions is drawn into the compressor where its temperature and pressure is raised. The high – pressure air proceeds into the combustion chamber fuel is burned

A. Wahab (✉)
Almustaqbal University College, Hillah, Iraq
e-mail: abdulkareemwahab78@yahoo.com

T. Ibrahim
Tikrit University, Tikrit, Iraq
e-mail: tahsen52@yahoo.com

V. Dynybyl et al. (eds.), *The Latest Methods of Construction Design*,
DOI 10.1007/978-3-319-22762-7_28

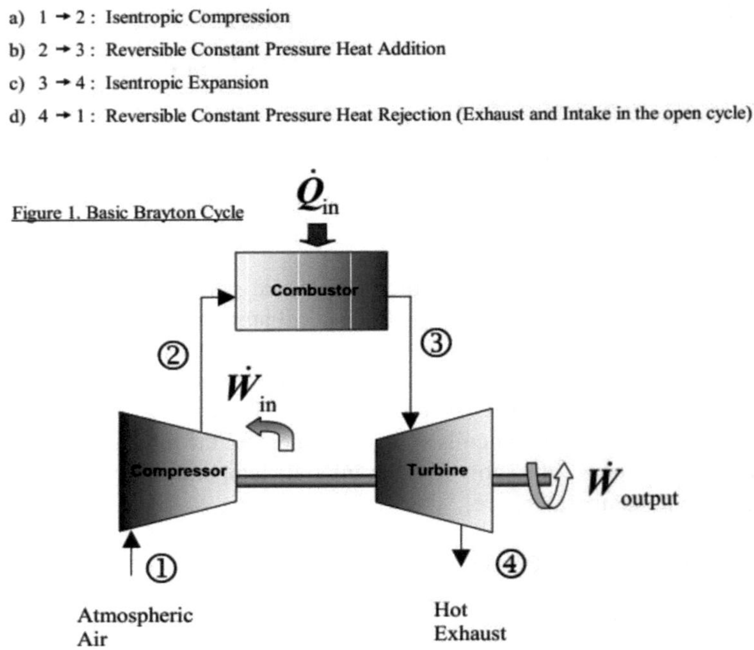

Fig. 1 Schematic diagram of Brayton cycle

at constant pressure. The resulting high temperature gases then enter the turbine where they expand to the atmosphere pressure, thus producing classified as an open cycle Brayton cycle. There is a drop in output power plant as the ambient air temperature increases. The aim of the present paper is to study the effects of ambient temperature. There are many methods for decrease the temperature of inlet air to compressor, but the method of evaporative cooling, which is widely used due to its low capital cost, uses a spray of water directly into the inlet air stream. Evaporative of the water reduces the temperature of the air, since cooling is limited to the wet bulb air temperature, is appreciably below the dry bulb (ordinary) temperature [1].

A schematic diagram for a simple gas turbine is shown in Fig. 1.

Evaporative cooling can assume large quantities of water, making it difficult to operate in arid climates. In a practical cycle with irreversibilities in the compression and expansion processes, the thermal efficiency depends on the maximum cycle temperature as well as on the pressure ratio. The maximum cycle temperature is limited by metallurgical consideration. The blades of the turbine are under great mechanical stresses and the temperature of the blade material must be kept to a safe working value [2].

2 Simple Gas Turbine Cycle

The ideal cycle for the simple gas turbine is the Brayton cycle is in Fig. 2, the thermal efficiency of the cycle is then readily shown to be given by:

$$\eta_{th} = 1 - \frac{1}{r_p^{(\delta-1)/\delta}} \quad (1)$$

The efficiency thus depends on the pressure ratio and of the gas specific heat ratio.

The work output, upon which the size of the plant for given power depends, is found to be a function not only of pressure ratio but also of maximum cycle temperature, thus.

$$W = T_1 C_{pg} \left(r_p^{(\delta-1)/\delta} - 1 \right) \quad (2)$$

3 Brayton Cycle with Regeneration

The temperature of the exhaust gas leaving the turbine is often considerably higher than the temperature of the air leaving the compressor. If these gases are allowed to pass into the atmosphere, then this represents a loss of available energy. Some of this energy can be recovered by passing the gases from the turbine through a heat exchanger (Regenerator), where the heat transferred from the gases is used to heat the air leaving the compressor and it enters to the compressor and it enters to the combustion chamber. The (T – S) diagram of regenerative cycle, shown in Fig. 2. The thermal efficiency of cycle increases as a result of regeneration, Eq. (3).

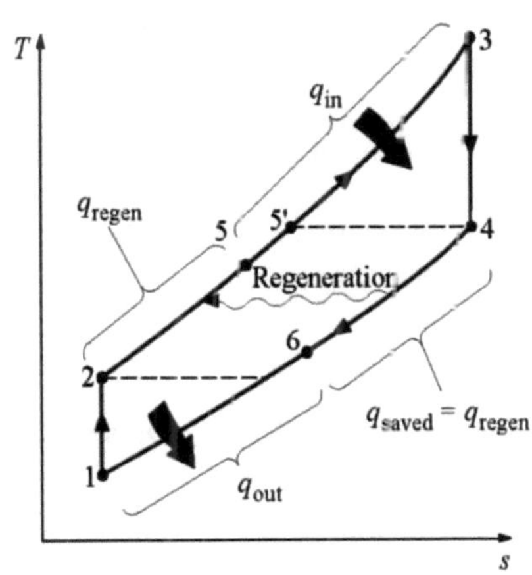

Fig. 2 (T – S) diagram of regenerative cycle

$$\eta_{\mathrm{th,\ reg}} = 1 - \left(\frac{T_1}{T_3}\right) r_p^{(\delta-1)/\delta} \tag{3}$$

Therefore, the η_{th} of an ideal Brayton cycle with Regenerative process depends on the ratio of the minimum to maximum temperatures as well as the pressure ratio.

4 Inlet Air Cooling System

The inlet air cooling system must be design to avoid icing at the compressor inlet air or any where in the air intake structure. When the air is drawn into the mouth of the compressor, its velocity of air increases and the it′s temperature drops further as air enthalpy is transformed into kinetic energy in an adiabatic process. Therefore, the design inlet air temperature should generally be at least 5 °C to avoid potential icing problems [3]. There are two methods for cooling inlet air, (1) Chiller system component, (2) Evaporative cooling.

4.1 Chiller System Component

The chiller coil system are designed as a module that is typically positioned downstream of the air filters. As airflow pass through the chilled coil, the air is cooled through an indirect heat exchange with the cooling fluid. The air then passes through it eliminator media and into the compressor, as shown in Fig. 3 [3].

The coil is cooled and therefore condensation is created. Condensate droplets are directed downward and collected in pans, then directed out of the system. Chiller coil tubes are typically made of copper and caged with aluminum fins that economically and effectively extend the cooling surface area, coil casing are strain steel for long life. For maximum turbine performance, the design temperature of the air leaving the cooling system and entering the compressor is typically no less than 7 °C.

The ambient air temperature and the altitude at the site are the major factors used in siting and designing a coil system. Cooling agents are usually water.

In this method, it is cool the inlet air regardless of ambient humidity. It is used in small or large system.

4.2 Evaporative Cooling

The evaporative cooler module is positioned down stream of the filter section, shown in Fig. 4 [3]. When the filtered air passes through the saturated evaporative

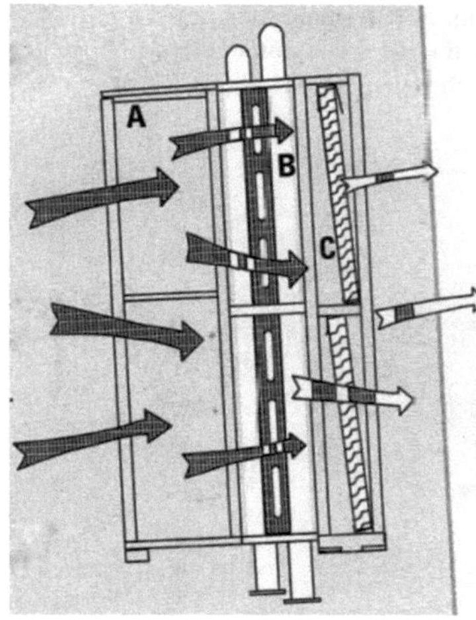

Fig. 3 Chiller coil system [3]

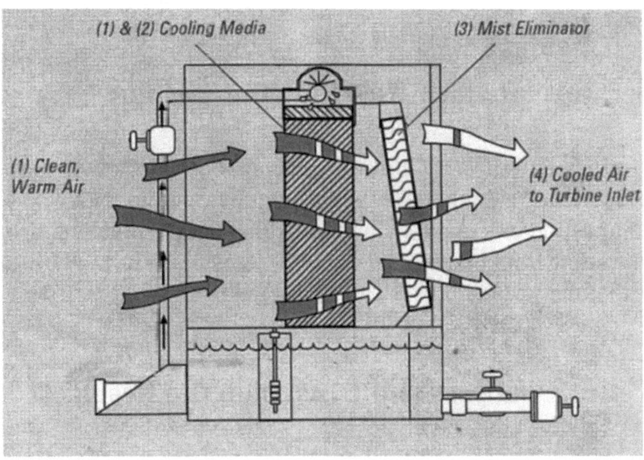

Fig. 4 Evaporative cooler module [3]

cooling media, water evaporates off the wet media. This evaporation is the process that reduces the air temperature. Cooled airs then passes through the integral mist eliminator, when left over water droplets are removed; clean cooled air is then directed into the compressor inlet. This method economical operation and uncomplicated system. But the cooling efficiency depends on relative humidity of ambient intake air and need source of make – up water.

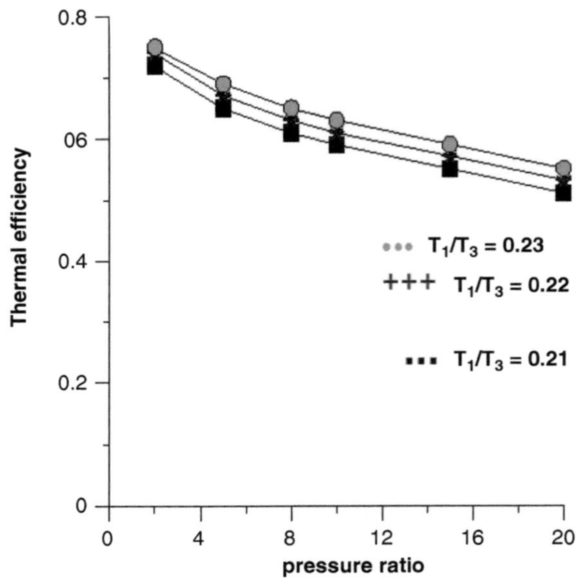

Fig. 5 Thermal efficiency of the ideal Brayton cycle with regeneration

5 Temperature of Hot Gases Inlet the Turbine

The thermal efficiency of a gas turbine depends on the allowable maximum gas temperature at the turbine inlet. The highest temperature in the cycle occurs at the end of the combustion process. Raising the turbine inlet temperature from 900 °C to 1200 °C increases the thermal efficiency by 20 %, and it is limited by the maximum temperature that the turbine blades can withstand [4].

6 Theoretical Analyses and Discussion the Result

Following typical operation parameters are considered for gas – turbine in this work, Brayton cycle with regenerative process.

$$r_p = 8, T_1 = 300\,\text{k}, T_3 = 1300\,\text{k}, \eta_c = 0.8, \eta_T = 0.85, \zeta = 0.8, \delta_{air} = 1.4$$

First, calculate the η of the cycle without regenerative process for several of pressure ratio by use Eq. (1).

Figure 5 represents the relationship between the η_{th} and the three values of the T_1, that is mean three values of ratios (T_1/T_3) with the several values of r_p, using Eq. (3). This Fig. shows the η_{th} of the cycle increase with decrease the T_1, and that regenerative process is most effective at lower values of r_p, because each gas turbine has a constant volumetric flow rate of air, as the air temperature decreases the air density and consequently, it mass flow rate increases. The increases air mass flow rate directly cases the gas turbine to produce more output power.

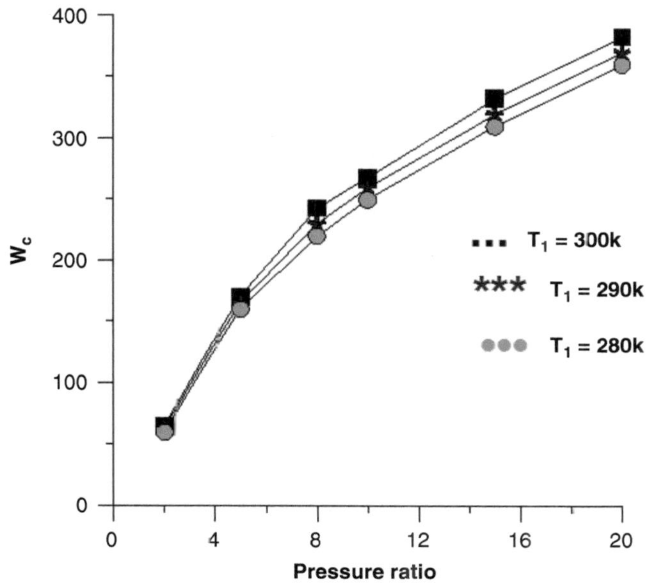

Fig. 6 Relationship between the compressor work and pressure ratio for several inlet air temperatures

Figure 6 represents the variation of work compressor W_C with several values of pressure ratio r_p for cycle with regenerative process for three values of temperatures T_1, that is mean the ratio of temperature (T_1/T_3) decrease. In this Fig. it is show the W_c is decrease with the T_1 of air inlet decreased, it is mean the thermal efficiency of the cycle is increased. Also it show the W_c is increased with increase the r_p, using Eq. (2).

Figure 7 represents the variation of η_{th} with the T_3 for several values of r_p, it is show the η_{th} increases with increase the T_3. The η_{th} increase with decrease of r_p, because when r_p increase the T_4 is decreased and T_2 is increased, so the recovered thermal energy in regenerator falls until zero corresponding to the r_p, that at this point the efficiency of simple cycle equal the efficiency of the regenerative cycle, using Eq. (3).

Figure 8 represents the variation of W_T with T_3 for several values of r_p. It is show the WT is increase with increase the T_3, but the W_T is increased with decrease of r_p. The value of optimum temperature would change slightly if the effects of blade cooling, using [5].

$$W_T = C_{pg} T_3 \left[1 - \frac{1}{r_p^{(\delta-1)/\delta}} \right] \tag{4}$$

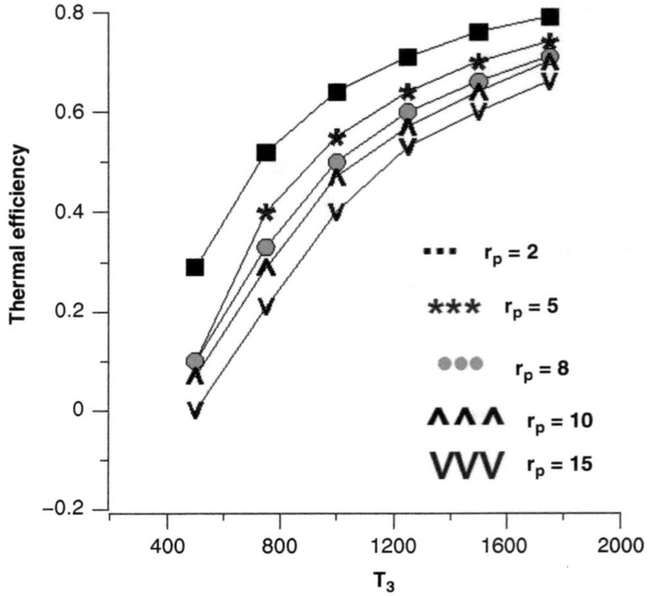

Fig. 7 Variation of thermal efficiency with the variation of values T_3 for many values of pressure ratio

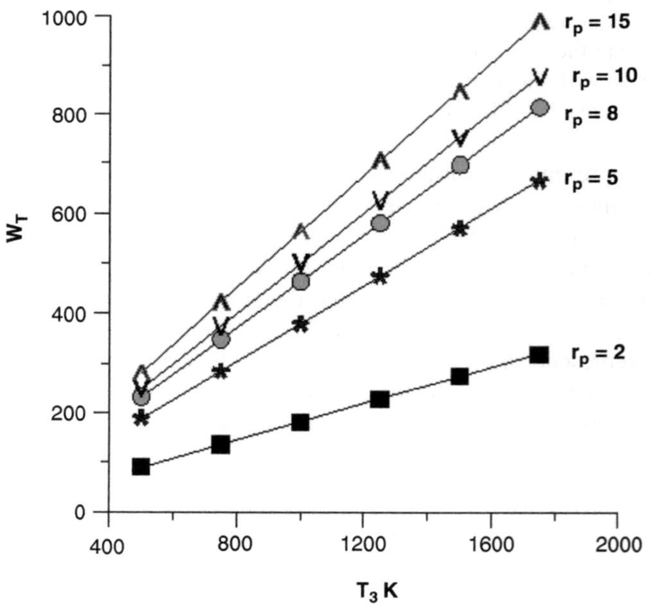

Fig. 8 Variation of turbine work with temperature of gases inlet to turbine at many values of r_p

7 Conclusions

The analysis of result from the modeling of the influence of parameter, showed that, the pressure ratio, inlet air temperature, turbine inlet temperature, effect on the performance of gas turbine power plant. The conclusions were summarized as follow:

1. The pressure ratio, inlet air temperature are strongly influence on the thermal efficiency of the gas turbine power plant.
2. The variations of thermal efficiency at higher pressure ratio, turbine inlet temperature, inlet air temperature are very important.
3. The thermal efficiency in the simple cycle depends on the pressure ratio only, but in the regenerative cycle depends on the pressure ratio and the inlet air temperature and inlet gas temperature and inlet gas temperature.
4. The work of compressor decreases when the inlet air temperature decreases, so the net work is increase.
5. The work of turbine increases when the temperature of the gases inlet the turbine increased.
6. The thermal efficiency of the regenerative cycle increase when the temperature of the gases inlet the turbine increased and when the pressure ratio is low.

References

1. H. Cohen, G.F.C. Rogers, H.I.H. Saravanamuttoo, *Gas Turbine Theory* (Wiley, New York, 1972)
2. T.D. Eastop, A. Meconkey, *Applied Thermodynamic for Engineering Technologist* (Pearson Education Limited, 1984)
3. Donaldson Company, *Cooling Inlet Air Improves Turbine Output* (Donaldson Company, Mineapolis, 2005) product letter
4. M.M. Rahman, T. Ibrahim, A.N. Abdalla, Thermodynamic performance analysis of gas turbine power plant. Int. J. Phys. Sci. **6**(14), 3539–3550 (2011)
5. B. Dawoud, Thermodynamic of power requirements. Appl. Therm. Eng. **25**, 1579–1598 (2005)

Part IV
Engineering Analyses

J. Andrus, J. Chmelar, J. Dvorak, V. Dynybyl, J. Hoidekr, S. Hosnedl, S. Hrcek, K. Kaláb, V. Kloucek, R. Kohar, J. Krepela, E. Kronerova, Z. Matuszak, Z. Nemec, J. Nevrly, K. Petr, M. Sasiadek, M. Spaniel, and V. Zoul

Innovation of Methodological Support of Risk Analyses of Technical Products in Their Life Cycle During Their Designing

J. Dvorak, M. Sasiadek, and S. Hosnedl

Abstract Concept of enhanced methodological support of risk analyses of technical products looked upon technical systems (TS) in the context of their Life Cycle (LC) is described in this paper. Within the methodological framework for innovation of risk analyses well proven method FMEA was used. The mentioned innovation of methodology consists of increasing of its complexity, systematic and also of increasing of its flexibility by application of key knowledge of Engineering Design Science, mainly from the Theory of Technical systems (TTS). The proposed methodology was verified and validated on specific example from industrial practice.

Keywords Risk • Analyse • Technical system • Life cycle

1 Introduction

It is very important for a design engineer to find out/predict as soon as possible, for the designed products their (future) properties and from them the resulting competitive advantages or disadvantages (strengths and weaknesses) and to make prompt efforts to their timely improvement or their elimination if necessary. During endeavours towards quality and competitiveness of a designed product, other nowadays contentious issues risks of product, process or organization which hides considerable potential to increase quality are being revealed. There are many approaches or methods for risk analyses of technical products. Those methods appeared especially during a second half of twentieth century when people began to appreciate a danger of consequences of their activities and fact that they can damage themselves apart from environment. There are many definitions of risks in available literature, risks are defined miscellaneous.

J. Dvorak (✉) • S. Hosnedl
University of West Bohemia, Plzeň, Czech Republic
e-mail: dvorakj@kks.zcu.cz; hosnedl@kks.zcu.cz

M. Sasiadek
University of Zielona Góra, Zielona Góra, Poland
e-mail: m.sasiadek@iizp.uz.zgora.pl

V. Dynybyl et al. (eds.), *The Latest Methods of Construction Design*,
DOI 10.1007/978-3-319-22762-7_29

So generally recognized definition obviously does not exist. Risks are defined in [1] as effect of uncertainty on objectives. In this definition, uncertainties include events (which may or not happen) and uncertainties caused by a lack of information or ambiguity. But there are many definitions e.g. [2]:

- simply future issues that can be avoided or mitigated
- the probable frequency and probable magnitude of future loss
- probability of a hazard resulting in an adverse event, times the severity of the event
- probability or possibility of inception of loss (generally failure)
- variability of possible results or uncertainity of their achievement
- etc.

Risk is mathematically defined as [3]:

$$\begin{aligned} Risk = & (probability\ of\ an\ accident\ ocurring) \\ & \times (expected\ loss\ in\ case\ of\ accident) \end{aligned} \quad (1)$$

Probability of accident occurring is measured in (1/year, 1/km, 1/km^2.year). Expected losses (consequences) are measured in (currencies, sudden deaths, subsequent deaths). There are many methods being used for risk indication and identification [2]. Methods used for risk evaluation can be divided into several groups according to the level of detail risk analysis capabilities and quantification of risks [1, 4]:

The level of detail of analyses

- comparative methods
- analytical methods based on deterministic approach
- analytical methods based on probability approach

According ability to quantify the level of risk

- qualitative methods
- semi – qualitative methods
- quantitative (probability) methods

The differentiation into such groups is relative to each other and the groups may overlap. Analytical methods based on a deterministic approach can be qualitative (FMEA) and semiquantitative (FMECA).

Comparative methods are focused on identification of risks. They are mostly based on comparison and application of operational experiences acquired from operation of danger machines added by inspection of machine [2, 4]. Those are e.g. safety review, checklist analyses, relative ranking. Analytical methods based on deterministic approach are focused on identification on source of risks e.g. preliminary hazard analysis (PHA), what if? (WI), hazard and operability analysis (HAZOP), failure mode and effect analysis (FMEA), fault tree analysis (FTA) [5], event tree analysis (ETA), cause-consequence analysis (CCA), human

Table 1 Differentiation of current methods regarding to their usability during LC individual phases (black pot means suitable method, gray pot means less suitable method) [4]

	SR	CL	RR	PHA	WI	WI/CL	HAZOP	FMEA	FTA	ETA	CCA	HRA
Research and development	○	○	●	●	●	○	○	○	○	○	○	○
Conceptual design	○	●	●	●	●	●	○	○	○	○	○	○
Prototype plant	○	●	○	●	●	●	●	●	●	●	●	●
Detailed engineering	○	●	○	●	●	●	●	●	●	●	●	●
Design/Start-up procedure	●	●	○	○	●	●	○	○	○	○	○	●
Regular Operation	●	●	●	●	●	●	●	●	●	●	●	●
Diversification/Modification	●	●	●	●	●	●	●	●	●	●	●	●
Investigation of events	○	○	○	○	●	○	●	●	●	●	●	●
Taking out of service	●	●	○	●	●	○	○	○	○	○	○	○

reliability analysis (HRA). Analytical methods based on probability approach are able to analyse causes of danger event and scenario of expansion of danger event e.g. reliability block diagram (RBD), Markov chains (MC).

Methods/approaches from analysed literature are also compared regarding to their usability to analyse risks during the phases of LC of any TS. After analyses which was mentioned it was found out that only What-if? (WI) method could be used for whole LC depicted in left column of Table 1 [4]. It is also clear from the picture that all presented methods are able to use during Regular Operation Phase and Diversification/Modification phase. From reasons of lack of usage during LC and absence of Theory Based Strategy endeavours for development of complex method started. With regard demands on quality and truncating of the engineering design process it is necessary to predict risks in all phases of LC. As it was mentioned What if analyses looks the most suitable when predicting all LC stages. However this method lacks systematic and is randomly used.

2 Application of a Traditional Risk Analysis with Use of FMEA

For our case (Fig. 1) the most known risk identification approach Failure Mode and Effect Analysis (FMEA) is used. This method is based on a deterministic approach and is also qualitative. At first step the technical system is analysed part by part and possible risks are stated including sources. Then the key indicators of RPN (Risk priority number) were determined (Severity, Detection, Occurrence). The selected crucial consequences are depicted in Table 2. The disadvantage of this approach is fact that consequences are stated more or less intuitively. The recommended FMEA procedure is described in e.g. [6].

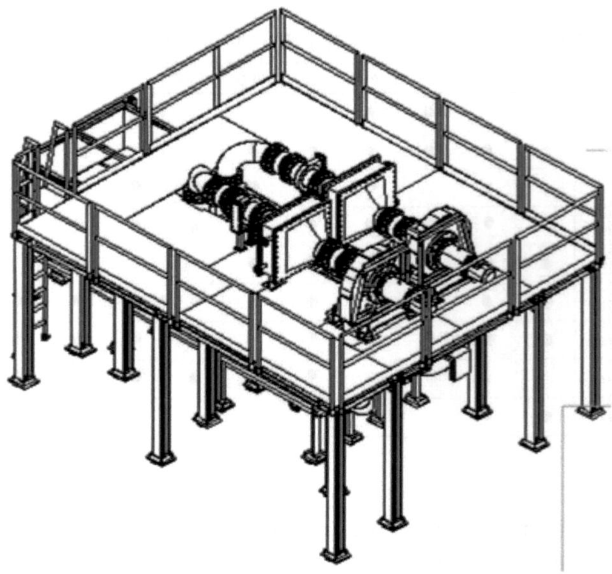

Fig. 1 Rapid cooler analysed by FMEA method

3 Systematic Risk Analyses Based on TTS

For brief introduction it is necessary to consider requirements as a TS property [7, 8] of a product to be designed should meet. TS Property is any TS 'feature, characteristics, attribute, etc.', such as: power, form, size, stability, durability, colour, manufacturability, transportability, suitability for storage, etc., which characterize TS. TS Properties can be classified in a number of ways. The described consistent and transparent hierarchical taxonomy stems from and our research supported by inquiries and by our large university and industry related experience.

All TS properties are split into three qualitatively different Domains: Descriptive, Reactive and Reflective ones. All their TS Property classes are split into the few subclasses each, their relevant properties, and their property indicators defined by their values (Fig. 2.) [9]. TS property classes of all Domains generally anyway differently relate to all Life cycle phases of TS. The risks were however again "derived" from TS properties. Conditions that each particular TS property does not meet requirements can evoke potential risk were considered although individual consequences could be caused by many property indicators.

If the part of current FMEA procedure (critical phase of intuitive process of finding, recognizing and describing risks) is replaced by risk analyses based on hierarchical taxonomy system for properties of technical products the system of investigated possible risks become systematic and comprehensive. The reason is easy. This system considers possible risks not only for main phases e.g. Operational

Table 2 Part of FMEA worksheet for the rapid cooler

Main system / Subsystem			Part/Operation name	Drawing/Part No.	Machine No.					Functional Specification-Technical Regulation	
Main Function / Operation			Date Performed	Date Time Updated 29.5.2014 16:22	Status - Hardware/Digital model						Issued by
PART		CHARACTERISTICS OF FAILURE				1st RATING				ACTIONS	
No	Function / Part / Operation	Failure mode	Causes of failure	Undesirable customer effects Effects of failure on syst. / part / operation	Testing - Simulation	Po	S	Pd	RPN	Recommended action	Planned action
1	Housing	cracks/fractures/raptures of the housing welded connections	penetration/infiltration of the temperature outside	increase of the environment temperature		3	3	4	36	checking 100% of the fusioon weld length	checking 100% of the fusioon weld length
			possibility of cooling water effluent	decrease of amount of the cooling water		3	3	4	36	checking 100% of the fusioon weld length	checking 100% of the fusioon weld length
		leak/leakage of the housing	penetration/infiltration of the temperature outside	increase of the environment temperature		3	3	4	36	check the leakproofness of the housing after welding	check the leakproofness of the housing after welding
			possibility of cooling water effluent	decrease of amount of the cooling water		3	3	4	36	check the leakproofness of the housing after welding	check the leakproofness of the housing after welding
2	Charge transport	non-rectilinear/not straight fixing of the roller tracks (bigger and	non-rectilinear/not straight travelling/shift of the chain and the	possibility of chain/charge/stock damaging		2	6	2	24	check the rectilinearity of the roller tracks fixing	check the rectilinearity of the roller tracks fixing
2.1	Big roller track	seizure/galling of the roller bearings	increase of the friction resistance between the roller and the chain	possibility of the chain break/rupture with larger amount of the seized rollers, process stoppage		2	7	6	84	selection of higher quality elements	selection of higher quality elements
		non-rectilinear/not straight chain guiding/leading	non-rectilinear/not straight travelling/shift of the chain and the	possibility of chain/charge/stock damaging		2	4	4	32	check the rectilinearity of the roller tracks	check the rectilinearity of the roller tracks
2.2	Small roller truck	seizure/galling of the roller bearings	increase of the friction resistance between the roller and the chain	possibility of the chain break/rupture with larger amount of the seized rollers, process stoppage		2	7	6	84	selection of higher quality elements	selection of higher quality elements

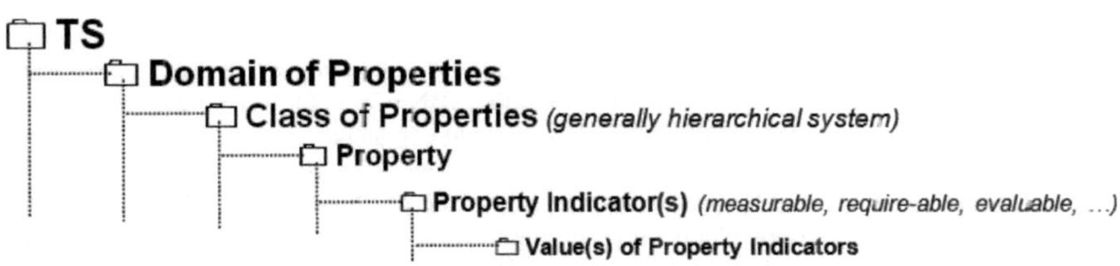

Fig. 2 Hierarchical taxonomy system for systematic classification and specification of any TS Property which could be generally source of risks [9]

phase of TS but also other LC phases are respected. Then the most dangerous risks are evaluated and calculated and subsequent control (measure that is modifying risk) [1] is performed.

4 Conclusions

The concept of methodological support of risk analyses of technical products in the context of their LC is described in this paper. This concept was successfully validated in cooperation of University of West Bohemia with company MEGA a. s., Stráž pod Ralskem (CZ) during development design project of electrodeionisation (EDI) module for preparation of highly pure water with use of electrodeionisation membrane processes. It contributes to the development of innovative design in the sense that generally design usually bears weak points which may annihilate their general value. In this paper it is believed that the paper makes a contribution to the innovative design reality.

Acknowledgments This paper includes partial results of the Project SGS-2013-050 "Complex support of design engineering of technical products to improve their properties and competitiveness II".

References

1. ČSN ISO 31000, *Management rizik – Principy a směrnice* (Český normalizační institut, Praha, 2010)
2. J. Hlinovský (in memoriam), J. Marek, P. Blecha, E. Krčálová, J. *Mareček, Management rizik výrobních strojů: MM Průmyslové spektrum* (2009). ISSN 1212-2572
3. W.G. Ramroth, *Risk Management for Design Professionals* (Kaplan Publishing, New York, 2007). ISBN 978-1-4277-5476-9
4. P. Fuchs, L. Vališ, Metody analýzy a řízení rizika. Available via http://www.pirozek.com/usi/lib/exe/fetch.php?media=analyza_rizeni_rizika_skriptum. Accessed 21 May 2014
5. ČSN EN 61025, *Analýza stromu poruchových stavů* (Český normalizační institut, Praha, 2007)
6. ČSN EN 60812, *Techniky analýzy bezporuchovosti systémů: Postup analýzy způsobů a důsledků poruch (FMEA)* (Český normalizační institut, Praha, 2007)
7. W.E. Eder, S. Hosnedl, *Design Engineering, a Manual for Enhanced Creativity* (CRC Press/Taylor & Francis Group, Boca Raton, 2008), p. 600. ISBN 978-1-4200-4765-3
8. ČSN EN ISO 9000 (idt ISO 9000:2005), *Systémy managementu (jakosti – 2002) kvality – Základní principy a slovník (Quality Management Systems – Fundamentals and Vocabulary)* Český normalizační institut, Praha, 2002, (Czech Institute for Standardisation, Prague, 2006)
9. S. Hosnedl, Z. Srp, J. Dvorak, Cooperation of engineering & industrial designers on industrial projects, in *Proceedings of the 10th International Design Conference – DESIGN 2008*, Zagreb, 2008, pp. 1227–1234

The Mathematical Description of Sunflower Seed's Hulling Process Using Impact Huller

J. Hoidekr, J. Andrus, and V. Dynybyl

Abstract This work aims to describe mathematical model of Sunflower Seed's Hulling process. Exactly, the movement of sunflower seeds inside rotor is described by equations of motion as a function of time.

Keywords Sunflower seeds hulling machine • Equations of motion

1 Introduction

The most of sunflower seeds hulling machines are based on the principle of using kinetic energy of thrown seed. Then the impact causes the seed is hulled. Seeds are accelerated by centrifugal force in rotating rotor with blades. There are three coordinate systems for mathematical description of seeds movement over the blades, like Fig. 1. The global system "*XY*" is fixed to machine frame. There are local Cartesian "*xy*" and Polar "$r\varphi$" coordinate systems, which are fixed to rotor blade [1].

2 Equation of Motion

Kinematic diagram of moving seed in the rotor can be thought of as a point slidably set on rotating rod with constant angular velocity "ω", like Fig. 2. Absolute velocity "v" can be broken into components "v_u" and "v_r".

There are several forces acting on a seed while it is moving over rotor blade. At first, there are gravity force "G" and its reaction "N". There is reaction force "F_C" between sunflower seed and rotor blade, also known as Coriolis force. When the

J. Hoidekr (✉) • J. Andrus
Czech Technical University in Prague, Prague, Czech Republic
e-mail: jan.hoidekr@fs.cvut.cz; jan.andrus@fs.cvut.cz

V. Dynybyl
Faculty of Mechanical Engineering, Czech Technical University of Prague, Prague, Czech Republic
e-mail: vojtech.dynybyl@fs.cvut.cz

V. Dynybyl et al. (eds.), *The Latest Methods of Construction Design*,
DOI 10.1007/978-3-319-22762-7_30

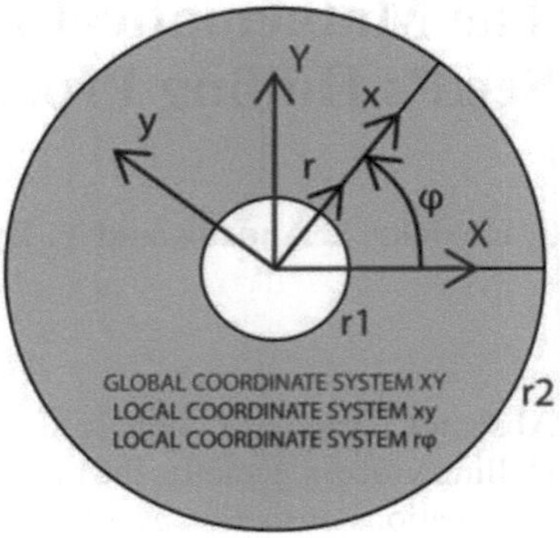

Fig. 1 Global and local coordinate systems of rotor

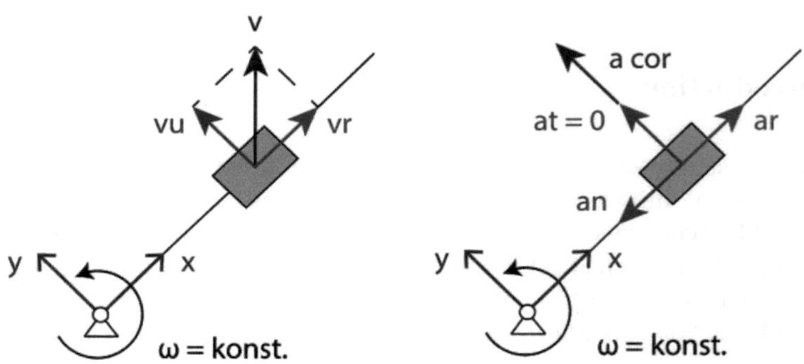

Fig. 2 Kinematic diagram of seed moving over rotor blade

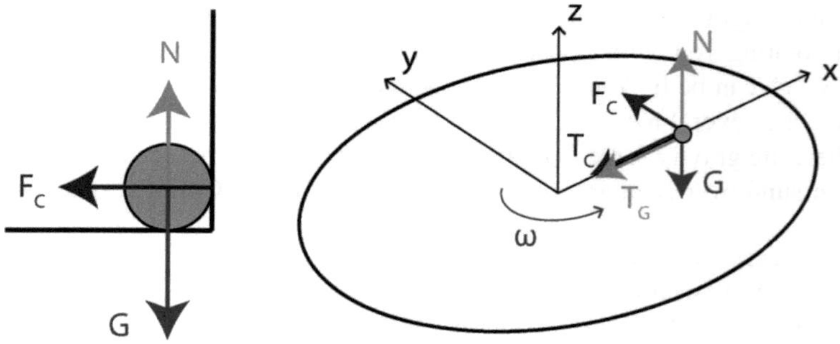

Fig. 3 Force conditions on seed while moving over rotor blade

seed is moving, there are frictional forces "T_C" and "T_G" caused by "N" and "F_C" forces (Fig. 3).

For set up equations of motion is used Newton's second law of motion. Forces and accelerations are expressed as 3D vectors in coordinate system "xy".

$$m \cdot \left(\begin{bmatrix} -\omega^2 \cdot x \\ 0 \\ 0 \end{bmatrix} + \begin{bmatrix} \ddot{x} \\ 0 \\ 0 \end{bmatrix} + \begin{bmatrix} 0 \\ 2 \cdot \omega \cdot \dot{x} \\ 0 \end{bmatrix} \right) = \\ = \begin{bmatrix} 0 \\ F_C \\ 0 \end{bmatrix} + \begin{bmatrix} 0 \\ 0 \\ N \end{bmatrix} + \begin{bmatrix} 0 \\ 0 \\ -m \cdot g \end{bmatrix} + \begin{bmatrix} -F_C \cdot \mu \\ 0 \\ 0 \end{bmatrix} + \begin{bmatrix} -N \cdot \mu \\ 0 \\ 0 \end{bmatrix} \tag{1}$$

Solving this system of equations we can get one second-order differential nonhomogeneous equation with constant coefficients in form:

$$\ddot{x} + k_1 \cdot \dot{x} + k_2 \cdot x = C \tag{2}$$

$$k_1 = 2 \cdot \omega \cdot \mu \tag{3}$$

$$k_2 = -\left(\omega^2\right) \tag{4}$$

$$C = -g \cdot \mu \tag{5}$$

Equation (2) can be solved analytically or using software MATLAB. Solving this equation we get the dependence of the position (6) and velocity (7) as a function of time. Coefficients "λ_1" and "λ_2" are known by solving (2). Coefficients "C_1" and "C_2" are known by solving Cauchy problem with specified initial conditions [2].

$$x_{(t)} = C_1 \cdot e^{\lambda_1 \cdot t} + C_2 \cdot e^{\lambda_2 \cdot t} + \frac{C}{k_2} \tag{6}$$

$$\dot{x}_{(t)} = C_1 \cdot \lambda_1 \cdot e^{\lambda_1 \cdot t} + C_2 \cdot \lambda_2 \cdot e^{\lambda_2 \cdot t} \tag{7}$$

Now we can write equations for position (8) and velocity (9) in global coordinate system "XY" fixed to machine frame. Trajectory of motion of the seed in rotor is illustrated on Fig. 4.

$${}^{XY}\underline{x} = \begin{bmatrix} x_{(t)} \cdot cos\,(\omega \cdot t + \varphi_0) \\ x_{(t)} \cdot sin\,(\omega \cdot t + \varphi_0) \\ 0 \end{bmatrix} \tag{8}$$

$${}^{XY}\underline{v} = \begin{bmatrix} \dot{x}_{(t)} \cdot cos\,(\omega \cdot t + \varphi_0) - \omega \cdot x_{(t)} \cdot sin\,(\omega \cdot t + \varphi_0) \\ \dot{x}_{(t)} \cdot sin\,(\omega \cdot t + \varphi_0) + \omega \cdot x_{(t)} \cdot cos\,(\omega \cdot t + \varphi_0) \\ 0 \end{bmatrix} \tag{9}$$

A direction of absolute velocity can be specified due to knowledge of sizes of velocities "v_u" and "v_r". This direction is important to set up outer stator impact areas for transfer a maximum of kinetic energy thrown seed into hulling process.

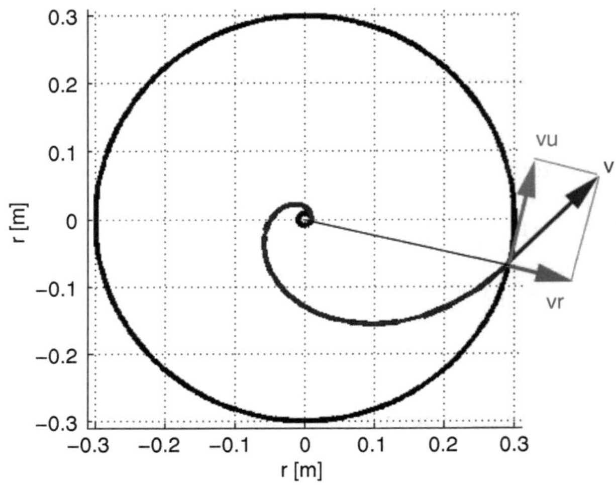

Fig. 4 Trajectory of motion of the seed in rotor

3 Conclusion

This is a base description of movement sunflower seeds over rotor blades inside mechanical impact huller. It can be used for setting ideal shape of outer stator impact area or describing energy condition of hulling process.

References

1. J. Hoidekr, Analýza loupání slunečnicových semen, in *Sborník konference STČ* (ČVUT v Praze, Fakulta strojní, Praha, 2014), s. 12. ISBN 978-80-01-05484-0. Dostupné z http://stc.fs.cvut.cz/pdf14/4522.pdf
2. J. Hoidekr, Analýza loupání slunečnicových semen, Diplomová práce, ČVUT v Praze, Praha, 2014, Vedoucí práce, Ing. Jan Andruš

Roller Bearing Fatigue Life Calculation with Inclusion of the Fits Parameters

S. Hrcek and R. Kohar

Abstract This article highlights the possibility of including the effect of fits on the accurate calculation of the fatigue life of rolling bearings and limits of ISO 281, which provides the calculation of service life for over 60 years.

Keywords Rolling bearings • Fatigue life • Lundberg – Palmgren theory

1 Introduction

Rolling bearing calculation now establishes international standards organization ISO 281 and its characterized by using the basic dynamic load rating C_r Based on this parameter and equivalent dynamic load that acts on the bearing is calculated fatigue life in millions of revolutions of one bearing ring to the other, which should reach the roller bearing at 90 % reliability. However, this method of calculation is greatly simplified and does not include a sufficient number of impacts affecting the resulting durability of rolling bearings.

This article builds on article *The effect of fits stiffness on distribution of internal loading in rolling bearings* and points to the rolling bearing life calculation, which includes a number of influences such as fits stiffness, direction of external forces, the radial internal clearance and fits of bearing rings on the shaft and housings.

To make this feasible, it is necessary to use basic theory to calculate the life of rolling bearings, written by Lundberg – Palmgren in 1947 [1] and which is based on the current ISO 281.

To determine the durability with the inclusion of the fits stiffness was used cylindrical roller bearing NU206 with parameters: basic dynamic load rating $C_r = 36,8\ kN$, diameter of the inner ring 30 mm, diameter of the outer ring 62 mm, width 16 mm, number of rolling elements 12, normal radial clearance. Fits of the inner ring on the shaft was considered standard with an interference – ∅ 30 k5 and fits of the outer ring in the housing was standard with clearance – ∅ 62 H7.

S. Hrcek (✉) • R. Kohar
University of Zilina, Žilina, Slovak Republic
e-mail: slavomir.hrcek@fstroj.uniza.sk; robert.kohar@fstroj.uniza.sk

V. Dynybyl et al. (eds.), *The Latest Methods of Construction Design*,
DOI 10.1007/978-3-319-22762-7_31

2 The Calculation of Fatigue Life According to ISO 281

Formula for calculating dynamic load rating according to ISO 281 [2]:

- for ball bearings the ball diameter $d_{we} \leq 25,4\ mm$:

$$C_r = b_m f_c (i\ \cos\alpha)^{0,7} Z^{2/3} {d_{we}}^{1,8} \tag{1}$$

- for ball bearings the ball diameter $d_{we} > 25,4\ mm$:

$$C_r = 3,647\ b_m f_c (i\ \cos\alpha)^{0,7} Z^{2/3} {d_{we}}^{1,4} \tag{2}$$

- for roller bearings:

$$C_r = b_m f_c \left(i\ l_{ef} \cos\alpha\right)^{7/9} Z^{3/4} {d_{we}}^{29/27} \tag{3}$$

From the above formula it is clear that the basic dynamic load rating of rolling bearings according to ISO 281 is dependent on pitch diameter D_{pw} (determination of coefficient f_c), the rolling elements diameter d_{we}, the number of rolling elements Z, nominal contact angle α, the number of rolling element rows i and the effective contact length of the roller l_{ef}.

Dynamic equivalent radial load P_e for ball, roller and angular contact ball bearings at constant radial load F_r and axial load F_a is given by:

$$P_e = X\,F_r + Y\,F_a \tag{4}$$

where X is called the radial factor and Y is called the thrust factor. The effect of radial load only on the bearing is equivalent to a load equal to:

$$P_e = F_r \tag{5}$$

Then ISO 281 defines the nominal life in millions of revolutions:

$$L_{10} = \left(\frac{C_r}{P_e}\right)^p \tag{6}$$

where the exponent p is for bearings with the point contact $p = 3$ and for bearings with line contact $p = 10/3$. Lundberg and Palmgren modified their value of the load-life exponent p for roller bearings from 4 to 10/3. Computer codes for rolling-element bearings incorporate $p = 4$ [3].

On the bearing N206 we assume purely radial load with size equal to half the basic dynamic load acting in any radial direction with the center in the bearing axis. Then according to Eq. (6) the basic life:

$$L_{10} = \left(\frac{36,8}{18,4}\right)^{10/3} = 10,08 \text{ or } L_{10} = \left(\frac{36,8}{18,4}\right)^4 = 16,0\ mil.\ ot. \tag{7}$$

3 Calculation of Life According to Lundberg: Palmgren Theory

The Lundberg – Palmgren theory states that for bearing rings subjected to N cycles of repeated loading the probability of survival S is given by:

$$ln\frac{1}{S} = A\frac{N^e \tau_o^c V}{z_o^h} \quad (8)$$

Where τ_o is the maximum orthogonal shear stress in the contact, z_o is the corresponding depth at which this stress occurs, and V is the stressed volume of material. The parameters A, c, and h are material characteristics that are determined experimentally. The parameter e is the Weibull slope for the experimental life data plotted on a Weibull probability paper. The stressed volume of material V was assumed to be:

$$V = 2a\, z_o(2\pi r_r) \quad (9)$$

$$V = l_{ef} z_o(2\pi r_r) \quad (10)$$

where the dimensions $2a$, z_o, and r_r correspond to the width, depth, and length of the volume. For bearings with a line contact applies Eq. (10) in which is the width stressed volume is equal to the effective contact length of the roller l_{ef}.

Adjusted Lundberg – Palmgren (8) formula for the calculation of life in millions of cycles when crossing the rolling element above one point of the raceway ring is:

$$N = \left(\frac{\ln\frac{1}{S}{z_0}^h}{A\ {\tau_0}^c\ V}\right)^{1/e} \quad (11)$$

In order to determine the durability of the rolling bearing, it is necessary to calculate the size of orthogonal shear stress τ_0 and depth in which the stress occurs z_0. Unless it is not possible to calculate the values from the real model of bearing, for example using finite elements method, than it can be based on the Hertz theory about the contact of two bodies. Hertz established formulas to calculate the width of the contact area and the size of the contact stress of two cylinders according to the following formulas:

- width of the contact surface for line contact of two cylinders:

$$2b = 2\sqrt{\left(\frac{8Q}{\pi l_{ef} E \sum \rho}\right)(1 - \nu^2)} \quad (12)$$

– contact stress for a line contact of two cylinders:

$$p_o = \sqrt{\frac{Q\,E\sum\rho}{2\pi l_{ef}(1-\nu^2)}} \tag{13}$$

After the calculation of the width of the contact area and the size of the contact pressure according to the Eqs. (12) and (13) it is possible to calculate the magnitude of the maximum orthogonal shear stress τ_0 and depth at which the stress occurs z_0. From the curve of subsurface stress [4] at the point of contact of two bodies (Fig. 1) it is clear that the size of the orthogonal shear stress is for line contact given by:

$$\tau_0 = 0,256 * p_o \tag{14}$$

and the depth at which this stress occurs:

$$z_0 = 0,25 * 2b \tag{15}$$

From the result of the load distribution on the rolling elements of NU206 bearing were taken the values of maximum rolling element load depending on the direction of radial force for the two boundary conditions, where the radial clearance in the bearing and the parameters of fits for durability are the best and worst [5]. From these load was calculated, according to the Hertz theory (Eqs. 12 and 13), the width of the contact surface and the size of the contact pressure between the rolling element and the inner ring of bearing. From these was calculated maximum orthogonal shear stress τ_0 and depth z_0, in which this stress occurs (Eqs. 14 and 15). Values of durability calculated according to the Lundberg – Palmgren theory are summarized in Tables 1 and 2 for the boundary of the fits.

While the calculated durability based on L-P theory is the number of cycles, in which is the place exposed to shear stress τ_0, it is still needed to recalculate the value of durability in millions of revolutions. For this it is necessary to calculate how many rolling elements passes during one rotation of one bearing ring to the other over one point on the raceway of the inner ring. For this purpose, you can use the bearing frequencies on which it is possible to detect damage to the particular components [6]. The inner race defect frequency is calculated according to:

$$BPF_i = \frac{Z}{2} n \left(1 + \frac{d_{we}}{D_{pw}} \cos\alpha\right) \tag{16}$$

The required frequency is calculated if speed n ($1/s$) is equal to 1. Then it is possible to divide the number of calculated cycles by error rate and get durability in millions of revolutions (right columns in Tables 1 and 2).

Summary of the results of Tables 1 and 2 is shown in Fig. 2.

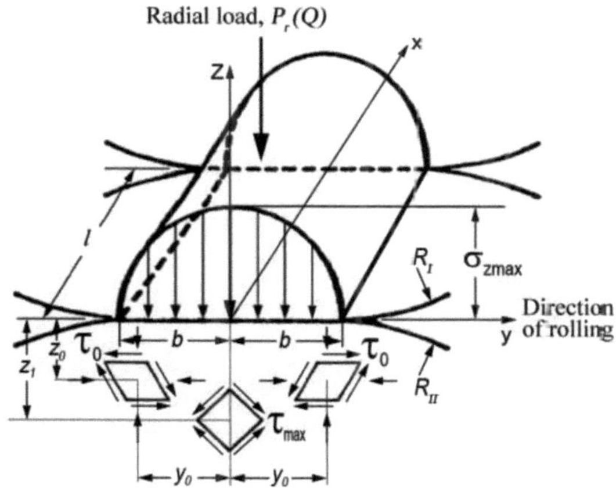

Fig. 1 Location of shear stresses below the surface of rolling contact

Table 1 NU206 bearing life depending on the direction of radial force for best condition of fits

dsrot	Qmin	2b	pmax	τo	zo	V	N	N
°	N	mm	MPa	MPa	mm	mm3	cykly	mil.ot
0	4776	0,291177	2321	594	0,072794	76,15384	339,74	47,38
20	4803	0,292007	2327	596	0,073002	76,37077	332,09	46,32
40	4762	0,290734	2317	593	0,072684	76,45421	343,90	47.96
60	5599	0,315267	2513	643	0,078817	82,45421	179,88	25,09
80	6851	0,348744	2779	712	0,087186	91,20957	80,23	11,19
100	7107	0,355182	2813	725	0,088796	92,89352	69,31	9,67
120	6542	0,340792	2716	695	0,085198	89,12988	96,49	13,46
140	6465	0,338772	2700	691	0,084693	88,60162	101,19	14,11
160	5643	0,316492	2522	646	0,079123	82,77452	174,38	24,32
180	5123	0,301571	2403	615	0,075393	78,87214	256,62	35,79

Table 2 NU206 bearing life depending on the direction of radial forces for the worst condition of fits

dsrot	Qmin	2b	pmax	τo	zo	V	N	N
°	N	mm	MPa	MPa	mm	mm3	cykly	mil.ot
0	5281	0,306184	2440	625	0,076546	80,07864	227,27	31,70
20	5265	0,305717	2436	624	0,076429	79,95659	230,06	32,09
40	4965	0,296883	2366	606	0,074221	77,64613	290,88	40,57
60	6116	0,329484	2626	672	0,082371	86,17246	126,39	17,63
80	7611	0,367575	2929	750	0,091894	96,13456	52,68	7,35
100	7493	0,364701	2906	744	0,091175	95,38298	56,09	7,82
120	7453	0,363736	2899	742	0,090934	95,13057	57,29	7,99
140	7561	0,366351	2920	747	0,091588	95,81443	54,10	7,55
160	6773	0,34674	2763	707	0,086685	90,68545	84,02	11,72
180	5968	0,325478	2594	664	0,081369	85,12467	139,39	19,44

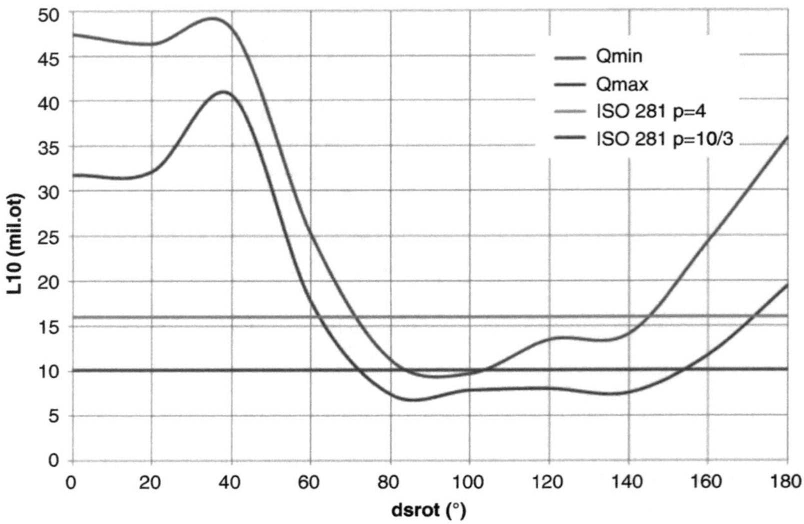

Fig. 2 Calculated bearing life of NU206 depending on the direction of radial force and parameters of fits

4 Conclusion

The aim of this paper was to show the possibility of accurate determination of the basic life of rolling bearings, as it allows the calculation according to ISO 281, which is largely simplified. The theory on which this standard is based allows to specify more accurate basic life of rolling bearings according to force field analysis that exist between the components of rolling bearings and tension in the contact of rolling elements with the raceways. Traditional life calculation is today well established since 1952 [1] and simplification of the original L-P theory resulted from those times when it was very difficult to include into analytical calculations various influences that affect the durability of rolling bearings. At present, thanks to modern computing and simulation software tools and powerful hardware it is possible to study and analyze action of force and stress ratios not only in individual bearings, as well to review these effects in specific applications. This was not possible in past, but today it is possible.

Acknowledgments This work was supported by the Slovak Research and Development Agency under the contract no. APVV-0419-11 – Adaptation of modern computer-simulation methods to the development of rolling bearings and their verification in real conditions.

This work was supported by the Scientific Grant Agency of the Ministry of Education, Science, Research and Sport of the Slovak Republic under the contract no. V-1/0396/14 – Research of the influence of construction and technological parameters of rolling bearings on the service life thereof.

This work was supported by the Scientific Grant Agency of the Ministry of Education, Science, Research and Sport of the Slovak Republic under the contract no. V-1/0844/13 – Research on guidance elements of rolling bearing and their design.

References

1. F. Sadeghi, B. Jalalahmadi, T.S. Slack, N. Raje, N.K. Arakere, A review of rolling contact fatigue. J. Tribol. **131**/041403-1 – 041403-15 (2009)
2. ISO 281:2007 - Rolling bearings – Dynamic load ratings and rating life. International Organization for Standardization
3. E.V. Zaretsky, *Rolling Bearing Life Prediction, Theory, and Application*, NASA/TP-2013-215305, 2013
4. A.T. Harris, M.N. Kotzalas, *Advanced Concepts of Bearing Technology*, Rolling Bearing Analysis, 5th edn. (CRC Press, Boca Raton, October 9, 2006). ISBN 9780849371820 - CAT# 7182. https://www.crcpress.com/Advanced-Concepts-of-Bearing-Technology-Rolling-Bearing-Analysis-Fifth/Harris-Kotzalas/9780849371820
5. J. Brònček, M. Dzimko, B. Hadzima, J. Takeichi, Experimental investigation of aluminium alloys 2024-T3 form in terms of tribocorrosion characteristics. Acta Metall. Slovaca **20**(1), 97–104 (2014). ISSN 1335-1532
6. L. Kucera, M. Lukac, L. Jurak, F. Brumercik, Hydromechanical automatic transmission. Komunikacie **11**(2), 33–35 (2009)

Application of Statistical Methods to Analysis of Parallel Structure of Gearbox Test Rig

J. Chmelar, K. Petr, and V. Dynybyl

Abstract Paper presents usage of statistical methods for an analysis of run of parallel structure. There was reported a problem that actuators of vertical axis are not uniformly loaded. Variance of acceleration record from each actuator were analysed and compared. There was performed normality test with Chi-Square. To compare the variance of acceleration there was used a non-parametric Kruskal–Wallis test. To identify a problematic actuator there was used a multiple comparison test and matrix of correlation coefficients with graphic output was constructed. The analysis pointed out on the most problematic actuator. Gathered information was used for further optimization and improvement of structure.

Keywords Parallel structure • Test rig • Gearbox • Statistics • Chi-square • Kruskal–Wallis

1 Introduction

Testing device for testing of gearboxes of rail vehicles was developed and implemented in testing facility of company Wikov MGI a.s., Hronov, Czech Rep. Testing device is supposed to test fatigue properties of gearboxes loaded with torque and impacts.

It compounds of two main electric motors that load gearbox with torque, they work in closed electro-mechanical loop. Basic principle of torque load is described in [1]. Tested gearbox is also loaded with shocks and impacts generated by hexapod structure. It consists of six hydraulic actuators that can load gearbox with impacts from three directions. Description of such device is presented in [1].

During final stage of implementation of device when all parts were tested there appeared a problem that one of used hydraulic actuators in axis Z according to Fig. 1 is significantly more loaded than the others. It caused problems with control setup

J. Chmelar (✉)
Czech Technical University in Prague, Prague, Czech Republic
e-mail: jakub.chmelar@fs.cvut.cz

K. Petr • V. Dynybyl
Faculty of Mechanical Engineering, Czech Technical University of Prague, Prague, Czech Republic
e-mail: karel.petr@fs.cvut.cz; vojtech.dynybyl@fs.cvut.cz

V. Dynybyl et al. (eds.), *The Latest Methods of Construction Design*,
DOI 10.1007/978-3-319-22762-7_32

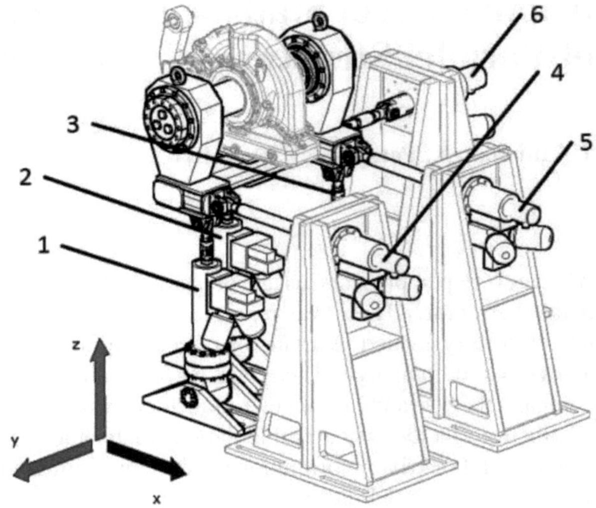

Fig. 1 Hexapod structure

of the system, when after some time of run the system suffered from instabilities. Even when this problem was partly solved by control strategy setup there still was the fact, that one actuator is more active than the others that could from long – term point of view lead to premature failure.

It was suggested to investigate this problem and identify the problematic actuator. This paper presents used procedure of investigation based on statistical analysis of recorded acceleration data of actuators.

2 State Description

The investigation is based on analysis of records of acceleration of actuators from control system. It was assumed that the problematic actuator should have wider variance of acceleration than other ones. For analysis were used 22 samples of records (one sample consists of records of acceleration from all six actuators).

During test cycle there were impacts with acceleration between 10 and 15 g that are rare and usually appears once per load cycle of each axis. The most impacts are expected to be up to 4 g, because in test cycle there are impacts with value of acceleration 2–4 g with frequency 2 Hz.

Cylinder six according to a Fig. 1 was not used in test cycle, it can be considered for this analysis that its role was to stabilize the system.

3 Data Description

For analysis all values of acceleration were transformed to its absolute values. Below there are presented examples of histograms of actuators 1, 2, 3 and 4 (Fig. 2).

It is obvious that actuator 2 has wider variance of acceleration than the others.

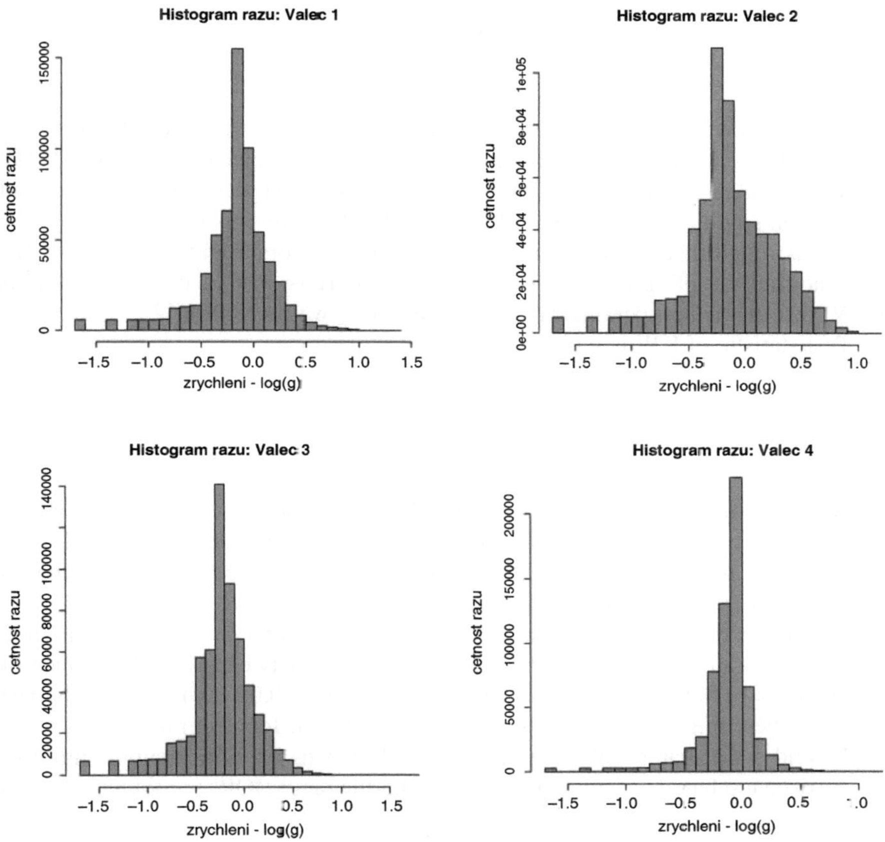

Fig. 2 Histograms of actuator acceleration. Actuator designaton *1*, *2*, *3* and *4* according to Fig. 1

4 Data Analysis

To quantify the results there were computed the variances of acceleration of all actuators. It showed that acceleration of actuator 2 is in each recorded sample the highest one.

To be the result statistically proven it was suggested to use multiple sample test. It is expected to show that the variance of acceleration between actuators is more significant than the variance of acceleration of actuator as such.

For further analysis it was checked whether recorded data agree with normal sampling distribution. It was used the Chi-square [2] test according to formula Eq. (1)

$$Q^2 = \sum_{i=1}^{K} \frac{(O_i - E_i)^2}{E_i} \tag{1}$$

Where: Oi observed frequency, Ei expected frequency, k number of intervals.

Hypothesis that data agree with normal distribution is rejected when $Q^2 > 9{,}448$ which is a value given by Chi-square distribution with four degrees of freedom and the chosen significance level $\alpha = 0.05$. Table 1 below presents the result.

It was found that acceleration distribution of actuators 3 and 6 does not agree with normal sampling distribution for further analysis was used Kruskal–Wallis (KW) test that is non-parametric alternative to one-way ANOVA [2].

Test statistic of KW is given below

$$H = \frac{12}{N(N+1)} \sum_{i=1}^{K} \frac{r_i}{n_i} - 3(N+1) \tag{2}$$

Where: N amount of samples, r_i sum of ranks of order in group i, ni sum of samples in group. Whole procedure of presented KW test is described in [2].

Rejection region for KW test hypothesis is when $H > 9{,}448$ according to Chi-Square distribution on significance level $\alpha = 0.05$.

Result for described case was $H = 103{,}233$, that falls into rejection region. Therefore it can be stated that there exist difference between size of variance of acceleration between actuators on level of significance 0.05. Other words the test said that there might be an actuator that significantly differs from the others.

To find out actuator with biggest variance there was performed simple multiple comparison test based on comparison of Interquartile range (IQR) according to Eq. (3)

$$IQR = Q_3 - Q_1 \tag{3}$$

In Table 2 there is highlighted highest value of IQR that belongs to actuator 2. It is also supported by graphical representation of multiple comparison test on a Fig. 3. That shows actuator 2 has significantly bigger variance over other used actuators.

Table 1 Results of Chi-Square test according to Eq. (1)

Actuator no.	1	2	3	4	5	6
Q^2	6.51	2.72	47.94	6.49	1.3	12.17
Result	Ok	Ok	Rejected	Ok	Ok	Rejected

Table 2 Multiple comparison test

Actuator no.	1	2	3	4	5	6
Median	0.515075	1.081763	0.383403	0.139142	0.083158	0.023331
Q1	0.447955	0.847845	0.28825	0.121384	0.075627	0.016026
Q3	0.64417	1.507605	0.470231	0.157666	0.095402	0.256708
IQR	0.196216	0.65976	0.181981	0.036281	0.019775	0.240681

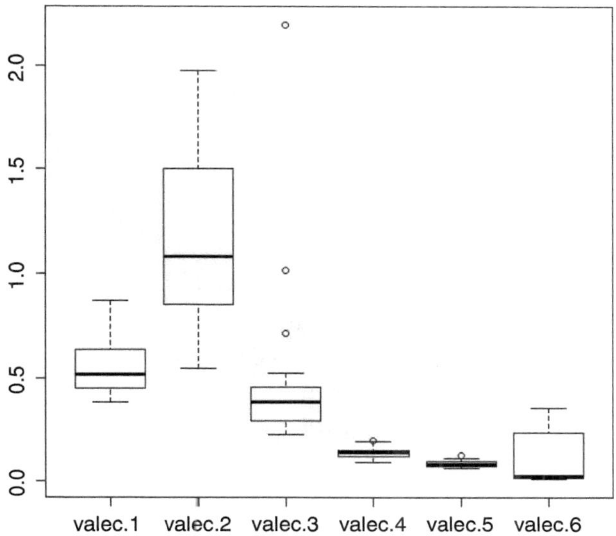

Fig. 3 Box-plot of variances of acceleration of actuators

5 Correlation Analysis

Last step of investigation was to find out, whether there exist any significant correlation between acceleration of actuator 2 and others in acceleration.

There were computed Pearson's correlation coefficients [2] between all actuators that are graphically presented on Fig. 4. The significance of the analysis is proved by significance test of correlation coefficients. Computed p-value was in all cases smaller then defined level of significance $\alpha = 0.05$.

The Heat-map shows that there exist linear dependency of actions of actuator 2 and 1 and 3 that is because these work in one axis. Less reacts actuator 2 on actions of actuator 4 and 5. Interesting is the dependency between 2 and 6 that could be due to unbalanced mass distribution on the frame with gearbox.

There wasn't found any special unexpected dependency between acceleration of actuators.

6 Conclusion

Analysis of variance of acceleration pointed on an actuator 2 that is significantly ($\alpha = 0.05$) more loaded than the others in the hexapod structure. For an investigation were used: normality test by Chi-Square, non-parametric variance analysis

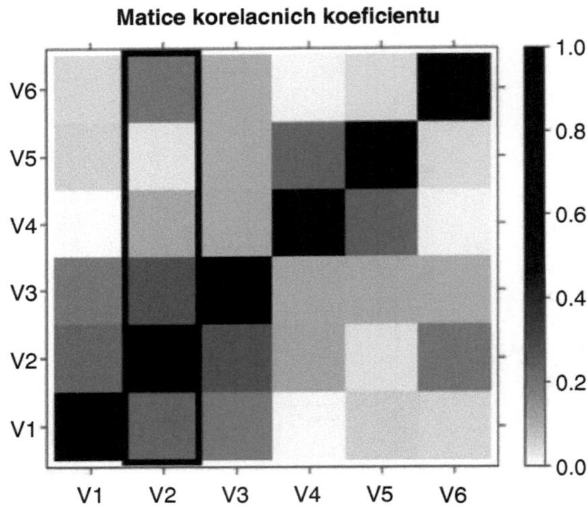

Fig. 4 Heat-Map of Pearson's correlation coefficients

Kruskal–Wallis and multiple-comparison test by comparing inter-quartile range. Also there was performed analysis of correlation of acceleration records between actuators. It does not showed any unexpected dependency of actuator 2 to the others. Results from this analysis were used as a base for design of new supporting frame takes more into account mass distribution of the clamped gearbox and its dynamic properties.

References

1. J. Chmelař, K. Petr, V. Dynybyl, J. Křepela, Design and implementation of testing stand for gearboxes of railway vehicles. Appl. Mech. Mater. [online]. 732, 283–286 (2015). DOI:10.4028/www.scientific.net/amm.732.283 [online]
2. K. Ramachandran, C.P. Tsokos, *Mathematical Statistics with Applications* (Academic, Boston, 2009), xxiii, p. 824. ISBN 01-237-4848-8

Driving Shaft Fracture Analysis

K. Kaláb

Abstract The article deals with the determination of the causes of failure of the driving shaft of the conveyor belt. The detection of the causes of this damage is based on the appearance of the shaft fatigue fracture and subsequent fracture surface analysis, the calculation of the stress for actual load and information about the belt operation.

Keywords Fracture face • Surface analysis • Fatigue • Conveyor belt

1 Introduction

The conveyor belt was driven by the station 2×250 [kW] and was operated 10 months (20 h per day), i.e. about $2 \cdot 10^7$ shaft revolutions. The broken shaft of the first driving drum pulley of the conveyor is shown in Fig. 1.

The shaft material was medium-alloyed stainless chromium – molybdenum – vanadium steel for heat treatment 31CrMoV9 according to EN 10085 (DIN 17221). After finishing the $R_{eMIN} = 650$ [MPa] and $R_m = (850 \div 1050)$ [MPa]. It was suitable for heavy-duty and large parts. Identifying the causes of the shaft failure based on macro fractography, personal inspection, photographs, original unprocessed fracture surface and profile damaged shaft, hardness measurement HB 329, HB 311, HB 296, HB 313, HB 312, HB 316 and by chemical analysis certificates. Metallographic analysis (micro fractography) has not been implemented.

2 Fracture Analysis

The next figures of the fracture show the major surface features seen on almost every fatigue face [1]. The arrows point to the initiation sites – origins of the fatigue crack denoted by capital letters (Figs. 2, 3, 4, 5, 6, and 7).

K. Kaláb (✉)
VSB-Technical University Ostrava, Ostrava, Czech Republic
e-mail: kvetoslav.kalab@vsb.cz

V. Dynybyl et al. (eds.), *The Latest Methods of Construction Design*,
DOI 10.1007/978-3-319-22762-7_33

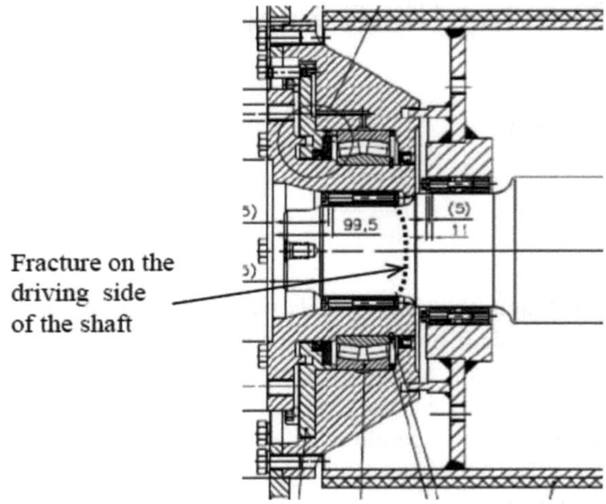

Fig. 1 Location of the shaft fracture on the TOLLOK locking assembly edge

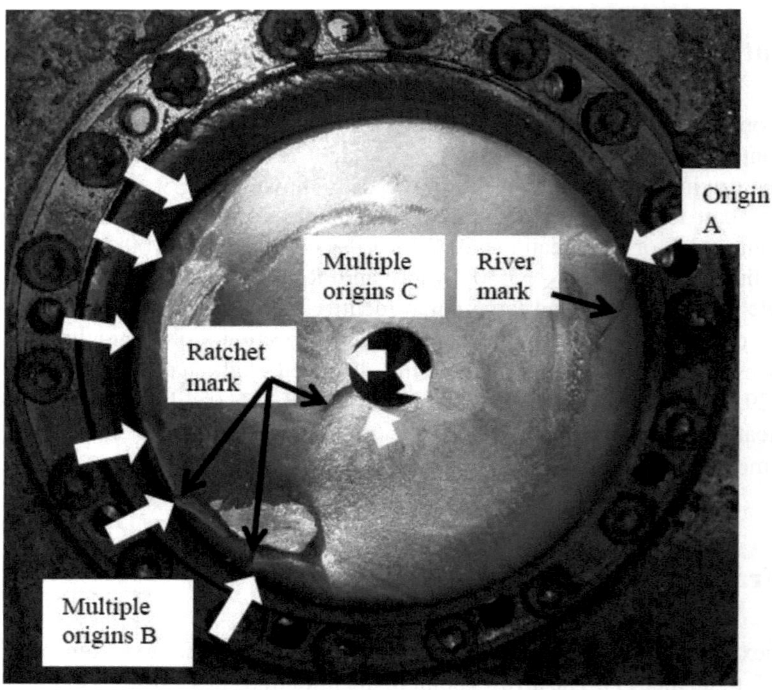

Fig. 2 Location of the shaft fracture on the TOLLOK locking assembly edge and description

The macroscopically smooth surface without significant progression marks shows that during the development of fatigue crack section, there were no significant changes in size of the operating stress or overloading of the shaft, such as

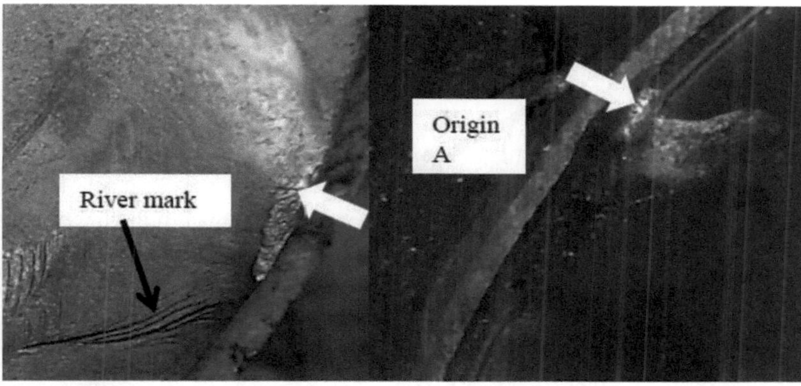

Fig. 3 Detail of the origin A

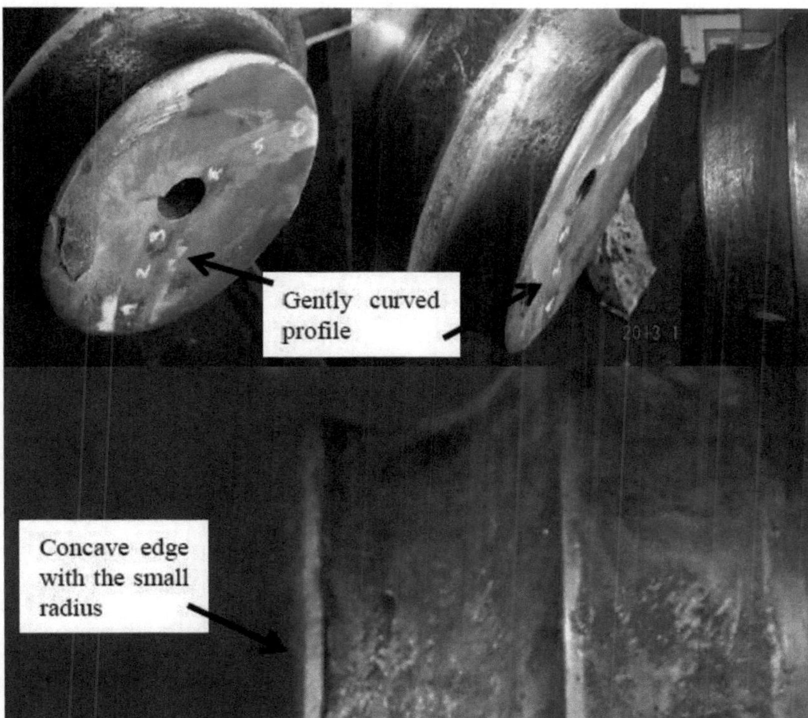

Fig. 4 View of the second side of the shaft fracture

irregular starting the conveyor. The fatigue fracture had multiple origins. The crack was initiated primarily at the flaw A. The river mark shows the direction of progression of the fatigue crack. River marks show up in the relatively fast-growing sections of the fatigue zone. During this crack growth, formed next multiple origins

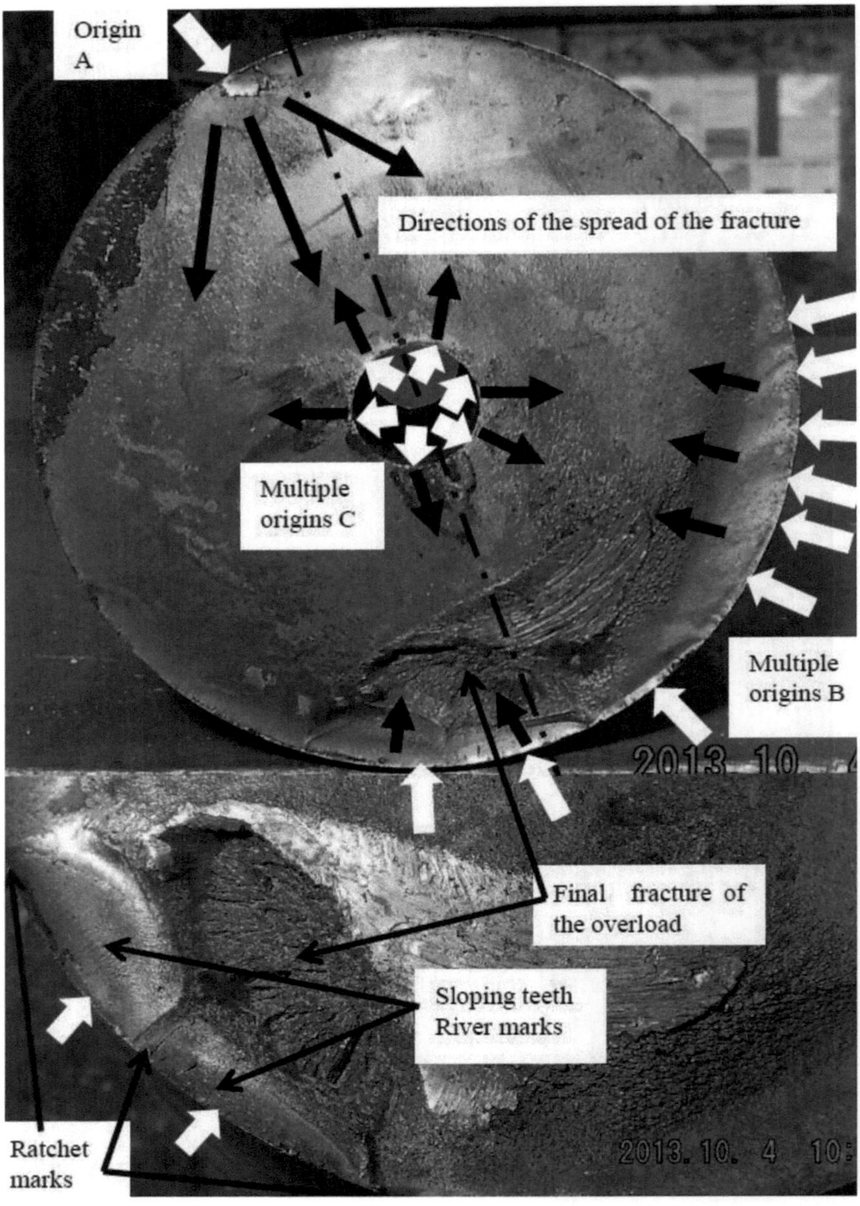

Fig. 5 Description of the cross-sectional of the fracture after removing the locking assemble

B and C from stress risers. The cross-sectional view in Fig. 4 shows that the piece is domed with the smallest radius near the outer edge. This small radius testifies to a high stress concentration. The high stress concentration also indicates the presence of multiple origins. The presence of ratchet marks indicates multiple origins and

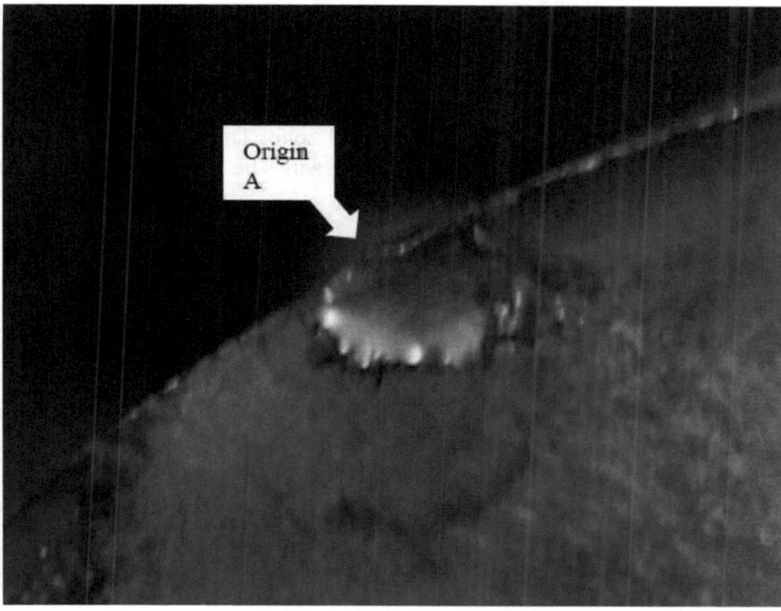

Fig. 6 Detail of the origin A – material, respectively manufacturing or assembly defect

Fig. 7 Rough- machined center hole of the shaft

relatively high total stresses. Ratchet marks can result from either high stress on the part or from high stress concentrations. The combination of many ratchet marks and a small overload zone indicates that the load was light, but there were high stress concentrations and that the stress concentration was the major cause of the fracture.

The calculated rated stresses in the cross section of the fracture are 214 MPa – the outer surface of the shaft and 323 MPa – surface of the axial bored hole. The size of the operating stress was near the fatigue limit and the number of cycles to fracture near the base N_C, in the investigated shaft that is about $2 \cdot 10^7$ cycles.

Also of interest is the shape of the overload zone. The fact that it is elongated indicates some plane bending loads were present. The subsequent check of the tensioning device revealed a cylinder jam.

3 Conclusion

The fracture of the shaft originated and spread at low nominal stresses and high notch effects of stress concentration. So the main causes of the fatigue fracture of the shaft were rotational stress concentrations that were located in the outer surface of the shaft at the edge of the clamping ring TOLLOK. Fatigue was the result of the improper assembly of the locking assembly and failure of the tensioning device.

Acknowledgments The research work reported here was made possible by the SGS Project 2014/24.

Reference

1. N.W. Sachs, Understanding the surface fractures of fatigue fractures: how they describe the failure cause and the failure history. J. Fail. Anal. Prev. **5**(2), 11–15 (2005), ASM International

Force Conditions at an Antibacklash Designed Gear Mechanism

V. Kloucek

Abstract The paper contains a description of one of principles of the backlash-free transmissions. Specially is resolved design of the zero-backlash gearing with countershafts. The backlash elimination in gears and bearings is obtained by the preloaded torsion-bar spring. There is determined dependence of reaction forces, acting on gears and bearings, to magnitude of the torsion-bar spring preload. Using the computer program, created in the Maple environment, is simulated running of loaded gear with reversation of rotating direction. The outcomes are used for control of strength, fatigue of material and service life of gearing.

Keywords Anti-backlash transmissions • Gear trains • Precision position control • Preload • Torsion-bar spring

1 Introduction

In engineering practice there is often a need for precise positioning of physical objects such as workpieces, tools, assembled parts, transported materials, finished products etc. Positioning device can be a variable-axis of machine tool, rotary positioning table, a robotic manipulator and many other cases. In such cases it is necessary to accelerate and brake considerable weight objects. It follows that drives of positioning devices operates with a relatively low speeds and relatively large forces and torques.

In terms of type of motion drives can be categorized into (a) rotary and (b) linear. The dominant rotary drive are electric motors of various design. However, they have an optimum efficiency and loading characteristics at higher speeds and lower torques than the positioning drives require. Therefore, between the electric motor and positioning device is necessary to insert the appropriate transmission or gearbox. Gearbox, as well as any kinematic mechanism, is necessarily made with clearances and dimensional tolerances. Still current problem is the suppressing of these backlashes in reversing the shaft rotate direction. This article describes one

V. Kloucek (✉)
VÚTS Liberec, a.s., Liberec, Czech Republic
e-mail: vojtech.kloucek@vuts.cz

V. Dynybyl et al. (eds.), *The Latest Methods of Construction Design*,
DOI 10.1007/978-3-319-22762-7_34

way of suppressing the backlash in gears meshes and bearings of gearbox with spur gears and the load simulation of the gearbox.

2 Design

2.1 Principle of the Backlash-Free Gearbox

Figure 1a schematically illustrates the principle of backlash-free gearing with countershafts. Torsion bar spring is mounted at a determined preload torque, and thus act on meshing pinion wheels in opposite directions. When transmission a rotary motion, then under the direction of rotation of input shaft performance by either one or the other path of gearing [1, 3].

2.2 Mounting Conditions

The gear by the Fig. 1a is closed loop of gears. Therefore, it is necessary in design to comply with geometric and kinematic constraints, which relate to axis distances and gear circumferential speeds. In general case (Fig. 2a) must be satisfied condition for pitch diameters

$$\frac{d32}{d31} = \frac{d22}{d12} \cdot \frac{d11}{d21} \tag{1}$$

Assuming the same module of all the gears, pitch diameters can be replaced by the number of teeth

$$\frac{z32}{z31} = \frac{z22}{z12} \cdot \frac{z11}{z21} \tag{2}$$

The advantage of this type of gearing is high variability in dimensions and therefore large area for optimization of specific applications. For the prototype design and subsequent calculations was chosen case, when $d11 = d12 = d31 = d\,32 = d1$ and $d21 = d22 = d4 = d2$ (Fig. 2b).

Table 1 lists technical parameters of the gearbox, which are used for subsequent calculations [2].

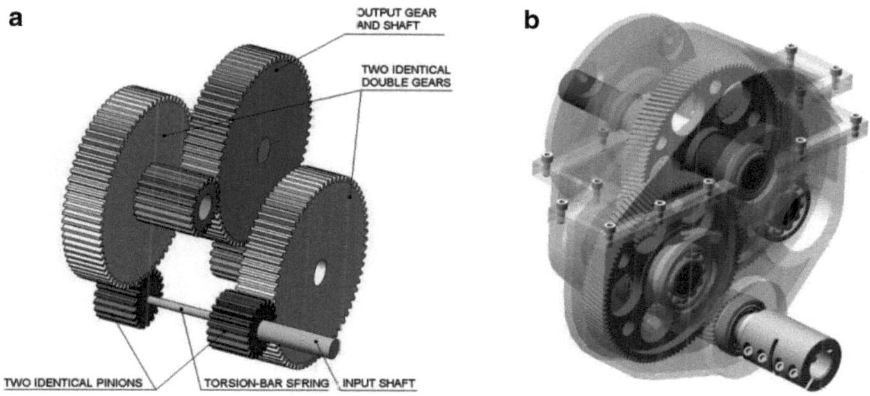

Fig. 1 (**a**) Anti-backlash gear assembly, (**b**) prototype design of the backlash-free gearbox

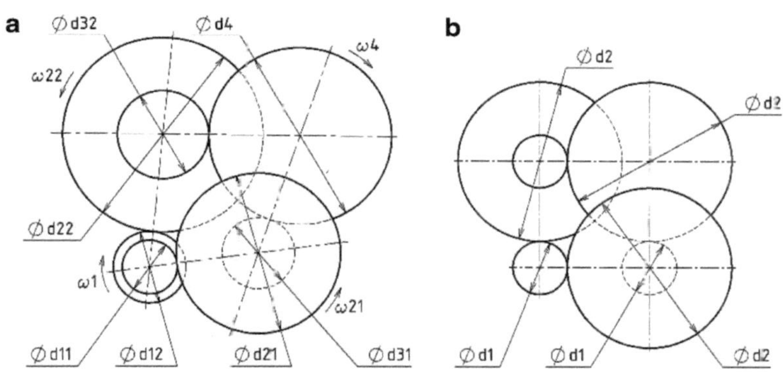

Fig. 2 Mounting conditions (**a**) general case, (**b**) prototype design

Table 1 Technical parameters of gearbox

Rated power	P	2.9	[kW]
Input torque	$M1$	18.6	[Nm]
Input speed	$n1$	1500	[rpm]
Gear module	m	1.5	[mm]
Tooth number	$z1$	31	[–]
Tooth number	$z2$	108	[–]
Reduction ratio	i	12.1	[–]

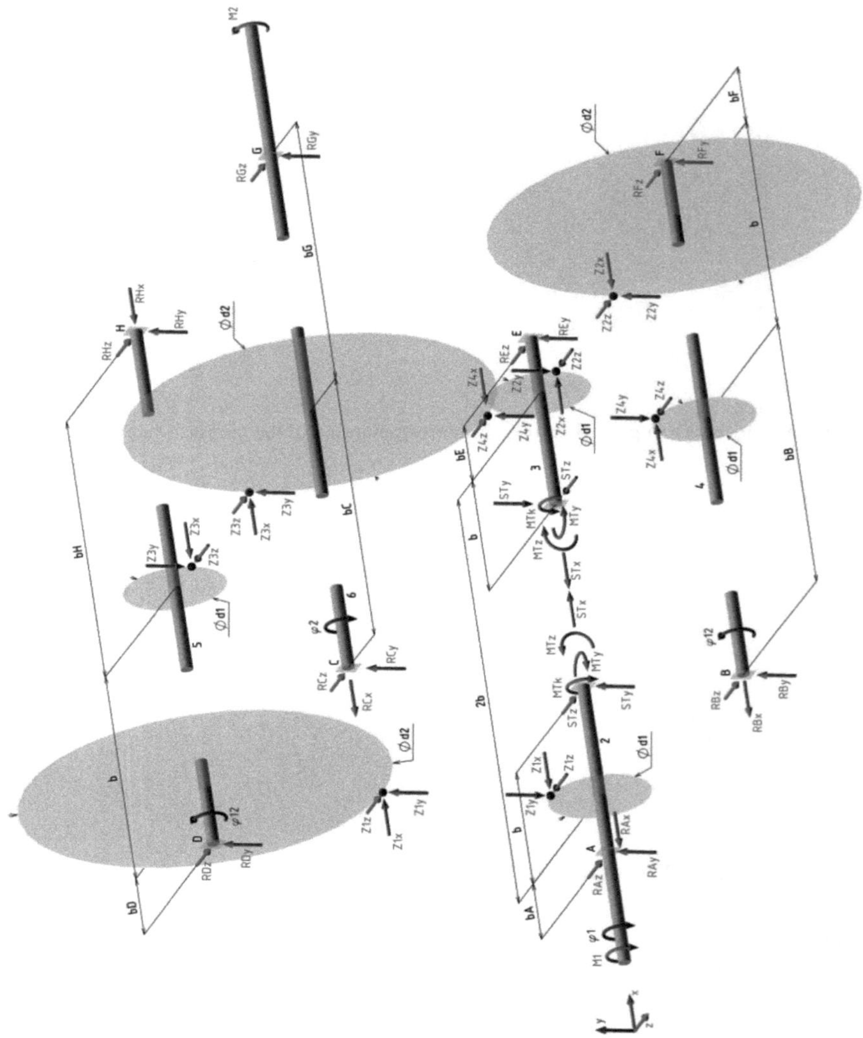

Fig. 3 External forces and forces on bearings, gear meshes and within torsion-bar spring

3 External and Internal Forces

3.1 Static Equilibrium

The gearing on Fig. 1a is a combined six-member mechanism (including frame). On Fig. 3 is release of each member. It should be determine 38 unknown forces and torques, torques *Mtk* (torsion-bar spring preload) a *M*2 (load on output shaft) are parameters of equation system. There are 30 equilibrium equations, the remaining

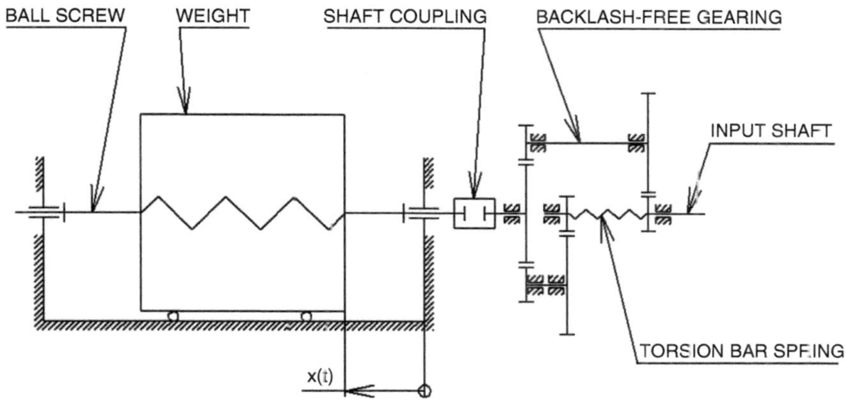

Fig. 4 Dynamic model

8 equations resulting from the geometry of toothing (4 meshes, for each one 2 equations).

3.2 *Dynamic Analysis*

For the dynamic analysis was created model by Fig. 4. The gear unit is connected by a ball screw with a weight, which is constrained by a ball linear guide. The input of model is known desired kinematics of weight. Subsequently, the reaction forces within the gearbox and input torque course are calculated. Mainly are controlled forces on meshes of pinions. To correct function of gearbox, they must be positive at all events.

For dynamic solution is necessary to know geometric characteristics as well as characteristics of mass. Weight mass was chosen of 100 kg. Inertia of gears, shafts and ball screw were determined from the 3D model in SolidWorks environment. For computing was created universal program in the Maple environment. All parameters are a variable, so the algorithm is applicable to any gear unit of the same design (Figs. 5, 6, and 7).

4 Conclusions

Presented procedure is applicable for optimizing the dimensions of the gearboxes with same design. Created algorithm is universal, all parameters are implemented as a variable. It, therefore, is useful for e.g. design of gearbox range. In the dynamic calculations can be any reaction force expressed as a function of time and therefore

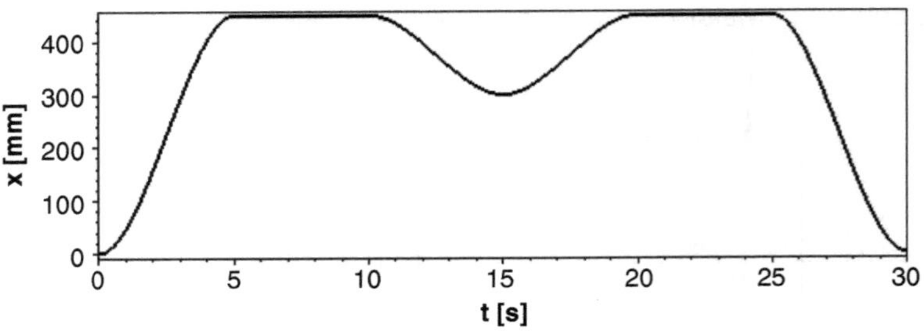

Fig. 5 Time dependence of weight position

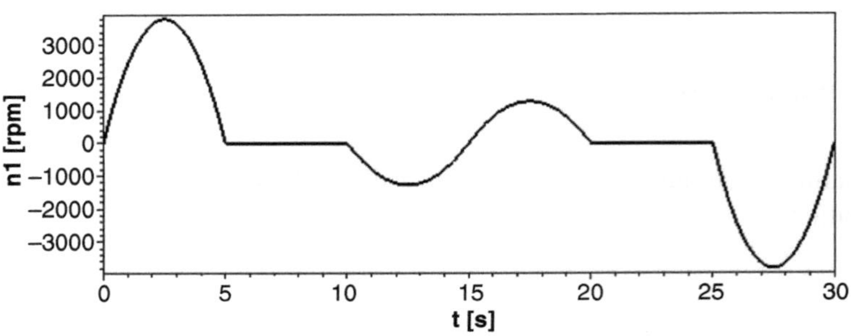

Fig. 6 Time dependence of gearbox input shaft speed

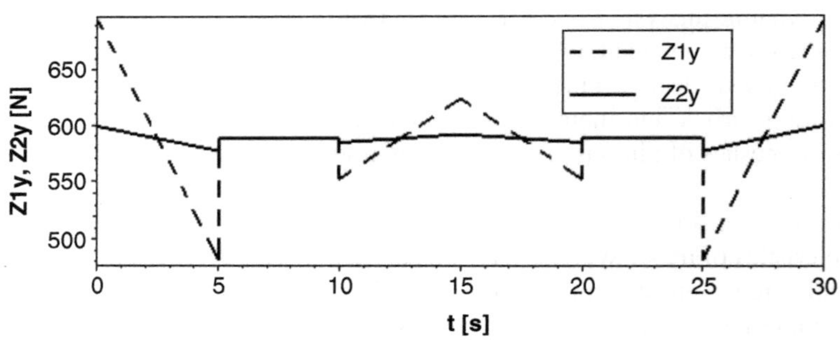

Fig. 7 Time dependence of reaction forces on pinion meshes

the results are important for monitoring fatigue of material of cyclically loaded parts. Kinematics of output motion is also freely modifiable.

Acknowledgements This work was supported by MPO ČR, research project TIP, no. FR–TI1/594, Research of sophisticated methods of design and development of special-purpose machines, components and peripherals of industry machines.

References

1. L.C. Hale, A.H. Slocum, Design of anti-backlash transmissions for precision position control systems. Precis. Eng. **16**(4), 244–258 (1994)
2. K. Baumgarten, K. Schloeglmann, Mechanical Gear Drive, U.S. Patent 4,953,417, 4 Sept 1990
3. C.L. Hannel, Anti-Backlash Gear Assembly, U.S. Patent 4,805,475, 21 Feb 1989

The Effect of Fits Stiffness on Distribution of Internal Loading in Rolling Bearings

R. Kohar and S. Hrcek

Abstract This article discusses the effect of fits stiffness on distribution of internal loading in rolling bearings. Investigating effect of stiffness was performed by using the finite element method, because with the analytic method, this effect is impossible to include in the load distribution calculations on body parts, which transmit load in rolling bearings.

Keywords Rolling bearings • Distribution of internal loading • Rolling elements • Finite elements method

1 Introduction

Calculation of the forces distribution in roller bearing operating at the outer radial and axial loads is very important, because maximum rolling element load is one of the determining factors for the static case, as well as to calculate the dynamic load rating of the rotating bearing.

Because the bearings are usually loaded with more than two rolling elements, the structure is statically indeterminate and distribution of internal loading of rolling elements is dependent on the elastic deformation of the contact points of rolling elements with the raceways of rings.

Calculation of load distribution on the rolling elements from the external forces is based on the known radial internal clearance or preload, which is in rolling bearings. From the values of radial internal clearance is determined, respectively load-zone calculated, which is characterized by the load-zone parameter ε (Fig. 1).

The procedure of calculation the load zones was defined by Stribeck already in 1900 [1]. Load-zone parameter is calculated in an iterative way, and compares the estimated value of the load-zone parameter ε with a calculated parameter ε' if their difference is minimal.

From the load-zone parameter ε is calculated the $J_r(\varepsilon)$ integral according to Eq. (1) – needed to calculate the maximum rolling element load Q_{max}.

R. Kohar (✉) • S. Hrcek
University of Zilina, Žilina, Slovak Republic
e-mail: robert.kohar@fstroj.uniza.sk; slavomir.hrcek@fstroj.uniza.sk

V. Dynybyl et al. (eds.), *The Latest Methods of Construction Design*,
DOI 10.1007/978-3-319-22762-7_35

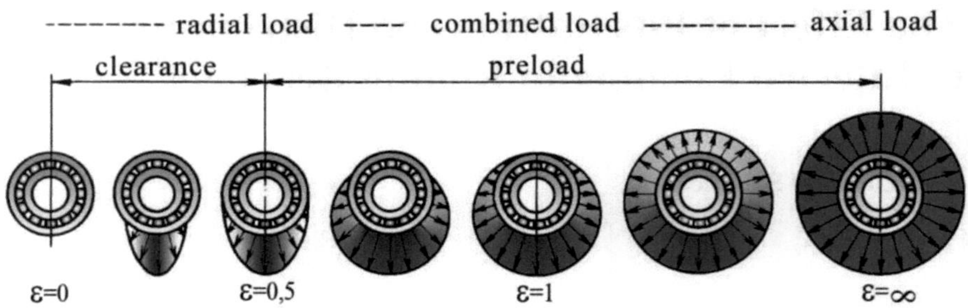

Fig. 1 Load zone of rolling bearings

$$J_r(\varepsilon) = \frac{1}{2\pi} \int_{-\psi_1}^{+\psi_1} \left[1 - \frac{1}{2\varepsilon}(1 - \cos\psi)\right]^n \cos\psi \, d\psi \tag{1}$$

Most loaded rolling element is calculated from the radial force applied to the bearing F_r, the number of rolling elements Z, from the values $J_r(\varepsilon)$ radial integral and contact angle α according to Eq. (2):

$$Q_{max} = \frac{F_r}{Z\, J_r(\varepsilon) \cos\alpha} \tag{2}$$

Other rolling elements load is calculated according to Eq. (3):

$$Q_\psi = Q_{max}\left[1 - \frac{1}{2\,\varepsilon}(1 - \cos\psi)\right]^n \tag{3}$$

while n is load deflection exponent ($n = 3/2 = 1{,}5$ for ball bearings and $n = 10/9 \cong 1{,}11$ for roller bearings) and ψ is azimuth angle [2].

This method of calculating distributions of internal loading on the rolling elements does not reflect the fits stiffness of bearing rings in the housing and the shaft, so the calculation is approximate and idealized. If there is the need to include all the effects in the calculation of distribution of internal loading in loaded bearing, it is necessary to take into account the stiffness of the shaft and housing and fits tolerances.

2 Model of Bearing

For the simulation of load distribution was used cylindrical roller bearing NU206 with parameters: diameter of the inner ring 30 mm, diameter of the outer ring 62 mm, width 16 mm, number of rollers 12, normal radial clearance (20–45 μm).

Fig. 2 Standard model of bearing housing used in bearing units by bearing manufacturers

Fits of the inner ring on shaft was considered standard with an interference, i.e. shaft ∅ 30 k5 and fits of the outer ring in the housing standard with the clearance, i.e. ∅ 62 H7. The shaft was considered as solid model, for housing was used standard bearing unit from leading global bearing company (Fig. 2).

On this model of bearing housing that has different cross-section stiffness, the finite elements method shows what impact has this cross-sectional stiffness on the distribution of internal loading acting in different directions to the rolling elements.

3 Static Structural Analysis in ANSYS/Workbench

ANSYS/Workbench was used for the simulation of distributions of internal loading in loaded cylindrical roller bearing NU206. 3D model of the entire body and the bearing arrangement (Fig. 3) was created in the CAD system PTC/Creo 2 and then imported into ANSYS/Workbench.

Static structural analysis was defined in software environment, the 3D model was defined in half symmetry passing through the center and perpendicular to the axis of the bearing. Between the rolling elements and raceways were defined frictionless contacts ignoring friction, fits of bearing rings on the shaft and housing was defined by the same type of contact. The model was meshed using standard finite elements of ANSYS library, for the volume was used Solid185 element and a pair of contact elements Conta174 and Targe170. On the lower surface of the housing was defined displacement simulating the fixation of housing [3]. Radial load was defined to the center of the shaft that was affecting the bearing at different angles from the vertical plane passing through the axis of the bearing (Fig. 4).

Fig. 3 Model of simulated cylindrical roller bearing NU 206 with seating

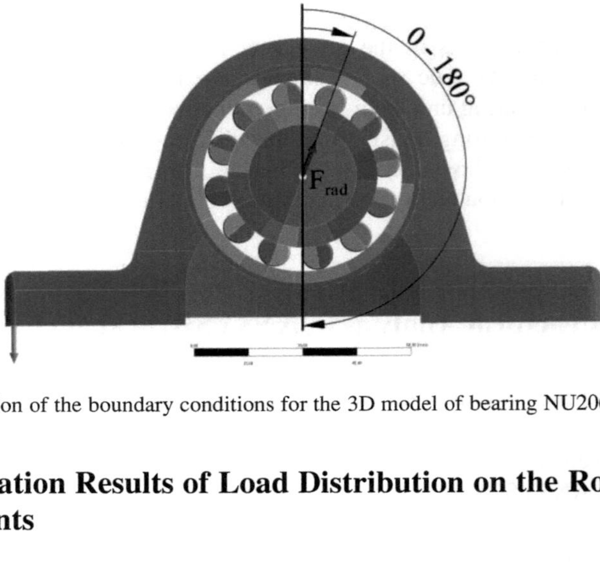

Fig. 4 Definition of the boundary conditions for the 3D model of bearing NU206 arrangement

4 Calculation Results of Load Distribution on the Rolling Elements

To the FEM analysis was included the load distribution size of radial internal clearance of the bearing NU206, fits stiffness, effect of the interference size of inner ring on the shaft and the effect of the clearance size of the outer ring in the housing. Figure 5 shows distributions of internal loading on individual rolling elements depending on the direction in which bearing carries the radial load.

From the figure it is clear that the load distribution affects the cross-section stiffness of housing, moreover, if the force is applied at an angle of 60–100° from the vertical plane, the maximum rolling element load is not within this direction but on the side. Distributions of internal loading, which is shown in Fig. 5 is for the most favorable effect of overall fits, when the radial internal clearance is minimal, interference on the shaft is minimal and the clearance of the outer ring in the housing is zero.

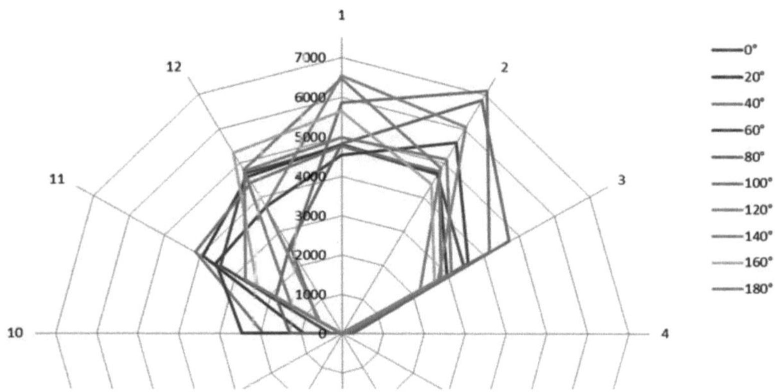

Fig. 5 Load distributions on the individual elements based on the direction of radial force

Figure 6 shows the dependence and the range of the maximum rolling element load depending on the direction of radial force. Boundary of the range of load is depending from the size of the bearing radial clearance and min. – max. values in the fits of bearing rings on the shaft and housing. Furthermore, Fig. 6 displays the maximum rolling element load values calculated analytically based on Stribeck elements of the maximum and minimum radial clearance. As can be seen, analytical calculation of load does not include any effect by the fits or stiffness of the individual components.

Figure 7 shows the correlation between the cross-section stiffness of housing and maximum rolling element load, for the most favorable effect of fits [4]. This dependence clearly proves that, the fits stiffness greatly affects distributions of internal loading and on this basis the fatigue life of rolling bearings.

5 Conclusion

Analysis results have shown a connection between distributions of internal loading in loaded bearing, parameters of fits and stiffness. In today's competitive environment it is important to design structures, resp. machine elements with high efficiency and with controlled or as closely as possible service life. The established procedure to calculate the nominal life of rolling bearings as the basic machine elements, which describes ISO 281 is only to some of these requirements sufficient. Therefore today's usage of modern computing and simulation methods can accurately calculate the forces acting in roller bearings and thus accurately determine their durability.

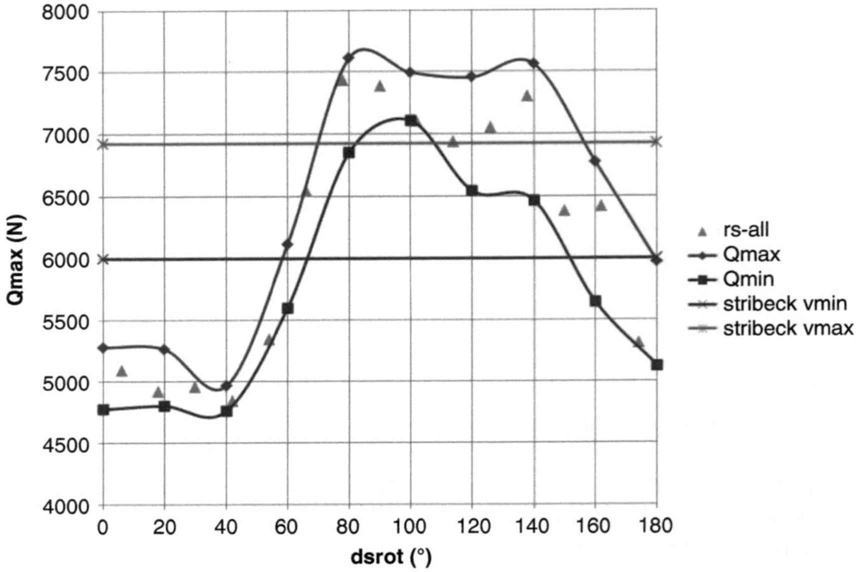

Fig. 6 Dependence of the size of the most loaded element on direction of radial force

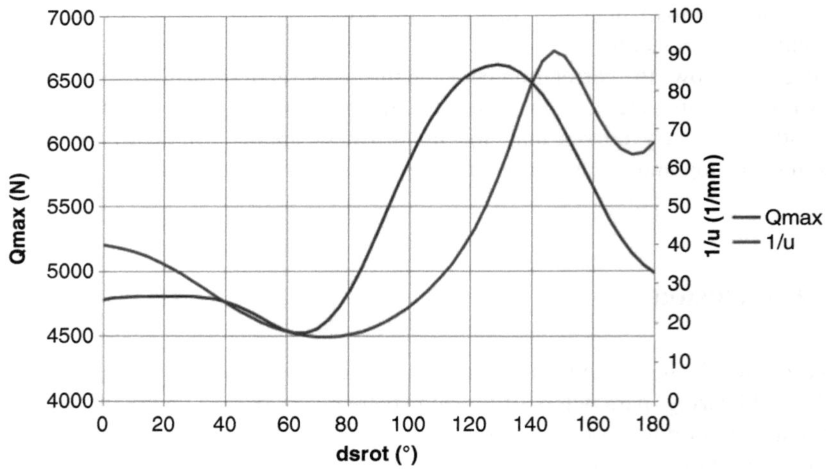

Fig. 7 Correlation between the cross-section stiffness of housing and max. rolling element load

Acknowledgments This work was supported by the Slovak Research and Development Agency under the contract no. APVV-0419-11 – Adaptation of modern computer-simulation methods to the development of rolling bearings and their verification in real conditions.

This work was supported by the Scientific Grant Agency of the Ministry of Education, Science, Research and Sport of the Slovak Republic under the contract no. V-1/0396/14 – Research of the influence of construction and technological parameters of rolling bearings on the service life thereof.

This work was supported by the Scientific Grant Agency of the Ministry of Education, Science, Research and Sport of the Slovak Republic under the contract no. V-1/0844/13 – Research on guidance elements of rolling bearing and their design.

References

1. F.B. Oswald, E.V. Zaretsky, Effect of internal clearance on load distribution and life of radially loaded ball and roller bearings, NASA/TM-2012-217115 (2012)
2. A.T. Harris, M.N. Kotzalas, *Advanced Concepts of Bearing Technology*, 5th edn. (CRC Press, Boca Raton, 2007). ISBN 0-8493-7182-1
3. J. Brònček, M. Dzimko, B. Hadzima, Y. Takeichi, Experimental investigations of aluminium alloys 2024 -T 3 form in terms of tribocorrosion characteristics. Komunikacie **20**(1), 97–104 (2014)
4. L. Kucera, M. Lukac, L. Jurak, F. Brumercik, Hydromechanical automatic transmission. Komunikacie **11**(2), 33–35 (2009)

Design Calculation of Spur Gear Module

E. Kronerova

Abstract Spur gearset has been designed and verified using three software programs and an analytical approach. The programs used in the experiment included Inventor, KISSsoft and MITCalc. By comparing the outcomes achieved with identical input data, one can assess the suitability and potential of the software programs for industrial and educational uses.

Keywords Module • Spur gear • Dimension

1 Introduction

This paper was inspired by the need for comparing gear module design calculations according to Bach formula and the Czech Standard ČSN 01 4686. Eventually, the comparison was extended to include several software programs as well. The programs selected for comparing the module calculations were those available at the Department of Machine Design of University of West Bohemia: Inventor, KISSsoft and MITCalc.

2 Design Calculation of Module

2.1 Calculation Formulas

The input data for Bach formula calculation include the following [1]:

Power	P [W]
Speed (revolutions)	n [s^{-1}]
Number of teeth of the smaller gear (pinion)	z_1 [−]
Helix angle	β [°]
Permitted flexural stress coeff. for the material	$c = (0.03–0.08) \cdot \sigma_{Do}$ [MPa]
Gear width coefficient	$\psi = b/m = 10–30$ [−]

E. Kronerova (✉)
University of West Bohemia, Plzeň, Czech Republic
e-mail: kronero@kks.zcu.cz

V. Dynybyl et al. (eds.), *The Latest Methods of Construction Design*,
DOI 10.1007/978-3-319-22762-7_36

$$m = 8.6(7.5) \cdot \sqrt[3]{\frac{M_t \cdot \cos\beta}{c \cdot \psi \cdot z_1}} = 8.6(7.5) \cdot \sqrt[3]{\frac{\frac{P}{2 \cdot \pi \cdot n} \cdot \cos\beta}{c \cdot \psi \cdot z_1}} \quad (1)$$

The input data for the calculation according to Czech Standard ČSN 01 4686 are as follows:

Power	P [W]
Speed (revolutions)	n [s-1]
Number of teeth of the smaller gear (pinion)	z_1 [−]
Coefficient of external dynamic forces	K_A [−]
Coeff. for non-uniform load on teeth across the width in contact	$K_{H\beta}$ [−]
Flexural fatigue strength of pinion material	σ_{Flim1}[MPa]
Gear width coefficient	$\psi_F = b_{WF}/m$ [−]
Active gear width for flexure calculation	b_{WF} [mm]

$$m = 18 \cdot \sqrt[3]{\frac{K_F \cdot M_{t1}}{\sigma_{FP} \cdot \psi_F \cdot z_1}} = 18 \cdot \sqrt[3]{\frac{K_A \cdot K_{H\beta} \cdot \frac{P}{2 \cdot \pi \cdot n}}{\sigma_{Flim1} \cdot 0.6 \cdot \psi_F \cdot z_1}}, \quad (2)$$

2.2 *Effect of Material on Module*

As illustrated in Fig. 1, the dependence of the module on material is dictated by the calculation formula. With both formulas, however, one can obtain the same module even with materials with lower permitted stresses, provided that the material and gear width are selected appropriately. This is more readily apparent in the case of the Bach formula.

2.3 *Effect of Gear Width on the Module*

In the module calculation formulas, the gear width is characterized by means of the gear width coefficient. As the figures show, the module decreases with increasing gear width. With materials that exhibit higher permitted stresses the module varies less (Fig. 2).

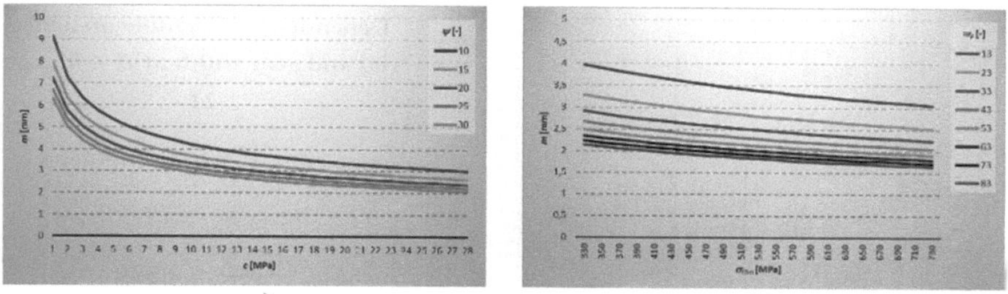

a) b)

Fig. 1 Dependence of the module of the gear material at (**a**) $z_1 = 21$, $\beta = 15°$ and (**b**) $z_1 = 21$, $K_A = 1.625$, $K_{H\beta} = 1.25$

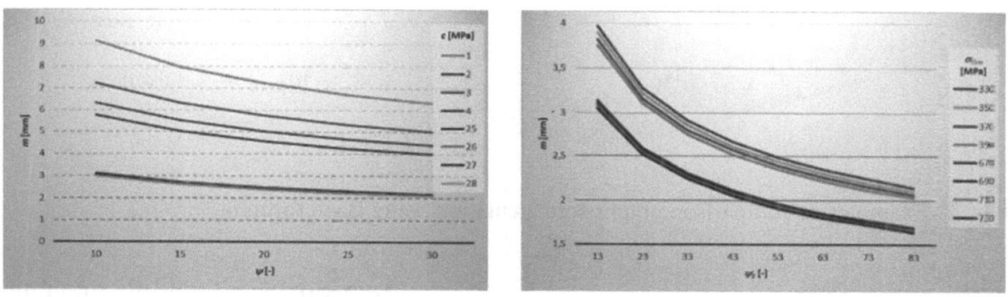

Fig. 2 Dependence of modulus on gear width at (**a**) $z_1 = 21$, $\beta = 15°$ and (**b**) $z_1 = 21$, $K_A = 1.625$, $K_{H\beta} = 1.25$

2.4 General Comparison of Dependence of Modulus on Material and Gear Width at Various Numbers of Teeth and Helix Angles

The comparison of the dependence of module on the number of teeth and the helix angle shown in the graphs suggests that with increasing number of teeth and increasing helix angle the module remains within a narrower interval across the entire range of materials and gear widths (Fig. 3). Individual ranges of module magnitudes are shown in various colours on the surface (Fig. 4).

It was also found that the module increases with the increase in the coefficient of external dynamic forces K_A [−] and the coefficient for non-uniform load on teeth across the width in contact $K_{H\beta}$ [−]. Even with increasing number of teeth, the module remains constant across a wider range of materials and gear widths.

In both calculation methods, the resulting module value remains constant across a wider range of materials and gear widths despite the increasing number of teeth. The module distribution is finer in the Bach formula [2].

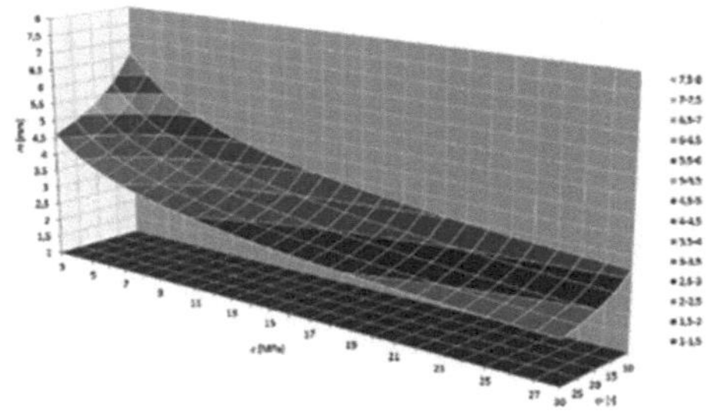

Dependence of module on material and gear width at $\beta = 0°$, $z_1 = 21$

$$m_{max} = 6.617\,mm$$

$$m_{min} = 2.179\,mm$$

$$\frac{m_{min}}{m_{max}} = 0.329$$

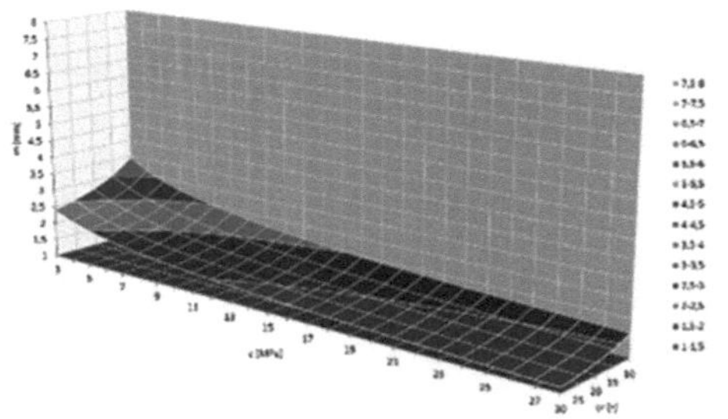

Dependence of module on material and gear width at $\beta = 0°$, $z_1 = 150$

$$m_{max} = 3.436\text{ mm}$$

$$m_{min} = 1.132\text{ mm}$$

$$\frac{m_{min}}{m_{max}} = 0.329$$

Fig. 3 Examples of comparison graphs for calculation using Bach formula

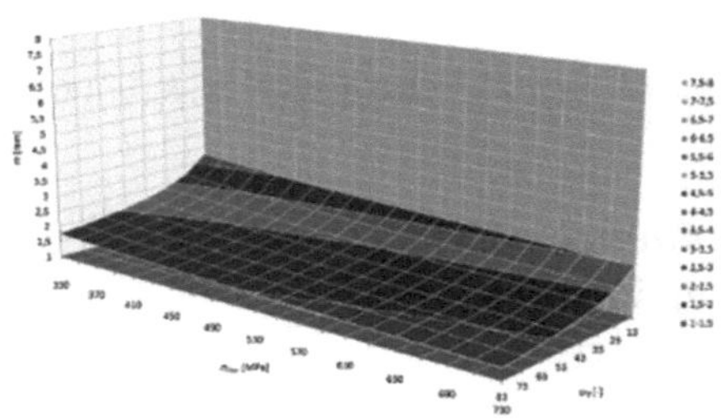

Dependence of module on material and gear width at $z_1 = 21$

$$m_{max} = 3.22\text{ mm}$$

$$m_{min} = 1.33\text{ mm}$$

$$\frac{m_{min}}{m_{max}} = 0.413\,mm$$

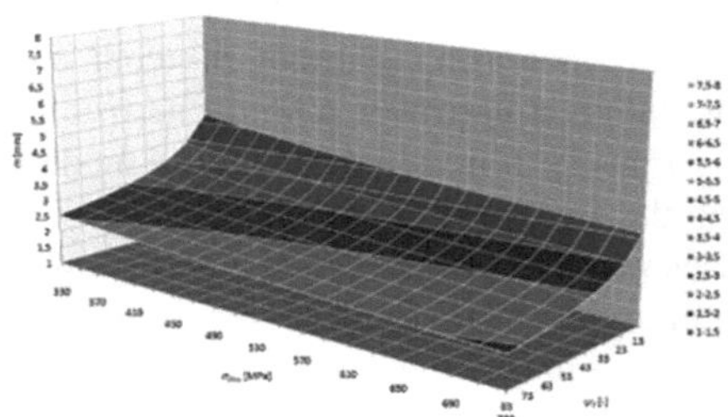

Dependence of module on material and gear width at $z_1 = 21$, $K_A = 2.25$, $K_{H\beta} = 1.50$

$$m_{max} = 4.71\,mm$$

$$m_{min} = 1.95\,mm$$

$$\frac{m_{min}}{m_{max}} = 0.414$$

Fig. 4 Examples of comparison graphs for calculation using the Czech Standard ČSN 01 4686

Table 1 Comparison between calculated module values

Software/calculation	Calculation type	Number of teeth	m_{Bach} [mm]	m [mm]
Analytical calculation	Strength-based	21	2.63	2.19
Analytical calculation	Geometry-based		–	2.96
Inventor			3	3
KISSsoft			–	3
MITCalc			–	3.5

3 Comparison Between Module Calculations Using Different Software Programs

The analytical calculation of gear module is based on strength calculation involving the flexural stress and contact load on the tooth. The computation using software programs (as indicated in Table 1) is based on the gear geometry. It requires the knowledge of centre-to-centre distances, the gear ratio and the formula for module calculation [3].

$$m = \frac{2 \cdot a_w \cdot \cos\beta}{43 + 14 \cdot i} \tag{3}$$

The comparison between outcomes of the software programs reveals that the Inventor package offers a single module value for a particular set of input data. KISSsoft calculates a module for certain ranges of the number of teeth and centre-to-centre distances and other parameters, yielding a range of module values. MITCalc yields a single module value, which is the highest of the results of all three software programs. Consequently, the gearset calculated using MITCalc will be the largest one for a particular set of parameters.

All software programs also perform strength calculations as part of this computation. The results thus include flexural and contact stress safety coefficients for the given gearset parameters.

4 Conclusion

Analytical calculations using Bach formula and the procedure according to Czech Standard ČSN 01 4686 were compared. The calculation according to the standard takes into account more factors affecting the module. Both calculations were also compared with calculations using software programs. The software calculation relies on the gearset geometry. The Bach formula calculation is comparable with the gearset geometry-based calculation.

Acknowledgments This contribution was created under the project no. SGS-2013-050.

References

1. R. Kříž et al., *Strojírenská příručka*, vol. 5 (Scientia, Praha, 1994), p. 241. ISBN 80-85827-59-X
2. J. Krátký, E. Krónerová, *Obecné strojní části 2* (Publishing House of University of West Bohemia, Plzeň, 2011). ISBN 978-80-261-0066-9
3. Czech Standard ČSN 01 4686-4, *Pevnostní výpočet čelních a kuželových ozubených kol.: Návrh a zjednodušený kontrolní výpočet čelních ozubených kol* (Úřad pro technickou normalizaci, metrologii a státní zkušebnictví, Praha, 1989). Available at https://csnonline.unmz.cz/Vysledky.aspx. Accessed 28 May 2014

Application of the Bond Graphs to the Description of Technical Systems

Z. Matuszak

Abstract The basic information about bonds graphs is summarized. Basic models of components used in the description of technical systems with bonds graphs are characterized. Selected elements of mechanical and energy are described with bonds graphs. Discussion on the nature of the diagnostic parameters described by graphs is conducted. The analysis of bonds graphs suitability as a source of diagnostic information is made. Ways to use the information provided by bonds graphs, especially for mechanical power generation equipment and their components, are suggested.

Keywords Bond graphs • Technical systems

1 Introduction

Graphical model by the method of the bond graph reflects the structure of the dynamic object and can be modified and used for the creation and verification of a mathematical model in the form of equations of state. The equations of state can be installed even without the use of a graphical model. The advantages of the bond graph method consists in the elimination of errors and much easier subsequent modification of the model.

From the standpoint of the modeling graph is a concise method of recording elements connections of elements or interactions between the elements of the system, and hence the structure of the system. One of the major questions relating to the modeling of bond graphs is the maximum simplification of the procedure for formulating models. Therefore, the information about the characteristics of system elements can be expressed graphically or analytically, as will be presented later in this paper.

Graphical model in the form of bond graph shows in a clear form of the dynamic structure of the object and can be easily modified and can be used to create and verify the mathematical model in the form of equations of state. The essence of the bond graph method is subordinate identical mathematical formulas and symbols

Z. Matuszak (✉)
Maritime University of Szczecin, Szczecin, Poland
e-mail: z.matuszak@am.szczecin.pl

V. Dynybyl et al. (eds.), *The Latest Methods of Construction Design*,
DOI 10.1007/978-3-319-22762-7_37

Table 1 Parameters of energy for the various forms of energy [8]

System	Generalized potential e	Generalized flow f	Generalized resistance R
F / v progressive mechanical	Force F [N]	Speed v [m/s]	F/v [Ns/m]
M / w mechanical rotary	Torque M [Nm]	Angular velocity ω [rad/s]	M/ω [Nms/rad]
p / $\dot{Q}$ hydraulic	Pressure p [Pa]	Volume flow $\dot{Q}$ [m^3/s]	$\Delta p/\dot{Q}$ [Pa/m^3]
u / i electric	Voltage U [V]	Current intensity I [A]	R [Ω]
T / $\dot{S}$ thermodynamic heat exchange	Temperature T [K]	Entropy flow $\dot{S}$ [J/Ks]	–
i / $\dot{m}$ thermodynamic mass exchange	Appropriate entropy mass flow i [J/kg]	Mass flow $\dot{m}$ [kg/s]	–
p / $\dot{V}$ thermodynamic process	Pressure p [Pa]	Rate of change of volume $\dot{V}$ [m^3/s]	–

contractual elements of the system, regardless of their physical nature. It is a concept based on the publication Pyntera of 1961 [1–5]. It was later extended by Karnoppa, Rosenberg and Margolis [6, 7]. Although this method has a limited field of efficient use of its rapid growth was observed.

2 Application of the Bond Graphs

Energy parameters for the various forms of energy are shown in Tables 1 and 2. Selected forms of energy are assigned to the basic parameters of generalized flow and a generalized capability in the form of physical quantities. Characteristic forms of energy associated with heat transfer and thermodynamics is no physical counterparts generalized momentum p generalized displacement q.

Diagram of the compressed air system in the engine room of ship is shown in Fig. 1. On this basis the functional structure of the compressed air system was made (Fig. 2).

Modeling of physical systems is a common approach to modeling energy processes regardless of their physical nature. Sources of energy storage elements, processing elements energy dissipation elements of energy are elements of a model in which energy processes is occurred. Rules for the preparation of models of systems described graphs bond include procedures include in [6, 8]. Figure 3

Table 2 Parameters of energy for the various forms of energy [8]

System	Accumulation of the potential L	Accumulation of the flow C	Generalized momentum p	Generalized displacement q
F / v progressive mechanical	Mass m [kg]	Resilience $\Delta l/F = 1/k$ [m/N]	Momentum p [kgm/s]	Displacement s [m]
M / w mechanical rotary	Moment of inertia J [kgm^2]	Momentum $\Delta\alpha/M$ [rad/Nm]	Angular momentum h [kgm^2/s]	Angle α [rad]
p / $\dot{Q}$ hydraulic	Inertia unit m/f [kg/m^2]	Surface f [m^2]	Momentum unit Γ [kg/sm]	Volume V [m^3]
u / i electric	L [H]	C [F]	–	Q [C]

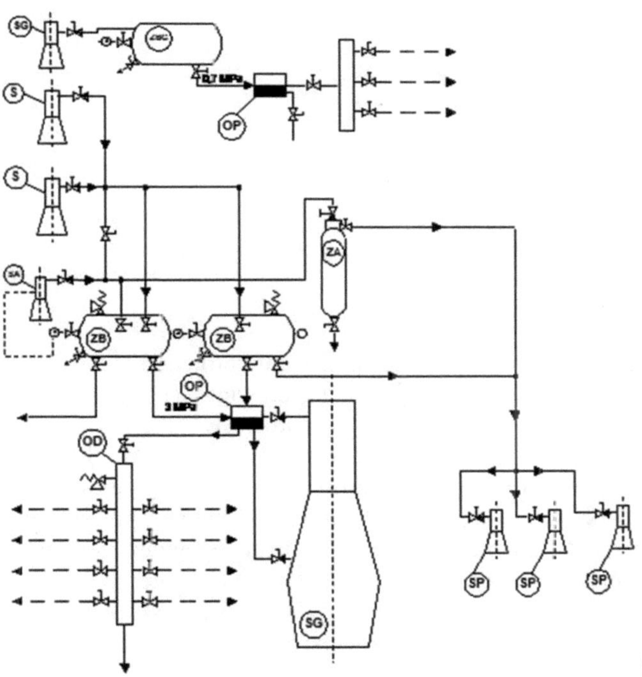

Fig. 1 Simplified diagram of the compressed air system in the ship's: *SG* – main engine; *SP* – auxiliary engines; *S* – compressor 30 bar; *SA* – emergency compressor; *SG* – economic compressor; *ZB* – starting air tanks; *ZA* – tank (emergency) starting air auxiliary engines; *ZBG* – economic air tank; *OP* – dehumidifiers

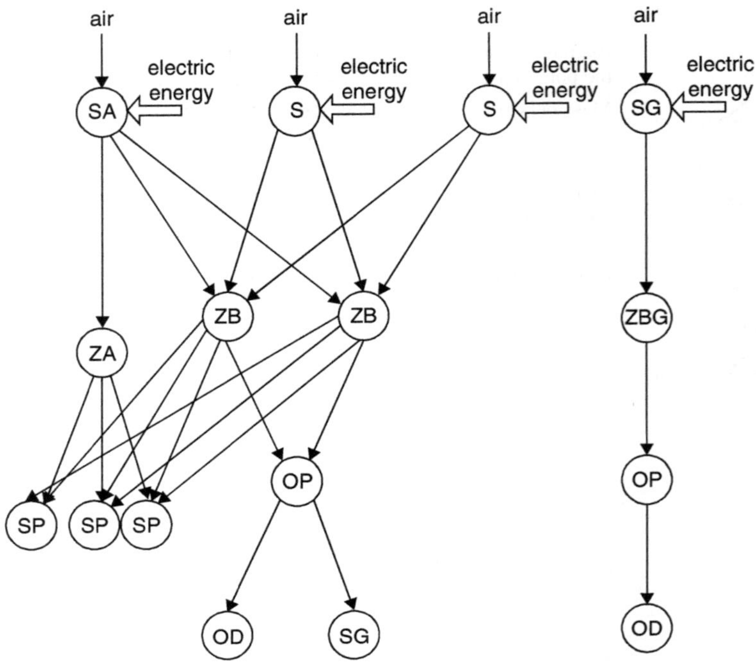

Fig. 2 Simplified diagram of the compressed air system in the ship's: *SG* – main engine; *SP* – auxiliary engines; *S* – compressor 30 *n C*; *SA* – emergency compressor; *SG* – Economic compressor; *ZB* – starting air tanks; *ZA* – tank (emergency) starting air auxiliary engines; *ZBG* – economic air tank; *OP* – dehumidifiers

shows graphs of compressed air bonding, and the structure of the diagram shown in the previous drawings. As shown on the drawings and principles of preparing graphs the symbols mean: SF – energy source; SE – the electric motor; SPRZ – Clutch; SA – emergency compressor; S – Main compressor; SG – air compressor economic; ZA – tank (emergency) starting air auxiliary engines; ZB – the main air tanks; ZBG – economic air tank; SP – auxiliary engines, power generation; OP – dehumidifiers; SG – the main engine; OD – load; ω _ angular velocity; M – torque; R_{SE} – mechanical losses of the electric motor; R_{ST} – material losses of the electric motor; RT – friction losses; and – entropy appropriate; $\dot{m}$_ mass flow; R_C – losses due to cooling of the compressor cylinders; R_{CM} – losses due to intercooling; R_{ZB} – loss of air in the tank; W_d – calorific value of the fuel; G_e – second fuel consumption; R_{CI} – heat loss; R_M – mechanical losses; R_{OP} – losses in the **dehumidifier**.

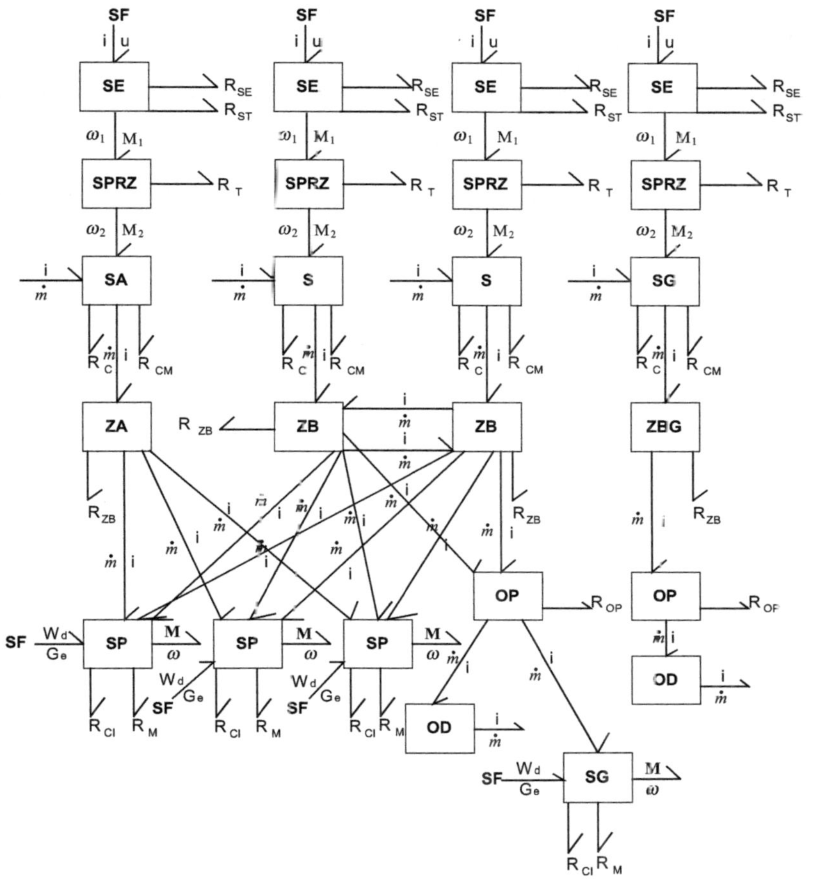

Fig. 3 Bond graphs compressed air system (described in text)

3 Conclusion

Modeling method using graphs and equations of state bond is particularly useful in situations where modeled object of energy contains elements of different physical nature. This allows use of unified modeling apparatus for modeling mechanical, hydraulic, electrical, and thermal elements. With this modified method, graphic model is particularly easy when simplifying both the model and at its advancement.

References

1. R. Diestel, *Graph Theory*, Electronic edn. (Springer, New York, 2000)
2. B. Korzan, *Elementy teorii grafów i sieci* (WNT, Warszawa, 1978)

3. N. Deo, *Teoria grafów i jej zastosowania w technice i informatyce* (PWN, Warszawa, 1980)
4. R. Wilson, *Wprowadzenie do teorii grafów* (Wydawnictwo Naukowe PWN, Warszawa, 1998)
5. www.bondgraphs.com. vid. 22 Mar 2014
6. K. Arczewski, Z. Goraj, J. Pietrucha, *Elementy modelowania w mechanice* (WPW, Warszawa, 1983)
7. R. David, H. Alla, *Petri Nets and Grafcet-Tools for Modelling Discrete Event Systems* (Prentice Hall, New York, 1992)
8. M. Cichy, *Modelowanie systemów energetycznych* (Wyd. Politechniki Gdańskiej, Gdańsk, 2001)

Assessment of the Validity of Technical Elements of the Complex Structure Using Flow Streams of Energy Factors

Z. Matuszak

Abstract In this paper have been characterized, reported in literature, the validity of measurement items (measures Birnbaum, the potential of improving the reliability, Lambert, ...), mainly relating to the related concepts with more reliability and structures of the more reliability. In this paper, an analysis of the direction and amount of flow streams of energy factors in the systems. The analysis will be performed for the so-called. dynamic functional structures. These are the structures in which during operation These are the structures in which during operation occurs change the direction of flow of streams of energy factors and the changing number of streams of energy factors and the number of working components. On the basis of these changes can extract the individual states of the system. For particular operating conditions shall be tables streams flow of energy factors for functional structure of the system. The validity and importance of the element in the system is evaluated based on the number of streams of energy factors and leave the water runs out of the item. The greater the number of these streams, the element is more responsible for the operation of the system in a given state.

Keywords Measurement validity • The flow of streams • Energy factors

1 Introduction

Historically, the first measure of the validity of proposed Birnbaum. Birnbaum measure of validity of the element c_i at time t is defined as follows [1–3]:

$$I_B^{(i)}(t) = h\big(l_i, \underline{R}(t) - h(0_i)\underline{R}(t)\big), \tag{1}$$

where:

$$h\big(l_i, \underline{R}(t)\big) = h(R_1(t), \ldots, R_{i-1}(t), \cdot, R_{i+1}, \ldots, R_n(t)). \tag{2}$$

Z. Matuszak (✉)
Maritime University of Szczecin, Szczecin, Poland
e-mail: z.matuszak@am.szczecin.pl

V. Dynybyl et al. (eds.), *The Latest Methods of Construction Design*,
DOI 10.1007/978-3-319-22762-7_38

T should be noted that the $h(l_i, \underline{R}(t))$ the reliability of the system the element c_i is fit; $h(0_i, \underline{R}(t))$ the reliability of the system, if c_i it is unsuitable (in time t). Thus, formula (1) determines loss of the reliability of the system caused by damage to the element c_i. It is also to be interpreted as the probability that damage to the element c_i in time t may cause damage to the system or as the probability that at time t the system is in a state for which the element c_i is critical.

Equivalent definition Birnbaum measure is as follows:

$$I_B^{(i)}(t) = \left.\frac{\delta h(R)}{\delta R_i}\right|_{R_j = R_j(t)}. \tag{3}$$

From the above equations it follows that the measure of the validity for element c_i does not depend on the reliability of the element, and only from the structure of the system reliability, time and reliability of the other components.

There are several measures of validity independent from time. Most of them are weighted measure of Birnbaum. The most famous of them is one measure bearing the name of Barlow-Proschan measure [1–3]:

$$I_{B-P}^{(i)} = \int_0^\infty f_i(t) I_B^{(i)}(t)dt, \tag{4}$$

where:

$f_i(t) = F_i'$ is the density of damage of the element c_i,

$$0 \le I_{B-P}^i \le 1 \tag{5}$$

$$\sum_{i=1}^n I_{B-P}^{(i)} = 1 \tag{6}$$

A measure of Barlow and Proschan is equal to the probability that cause damage to the system is the damage to the element c_i. This measure can also be regarded as an average measure of Birnbaum due to $F_i(t)$. A popular variation is the measure according to formula (4), which differs in that the integration is carried out in the interval $[0, t]$. This means that this measure expresses the probability that the system is corrupt to the time t that cause damage to the system will be element c_i. Another measure of validity of the proposed system components Natvig. According to this proposal, a measure of the validity of the element is the loss of the remaining time system is working correctly due to damage to the element in question. If the damage to the system components independently of each other, is a measure of Natvig is [1–3]:

$$I_N^{(i)} = k_N \int R_i(t)(-\ln R_i(t)) I_B^{(i)}(t)dt. \tag{7}$$

k_N factor is necessary to ensure summation $I_N^{(i)}$ to unity.

Another measure of validity, independent of time, is a measure of Bergman:

$$I_E^{(i)} = k_E \int_0^{\infty} t f_i(t) I_B^{(i)}(t) dt \tag{8}$$

whose usefulness is comparable to the measure of the Barlow-Proschan and Natvig

A more general importance is the measure $I_P^{(i)}$ defined as:

$$I_P^{(i)} = k_P \int_0^{\infty} t^P I_B^{(i)}(t) dF_i(t) \tag{9}$$

Measures of Barlow-Proschan and Bergman are particular cases of measure specified in the formula (6) because:

$$I_P^{(i)}\Big|_{p=0} = I_{B-P}^{(i)} \tag{10}$$

$$I_P^{(i)}\Big|_{p=0} = I_E^{(i)} \tag{11}$$

Using the concept of the so-called a critical element of the measure of the validity in the system of Lambert introduced.

In a system with all the elements of the serial structure are critical. In the case of other structures element becomes critical when they damage all the other elements of section unfitness to which the considered element.

The probability that at time t the system is in such a state that these damaged element to the time t is equal to [1–3]:

$$\{h(l_i, \underline{R}(t)) - h(0_i, \underline{R}(t))\} F_i(t) \tag{12}$$

Taking into account the above-defined event, provided that the damaged system time t, a measure of the validity of the Lambert form:

$$I_L^{(i)} = \frac{\{h(l_i, \underline{R}(t)) - h(0_i, \underline{R}(t))\} F_i(t)}{1 - h(\underline{R}(t))} \tag{13}$$

The measure validity described above are the most popular and described in the literature.

2 The Matrix Analysis Flow of Energy Factors

Functional structure (graphs of flow factors) fuel system is shown in Fig. 1 it was drawn up on the basis of the diagram of Fig. 2.

In schemes have been marked measurement points and alarm sensors whose task is to control the proper flow of refrigerant. In order to analyze whether the number of installed measurement points and alarm sensors is sufficient to provide full control of transport fuel is drawn up so-called matrix of flow.

The matrix contain dry the number of columns and rows, how many components of a functional diagram (1–18 for the installation of auxiliary fuel supply engines).

They contain information about the flow of fluid between the individual elements of the installation ("0" means no flow factor, "1" occurrence rate). The matrix also indicated location of detectors and alarms (Table 1).

The validity and importance of the element in the system is evaluated based on the amount of streams of energy factors that flow to and from the element. The greater the number of these streams, the more responsible is the element for the operation of the system in this state. The validity of an element in the structure is described by the total sum of streams of energy factors $I^{KM}(i)$ (*the stream measurement*), which flows to and from the element s_o. The dependence can be described as:

$$I^{KM}(i) = s_i(i) + s_o(i) \tag{14}$$

or for the representation of that measurement by a number from the interval <0, 1>:

$$I^{M}(i) = k_{KM} I^{KM}(i) = k_{KM}[s_i(i) + s_o(i)] \tag{15}$$

where:

$k_{KM} = [\sum_{i=1}^{n} I^{KM}(i)]^{-1}$ – is the factor to ensure summation to unity,

n – is the number of elements in the system.

3 Conclusion

The assumed research procedure is as follows: the choice of complex technical systems with variable functional structure to be analyzed; preparation of functional structures for different operating states; drafting tables of streams of energy factors flow in the functional structures of the system for different operating states of the system; mathematical description of the elements validity on the basis of tables of streams of energy factors flow; developing validity measurements of the system elements based on the streams of energy factors flow; estimation of selected validity

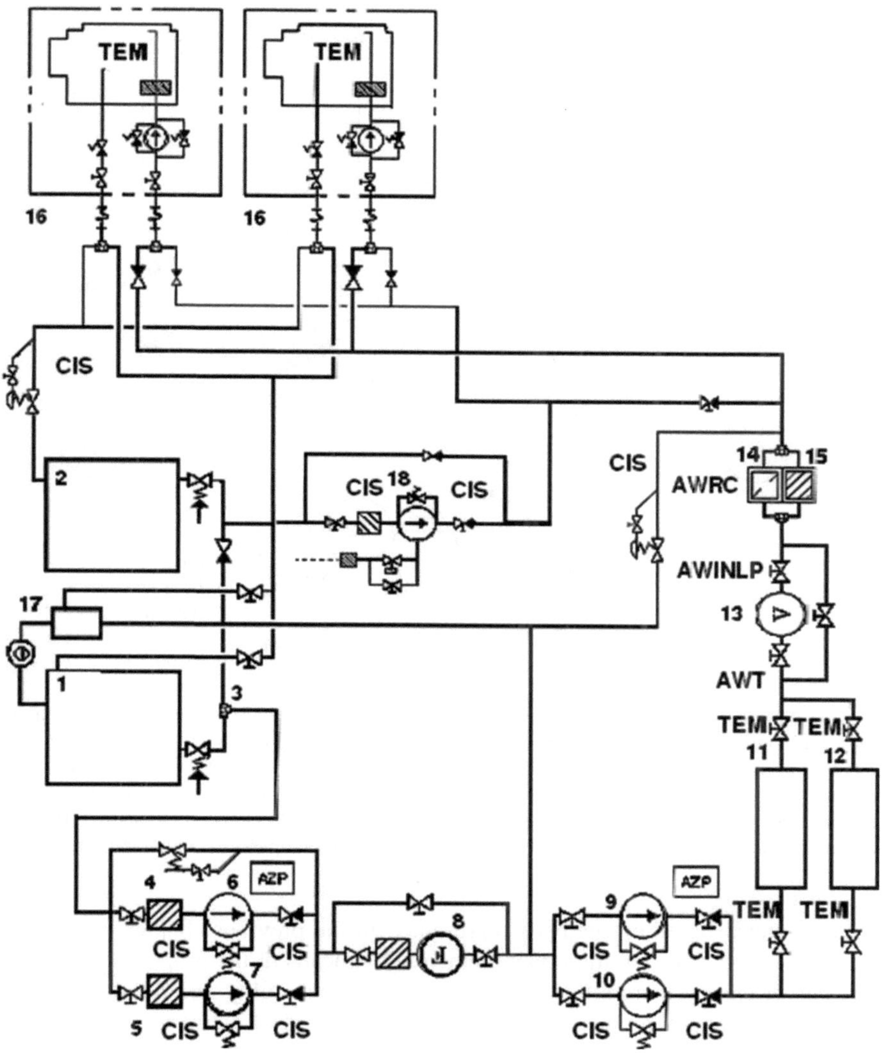

Fig. 1 Fuel supply auxiliary engines – flow chart: *1* – tank of heavy fuel; *2* – tank of light fuel; *3* – three way valve; *4, 5* – filters; *6, 7* – feeding pump; *8* – flowmeter; *9, 10* – circulation pumps; *11, 12* – heaters; *13* – Viscometer; *14, 15* – automatic filter; *16* – auxiliary engines; *17* – tank circulation; *18* – emergency fuel pump light; *TEM* thermometer, *CIS* manometer, *AZP* alarm actuating the backup pump, *AWT* high temperature alarm, *AWINLP* alarm high and low viscosity fuel, *AWRC* alarm high differential pressure across the filter

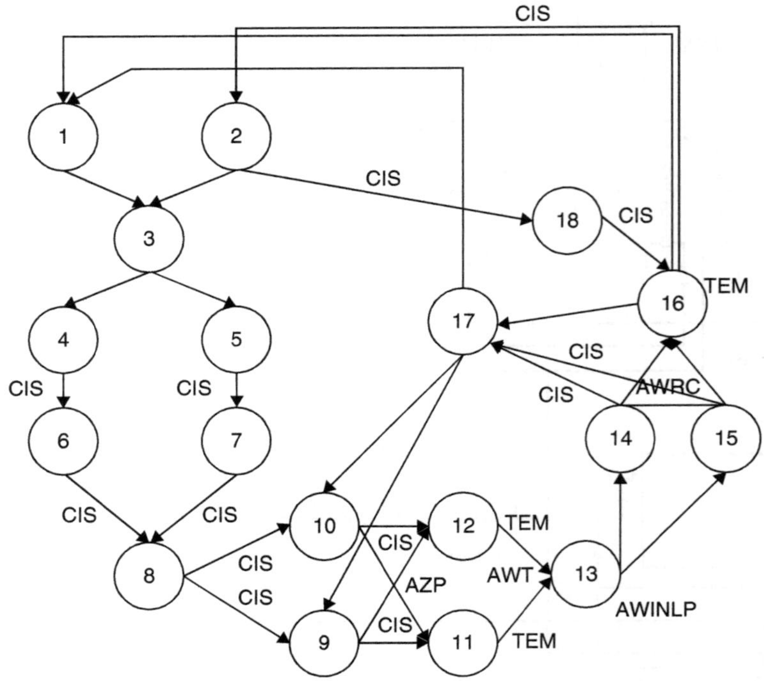

Fig. 2 Diagram of the fuel supply auxiliary engines: *1* – outgoing heavy fuel tank; *2* – outgoing light fuel tank; *3* – three way valve; *4, 5* – filters; *6, 7* – feeding pump; *8* – flowmeter; 9, 10 – circulation pumps; *11, 12* – heaters; *13* – Viscometer; *14, 15* – automatic filter; *16* – auxiliary engines; *17* – tank circulation; *18* – emergency pump of fuel light; *TEM* thermometer, *CIS* manometer, *AZP* alarm actuating the backup pump, *AWT* high temperature alarm, *AWINLP* alarm high and low viscosity fuel, *AWRC* alarm high differential pressure across the filter

measurements based on methods represented in the previous chapter; comparing the results obtained (the important elements) by means of the proposed method of streams of energy factors flow based on elements reliability and reliability structures of the system; an attempt to develop the method of elements validity measurement in a complex technical system, binding the method of streams of energy factors flow with the method based on elements reliability and reliability structures evaluation.

Preliminary results of the research of the described validity measurement, presented in [4–7] give an optimistic outlook on the further development of this method.

Table 1 Flow matrix factor in the installation of auxiliary fuel supply engines

Element	1	2	3	4	5	6	7	8	9	10	11	12	13	14	15	16	17	18
1	–	0	1	0	0	0	0	0	0	0	0	0	0	0	0	0	0	0
2	0	–	1	0	0	0	0	0	0	0	0	0	0	0	0	0	0	CIS
3	0		–	1	1	0	0	0	0	0	0	0	0	0	0	0	0	0
4	0	0	0	–	0	CIS	0	0	0	0	0	0	0	0	0	0	0	0
5	0	0	0	0	–	0	CIS	0	0	0	0	0	0	0	0	0	0	0
6	0	0	0	0	0	–	0	AZP	0	0	0	0	0	0	0	0	0	0
7	0	0	0	0	0	0	–	AZP	0	0	0	0	0	0	0	0	0	0
8	0	0	0	0	0	0	0	–	CIS	CIS	0	0	0	0	0	0	0	0
9	0	0	0	0	0	0	0		–	0	TEM CIS AZP	TEM CIS AZP	0	0	0	0	0	0
10	0	0	0	0	0	0	0	0	0	–	TEM CIS AZP	TEM CIS AZP	0	0	0	0	0	0
11	0	0	0	0	0	0	0	0	0	0	–	0	TEM AWT	0	0	0	0	0
12	0	0	0	0	0	0	0	0	0	0	0	–	TEM AWT	0	0	0	0	0
13	0	0	0	0	0	0	0	0	0	0	0	0	–	AWI NLP	AWI NLP	0	0	0
14	0	0	0	0	0	0	0	0	0	0	0	0	0	–		AWRC	CIS	0
15	0	0	0	0	0	0	0	0	0	0	0	0	0	0	–	AWRC	CIS	0
16	1	CIS	0	0	0	0	0	0	0	0	0	0	0	0	0	–	0	0
17	1	0	0	0	0	0	0	0	1	1	0	0	0	0	0	0	–	0
18	0	0	0	0	0	0	0	0	0	0	0	0	0	0	0	CIS	0	–
Σ	2	1*	2	1	1	1*	1*	2*	1+1*	1+1*	6*	6*	4*	1*	1*	3*	2*	1*

AZP alarm actuating the backup pump, *AWINLP* alarm high and low viscosity fuel, *AWT* high temperature alarm, *AWRC* alarm high differential pressures. Symbols with an asterisk denote the number of points of measurement of critical parameters. If there is no asterisk means the same flow without any measurement

References

1. B. Bergman, On reliability theory and its applications. Scand. J. Stat. **12**, 1–42 (1985)
2. B. Natvig, A suggestion of a new measure of importance of importance of system components. Stoch. Process. Appl. **9**, 319–330 (1979)
3. B. Natvig, New light on measures of importance of system components. Scand. J. Stat. **12**, 43–54 (1985)
4. M. Kołodziejski, Z. Matuszak, *Estimating the Importance of Element in Multicomponent Technical System*. (Computational Mechanics Publications – Marine Technology ODRA'95, Southampton 1995), pp. 525–532
5. M. Kołodziejski, Z. Matuszak, Importance assessment of ship power plant system components. Pol. Marit. Res., No. 2(8). **3**, 27–30 (1996)
6. Z. Matuszak, Estimating the reliability of a systems of fixed and variable operating structure. *Mieżdunarodnyj sbornik naucznych trudow „Nadiożnost i effektiwnost techniczeskich sredstw"*, (Kaliningradskij gosudarstwiennyj techniczeskij uniwersytet, Izdatelstwo KGTU, Kaliningrad, 2004), s. 103–108, (in Polish)
7. Z. Matuszak, Selected methods for assessing reliability of marine power systems. *Mieżdunarodnyj sbornik naucznych trudow "Nadiożnost i effektiwnost techniczeskich sredstw"*. (Kaliningradskij gosudarstwiennyj techniczeskij uniwersytet, Izdatelstwo KGTU, Kaliningrad 2004), s. 109–118, (in Polish)

Influence of Short-Circuit Valve on Hydraulic Recovery of Vehicle Kinetic Energy

Z. Nemec and J. Nevrly

Abstract The present paper deals with one of possibilities how to improve the efficiency of recovery processes using the experimental road roller AMMANN AP 240 H. The recovery of braking kinetic energy can significantly contribute to fuel consumption lowering of heavy horse vehicles working in start-stop regime. The paper presents and evaluates the results of parallel measurements performed on the experimental vehicle.

Keywords Energy • Recovery • Hydraulic • Vehicle • Efficiency

1 Introduction

The researchers from the company Rexroth Bosch suggested supplementing the existing hydraulic circuit of the hydrostatic module with a short-circuit valve, which operates in the regime Decel, i.e. during the vehicle deceleration. Prior to tests on the experimental vehicle, it is desirable to verify the usefulness of suggested modification by means of simulations with the developed mathematical models of vehicle system.

The expected contribution of the short-circuit valve implementation was a relief of hydro-generator operation and improvement of efficiency during the conversion of vehicle kinetic energy into pressure energy in accumulators. Three basic variants of system operation are presented and the simulated time courses of principal hydraulic, mechanic and electric quantities including the efficiency evaluation and the amount of converted energy are introduced as examples for presumable operation conditions.

Z. Nemec (✉) • J. Nevrly
Brno University of Technology, Brno, Czech Republic
e-mail: nemec@fme.vutbr.cz; nevrly@fme.vutbr.cz

V. Dynybyl et al. (eds.), *The Latest Methods of Construction Design*,
DOI 10.1007/978-3-319-22762-7_39

2 Principle of the Solution

The short-circuit valve is connected up between the inlet and outlet of the hydrogenerator and it is controlled by the electric signal from the recovery control unit.

The analysis of the time courses of physical quantities and efficiencies is based on simulations with mathematical models of energy recovery that have been developed and described in the previous research works. Fundamental are the simulations carried out repeatedly with different modifications of recovery conditions [1].

In this study, we deal only with the recovery process of deceleration (Decel) because in other possible processes (Free run and Accel – acceleration of the vehicle) the short-circuit valve is considered to be inactive (it is closed) – see chapters "Design of the Vibration Separatory Apparatus", "Designing of Mixing Equipment, Reactors and Bioreactors", and "Modern Trends in the Drive Wheelsets of Rail Vehicles".

A separate problem which was also solved by means of modeling is timing of control of all nine valves. These determine the mode of vehicle's regime and direction of its movement. It is a sequential control of their switching on and switching off whereas it is necessary to take in consideration dynamical properties and technical claims (e.g. to eliminate hydraulic short circuits, etc.). The time delays of control signals are set in the range of 0–145 ms (Fig. 1) [2].

3 Version "Without the Short-Circuit Valve", Setting of the Valves Changed Immediately

In this case, it is the simplest version, but from the view of operation and effectiveness, it is the least suitable one. Just after releasing the driver pedal (here in time of 8 s), a sequence of control voltage is generated for valves for change to Decel. The end of Decel is derived from derived hydraulic motor (HM) velocity limit at which oil flow between accumulators sinks to zero (here, to 390 rev/min for all three examples) [3].

The described regime with intensive energy wasting has also a negative influence on the recovery process efficiency. The efficiency value of 11.9 % obtained by simulation is considerably lower than that in other Decel versions described in following chapters.

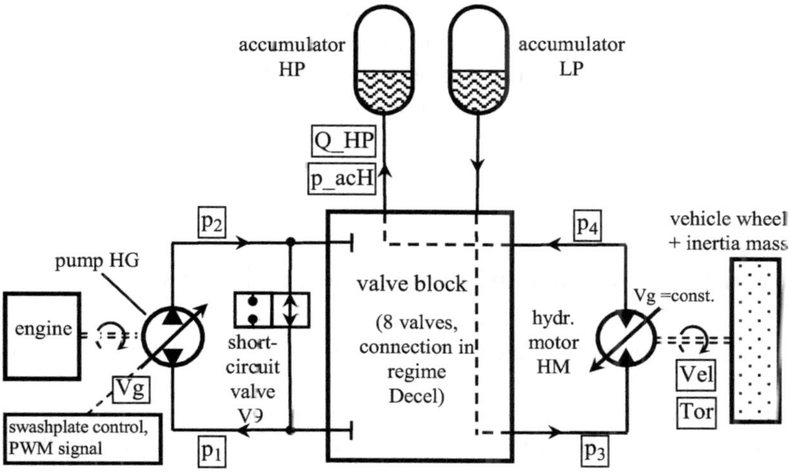

Fig. 1 Simplified scheme of recovery drive in the regime Decel, position of the short-circuit valve, notation of quantities

4 Version "Without the Short-Circuit Valve", Waiting for Drop of HG Flow

This version is a modification of the previous one; the difference is in a time shift of valves control at the beginning of Decel process. Immediately after releasing the pedal, according to the prescribed ramp, HG flow drops by means of the control plate in HG. Only in case of reaching a lesser value than the chosen limit of its position (here 20 %), the valves control will start to switch to Decel. Thus, the unfavorable regime described in the previous chapter will be avoided.

This more suitable regime without energy wasting will also favorably influence the efficiency, reaching 19.8 %. However, the possibility to reach a timely conversion of kinetic energy into pressure energy is lost due to waiting for suitable conditions.

A prospective improvement can be reached by using a quicker deceleration ramp of HG. For instance, with the ramp 1 s/100 % the efficiency will be improved to 22.9 % and the increase in the high-pressure accumulator will reach 2.09 MPa; with the ramp 0.5 s/100 % to 23.5 % and to 2.44 MPa [4].

5 Version "with the Short-Circuit Valve"

The core of this solution is a hydraulic short circuit of HG during the time interval when HG is disconnected from the system, which is suitable mainly at the beginning of Decel. For this purpose, it is necessary to enlarge the number of existing valves in the valve block (valves denoted inside of the valve block as V1–V8) by adding the valve V9, which is closed in the idle condition.

Electric control of the extra short-circuit valve V9 takes place in co-operation with control of disconnecting valves V1 and V2, i.e. a delay of control voltage at the beginning of Decel is adjusted in such a way that opening of V9 is realized simultaneously with closing of V1 and V2.

6 Results of Simulations, Comparison of Efficiencies

A number of experiments were performed on the developed computer model in Matlab/Simulink. Figure 2 shows the examples of time courses of the chosen quantities, including some explaining remarks, in for the described problem of choice of regimes of the process Decel. The number of modeled quantities is in reality about of 170 (Table 1).

However, its prospective use is conditioned by a quick ramp operation of HG, i.e. by quick lowering of oil flow on the side of HG and by starting of Decel not earlier than a sufficiently small flow is reached. This conception does not require a modification.

7 Conclusion

This working study was a basis to make a decision concerning the choice of a suitable variant for the recovery process Decel. Three basic variants were considered and the time courses of the principal quantities, including evaluation of efficiency and amount of converted energies are shown on the examples for expected operation conditions.

It follows from the analysis that considerably most suitable is the conception according to chapter "Modern Trends in the Drive Wheelsets of Rail Vehicles", i.e. supplementation of the existing hydraulic system with the short-circuit valve, which ensures the favorable regime of HG during the time interval of Decel, i.e. an idle run. The only disadvantage over prevailing advantages is the necessity of supplementation of the existing system with the above-mentioned valve.

The second most suitable variant is the one according to chapter "Designing of Mixing Equipment, Reactors and Bioreactors". The variant according to chapter "Design of the Vibration Separatory Apparatus" is not suitable both from the view of efficiency and of unfavorable operation conditions for HG. Courses of efficiency found during Decel by means of simulation reach relatively small values.

It is necessary to have in mind that the concept of efficiency offers more variants according to the losses taken into consideration for computation and that the introduced efficiencies represent the mean value from the time interval of Decel. At the beginning of the process, immediate values of efficiency are approximately

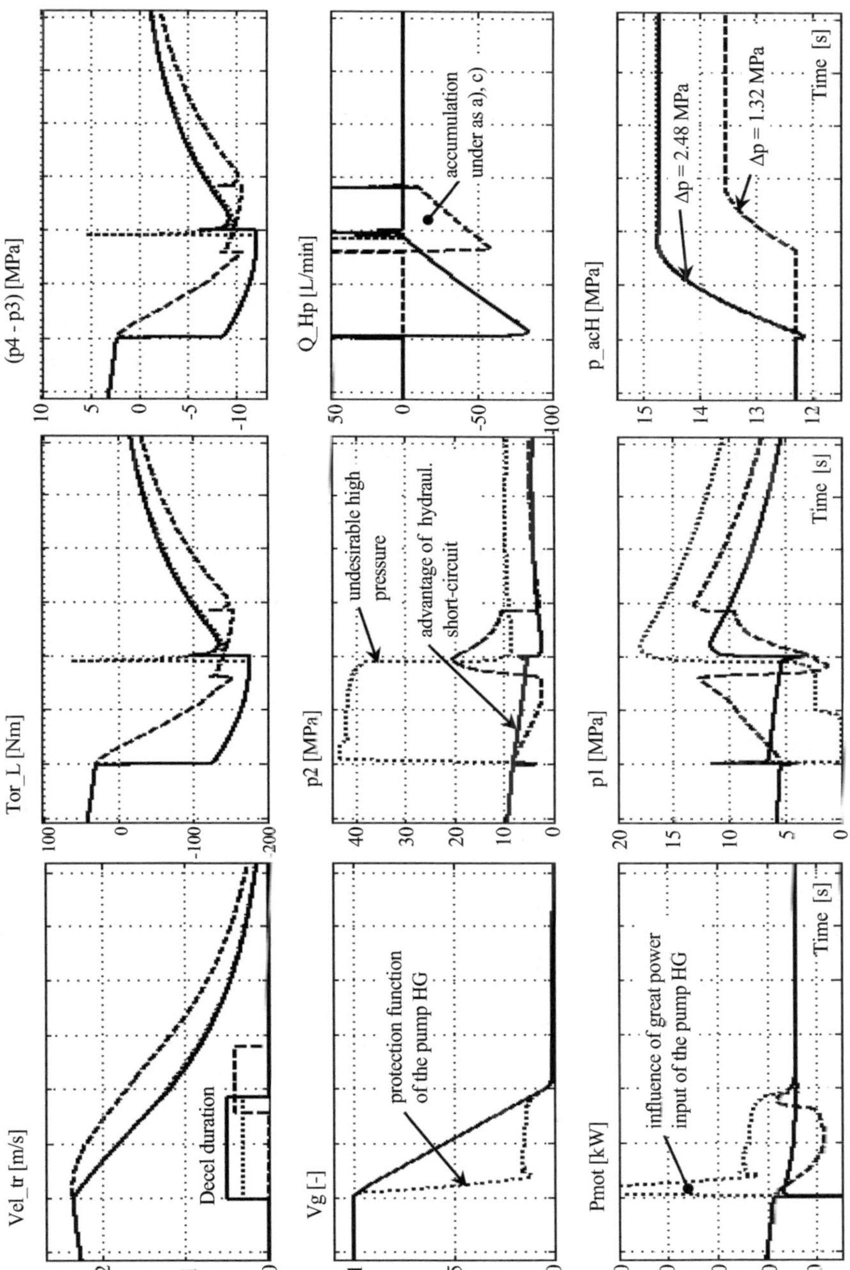

Fig. 2 The time courses of the main quantities during process Decel, comparison of versions – (**a**) dotted, (**b**) dashed, and (**c**) continuous

Table 1 Obtained values of efficiency for considered variants of the recovery regime Decel

Variant	Operation principle of Decel	Efficiency η_{dec} [%]
a	"Without the short-circuit valve", setting of the valves changed at once	11.9
b	"Without the short-circuit valve", with waiting for pump flow drop	19.8
c	"With the short-circuit valve"	22.1

doubled but in the course of time they fall to zero. This is related to HM falling velocity and oil flow between accumulators. At the end of Decel, a small influence of energy conversion can be seen and on the contrary the influence of losses prevails.

Acknowledgement This work is an output of research project Eureka (LF12029) "Hydrostatic system to energy recovery for commercial vehicles" supported by the Ministry of Education, Youth and Sports of the Czech Republic and NETME Centre, regional R&D centre built with the financial support from the Operational Programme Research and Development for Innovations within the project NETME Centre (New Technologies for Mechanical Engineering), Reg. No. CZ.1.05/2.1.00/01.0002 and, in the follow-up sustainability stage, supported through NETME CENTRE PLUS (LO1202) by financial means from the Ministry of Education, Youth and Sports under the "National Sustainability Programme I".

References

1. Z. Němec, Implementation of the short-circuit-valve in model of Decel recovery process. Partial research report to the project Hydrostatic system to energy recovery for commercial vehicles (Brno University of Technology, Faculty of Mechanical Engineering, 2014)
2. J. Nevrlý, Hydraulic energy recovery stands for vehicles, in *Proceedings of 53rd International Conference of Machine Design Departments,* Brno University of Technology, Faculty of Mechanical Engineering, Institute of Machine and Industrial Design, Mikulov, 2012, pp. 199–206. ISBN 978-80-214-4533-8
3. A. Pourmovahed, N.H. Beachley, F.J. Fronczak, Modeling of hydraulic energy regeneration system – part I: analytical treatment. J. Dyn. Syst. Meas. Control **114**, 155–159 (1992)
4. A. Pourmovahed, N.H. Beachley, F.J. Fronczak, Modeling of hydraulic energy regeneration system – part II: experimental program. J. Dyn. Syst. Meas. Control **114**, 160–165 (1992)

FEM Simulations on Gears with Different Carbonized Depth

K. Petr, V. Dynybyl, and J. Krepela

Abstract The article describes the possibilities of increasing of the load capacity of gearing by optimizing thickness of the surface layer of case-hardened tooth flanks, which have different thicknesses and pressure angles. Further is described the creation of FEM (finite element method) simulation model of gear-mesh with the application of various surface layers and then are described calculation results. Mechanical properties of the surface layers are determined from the material analysis.

Keywords Gears • Finite element method • Gear-mesh • Elastic–plastic deformation • Carbonized depth • Case hardened layer • Experimental tests • Material analysis

1 Introduction

The article describes the possibilities of increasing of the load capacity of spur gears from reasons of greater competitiveness of gearboxes on the market. Load capacity of gears means determination of the load gearing which does not cause exceedance of all safety against any kind of gear damage (failure). Increasing the load capacity can be achieved by various ways (better material, type of surface treatment, modification of flank shape, etc.) [1–3].

The article includes a description and creation of the FEM (finite element method) simulation model of gear-mesh (gearing) with the application of different thicknesses of the surface layers of case hardened (CHD) tooth flanks. In the article are compared two geometry of the gearing – for two different pressure angles (20° and 25°).

K. Petr (✉) • V. Dynybyl
Faculty of Mechanical Engineering, Czech Technical University of Prague, Prague, Czech Republic
e-mail: karel.petr@fs.cvut.cz; vojtech.dynybyl@fs.cvut.cz

J. Krepela
Wikov MGI a.s., Hronov, Czech Republic
e-mail: jkrepela@wikov.com

V. Dynybyl et al. (eds.), *The Latest Methods of Construction Design*,
DOI 10.1007/978-3-319-22762-7_40

Mechanical properties of the surface layers (with different distribution of hardness in these layers) has been taken from the material analysis. These analyses were carried out in parallel with experimental tests on identical samples of teeth.

2 Input Parameters for the Creation of the FEM Model

The most important input parameter is involute gearing geometry. Geometric data are exported from the program KISSsoft. Further important parameter is the distribution of hardness in normal direction from the tooth flank towards the core (see Fig. 2). Hardness was measured by device CSM at the load 0.5 N. This device worked with a diamond indenter – type Berkovich. Hardness was measured from the edge of the sample at points spaced 100 μm (see Figs. 1 and 2).

Parameters of tested gearing: module m = 6 mm, pressure angles 20° and 25°, number of teeth $z_1 = 16$ (pinion) and $z_2 = 50$ (gear).

In the Table 1 are indicated the thickness of the case hardened layer (CHD) and values of the critical shear stress under the case hardened layer (σ_{CHD_K}).

3 Creation of the FEM Model and Definition of Boundary Conditions

FEM model and calculation of gear-mesh is the difficult contact problem. The solution of the contact problem is solved in the part of gearing (2D FEM model – see Fig. 3) [1]. On pinion and gear have been created only three teeth. Three teeth are sufficient for the simulation of gear-mesh across the tooth (middle tooth).

The geometric model is divided into several parts (part of the layers – shells), which are shown in Fig. 3. Way of division of model on individual parts has been selected with respect to future generating of mesh.

As material for pinion and gear has been used case-hardened steel 18CrNiMo7-6 with elastic properties ($E = 210\ 000\ Nmm^{-2}$; $\mu = 0.3$; $\rho = 7\ 830\ kg/m^3$; $R_{p0.2} = 830\ Nmm^{-2}$).

Carbonized layers are created as elastic–plastic material model, with application of Ramberg–Osgood model of plasticity. For this plasticity model is characterized by a smooth elastic–plastic transition (especially for highly alloyed and hardened steel). Ramberg–Osgood model of plasticity is defined by (1), where ε_{el} is the relative deformation of the elastic zone (elastic deformation) and ε_{pl} is the relative deformation of the plastic zone (plastic deformation). Constants K and n are dependent on the material properties.

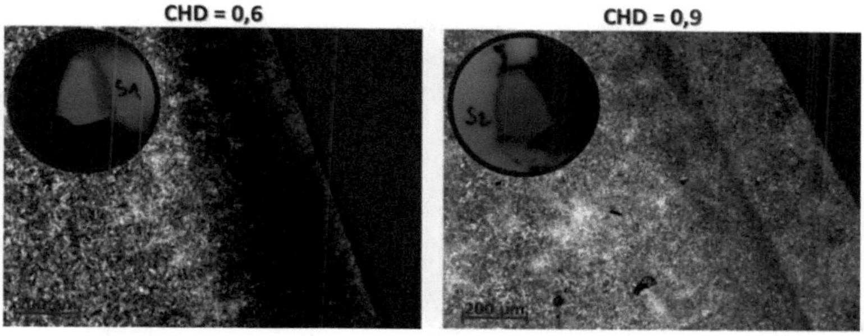

Fig. 1 Example of thin section for CHD samples (CHD = 0.6 mm and CHD = 0.9 mm)

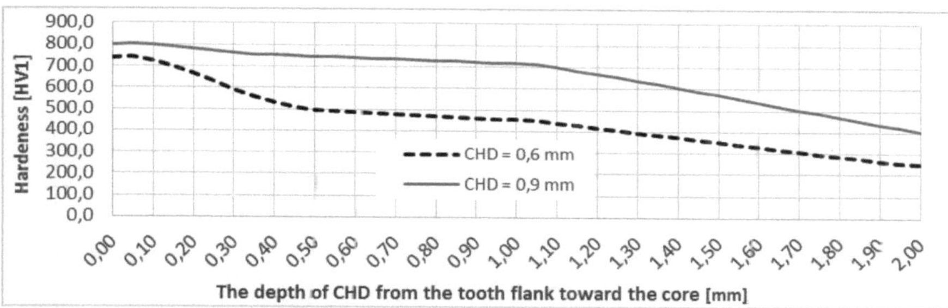

Fig. 2 Distribution of hardness for CHD = 0.6 and CHD = 0.9

Table 1 The thickness of the case hardened layer (CHD) and values of the critical shear stress

Pressure angles α [°]	Case hardened depth (CHD) [mm]	The critical shear stress under the case hardened layer σ_{CHD_K} [Nmm^{-2}]
20	0.6	368.29
	0.9	313.15
25	0.6	400.67
	0.9	342.88

$$\varepsilon = \varepsilon_{el} + \varepsilon_{pl} = \frac{\sigma}{E} + \left(\frac{\sigma}{K}\right)^{1/n} \tag{1}$$

The contact bonds (Surface – Surface) among tooth flanks is added to the model and meshed (friction coefficient f = 0.0718). The gear and pinion are connected with the self-centre point of rotation with help of bond type "Coupling". The subsequent rolling of the pinion and the implementation of moment on gear are applied through these points. Simulation of gear-mesh is performed quasi-statically during three steps (see Fig. 4) [2, 3].

Quality of FEM mesh is shown in Fig. 5 (for the pinion and gear). In the Fig. 5 is shown as the mesh in the tooth flank (contact bond) is very fine and towards the core

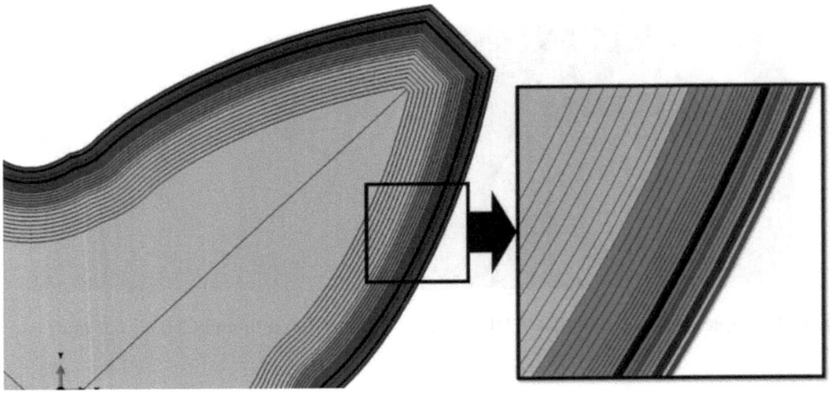

Fig. 3 Detail of the "part of the layers – shells" case hardened layer

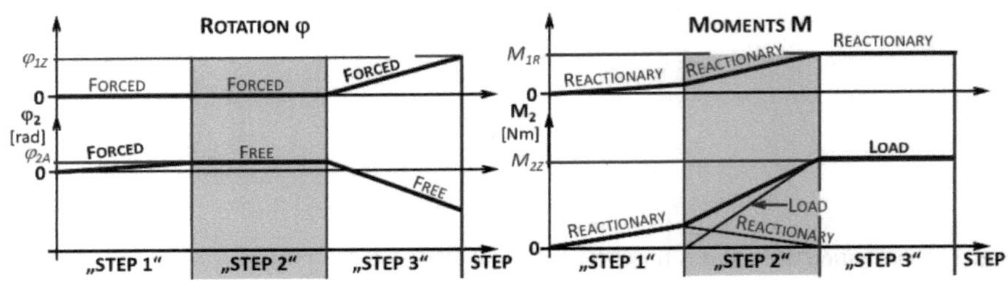

Fig. 4 The loading status during three steps of calculation (index 1 – pinion; index 2 – gear)

(respectively to the centre of the gears/pinion) is coarser. For FEM mesh was used combination of two types of elements for easier generating of mesh (*linear quadrilateral elements* and *linear triangular elements*).

4 The Results of the FEM Calculation of Gear-Mesh with Simulation of Case Hardened Layer

From the distribution of stress in the gear-mesh (Fig. 6) is obvious that the transition of stress from tooth flank towards tooth root is greater for pressure angle 20° than for 25°. It follows known conclusion that with greater pressure angle of gear increases resistance fracture of the tooth in the tooth root. For greater pressure angle we can see smaller Hertzian contact stress.

From distribution of main shear stress (S12) according Figs. 7 and 8 is obvious that with greater thickness of CHD is moved the maximal value S12 nearer to the tooth flank. For pressure angle 20° is the area with maximal shear stress greater than for pressure angle 25° and is placed deeper under case-hardened layer.

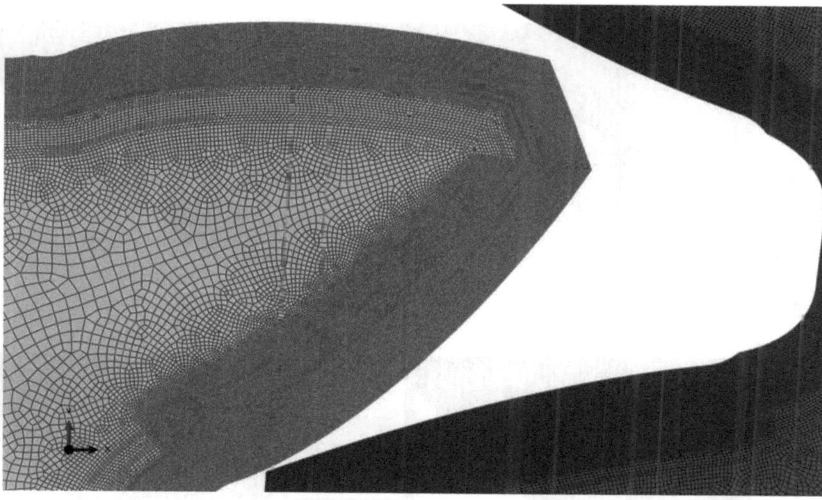

Fig 5 Detail of FEM mesh in the area of gear-mesh of teeth flank on the pinion

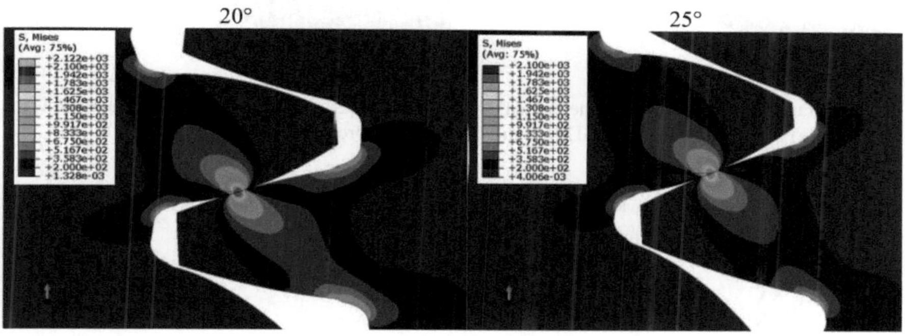

Fig 6 Distribution of (von Misess) stress in the gear-mesh

In the Fig. 7 are shown distribution of main shear stress (white area on right side) for different carbonized depth (CHD).

In the Fig. 8 is shown distribution of main shear stress over the depth (path of depth is in the direction of normal line) for $CHD_{20^\circ \text{and } 25^\circ} = 0$; $CHD_{20^\circ \text{and } 25^\circ} = 0.6$; $CHD_{20^\circ \text{and } 25^\circ} = 0.9$ mm).

5 Conclusion

From this analysis suggests that when using a greater pressure angle than angle 20°, is necessary to calculate with the fact that position of the maximum main shear stress is moved towards from tooth flank to core. On this problem is necessary to

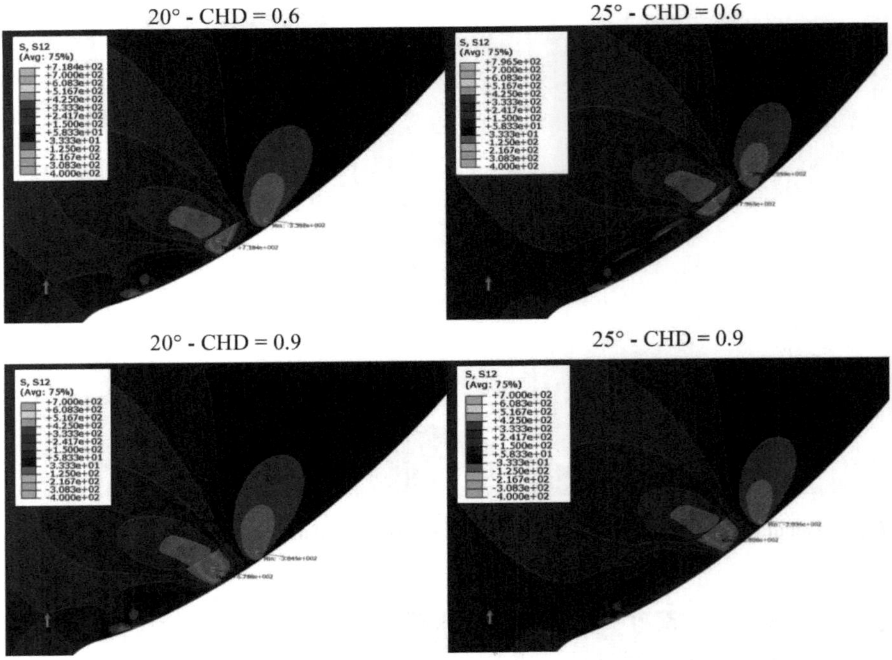

Fig. 7 Graphical comparison of the critical shear stress (S12) in the area of max. values (near the pitch circle) for thicknesses of CHD 0.6 and 0.9 mm and pressure angles 20° and 25°

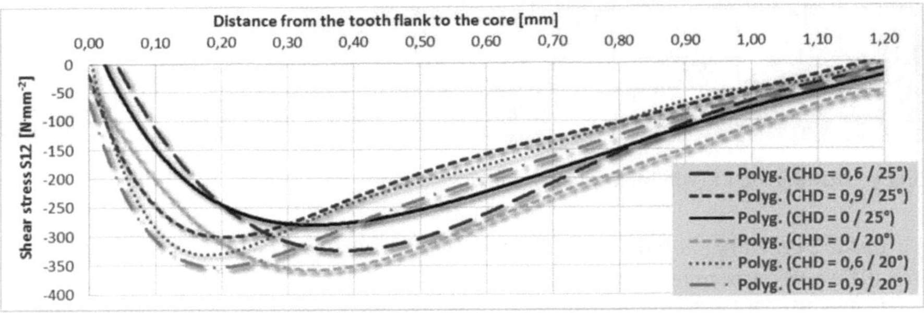

Fig. 8 Graphical comparison of the main shear stress values (distribution of main shear stress over the depth/path of depth is in the direction of normal line/ – CHD 0; 0.6; 0.9 mm – 20° and 25°)

respond by greater CHD. With sufficient increase of CHD, is reduced the area of critical values of shear stress. A positive aspect of a greater pressure angle is a greater resistance to bending stress.

Acknowledgments The research work reported here was made possible by Ministry of Industry and Trade of the Czech Republic grant No. FR-TI4/054.

References

1. K. Petr, V. Dynybyl, M. Dub, FEM simulation of flank breakage on tooth of gears and experimental photos, in *Proceedings of the 54th International Conference of Machine Design Departments* (Technická univerzita v Liberci, Hejnice, 2013), pp. 259–264
2. K. Petr, J. Kanaval, V. Dynybyl, Modification and non-standard methods of increasing tooth flank resistance of gears, in *Proceedings of International Conference on Gears*, VDI-Society for Product and Process Design and TUM/VDI Verlag GmbH, Munich/Düsseldorf, 2010, ed. by R.B. Höhn, pp. 1439–1142
3. K. Petr, P. Žák, V. Dynybyl, J. Kanaval, FEM optimization of gear flanks shape modifications, in *Proceedings of the 3rd International Conference Power Transmissions*, Sofia Publications Thessaloniki, Kallithea, 2009, ed. by A. Mihailidis, pp. 93–100

Numerical Simulation of Ductile Fracture

M. Spaniel

Abstract Brief introduction into ductile fracture material models based on damage mechanics approach suitable for FE simulation is given in the article following with two examples of calibration. The first example introduces calibration of steel commonly used in nuclear power plant industry that was based on wide portfolio of 15 experimental specimens. The second one deals with simplified calibration of armor steel based on shooting tests.

Keywords Ductile fracture • Fracture locus • Damage mechanics • Ballistic protection

1 Introduction

Numerical simulation of failure provides us with significantly deeper insight into behaviours of structure even before it was manufactured and tested. This is particularly important in cases of (1) exposed and expensive structures i.e. components of nuclear facilities, when full scale experiments are impossible; (2) exposed structures the failure of which may cause people death or serious injuries, i.e. safe vehicles should keep integrity if crash happens; (3) structures the function of which is based on partial failure, i.e. kinetic energy absorbers; (4) structures providing protection against impact, i.e. protection against fractured parts of high speed rotating machines, or ballistic protection.

Despite advances in the study and modeling of the microstructure of materials, nowadays in solving complex engineering problems general approach is to apply a phenomenological material models in the continuum. Damage mechanics – one of commonly applied engineering approaches modeling failures – is based on assumption of damage cumulation during loading process. In most cases the damage is described as abstract scalar parameter. Damage mechanics failure models differ by damage cumulation laws (usually incremental form) and critical value of damage (usually renormalized to be unity). Wide range of phenomena (fatigue, ductile

M. Spaniel (✉)
Czech Technical University in Prague, Prague, Czech Republic
e-mail: miroslav.spaniel@fs.cvut.cz

V. Dynybyl et al. (eds.), *The Latest Methods of Construction Design*,
DOI 10.1007/978-3-319-22762-7_41

fracture, composite fracture) has been more or less successfully solved in scope of damage mechanics concept.

2 Ductile Failure Models in Scope of Damage Mechanics

Ductile fracture is the process of progressive failure of metallic material in conditions of monotonic loading. Evolution of the damage depends on plastic straining and ends by material point failure that causes the loss of load carrying capacity in the material point. This and monotonically increasing loading results into overloading of other material points in the vicinity of the failed one. This process is commonly mentioned as progressive failure.

Phenomenological material models describing ductile failure mostly introduce extension of plasticity models. Two types can be distinguished – coupled models modifying plastic response in dependence on damage evolution; uncoupled models avoiding the damage influence on plastic response. Easier calibration process is an essential advantage of uncoupled material models, for which the calibration of plastic response and calibration of ductile damage can be separated.

The phenomenological models of ductile failure do not have physical meaning and try best capture the behaving of real material on base of empirically designed relations. Their parameters are designed based on experience, intuition or are deduced from micromechanical continuum models. By the latter way the substantial influence of stress triaxiality was demonstrated. Formulation of failure criterion describing efficiently the process of damage is another crucial point of phenomenological models. Phenomenological models provide us with acceptable description of response in case of materials they was designed and calibrated for. On the other hand the experimental tests are costly and their portability to other materials is limited.

Material model discussed in this paper is based on both classical incremental model of plastic response with isotropic hardening and phenomenological concept of damage in continuum mechanic. This model supposes isotropy, and for description of stress state uses Von Mises stress q, stress triaxiality η and Lode parameter ξ. These quantities are defined using hydrostatic stress, p and second, J_2 and third, J_3 invariants of deviatoric stress [1, 2].

$$q = \sqrt{3J_2},\ \eta = -\frac{p}{q},\ \xi = \frac{27J_3}{2\ q} \tag{1}$$

Failure criterion is based on phenomenological quantity damage ω defined as non-decreasing scalar parameter

$$\omega = \int_0^t \frac{\dot{\overline{\varepsilon}}_{pl} dt}{\overline{\varepsilon}_f(\eta, \xi)}, \tag{2}$$

\$10that depends on loading history (via accumulated plastic strain $\dot{\overline{\varepsilon}}_{pl}$), and can be understood as linear accumulation of incremental damage in process of monotonic loading. Fracture locus $\overline{\varepsilon}_f$ is function of stress triaxiality and Lode parameter and it has to be calibrated experimentally. The fracture locus has physical meaning of accumulated plastic strain at the instant of ductile failure initiation at the end of hypothetic monotonic loading with constant both triaxiality and Lode parameter. In such loading process the damage at failure reaches value $\omega_{crit} = 1$, so damage defined in (1) is normalized.

3 Calibration of Typical Nuclear Industry Steel Fracture Locus

This example describes the calibration of uncoupled material model of ductile damage by Johnson-Cook that is implemented in FE package Abaqus. As the work was supported by the grant "Identification of ductile damage parameters for nuclear facilities", calibration was done for typical steel used in nuclear power plant industry.

Several specimen types with different values of both stress triaxiality and Lode parameter at expected locations of ductile fracture should be used for successful ductile failure model calibration. As neither η neither ξ in most cannot be determined using analytic formulas, the specimens have to be analyzed via FE [5]. The calculated course of these quantities serves as input into calibration process. Critical extension at the instant of failure is directly determined experimentally. Fracture strain in expected location of failure is extracted from FE simulation based on the critical extension. The critical extension could be determined for example via direct surface observation and first individual cracks detection. Usually, critical extension is identified on base of sudden decrease of loading force in force-extension record.

Two approaches are commonly cited to be used in calibration process of uncoupled ductile damage model. The first one is based on averaged values of stress triaxiality and Lode parameter [6]. The main disadvantage of this approach is wide range of stress triaxiality and Lode parameter for some specimen types resulting into non-negligible error caused by averaging of these quantities. The second approach defines target as deviation of damage integrated up to fracture strain for each specimen averaged over all specimens.

The portfolio of 15 calibration specimen types for quasi-static loading was designed to calibrate fracture locus. Single specimens in space of stress triaxiality and Lode parameter are shown in Fig. 1a. It is based on averaged values of η_{av} and ξ_{av} at instant of expected fracture. The substantial subset of calibration portfolio is based on special specimen with double curvature (butterfly) that was designed by Mohr. Essential advantage is possibility to cover wide range of stress triaxiality and normalized Lode angle with only one specimen geometry.

The resultant fracture locus is shown in Fig. 1b. The figure introduces fracture strains of single specimens. Symbol associated with i – th specimen is plotted at position $\left[\eta_{av_i}, \xi_{av_i}, \overline{\varepsilon}_{f_i}\right]$ of averaged FE determined triaxiality, resp. Lode parameter

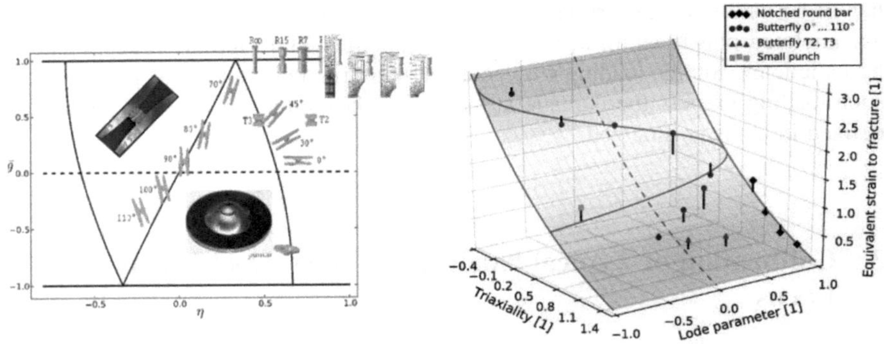

a) b)

Fig. 1 Calibration specimens (**a**), Johnson-Cook fracture locus (**b**)

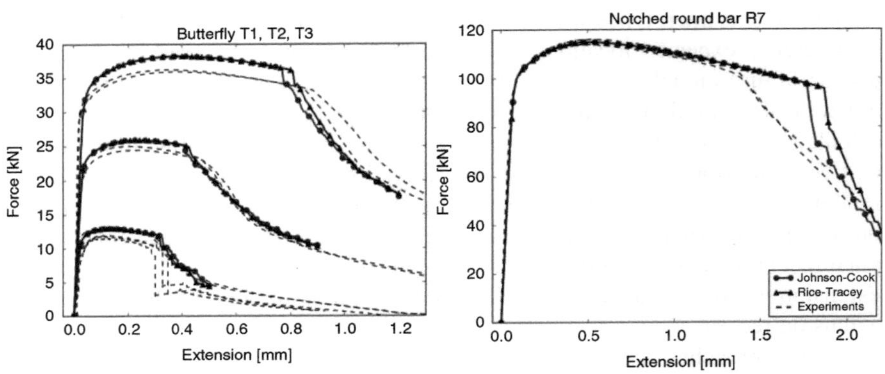

Fig. 2 The result of simulations with Johnson Cook and Rice Tracey model

and experimentally determined fracture strain. Comparisons of FE simulations to experiments for notched round bar and butterfly 0° with different thicknesses are shown in Fig. 2. Response of Johnson-Cook material model is more close to experiments for most of specimens. It is possible to say that both models reach acceptable agreement with experimental data.

4 Calibration of Typical Ballistic Protection Armor Steel

In this example some experimental results of a bullet impact on steel armor are presented together with numerical modelling and calibration approaches. The development of high-quality composite sandwiches for ballistic protection is the target of a grant project in terms of which the research is being conducted. Traditionally, pure metal-based armor which is heavy and thus negatively affects other vehicle parameters, is used. These days, composite or hybrid sandwiches represent an alternative. Numerical simulations may reduce excessive experiments needed to obtain appropriate design of ballistic protection, but they require verified modeling approaches and reliable material data [3, 4]. Therefore, appropriate failure model of steel based strike-face armor has been searched and calibrated as the first necessary step in modelling hybrid steel-composite armors.

Johnson-Cook material model including rate dependency was employed for description of fracture locus

$$\overline{\varepsilon}_f\left(\eta, \dot{\overline{\varepsilon}}_{pl}\right) = \left[d_1 + d_2 e^{-d_3\eta}\right]\left[1 + d_4 \ln\left(\frac{\dot{\overline{\varepsilon}}_{pl}}{\dot{\varepsilon}_0}\right)\right], \quad (3)$$

where $d_1 \ldots d_4$ are failure parameters. In our case, we neglected the dependence of damage on triaxiality (parameters d_2 and d_3). There are two remaining parameters. Parameter d_1 describes the size of the equivalent plastic strain at the beginning of the failure and parameter d_4 expresses dependency on the strain rate. These two parameters were fitted so the values of residual velocities for different impact velocities agree with the experimental shooting tests of the strike-face armor plate (Fig. 3) as well as possible. Parametric study utility in the software Abaqus/Explicit was used. The best agreement was reached with parameters $d_1 = 0.25$ and $d_4 = 0.024$ as shown in Fig. 4.

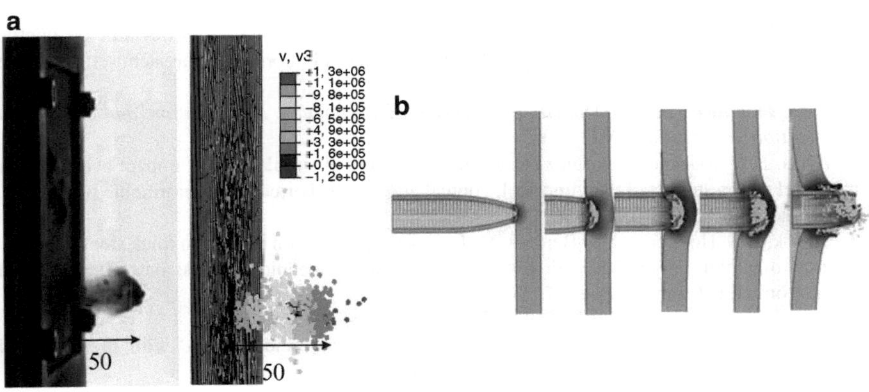

Fig. 3 Comparison of experiment and simulation 50 microseconds after the impact with velocity v = 1033 m/s (**a**). Five different moments during the simulation of a bullet impact with the initial velocity vi = 1022 m/s (**b**)

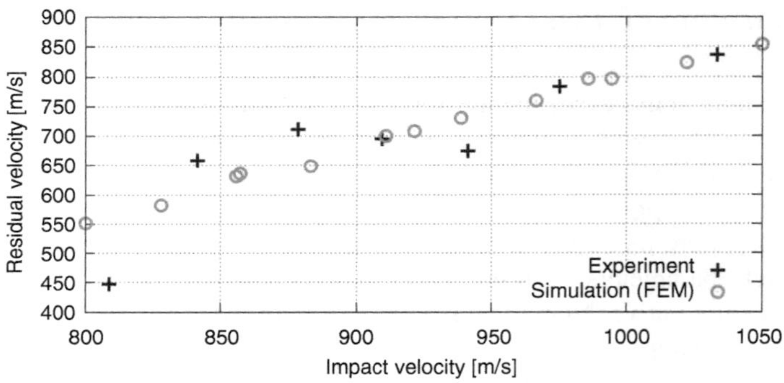

Fig. 4 The impact-residual velocity dependence

5 Conclusion

Calibration of damage mechanics based phenomenological material models of ductile fracture is complex procedure that requires both experimental and computational approach. As the costs of calibration experiments may be significant, models and approaches reducing experimental portfolio are of particular interest. Some experience with calibrations was discussed.

Acknowledgments The research work reported here was made possible by projects "Ductile damage parameters identification for nuclear power plants – FR-TI2/279" and "Development of the hi-tech composite sandwiches for ballistic protection – FR-TI4/317" sponsored by Ministry of Industry and Trade of The Czech Republic

References

1. T. Wierzbicki, L. Xue, On the effect of the third invariant of the stress deviator on ductile fracture, in *Technical Report* (Impact and Crashworthiness Laboratory, Massachusetts Institute of Technology, Cambridge, 2005)
2. L. Xue, *Ductile Fracture Modeling – Tudory, Experimental Investigation and Numerical Verification*, Dissertation, MIT, 2007
3. A. Arias, J.A. Rodriguez-martinez, A. Rusinek, Numerical simulations of impact behaviour of thin steel plates subjected to cylindrical, conical and hemispherical non-deformable projectiles. Eng. Fract. Mech. **75**, 1635–1656 (2008)
4. T. Borvik, O.S. Hopperstad, T. Berstad, M. Langseth, Perforation of 12 mm thick steel plates by 20 mm diameter projectiles with flat, hemispherical and conical noses: part II: numerical simulations. Int. J. Impact Eng. **27**, 37–64 (2002)
5. Dassault Systemes, Abaqus 6.12 Online Documentation, 20126.
6. Y. Bai, T. Wierzbicki. A new model of metal plasticity and fracture with pressure and Lodedependence. International journal of plasticity. 2007

Dynamic of Propulsion, New Challenges for Analysis

V. Zoul

Abstract Every propulsion system, especially when it contains piston machines, or engines, or compressors, needs already in project phase an analysis of propulsion dynamic to avoid serious problems or damages in operation. A basic part of such analysis is a torsional vibration calculation. Methods for such calculation are nowadays quite good known und used. Development of programs for these calculations concerns improvement of their user-friendliness. But development of modern propulsions, where new types of machines and more sophisticated control systems are used, brings new aspects which need to be respected. Some of these aspects are described in this paper.

Keywords Torsional vibration • Methods of calculation • Analysis • New aspects

1 Introduction

Torsional vibration calculation (TVC) and dynamic analysis of propulsion systems count among oldest disciplines of machine engineering. Starting with introduction of piston steam machines for ship propulsion systems the problems of torsional vibration arose almost immediately. The first known problem is probably paddle wheel steamer with three cases of shaft breaks already in 1872. After it the problems of torsional vibration (TV) spread (above all in connection with development of internal combustion piston engines) in other branches of machinery engineering aside from the shipbuilding, in manufacturing of all kind of vehicles, in power and chemical engineering. It could be said that already after the WW II, especially on the end of sixties, the common TVC were managed on a very good level based on plenty of practical experiences. To enable calculation some simplifications, tabular methods and classical algebraic formulas were used.

On the beginning of the seventieths with introduction of computer technique, the existing methods were transformed in computer programs. Important contribution was also a possibility to use a numerical integration for solution of nonstationary states and nonlinear problems. Nowadays a considerable amount of software

V. Zoul (✉)
Czech Technical University in Prague, Prague, Czech Republic
e-mail: vaclav.zoul@fs.cvut.cz

V. Dynybyl et al. (eds.), *The Latest Methods of Construction Design*,
DOI 10.1007/978-3-319-22762-7_42

packets for solution of machine dynamic exists. Some of them could be used also for TVC. It looks that it is nothing to solve on the field of TVC, all is finished. But in reality the new technique helps not only in calculations, but brings generally changes in the design of machines and their control too. These changes are also new challenges for engineers.

2 Present Methods for TVC

For current TVC which are nowadays required as a part of design of propulsion system of vehicles (mostly ships) and other machines the programs could be used based on traditional TVC computation methods. These methods are also accepted by different classification societies. In this situation the main problem is not to find a method for TVC, but to have programs to disposal, which allow to carry out the calculation cheap and quickly. In fact, without a good, user-friend program could be possible to make a TVC too. But at a cost and in time which are full unacceptable.

Programs which meet these more or less economic conditions must have, as a minimum, the following properties.

- Easy to insert and to change the input data (calculation of variants).
- The input data must be in form corresponding to the form of data which are usually provided by manufacturers of engines and other parts (propeller, gear box, flexible couplings, TV dampers) of the propulsion line. Ideal is to take for some inputs dimensions directly from drawings.
- Inputs can respect these nonlinear properties which depend on the speed of rotation, like change of the coupling's stiffnesses or properties which depend on the frequency of excitation, like stiffness or damping.
- The output data must be in form corresponding to convention of customers and it should be easy to include the results in report without necessity to adapt them.
- The results should be possible to compare immediately with permissible values.
- It should be possible to transfer the results of measurement in packets of results of calculation or in the opposite way to enable to compare both each with other and to use it for identification of the system.
- The calculation and results must be accepted and agreed by classification societies or similar organisation.
- It is recommended to have databanks of properties of standard parts (engines, couplings, damper) to disposal.
- It is recommended to store the results in a databank.

Everybody, who's intention is to deal with the TVC professionally, must solve the dilemma or to build such a system fulfilling these requirements

- with his own capacity or
- to buy a really good readymade professional program.

Further perspective in the field of present TVC programs is naturally to complete and to improve stepwise the programs with new piece of knowledge and new findings.

Last and not least it will be necessary to complete the classic TVC with calculation with load collectives to have a basis for calculation of lifetime.

3 Changes on the Field of TVC

Changes in the propulsion lines, which will be dealt here, are connected above all with spread of electronic and digital technology. An influence of these technologies is seen in more branches. Probably most noticeably it is in the propulsions with:

- piston engines managed by electronic engine management,
- electric motors with variable frequency control.

Cooperation of the electric power aggregates driven by diesel engines with electrical network is also a theme tight connected with the electronic control.

An electronic control brings plenty of useful advantages as a high efficiency, easier manoeuvrability and lower pollution. These topics have usually priority over problems which can bring dynamic of the system. An engineer dealing with machine dynamic has to use in such cases all his knowledge and cleverness to keep the propulsion free from dangerous vibrations.

Some examples which are presented here should show the actual problems which occur in praxis.

4 Working Cycle of the Internal Combustion Engines

Long time it was possible to use for definition of the excitation caused by working cycles in internal combustion engines data in form of tables or diagrams, compiled from indication of similar engines. It was possible as the working cycles in plenty of engines were simple and similar. The working cycles now could be very different due to use of more stage turbocharging and more stage fuel injection. The last generation of the diesels uses for example till four fuel injections in one working cycle. Such working cycle differs naturally from the old one. It could be said that the best way is to obtain the correct data directly from the manufacturer of the engine. In reality it is not very easy.

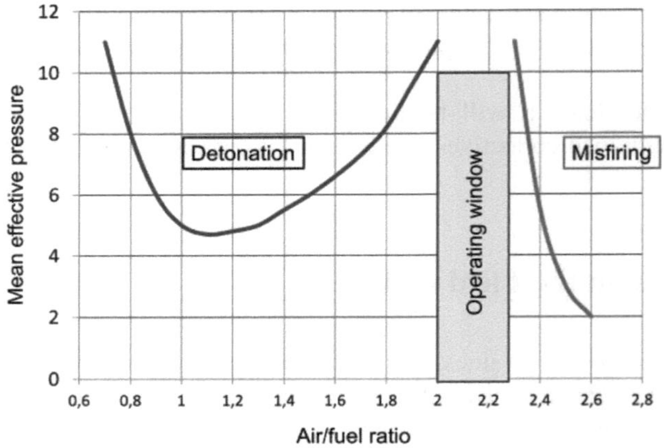

Fig. 1 Operating window of a dual fuel engine

5 Gas Engines

Gas engines, especially dual fuel gas engines, which use small amount of regular diesel fuel as an initiation for ignition, run efficiently in so called lean burning regime. This regime lies in narrow band of air/gas ratio. Outside this band the engine run unstable and disturbed (Fig. 1).

The engine is kept in the operating range by electronic control system. Not every time is the control able to react quickly enough, especially during change of speed, load or on change of gas characteristic (Fig. 2).

6 Possibility to Shift the Strong Resonances Out of Operation Range

Basic philosophy how to protect propulsions with piston engines against torsional vibration is to shift the strongest resonances out of operation speed zone, where only the resonances with low intensity ("minor" harmonics) can stay. Requirement for an existence of minor harmonic is that the excitations of single cylinders are eliminated by the excitation of other cylinders which acts in opposite direction (Fig. 3).

With introduction of common rail technology in engine's branch it is possible to adjust the fuel delivery in each cylinder individually. Fuel volume delivered in each cylinder could looks like an example on the Fig. 4.

The Fig. 4 shows clearly that the advantage which had the engines with a classical injection system controlled by one fuel rag, could not be applied, when the engine has common rail injection system and its SW does not respect balance between cylinders.

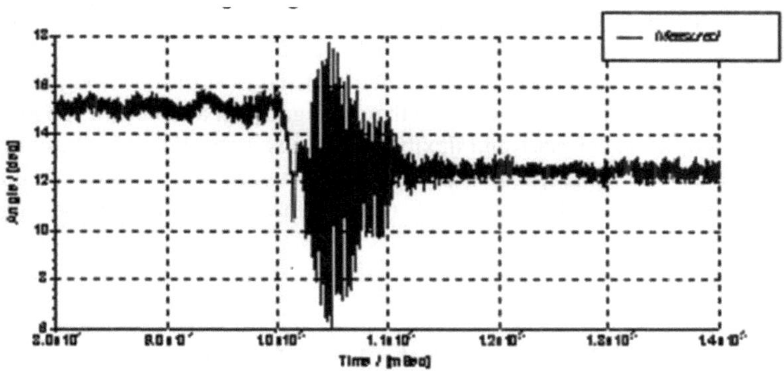

Fig. 2 Dual fuel engine. Coupling's reaction on change of load

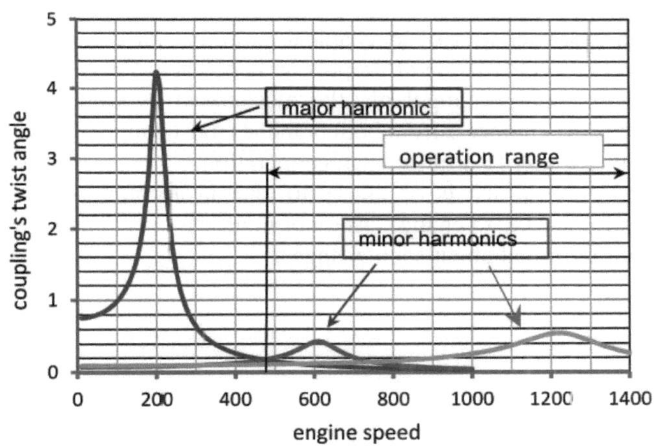

Fig. 3 Load of a coupling in a typical aggregate with a piston engine

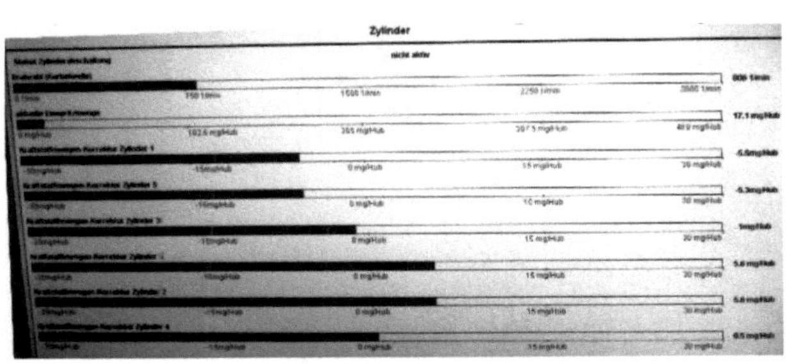

Fig. 4 Six cylinder diesel engine. Example, showing how the fuel volumes delivered in each cylinder could differ each between other, when the engine is equipped with common rail injection

Another example, how the new control methods can influence the propulsion dynamic are variable frequency drives which are often used for example in drive of piston compressors. Also here the principle was used to shift the strong resonances out of working zone. It was easy, when the aggregate run at constant speed. With a combination of flywheel mass and flexible couplings it was possible to achieve that the speed of the aggregate was located between resonances. Using the electromotors with variable frequency (speed) the speed zone of the aggregate is wide.

7 Influence of Electronic Control and Conclusions

As seen from the previous paragraphs a correct function of the propulsion and level of load of the parts of the propulsion line depends not only, like before, on properties and characteristics of mechanical parts, but on correct function of the electronic control. The dynamic of propulsion is nowadays necessary to solve as a complex system, incorporating both, the mechanical as well as the electronic system. It means the engineers from both branches must cooperate closely together. In practice it is not easy to arrange such cooperation.

The second point which is necessary to take in account is that the SW for electronic control is easy to change. And not in all cases is guaranteed that the all participants are informed about such changes.

Part V
Modern Material and Technology

M. Hasan, M. Khidhair, R. Madaj, P. Spišák, P. Maly, V. Dynybyl, J. Sojka, and F. Lopot

Corrosion in Some Iron Alloys

M. Khidhair and M. Hasan

Abstract This work contains many investigations to estimate the corrosion behavior of iron alloys such as (BC18, BC13, 25CrMo4) during exposure to inorganic acids (Hcl—HNO_3—H_2SO_4) and (NaOH). Chemical compositions of allays were analyzed by spectrometer type (ARL), in weight percent. Many heat treatments were done to show their effects on corrosion resistance. Investigated specimens in disc form were immersed in aggressive mentioned solutions, for defined times. The lost in weight of specimens due to corrosion were removed manually and registered. The corrosion rates ($g/m^2 \cdot h$) were calculated according the Eq. (2) referred to in the text, during exposure time of (120) h and (240) h for some alloys. This enables the conclusion that alloys with Cr% more than 12 % had more corrosion resistance than others. Also heat treatment improved corrosion resistance, and that solution of the acids was the most aggressive medium.

Keywords Heat treatment • Corrosion • Resistance • Iron alloys

1 Aim of Investigation

The goal of this work was to test the effect of H_2SO_4, Hcl, HNO_3 acids and their mixture with NaOH, on the corrosion of some iron base alloys with different Cr and C contents. Also to show the heat treatment influence on their corrosion rates.

2 Introduction

Corrosion is a natural process happened in the metals and alloys to return back to their least level of energy [1].

Corrosion is controlled by electrochemical and thermodynamic rules, depending on types of metals and the environments causing the corrosion. The period of exposure must be also determined – about 200 h – to get real values of corrosion

M. Khidhair (✉) • M. Hasan
Al-Mustaqbal College – Babylon, Hilla, Iraq
e-mail: khidhair_aljuboury@yahoo.com

V. Dynybyl et al. (eds.), *The Latest Methods of Construction Design*,
DOI 10.1007/978-3-319-22762-7_43

rates [2]. The field of corrosion science and engineering is of high importance in both life time and technological solutions [3].

It was found that a new and better understanding of corrosion and its effect has very high importance mainly in physical and chemical base of corrosion [4].

As we know that corrosion is the gradual destruction of material. Usually metals by chemical reaction with its surroundings. Also it may occurs in other materials such as ceramics and polymers [5].

The iron and steel are the most commonly used metals, corrode in many media including outdoor atmosphere. Alloying elements can play a dominant rule, mainly Cr and Sl. Massive Cr carbides present in microstructure are $M_{23}C_6$. which can be described as composts with large and hard carbides in Cr-Fe alloy matrix, which improve corrosion resistance [6].

We must know the increasing of Si with Cr improve corrosion resistance [7].

Corrosion of irons and steels is accelerated by salt water because it is an electrolyte solution and its ions are able to move freely between the sites of oxidation and reduction, other strong oxidants as acids, and others could be aggressive also. Corrosion was accelerated due to electrochemical reaction [8].

Japanizes researchers found that stainless steel was better than Al. alloys in corrosion resistance and they calculated its corrosion rates [9], others referred to the methods of resisting or prevention corrosion [10–12].

Another Ault hers tried to find relations between chemical compositions and corrosion rates [10]. (ISO – 3651/1) [13] Stated a relation between corrosion rate and the mass lost and surface area of the tested specimens as follow:

$$\text{Vc} = \text{corrosion rate} = \frac{87600\ \text{m}}{Sdt} \tag{1}$$

$$\text{Vc} = \text{corrosion rate} = \frac{10000\,m}{S.t} \quad \left[\text{gr/m}^2 \cdot \text{hr}\right]\ [13] \tag{2}$$

Other authors used the other forms to calculate (Vc) [9, 14].

Where: S = Surface area exposed – in Cm^2; d = density of alloy – in g/cm^2; t = exposure time in h.; M = Mass loose due to corrosion (g.)

In this work we used formula (2) to calculate the (Vc) in the iron alloys with Cr—C using different aggressive materials.

3 Practical Part

3.1 Materials Used in the Work

The chemical composition of the four iron base alloys were analysed by spectrometer of (ARL) type. The composition was shown in Table 1. The aggressive media

Table 1 Chemical compassion of investigated alloys in wt. %

Alloy name and symbol	Chemical elements									
	C	Si	Mn	Cr	Ni	Cu	P	S	Al	Mo
High Cr low carbon steel A	0.327	0.365	0.641	15.80	0.606	0.054	0.025	0.014	0.04	0.994
High Cr and high carbon steel (BC18) B	2.43	0.95	0.43	14.80	0.19	0.24	0.033	0.036	0.002	0.39
Fe-Cr steel (BC13) C	2.80	0.69	0.52	11.60	0.27	0.29	0.038	0.037	0.006	0.07
Low Cr Low carbon steel 25CrMo4 D	0.245	0.465	0.389	1.23	0.06	0.09	0.017	0.03	0.10	0.27

Table 2 Aggressive media used in this investigation.

Material number	Chemical symbol	Name nation	Notes
1	H_2SO_4	Sulpheric acid	It is concentrated (95–97%) by Wt.
2	HNO_3	Nitric acid	Const. (65 %)
3	HCl	Hydrochloric acid	Const. (37 %)
4	NaOH + HNO_3 + HCl	Mixture of acids and NaOH	NaOH (50 %) const.; HNO_3 (65 %) const.; HCl (37 %) const.

used in this work were mainly inorganic acids. NaOH, also used in the mixture of HNO_3 and HCL, in equal portions in weight. Table 2 contains those media.

3.2 Method Used

Specimens were prepared as discs by machining and cutting to the weights and surface area shown in Table 3. Also hardness of the alloys was fixed in that table.

Electronic balance of ±0.01 g and dimensions measuring device of ±0.01 mm, accuracy were used. The samples were distinguished by fixed numbers. Iron base alloys were put in glass containers each with definite aggressive media. Two specimens of each alloy were used in each media. The media were poured on the alloy in the containers, which were covered by drying papers. After exposure time of 120 and 240 h., alloys were washed and cleaned from corroded materials, then weighted to determine the loss in weight due to corrosion using the same balance. The results obtained were show in Table 4 in case of 120 and 240 h., exposure time, the corrosion rates were also calculated by using Eq. (2) referred to in the introduction, the calculated corrosion rates were shown in Table 5 and in (240) h. Table 6 for (Bc18) alloy.

3.3 Heat Treatment

In order to show the effect of heat treatment on the corrosion properties of those alloys, they were hardened in oil and then tempered.

The heat treatments used were as follow.

(a) Alloys of high Cr content were hardened in oil from about 1000 °C, and then tempered in still air from about 300 °C, holding time in hardening 75 min.
(b) Alloy of low Cr and low carbon, was hardening in oil from 910 °C, then tempered at about 300 °C and still air cooled. Holding time in hardening about 90 min.

The above mentioned informations were given in Table 5.

Table 3 The initial mass and surface area of tested alloys

Alloy	Aggressive medium	Initial mass [g]	Surface area [cm^2]	Hardness (HV) for heat treated
High Cr low C after rolling	HCl Mixture	23.43	12.56	500–559 (As rolled)
		24.09	12.50	
High Cr high C BC18-as cast	HCl Mixture	43.17	19.92	520–570
		41.85	19.32	
High Cr high C BC18-Heat treated	HCl Mixture	40.09	22.14	750–790
		52.93	22.71	
Fe-Cr (BC13) alloy – as cast	HCl Mixture	32.41	18.07	500–520
		31.94	17.63	
Alloy (BC13) heat treated	HCl Mixture	44.44	20.34	720–750
		44.02	19.76	
Low Cr Low C alloy-as cast (25CrMo4)	HCl Mixture	45.50	20.13	190–210
		49.70	20.65	

Table 4 The loss in mass due to corrosion in as cast and treated conditions of iron base alloys

Type of alloy	Aggressive material	In as cast state			Heat treated		
		Lose in mass [g]	Surface area [cm^2]	Corrosion rate [g/m^2. h]	Lose in mass [g]	Surface area [cm^2]	Corrosion rate [g/m^2. h]
High Cr Low C	HCl mixture	2.75	12.56	0.00182	2.46	12.56	0.00163
-A-*		2.50	12.50	0.0033	3.14	12.21	0.0021
High Cr High C	HCl mixture	2.53	19.92	0.00106	2.40	22.14	0.00090
-B-		3.55	19.32	0.0024	3.13	22.71	0.0011
High Cr BC13	HCl mixture	2.97	18.07	0.00137	3.0	20.34	0.00127
-C-		4.15	17.63	0.0020	3.95	19.76	0.00168
Low Cr Low C	HCl mixture	5.6	20.13	0.00240	6.62	18.21	0.0030
-D-		5.3	20.65	0.00214	5.25	19.19	0.0016

4 Results Obtained

We could conclude from this work the following important results:

1. The highest corrosion resistance in all the tested alloys was in concentrated H_2SO_4 and HNO_3 acids. Also that resistance was decreased by decreasing Cr content.

Table 5 Calculated corrosion rates of (Bc18) after (120) h

	As cast			Heat treaded		
Aggressive materials	Surface area [cm^2]	Mass lose [g]	Corrosion rate [$g/m^2 \cdot h$]	Surface area [cm^2]	Mass lose [g]	Corrosion rate [$g/m^2 \cdot h$]
H_2So_4	19.90	0.47	0.00020	17.85	0.05	0.000023
HNO_3	19.90	1.38	0.00058	17.85	0.07	0.000033
HCL	19.80	3.20	0.00135	17.70	2.35	0.0011
Mixture – Acids + NaOH	19.70	5.63	0.00238	17.70	4.66	0.00220

Table 6 Calculated corrosion rates of alloy (BC18) – after (240) h

	As cast			Heat treated		
Aggressive material	Surface area [cm^2]	Mass lost [g]	Corrosion rate [$g/m^2 \cdot h$]	Surface area [cm^2]	Mass lost [g]	Corrosion rate [$g/m^2 \cdot h$]
H_2SO_4	21.28	0.15	0.000030	17.86	0.09	0.000020
HNO_3	22.50	0.12	0.000022	17.8	0.09	0.000020
HCl	22.3	4.2	0.00078	16.2	2.8	0.00066
Mixture acids + NaOH	24.5	3.5	0.00060	18.31	2.8	0.00046

2. The highest corrosion resistance in HCl acid was in alloy of highest Cr and low carbon (A) as rolled, and the least was in lowest Cr and C contents.
3. The corrosion rates of the two highest Cr content alloys (A and B), increased by exposure to the mixture of acid and NaOH more than in HCl acid.
4. Corrosion resistance of the alloy (BC18) – with high Cr – increased in the mixture of acids and NaOH by heat treatment, also its resistance improved by heat treatment.
5. Corrosion resistance of alloys with less than 12 % Cr (alloys C and D) decreased when exposed to HCl or acids and NaOH mixture.
6. Heat treatments improved corrosion resistance of alloys (BC18) and (BC13) relative to as in as cast condition.
7. Alloys of low Cr and carbon contents had relative by higher corrosion rates after heat treatment.
8. Corrosion rates generally decreased by increasing exposure time to 240 h in alloy (Bc18) in both Hcl and mixture.

5 Discussion and Recommendations

Referring to the results obtained from this work, we could state that (H_2SO_4 and HNO_3) acids had less ionic activation than Hcl and the mixture. That resulted in higher corrosion resistance in the farmer acids. The salt water also accelerates rusting due to its freely electrolyte solution.

In other hand and according to many scientific facts, the increasing of (Cr%) mainly over 12 % Cr, such as in alloys (A and B), causes high corrosion resistance – as shown in the results. Decreasing of (Cr%) below (12 %) as in alloy (C) caused decreasing of corrosion resistance relative to (A and B) alloys. Even with high (C %).

Lowest Cr% and C% contents in alloy D was the main reason to decrease its corrosion resistance especially in Hcl acid. As the investigated alloys were hardened and tempered, their microstructures were improved because of better stability and less internal stresses, accordingly their corrosion resistances, mostly improved, when we tried to check atmospheric corrosion effects, we found that alloy (D) showed low rusting rate within (240) h – about 0.00011 $g/m^2.h$.

Finally we can recommend that in order to achieve best result, chemical compositions, without undesirable impurities with as less stresses, and high structure stability must be taken in the considerations.

References

1. F. Reynolds, L. Edward, Stress corrosion cracking – ASM, in *Metal Handbook,* vol 13
2. P.L. LaQue, H.H. UHLIG, An easy on pitting crevice – Corrosion and related potentials, vol 22 No 8. (1983)
3. The National Academies Press, *Research Opportunities in Corrosion Science and Engineering*, Washington, DC (2011)
4. The National Academies Press, *Assessment of Corrosion Education* (2009)
5. Wikipedia – corrosion, -Free encyclopedia
6. NACE International, Material selection, iron corrosion (2010)
7. NSW. Charles Sturdy University, Chemistry corrosion of Fe and its alloys, Foundry Rev. (2010)
8. R.G. Lim, Jeongyoun – Inst of Technology, Effect of Cr and Si on corrosion of iron alloys (Crakov, 2008)
9. Windorff Zofia Journal, in *Metallurgical of Heat Treatment* Wyd. ((8)), Testing of corrosion resistance, Warszawa (1970)
10. C.R. Brooks, in *Heat Treatment of Fe Alloys* (New York/London, 1979)
11. S.K. Hajra, in *Materials Science – Corrosion and Its Prevention* (Indian Book, Distribution comp., 1982)
12. M.G. Fontana, in *Corrosion Engineering*, 2nd edn., ed. by International student, Corrosion Prevention (Singapore, 1984)
13. ISO – 3651/1, International Stand. org
14. Lawrence, in *Handbook of Stainless Steel. Effect on the Corrosion Composition, Structure, and Heat Treatment on the Corrosion Resistance* (New York/London, 1977)

Overview on the Clinched Joint Method

P. Maly, V. Dynybyl, J. Sojka, and F. Lopot

Abstract This article aims to give the general overview of the recent state in the problematic of joining thin-walled products. Clinched joint represents one of the technologies that are used for connecting sheet-metal parts into a final assembly. Characterization of the technology, different types and possible application of the joints are also introduced in the text.

Keywords Clinched joint • TOX®-joining system • Sheet-metal • Mechanical joining • Clinching • Riveting

1 Introduction

In recent years, there has been an effort to produce lightweight constructions because of economic and environmental reasons. The reduction of used material together with the maintaining of load capacity is the main goal of such constructions. Therefore, many products are made of sheet-metal parts that are connected together to the final shape to meet all requirements. This requires the choice of suitable methods for joining thin-walled parts into a final complex assembly.

2 Joining Methods of Sheet Metal

There are many joining methods of sheet-metal plates used in the industrial production that can be divided into four main groups: welding, joining with aid of consumable, adhesive bonding and joining by forming. The example of

P. Maly (✉) • J. Sojka
Faculty of Mechanical Engineering, Department of Designing and Machine Components, Czech Technical University in Prague, Prague, Czech Republic
e-mail: pavel.maly@fs.cvut.cz; jiri.sojka@fs.cvut.cz

V. Dynybyl • F. Lopot
Faculty of Mechanical Engineering, Czech Technical University of Prague, Prague, Czech Republic
e-mail: vojtech.dynybyl@fs.cvut.cz; frantisek.lopot@fs.cvut.cz

V. Dynybyl et al. (eds.), *The Latest Methods of Construction Design*, DOI 10.1007/978-3-319-22762-7_44

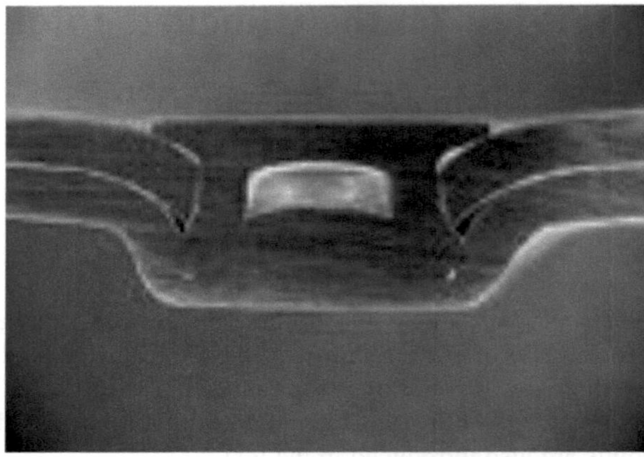

Fig. 1 Self-pierce riveting [1]

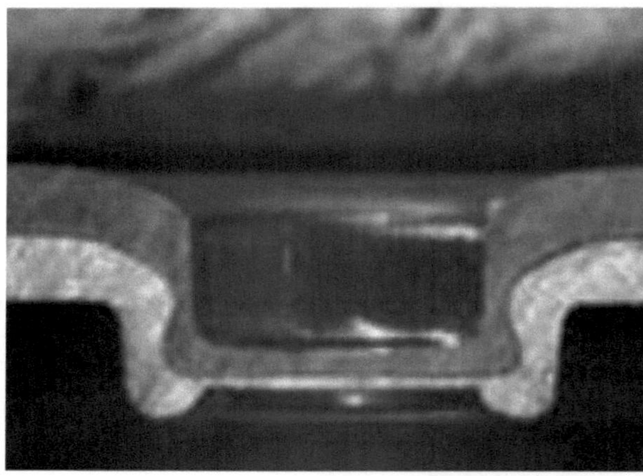

Fig. 2 TOX®-round joint [2]

connection where the additional connection aid is used can be represented for example by self-pierce rivet connection (see Fig. 1).

The joint by forming is a method of joining sheet metal by localized cold forming of materials and clinching process represents one of these technologies. There are many suppliers of the clinching technology (e.g. TOX PRESSOTECHNIK, BTM Corporation, Böllhoff, Norlok Technology, Eckold Clinching etc.). See Fig. 2 for example of such joint.

3 Clinched Joint Method

Different kinds of joining methods are used in a variety of industries and the choice of the suitable method is often crucial from many reasons. Focusing on the clinching method, there are different types of the joints depending mainly on the used tool, material and requirements. Round clinched joints (Fig. 2) is often used,

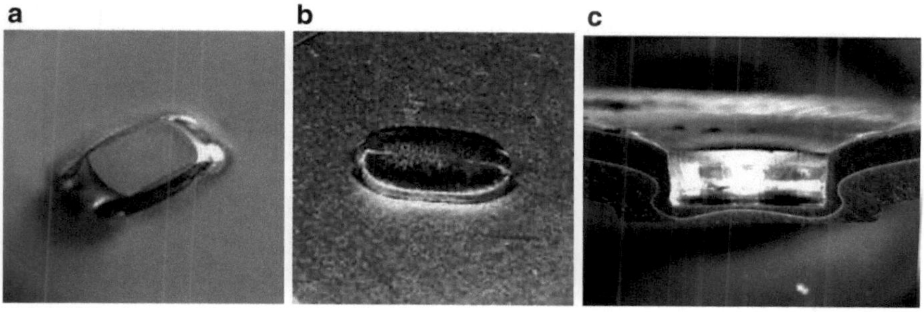

Fig. 3 Types of clinched joint: (**a**) joint with intersection [3], (**b**) oval joint [3], (**c**) dieless joint [5]

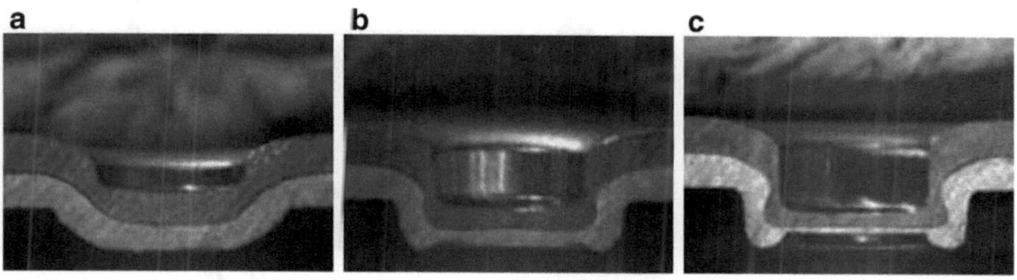

Fig. 4 TOX® joint forming: (**a**) penetration, (**b**) interlock formation, (**c**) completed joint [2]

other types of the clinched joint may be rectangular joint with intersection, oval joint [3], or dieless (flat) joint [5]. Examples are shown in Fig. 3. Main advantage of dieless joint is e.g. the possibility to local heating, therefore joining of brittle material is also possible [4].

3.1 Technology of Clinching

The process of forming the one stroke clinched joint can be divided into three main steps: penetration of the material, interlock formation of punch side sheet material and completed joint (see Fig. 4).

In general, the tool for clinching joints consists of two main components: punch and die. Modifications of the tool can occur depending on many factors, e.g. type of the clinched joint, or connected materials (Fig. 5).

Different types of machines are used for creating of clinched joints (see Fig. 6). Several types of stand-alone pneumatic presses are available. These include C-frame presses, small presses 2- and 4-column presses, specialty presses, or benchtop presses. Another type is a clinching unit that may be operated manually or automatically by a robot.

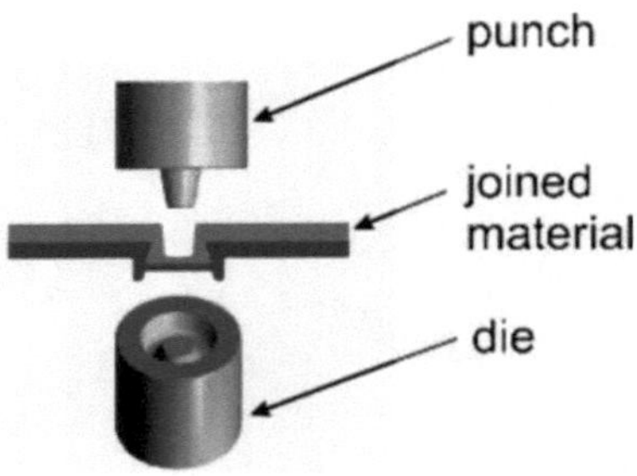

Fig. 5 Tool for clinching [2]

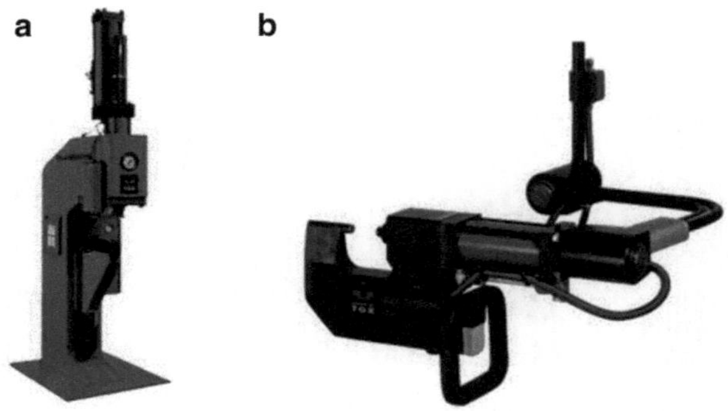

Fig. 6 Machines for clinching: (**a**) stand-alone press, (**b**) handheld clinching unit [2]

3.2 Application of the Clinched Joints

This joining method is used in many branches of production, especially where the fast, cheap and reliable connection of thin sheet-metal plates is needed. Automotive industry, HVAC, appliance manufacturing or electrical enclosures represent some main examples of clinched joints applications.

The specific application is the connection of two sheet-metal parts of the side stem of refrigerated rack that was subjected to a laboratory testing [6, 7].

4 Conclusion

The basic overview of the joining technology was introduced. It was focused on the connecting of thin-walled plates using modern clinching technologies. This joining method provides many advantages such as fast, economic, or clean technology without fumes and burrs. It also enables to connect dissimilar materials that may be pre-painted or coated. Therefore, the clinched joints can replace other methods of sheet-metal joining such as spot welding.

Acknowledgments The research work reported here was supported by an industrial cooperation.

References

1. RIVSET® Self-pierce riveting for perfect joints, Böllhoff, [online]. Available on http://www.boellhoff.com/en/de/assembly_systems/riveting/rivset.php. Accessed 14 Apr 2015
2. TOX®-Joining-Systems [online].. Available on http://www.tox-en.com/products/joining-systems/tox-clinch-procedure.html. Accessed 14 Apr 2015
3. http://www.btmcorp.com/v-loc.html
4. S. Coppieters, Experimental and numerical study of clinched connections. PhD Thesis, Katholieke Universiteit Leuven – Faculty of Engineering, Leuven, Belgium, 2012
5. R. Neugebauer, M. Todtermuschke, R. Mauermann, F. Riedel, Overview on the state of development and the application potential of dieless mechanical joining processes. Arch. Civ. Mech. Eng. **8**(4), 51–60 (2008). Wroclaw, ISSN: 1644–9665
6. P. Maly, F. Lopot, QUALISYS system applied to industrial testing, in *Applied Mechanics and Materials* (Trans Tech Publications, Uetikon-Zurich, 2014), pp. 135–140. ISSN 1660–9336. ISBN 978-3-03785-977-3
7. P. Maly, V. Dynybyl, F. Lopot, F. Stary, P. Kubat, P. Kubovy, Measurement report of side stem of refrigerated rack, Prague, 2011

The Process of Preparation Prototypes Bearing Cages for the Durability Testing

P. Spišák and R. Madaj

Abstract The aim of the article is to introduce readers to the topic of using modern technologies in prototyping, including the technology of vacuum casting. This article further describes a proposal of a special device that would be used for rapid prototyping of bearing cages, which would subsequently be a subject to a durability test.

Keywords Vacuum casting • Vacuum chamber • Bearing cage

1 Introduction

In today's dynamic market there is an increasing emphasis on a quick launch of a new product to consumers. The competition in the market and tendency to satisfy the customers' needs require new and new products. The speed of response to changes is an important factor in competitiveness. Companies therefore utilize the latest concepts and technologies already during development of prototypes.

One of these technologies is the technology of vacuum casting (VC). It is a technology of casting different material types (plastic, rubber, wax ...) into the silicone mold.

The fabrication of the mold and consequently the components themselves is very fast. For simple components it is possible to have a first sample within 5–6 h which is considerably quicker compared to traditional methods [1].

2 Casting Production Using Vacuum Casting

The manufacture process of components using VC can be divided into two stages namely the production of molds and the casting production.

P. Spišák (✉) • R. Madaj
University of Zilina, Žilina, Slovak Republic
e-mail: peter.spisak@fstroj.uniza.sk; rudolf.madaj@fstroj.uniza.sk

V. Dynybyl et al. (eds.), *The Latest Methods of Construction Design*,
DOI 10.1007/978-3-319-22762-7_45

2.1 Casting Form

In this case a casting form is a tool that allows us to make castings with the same shape, size and surface. The quality of the castings depends on the quality of casting form. Production of castings is divided into a number of operations.

- **Creation of the master model**, here a lot depends on attention to detail and perfect quality, because every little imperfection here will be reflected to the casting form, and later transferred to the individual castings.
- **Creation of casting form** from the master model. Important here is to properly place the master model and sink it in the silicone substance.
- **Removal of the master model** from the casting form means is to find an appropriate cut in order to remove master model from the silicone casting form.
- **Preparation of the silicon casting form** that deals with creation of a complete inlet system.

2.2 Casting Production

Quality of a casting is affected by the quality of casting form, proper preparation work for casting and casting process itself. It is important to follow instructions provided by manufacturers of casting form material, casting material, or equipment for the vacuum casting. The casting production process is subdivided into a number of operations.

- **Casting material preparation** includes weighing of the components and mixing them as recommended by the casting material manufacturer, it may also require pre-heating of the material.
- **The vacuuation and casting** consists of emptying the casting space, mixing the two components of the casting material, and filling the casting form with the casting material.
- **Tempering** will enable that material obtains characteristics and quality indicated by the manufacturer.
- **Removing of the casting form**
- **Finalisation of the casting**, these are the final touches like the removal of the inlet system or degreasing of the casting.

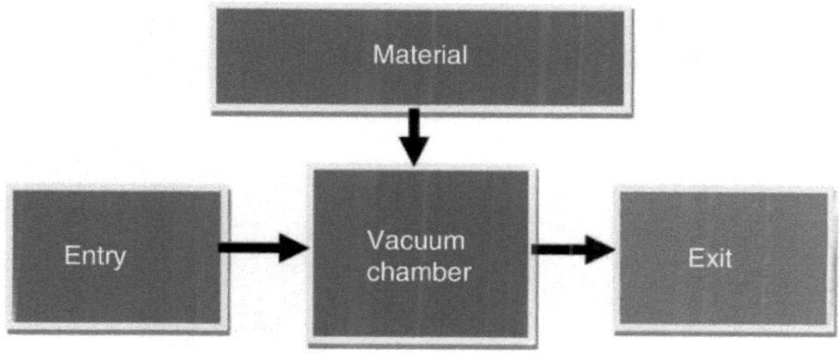

Fig. 1 Block diagram

3 Design of Special Equipment for Vacuum Casting Production of Prototypes

There are several manufacturers of equipment for vacuum casting on the market. However, there is a room for improvement to produce an equipment that would be faster and easier to use. These conditions are met by the vacuum casting equipment we've designed (Fig. 1).

The main parts of the equipment are:

- **Dosing device** – material is dispensed by dosing device Dekumed Unidos 200 which has a sufficiently large reservoir of material and is capable of smooth mixing ratio change (Fig. 2) [2].
- **Mixing equipment** – the material is blended by a static mixer, there are two input chambers for material and an output for homogeneous substance.
- **Vacuum chamber** – is made of 8 mm thick steel plates, designed for a vacuum value ≤ 0.5 mbar. There is a transparent polycarbonate with a thickness of 15 mm in front. Vacuum chamber is attached to a linear lead that is moved by a pneumatic cylinder (Fig. 3).
- **Rotary air-pump** – type SV25 1,090,249 is used for the extraction of air from the casting form. Detailed parameters listed in the Table 1.
- **Conveyor belt** – is a standardized conveyor with the driving gear inside the drum. It moves forms during the working process.

The complete 3D CAD model of the device can be seen in Fig. 4.

By using the device some tasks will take noticeably shorter (vacuuming space casting, mixing of the material), others will be even completely removed (weighing of the material) where errors occur frequently and thus the casting is flawed.

Fig. 2 Dekumed Unidos 200

Fig. 3 Vacuum chamber and conveyor belt

Table 1 Parameters of rotary pumps [3]

Vacuum power of the pump (m^3/h)	Voltage (V/Hz)	Power (KW)	Weight (kg)
25/$\leq$0.5 mbar	230/50	1.2	28

Fig. 4 Special equipment for vacuum casting

4 Use of a Special Device for Production of Bearing Cages Prototypes

This equipment can be used to manufacture prototypes of bearing cages that are destined for durability tests. This device would partially reduce time to manufacture the bearing cage and would also reduce faulty casting products due to incorrect material mix ratio.

The entire preparation of this kind of prototype includes a number of dimension corrections. Based on required characteristics it is important to compile a dimensional chain/matrix and follow it from production preparation phase. The final dimensions are determined mainly by shrinkage of polyurethane during solidfition, by the casting form dilatation during casting or also by chemical reaction, by the dimensions accuracy of the printouts for the casting form production.

Only after mastering and testing of these corrections it is possible to prepare components of a large-sized bearing in required quality and durability with geometric tolerances allowing passing durability test reflecting real situation at the specified load (Fig. 5).

Components produced by using the described processes with the use of modern technologies have been applied in durability tests. These tests were carried out as a part of the bearings testing framework that took place at the Faculty of Mechanical Engineering, University of Žilina.

Fig. 5 Silicone molds and casting bearing cage with rolling elements

Acknowledgments This work was supported by the Slovak Research and Development Agency under the contract no. APVV-0419-11 – Adaptation of modern computer simulation methods to the development of rolling bearings and their verification in real conditions.

This work was supported by the Scientific Grant Agency of the Ministry of Education, Science, Research and Sport of the Slovak Republic under the contract no. V-1/0396/14 – Research of the influence of construction and technological parameters of rolling bearings on the service life thereof.

This work was supported by the Scientific Grant Agency of the Ministry of Education, Science, Research and Sport of the Slovak Republic under the contract no. V-1/0844/13 – Research on guidance elements of rolling bearing and their design.

References

1. M. a kol Žarnay, Technologickosť konštrukcií strojov (EDIS, Žilina, 2000), str. 150. ISBN 80-7100-664-5
2. Dekumed. [Online] (2013). Available on http://www.dekumed.de/U200e.html. Accessed 19 Apr 2013
3. Mcae. [Online] (2013). Available on http://www.mcae.cz/vakuove-lici-sys. Accessed 25 Apr 2013

Part VI
Optimization and Design

V. Bauma, M. Bozca, J. Drahokoupil, J. Dvoracek, V. Hanus, J. Homišin, J. Hosek, L. Komenda, P. Lepsik, M. Moravic, L. Nohal, O. Novák, P. Oganesyan, M. Petru, V. Rada, S. Shevtsov, Z. Sika, F. Stary, P. Svatos, M. Valasek, J.-K. Wu, I. Zhilyaev, and J. Zicha

Optimisation of the Geometric Design Parameters of a Five Speed Gearbox for an Automotive Transmission

M. Bozca

Abstract Optimisation of the geometric design parameters of a five speed gearbox for an automotive transmission is studied. The bending stress is considered as the objective function, and the design parameters are optimised under several constraints, including contact stress and constant distance between gear centres. During optimisation, the contact ratio changes with respect to pressure angle, and the effects of the contact ratio and profile modification on tooth bending stress are analysed. By optimising the geometric design parameters of a gearbox, including the module, number of teeth, helix angle and face width, it is possible to reduce tooth bending stress and obtain a light-weight-gearbox structures. It can be concluded that increasing the contact ratio results in a reduced tooth bending stress; however, in contrast, increasing the pressure angle caused a reduction in the contact ratio and an increase in tooth bending stress and contact stress. In addition, it can be concluded that positive profile modification reduces tooth bending stress. All of the geometric design parameters determined by optimisation satisfy all constraints.

Keywords Gears • Contact ratio • Bending stress • Contact stress

1 Introduction

Optimisation of the geometric design parameters of a five speed gearbox for an automotive transmission is studied. The purpose of this study is to determine the geometric design parameters of a five-speed gearbox for an automotive transmission by minimising the tooth bending stress.

By optimising the geometric parameters of the gears, such as, the module, number of teeth, helix angle, and face width, it is possible to obtain light-weight gearbox structures. Optimised geometric design parameters satisfy all constraints, and the best solutions are selected from the obtained optimum solutions for each given speed.

M. Bozca (✉)
Yildiz Technical University, Istanbul, Turkey
e-mail: mbozca@yildiz.edu.tr

V. Dynybyl et al. (eds.), *The Latest Methods of Construction Design*,
DOI 10.1007/978-3-319-22762-7_46

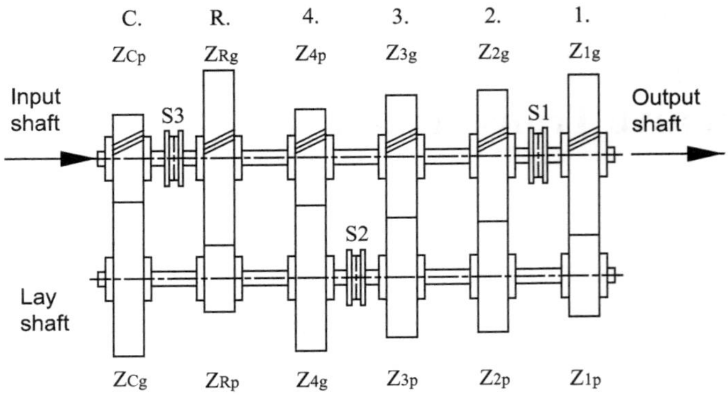

Fig. 1 A five speed gearbox for an automotive transmission

1.1 Gearbox Mechanism

The gearbox mechanism includes pinion gears, wheel gears, an input shaft, an output shaft, a lay shaft, a bearing support, and synchronisers, as shown in Fig. 1.

All pinion and wheel gears are helical, and all gears are made of 16MnCr5.

2 Contact Ratio

The average number of teeth in contact as the gears rotate together is the *contact ratio* (CR) [1].

The total contact ratio, ε_γ, is calculated as follows.

$$\varepsilon_\gamma = \varepsilon_\alpha + \varepsilon_\beta \tag{1}$$

where ε_α is the transverse contact ratio and ε_β is the overlap ratio.

3 Profile Modification

Profile modification is given as follows.

$$V = x.m \tag{2}$$

where x is the profile modification factor [−] and m is the module [mm]. When x is positive, it is called positive profile modification, and when x is negative, it is called negative profile modification [2].

4 Calculating the Load Capacity of Helical Gears

4.1 Tooth Bending Stress

The real tooth-root stress, σ_F, is calculated as follows [2–4]

$$\sigma_F = \frac{F_t}{bm_n} Y_F Y_S Y_\varepsilon Y_\beta K_A K_V K_{F\beta} K_{F\alpha} \tag{3}$$

where F_t is the nominal tangential load [N], b is the face width [mm], m_n is the normal module [mm], Y_F is the form factor [−], Y_s is the stress correction factor [−], Y_ε is the contact ratio factor [−], K_A is the application factor [−], K_V is the internal dynamic factor [−], $K_{F\beta}$ is the face load factor for tooth-root stress [−] and $K_{F\alpha}$ is the transverse load factor for tooth-root stress [−].

The safety factor for bending stress S_F is calculated as follows [2–4]

$$S_F = \frac{\sigma_{Fp}}{\sigma_F} \tag{4}$$

where σ_{Fp} is permissible bending stress.

4.2 Tooth Contact Stress

The real contact stress, σ_H, is calculated as follows [2–4]

$$\sigma_H = \sqrt{\frac{F_t}{bm_n} \frac{u+1}{u}} Z_H Z_E Z_\varepsilon Z_\beta \sqrt{K_A K_V K_{H\beta} K_{H\alpha}} \tag{5}$$

where d_1 is the reference diameter of the pinion [mm], u is the gear ratio [−], Z_H is the zone factor [−], Z_E is the elasticity factor [$\sqrt{N/mm^2}$], Z_ε is the contact ratio factor [−], Z_β is the helix angle factor [−], $K_{H\beta}$ is the face load factor for contact stress [−] and $K_{H\alpha}$ is the transverse load factor for contact stress [−].

The safety factor for contact stress, S_H, is calculated as follows [2–4]

$$S_H = \frac{\sigma_{Hp}}{\sigma_H} \tag{6}$$

where σ_{Hp} is the permissible contact stress.

5 Optimisation of Gearbox Design Parameters

Constrained optimisation approaches are applied to the gears system. All programs are developed using MATLAB and in all optimisation studies, the sequential quadratic programming (SQP) method is employed.

To find the optimum design parameter, the initial design parameters of the gear system including m, z, β, and b, are varied. Four design parameters are optimised simultaneously using the programs developed. During optimisation, different initial value vectors are used to identify the global minimum solution of the objective function, $\sigma(m, z, \beta, b)$.

The following objective function was employed:

$$F = \min(\sigma) \tag{7}$$

Minimum tooth bending stress $\min(\sigma)$, is defined as follows:

$$\min(\sigma_F) = \min\left(\frac{F_t}{bm_n} Y_F Y_S Y_\varepsilon Y_\beta K_A K_V K_{F\beta} K_{F\alpha}\right) \tag{8}$$

The tooth contact stress is considered as the constraint functions during optimisation. The following are considered as the constraint functions.

$$\sigma_H - \sigma_{Hp} \leq 0 \tag{9}$$

6 Numerical Example

The contact ratio and bending stress relations are shown in Fig. 2. Increasing of the contact ratio, results in reduced tooth bending stress and reduced contact stress.

The contact ratio and pressure angle relations are shown in Fig. 3. Increasing of the pressure angle, result in reduced the contact ratio and increased the tooth bending stress and contact stress.

The profile modification factor and bending stress relations are shown in Fig. 4. Increasing of the profile modification factor, results in a reduction in tooth bending stress.

7 Conclusion

Optimisation of the geometric design parameters of a five speed gearbox for an automotive transmission is studied. The following conclusions are drawn.

By optimising the geometric design parameters of the gearbox, including the module, number of teeth, helix angle and face width, it is possible to reduce the tooth bending stress and obtain a light-weight gearbox structures.

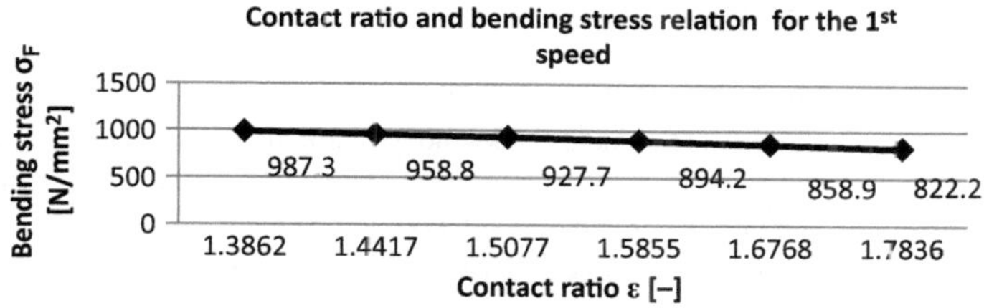

Fig. 2 Contact ratio and bending stress relation

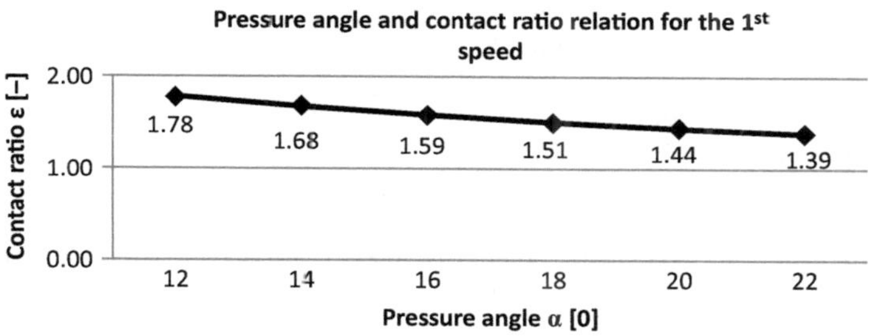

Fig. 3 Pressure angle and contact ratio relation

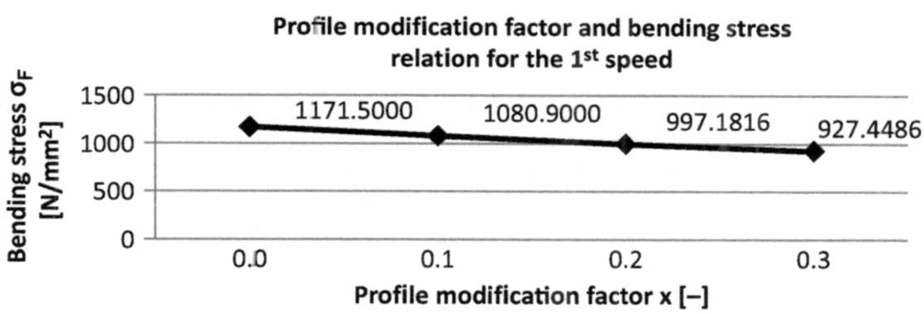

Fig. 4 Profile modification factor and bending stress relations

Increasing of the contact ratio, results in reduced tooth bending stress and contact stress. In contrast, increasing the pressure angle reduces the contact ratio and increases the tooth bending stress and contact stress.

Profile modification is an effective parameter to reduce tooth bending stress. Increasing of the profile modification factor, results in a reduction in tooth bending stress.

All geometric parameters can be selected independently for each speed inside the obtained optimum solutions. The best solutions are selected from the obtained optimum solutions for each speed (Table 1).

Table 1 Optimisation results

	1st Pinion	2nd Pinion	3rd Pinion	4th Pinion	Constant Pinion	Rear Pinion
	Sol. no 1	Sol. no 3	Sol. no 3	Sol. no 6	Sol. no 5	Sol. no 1
	$\alpha = 12°$	$\alpha = 16°$	$\alpha = 16°$	$\alpha = 22°$	$\alpha = 20°$	$\alpha = 12°$
Module m	3.4442	3.3707	3.2281	2.8267	3.6172	2.7141
Number of teeth z	14.0000	19.0000	19.0000	19.0000	19.0000	19.0000
Helix angle β	32.0000	30.7541	30.7453	30.7284	30.7733	31.6943
Face width b	34.0000	33.0000	32.0000	32.0000	32.0000	44.0000
Pressure angle α_t	14.0709	18.4523	18.4507	25.1743	22.9583	14.0261
Centre distance a	80.0000	80.0000	80.0000	79.9735	79.9885	80.0000
Transverse contact ratio ε_α	1.7836	1.6184	1.6296	1.4321	1.4535	1.8913
Overlap ratio ε_β	1.6651					
Bending stress σ_F	822.2394	738.8489	754.3754	680.0000	366.1000	847.6631
Safety factor for bending stress S_F	1.2162	1.3535	1.3256	1.4706	2.7314	1.1797
Contact stress σ_H	921.600	828.5000	772.0000	801.0000	610.7000	1141.1000
Safety factor for contact stress S_H	1.5191	1.6897	1.8134	1.7478	2.2926	1.2269

References

1. R.C. Juvinall, K.M. Marshek, *Fundamentals of Machine Component Design*, Asia (John Wiley & Sons, Inc., 2006)
2. M. Roloff, *Maschinenelement* (Vieweg & Sohn Verlag/Fachverlage GmbH, Wies-baden, 2005)
3. Decker. *Maschinenelement* (Carl Hanser Verlag, München, 2009)
4. H. Naunheimer, B. Bertsche, J. Ryborz, W. Novak, *Automotive Transmissions* (Springer Verlag, Berlin/Heidelberg, 2011)

Modification of Experimental Station AXMAT

J. Dvoracek, L. Komenda, and L. Nohal

Abstract In this article is described modernization of experimental station AXMAT, which is used for testing of contact fatigue of small axial bearings and flat specimens. The test-rig is placed in a laboratory of Institute of Machine and Industrial Design of Brno University of Technology. Original setup of the test-rig was based on applying of axial force at a specimen by mechanical lever with additional weight. This stand is designed for determination of resistance of materials and bearings against the contact damage. This is achieved with use of vibrodiagnostic. Essential changes were made on the test-rig by its modification. A new hydraulic load system for set up of axial force on a specimen was developed. A control program was created in SIMULING for operating of the hydraulic system and its force. This program gives a chance to adjust applied axial force on a specimen in optional modes. Moreover a feedback for record of vibration, acoustic emission (AE) and temperature signals was created. An analysis of AE signal enables a detection of contact fatigue formation before pitting occurs. Due to that it is possible to detect defects on a specimen earlier than by standard vibrodiagnostic's method. Because of a new modifications and improved measurement system the test-rig provides more complex and accurate detection of a contact fatigue.

Keywords Acoustic emission • AXMAT • Axial bearing

1 Introduction

Contact fatigue is a process which causes defects onto the surfaces of interacting elements. One of the most significant examples where fatigue occurs is a bearing contact. There are several forms of defects caused by contact fatigue, for instance pitting, spalling act.

Bearings are tested for a resistance against the contact fatigue in laboratory conditions with use of experimental stands. There are two basic methods for

J. Dvoracek (✉) • L. Komenda • L. Nohal
Brno University of Technology, Brno, Czech Republic
e-mail: dvoracek@fme.vutbr.cz; komenda@uk.fme.vutbr.cz; nohal@fme.vutbr.cz

V. Dynybyl et al. (eds.), *The Latest Methods of Construction Design*,
DOI 10.1007/978-3-319-22762-7_47

non-destructive detection of defects on specimens during run: Vibro-diagnostic and Acoustic emission analysis.

There is particular experimental station "AXMAT" in laboratory of Institute of Machine and Industrial design at Brno University of Technology. The station AXMAT is design for testing of small axial bearings and flat specimens. A load applied on the specimen/bearing is created by manual lever with possibility of additional weight. This setup results in constant load and axial force acting at the specimen in run. However there are almost no operation conditions with constant load in real situations. Due to that it was decided to modify AXMAT in a way where loading system would be adjustable during operation [1].

2 Materials and Methods

For new loading system were considered pneumatic and hydraulic systems. After comparison its advantages and disadvantages the hydraulic principle was chosen.

The new loading system is based on hydraulic piston which creates force in axial direction to the specimen (Fig. 1). The whole system is operated by PLC. There was also placed a feedback regulation of axial force on the specimen in the system. The feedback is based on a force sensor, which provides information about acting force and bring the signal to the operation unit (PLC).

Figure 2 performs a realization of new modifications of experimental station AXMAT [2].

For higher quality of operational conditions of running experiments, PC driver provides more variations of loading set up. This was achieved with operational modes which were exactly defined in PC software for control of hydraulic system.

Each mode was programmed for its own run conditions. This means setting of upper and lower limits for specification of applied axial force. That the axial force can change by specific mode definition between those limits.

Another way for control of hydraulic system is provided by software MATLAB – SIMULINK. In this software was created program for dynamic control of the system. It is possible to adjust running conditions of the test for arbitrary needs of user.

There are signals from vibration and acoustic emissions captured in the system. Vibration signal is manly used for detection of the end of the test. When defect occurs on the specimen/bearing the vibration signal exceeds acceptable limit and the whole test stops. Acoustic emission signal is used for detection of early stadium of a defect. It is more sensitive method for defect analysis that vibration method [3].

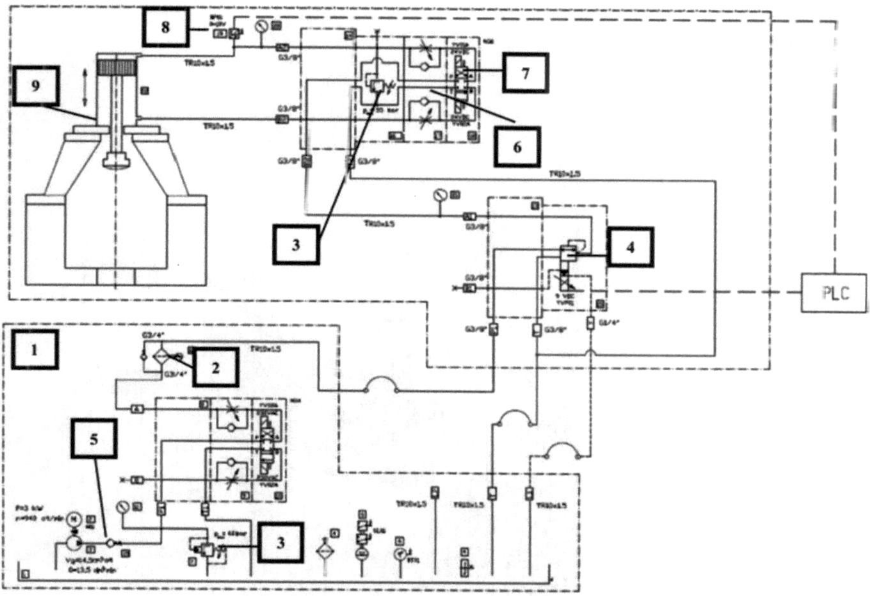

Fig. 1 Schema of hydraulic loading system
1. Hydraulic unit
2. Filter
3. Cylinder valve
4. Proportional reduction valve
5. One-way valve
6. Throttle valve
7. Hydraulic switchboard
8. Pressure sensor
9. Hydraulic motor

3 Experiment

The system for dynamic loading, which was designed and then realized, was tested to verify its functionality. Results of parameters of acoustic emission signals of contact fatigue test with dynamic loading are plotted on Fig. 3.

Comparison of signals from AE sensor and vibration sensor are plotted on Fig. 4.

4 Conclusion

Experimental station AXMAT was modified for possibility of defect detection of axial bearing running under variable loading conditions. A new hydraulic system was used for the aim of the AXMAT reconstruction whit possibility of PC driving.

Fig. 2 Realization of station modifications

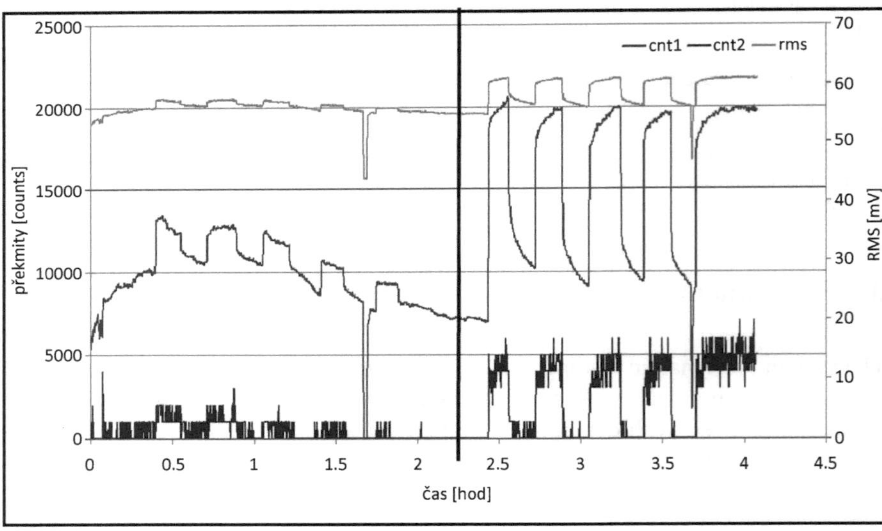

σ_{Hertz} = 4500/4000 Mpa
po 10 min

σ_{Hertz} = 5000/4000 Mpa
po 10 min

Fig. 3 Parameters of AE signal under dynamic loading

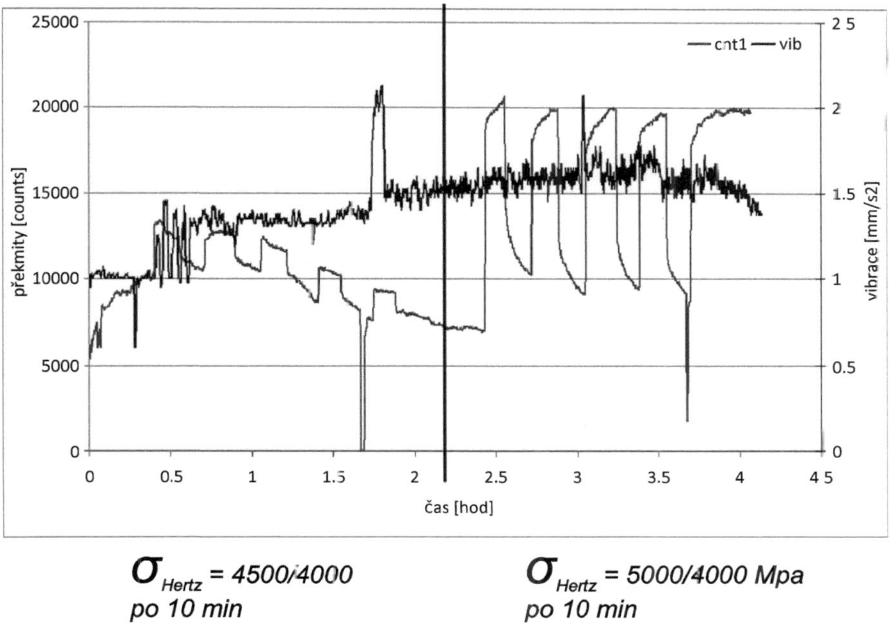

Fig. 4 Comparison of AE and vibration signals

Functionality of new loading system and measurement system was verified during an experimental test and also importance of monitoring of AE parameters for experiments running under dynamic contact loading conditions was shown.

Acknowledgments This work reported here was made possible by support of Institute of machine and industrial design at Brno University of Technology.

References

1. J. Dvořáček, Vývoj nové generace zařízení s pokročilou diagnostikou pro stanovení kontaktní degradace, Brno. Dissertation thesis. Brno University of Technology, 2014
2. Reconstruction of AXMAT testing station, Brno, Diploma thesis, Brno University of Technology, 2014
3. Z.A. Khan, M. Hadfield, Y. Wang, Pressurised chamber design for conducting rolling contact experiments with liquid refrigerant lubrication [online] 2005. Available http://www.sciencedirect.com/science/article/pii/S026130690400202X. Accessed 22 Feb 2013

Basic Conditions of the Implementation of Extremal Regulation

J. Homišin and M. Moravič

Abstract Extremal regulation is an optimization method that does not require knowledge of mathematical model because it is practiced by experimentation on device directly in operation. The first part of this article contains brief characteristic of extremal regulation and presentation of controlling of torsional vibration at the system. The second part includes illustration of control algorithm of extremal regulation system at the implemented mobile torsionally oscillating mechanical system at steady state. The main aim of this article is to summarize the way of application of this method in torsionally oscillating mechanical systems with usage of pneumatic flexible shaft couplings being generated at our department.

Keywords Method of extremal regulation • Automatic control circuit • Control algorithm of extremal regulation system • Torsional vibration • Pneumatic flexible shaft coupling

1 Introduction

Mechanical systems with combustion engines, compressors, pumps and ventilators may be characterized as torsionally oscillating mechanical systems (TOMS) [1]. In consequence there is emergence of dangerous torsional vibration. The function of optimization of TOMS is to minimize the size of torsional vibration at stabilized state of system. With the usage of this method we are capable tuning of TOMS and providing for the system to operate most effectively from viewpoint of the size of dangerous vibrations. It is provided for by finding optimal air pressure in pneumatic flexible coupling that is minimum of objective function. By means of convenient regulation of pressure of gaseous medium in compression space of couplings, their dynamic torsional stiffness can be changed, hence there is a change of natural frequency of the system [2].

J. Homišin (✉) • M. Moravič
Technical University of Košice, Košice, Slovak Republic
e-mail: jaroslav.homisin@tuke.sk; marek.moravic@tuke.sk

V. Dynybyl et al. (eds.), *The Latest Methods of Construction Design*,
DOI 10.1007/978-3-319-22762-7_48

2 Characteristic of Extremal Regulation

Extremal regulation is an optimization method that does not require knowledge of mathematical model. Static optimization can be applied during the steady-state operation of mechanical systems. In Fig. 1a and b there is an illustration of presentation of controlled torsional vibration mechanical system at the system at steady-state operating mode, it means at constant operating speed characterized by position *(1)* to *(10)*. Characteristics represent the results of the simulation of extremal regulation at implemented TOMS and a change in operating mode is implemented by skip. These results are illustrated by regulation of gaseous medium pressure p_S (Fig. 1a) and by regulation of the size of torsional vibration in system A_T (Fig. 1b). The change in operating mode of system causes an overpressure adjustment of gaseous medium from pneumatic tuner of torsional vibration. By Fig. 1b influence of overpressure of gaseous medium over the size of mechanical vibration in system A_T can be discovered. [3–6].

2.1 Input and Output Quantities of the Controlled System

Controlled system includes input and output quantities (Fig. 2). The input quantities encompass action quantities x, which are used to manage the status of the system, parameters p influencing the behaviour of the mechanical system, their value is predetermined, disturbances z affecting the status of the system are random and cannot be controlled. Output quantities include state variables F, from which a objective function is made up. Also technological and economic variables belong here, but they are not used for regulation to establish the objective function [7].

2.2 The Objective Function of the System

In our case input quantity is characterized by an air pressure in the pneumatic-flexible elements of used flexible coupling, which the stiffness and damping properties directly depend on. Output quantity value can be dynamic component of load torque M_{dyn}, effective value of dynamic component of load torque *RMS* and dynamic component of angle of twist of coupling φ_{dyn}. In Fig. 3 there is a functional dependence of effective value of dynamic component of load torque *RMS* transmitted by the coupling from the pressure of the gaseous medium in the compression space p_S [2, 4] as the objective function.

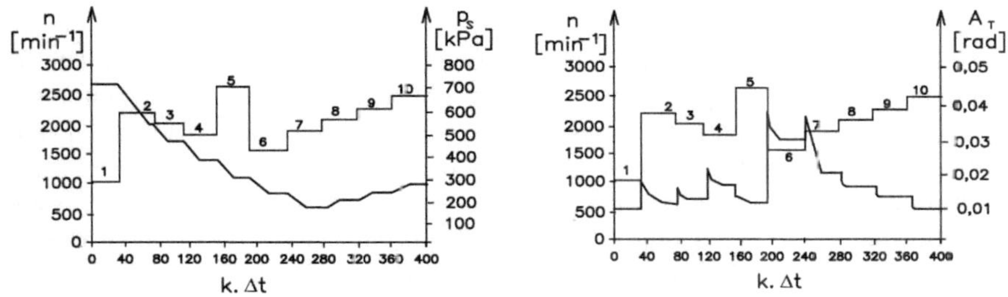

Fig. 1 (**a**) The result of static optimization of implemented system characterized by the change of the action quantity p_S in the range of operating speed n depending on the discrete time $k.\Delta t$ and (**b**) the result of static optimization of implemented system characterized by the change of the controlled quantity A_T in the range of operating speed n depending on the discrete time $k.\Delta t$

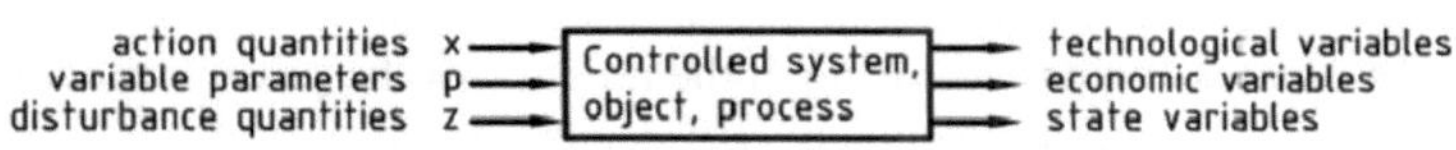

Fig. 2 Diagram of the controlled system

Fig. 3 The objective function as dependence of *RMS* on the pressure p_S in pneumatic coupling

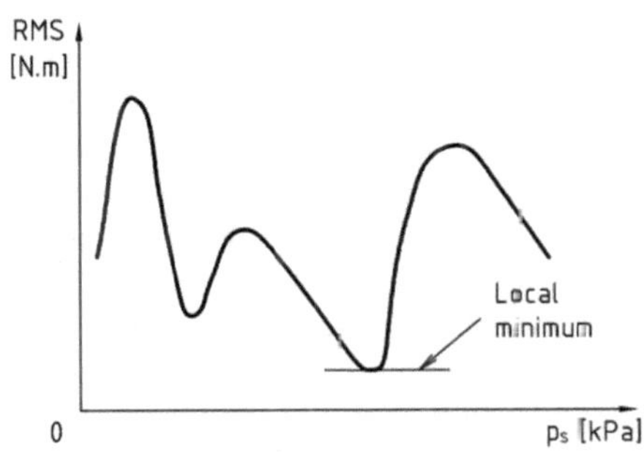

2.3 Implementing Conditions of Extremal Regulation

Pneumatic coupling, ie pneumatic tuner torsional vibrations must be part of the mechanical system. The basic inevitable conditions for application of method of extremal regulation are as follows:

- TOMS has to satisfy the basic requirements imposed to implementation of continuous tuning, i. e. tuning of system during operation in steady state [1],
- implementation of static optimization TOMS does not require knowledge of its mathematical model, but it is necessary to know, whether the particular objective function has extreme (local minimum) [2–5].

3 Automatic Control Systems and Circuits

Extremal regulation of system can be practised by the automatic control system for continuous change of pneumatic coupling characteristics or by the automatic control system for continuous tuning of mechanical system. There is a patent protection [8, 9] granted for these automatic control systems.

Automatic control system (Fig. 4a) (*1-position*) consists in a control (*2*) and actuating subsystem (*3*). Control subsystem (*2*) is characterized by a microprocessor (*6*), inductive sensor built directly in the pneumatic-flexible element (*10*) or at another suitable place of pneumatic flexible coupling and contact or contactless portable device (*5*) of electrical signal placed on the rotating shaft of the mechanical system. Actuating subsystem (*3*) consists in a two-position electromagnetic distributor (*9*), which is connected to the gaseous medium container (*7*) through the control valve (*8*). Microprocessor (*6*) directly controls distributor (*9*) after measured signal processing. The continuous tuning of mechanical system is accompanied by a continuous increasing or decreasing of the pressure of the gaseous medium in the pneumatic-flexible coupling elements (*10*) depending on position, in which electromagnetic distributor currently is. Automatic control system (Fig. 4b) (*1*) consists in a control (*2*) and actuating subsystem (*3*). Control subsystem (*2*) is characterized by a microprocessor (*5*) and the piezoelectric accelerometer of mechanical vibration (*4*). Sensor (*4*), which is placed on the mechanical system sends out an electrical signal to the microprocessor (*5*) proportional to the mechanical vibrations of the system. Actuating subsystem (*3*) consists in an electromagnetic distributor (*7*), which is connected to the gaseous medium container (*8*) through the control valve (*6*). Microprocessor (*5*) acting as the basic control component of the whole regulatory system (*1*) causes its shift by closing the electromagnetic two-position distributor (*7*), therefore the inlet branch *P* or the outlet branch *T* increases or decreases the pressure of gaseous medium in the compression volume of pneumatic flexible shaft coupling [2].

Some of automatic control systems can be used for extremal regulation implementation. Automatic control circuit according to control system [8] (Fig. 5a) consists in regulated TOMS (*1*), which in operation can be affected by disturbance quantities *z*. Optoelectronic sensor (*2*), with help of which quantities *F* values of objective function are sensing, sends out a signal in extremal regulator – microprocessor (*4*) by contactless transmission (*3*). Microprocessor on basis of processed electrical signal with help of generated algorithm directly controls actuator, and that is electromagnetic distributor (*5*), by which input quantity *x* – pressure in compression space of pneumatic coupling is regulated, namely by filling or draining of gaseous medium. Automatic control circuit illustrated in Fig. 5b is different from the former in a type of sensor used (piezoelectric accelerometer of mechanical vibration) and missing contactless portable device of electrical signal [2, 3].

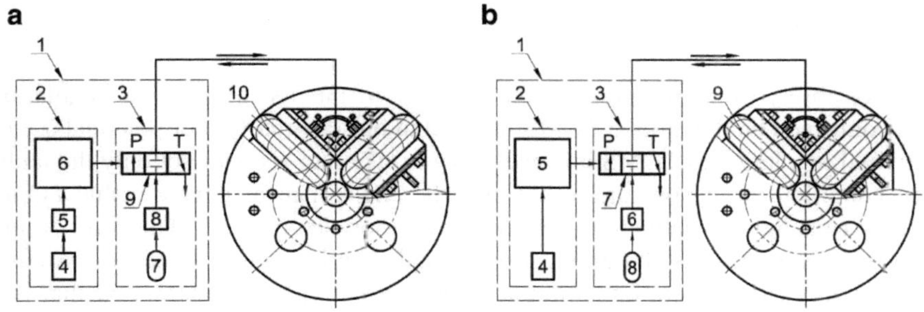

Fig. 4 (**a**) Automatic control system for continuous tuning of mechanical system and (**b**) automatic control system for continuous change of pneumatic coupling characteristics

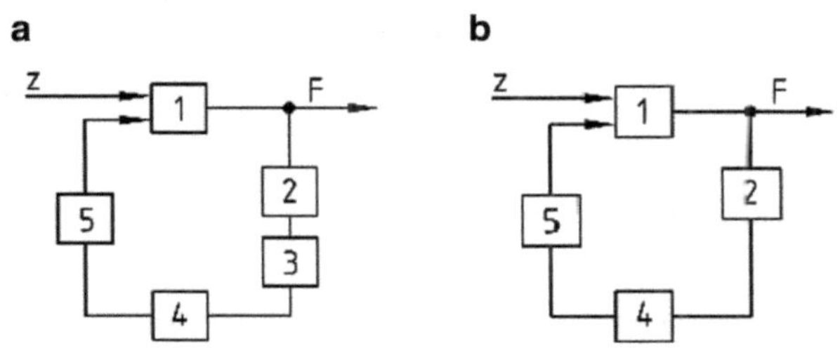

Fig. 5 Automatic control circuits of extremal regulation: (**a**) according to automatic control system [8] and (**b**) according to automatic control system [9]

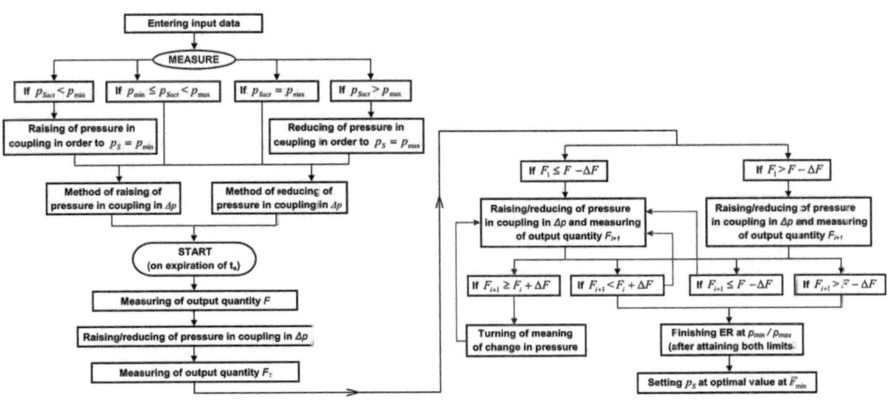

Fig. 6 Diagram of control algorithm of extremal regulation TOMS

4 Control Algorithm of Extremal Regulation

Figure 6 illustrates control algorithm of extremal regulation system at implemented TOMS.

Given algorithm corresponds to extremal regulation with stepping change of action quantity. Extremal regulation was practiced on mobile TOMS at constant operating speed and load.

5 Conclusion

The possibility of static optimization by usage of method of extremal regulation was confirmed on the mobile TOMS realized in our laboratory. For practical use of extremal regulation in TOMS it is necessary to take into account not only the fault-free condition during operation, but also the emergence of unpredictable failure effects causing the dangerous torsional vibration in systems. The aim of my doctoral dissertation is the application of static optimization based on extremal regulation during system operation at changing external influences, and that is a change in working mode (operating speed), irregularity of the working machine and falling of one cylinder of the engine out of operation.

Acknowledgments This paper was written in the framework of Grant Project VEGA: "1/0688/12 – Research and application of universal automatic control system in order to master the source of mechanical systems excitation".

References

1. J. Homišin, *Mechanická sústava vhodná pre realizáciu jej plynulého ladenia*. Patent file No. 276926/92
2. J. Homišin, *Nové typy pružných hriadeľových spojok. Vývoj – Výskum – Aplikácia* (Vienala, Košice, 2002). ISBN 80-7099-834-2
3. J. Homišin et al., *Súčasné trendy optimalizácie strojov a zariadení* (C-PRESS, Košice, 2006). ISBN 80-8073-656-1
4. J. Homišin, Možnosť realizácie extremálnej regulácie v torzne kmitajúcich mechanických sústavách. Automatizace **40**(5), 247–251 (1997)
5. J. Homišin, Príspevok k statickej optimalizácii torzne kmitajúcej mechanickej sústave. Strojnícky časopis. **4**(96), 272–279
6. J. Homišin, R. Grega, Príspevok k statickej optimalizácii mechanických sústav na základe metódy extremálnej regulácie. Acta Mech. Slovaca **9**(3-B), 19–26 (2005)
7. P. Kaššay, *Optimalizácia torzne kmitajúcich mechanických sústav metódou extremálnej regulácie* (Doktorandská dizertačná práca, Košice, 2008)
8. J. Homišin, *Regulačný systém pre realizáciu plynulého ladenia mechanickej sústavy*. Patent file No. 276927/92
9. J. Homišin, *Regulačný systém pre zabezpečenie plynulej zmeny charakteristiky pneumatickej spojky*. Patent file No. 259225/87

Design of a Disc Grinder for EDM Electrodes Machining

J. Hosek and J. Drahokoupil

Abstract This paper gives a review of possibilities for dressing rotational electrodes by electro discharge machining (EDM) technique. We designed a vertical EDM disc grinder as an alternative to available sacrificial block electrode dressing technique. We performed experimental machining tests with this vertical grinder and we have explained the effects, which affect the stability of this electro discharge machining technique.

Keywords Electro discharge machining • Electrode machining • Disc grinder • Erosion • Stability

1 Introduction

One of the main tasks of preparation stage for the workpiece EDM machining is the formation of an electrode of appropriate shape. There are available standard commercial electrodes of different materials, shapes and dimensions, but its limited variety of dimensions and shapes restricts the range of patterns which could be machined. This is a serious problem, especially in micro EDM, where the electrode dimension needs to be chosen according to the dimension of the spark gap distance, typically 5–20 μm. The microelectrodes of such dimensional steps necessary for optimal product dimensions EDM machining are not commercially available.

To tackle this problem, a few methods for preparation the electrode of any necessary dimension were developed. We have made a review of the most common possibilities for machining rotational electrodes or shafts by electro discharge machining (EDM) technique. Based on this review we designed a vertical EDM disc grinder as an alternative to the most common electrode machining technique by sacrificial block. We performed experimental machining with this vertical disc EDM grinder and we have explained the effects, which affect the stability of this electro discharge machining technique concept.

J. Hosek (✉) • J. Drahokoupil
Czech Technical University in Prague, Prague, Czech Republic
e-mail: Jan.Hosek@fs.cvut.cz

V. Dynybyl et al. (eds.), *The Latest Methods of Construction Design*,
DOI 10.1007/978-3-319-22762-7_49

2 Methods Used for the Electrode Dressing

A choose of electrode machining method is mainly affected by the electrode material and its dimension. Large enough electrodes, made from well machinable material, as graphite, can be simply turned or milled. Small electrodes for micro EDM are usually made of tungsten and the only way to form it into the desired shape and dimension is the use of EDM machining.

The most common method of rotational microelectrodes production is its EDM dressing by a precisely grinded AgW sacrificial block [1, 2]. The rotating electrode is pushed towards the sacrificial block by appropriate path and velocity till it erodes itself to desired dimension, as it is shown in the Fig. 1, left. This simple technique has few practical disadvantages. The first, it is a compromise between machining time and precision. The precise diameter can be reached under fine machining condition, but this condition extends the machining time. The second disadvantage is a local wear of the AgW sacrificial block. This local wear, see Fig. 1 in the middle, affects the precision of the electrode final diameter. If a new electrode will be dressed at the same place as previous, the electrode reaches other diameter. This is a limiting factor affecting the sacrificial block lifetime.

The worn area of the sacrificial block can be minimized by electrode dressing by its sinking into the sacrificial block [3, 4], as it is shown in the Fig. 1, right. In this case the sacrificial block can be made of easily EDM worn and cheap material, compared to AgW sacrificial block.

Another electrode dressing methods eliminate local sacrificial block wear by using moving elements like a wire or disc [1, 2]. Disadvantage of use wire electrode dressing can be a low stiffness of dressing wire, given by its tension. This drawback can be overcome by using a wire guide [5]. Based on this review we decided to design a disc grinder for micro electrodes dressing, where the wear is applied along large area of the disc and the impact of the wear on the disc dimensions could be minimized.

3 Design of the Disc EDM Grinder

The design of the disc EDM grinder has to fulfil a few limiting conditions. The experimentally verified optimal disc rotational speed is between 50 to 150 rounds per minute. It needs to use a gear to reduce drive engine rotational speed and increase its torque. Another critical condition comes from the limitation of disc wobbling, which has to be much smaller than the spark gap distance, typically 5–20 μm. The whole system design has to be compact to be fixed to the magnetic table of our EDM machine SODICK AP1L and to be able submerged into the dielectric oil.

The minimum wobbling of the EDM grinding disc was achieved by the use of precise ball bearings directly pressed on the grinder vertical shaft. An electrical

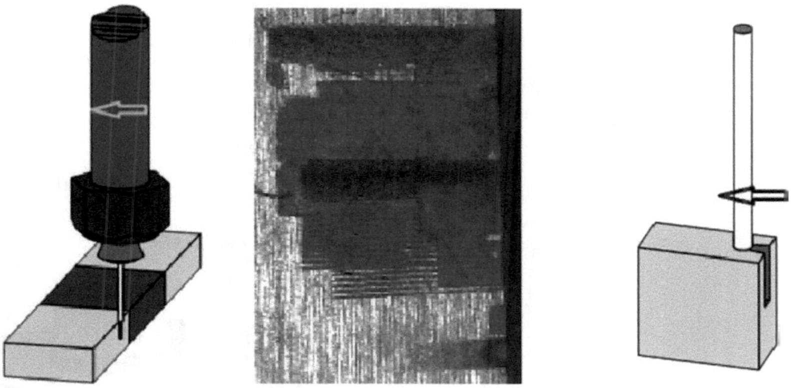

Fig. 1 Scheme of electrode dressing by sacrificial block – *left*; local wear of the sacrificial block by dressed electrode erosion – *middle*; and sinking the electrode into the sacrificial block – *right*

contact was assured by a preloaded career bar and the plate insulation by a ceramic wheel coaxially cemented to the shaft. A drive part of the system was designed separately and the motion transmission is performed by preloaded tooth belt to eliminate clearances and vibrations. There was used a DC motor Black&Decker 400 W with a planetary gear as a drive. The design of the system is shown in the Fig. 2.

4 Experimental Results

We performed series of EDM machining tests with this EDM disc grinder. We used an readily worn material (aluminium) for a grinding wheel to easily observe the wear effects caused by EDM process. Machining goes well if the disc has no defects or wobbling. If any defect appears on the disc, this defect starts to increase itself. Compared to common grinding, where the grit with a greater radius is more easily broken up and the wheel self dresses (a stable system with negative feedback) the ED machining behave as a positive feedback system, i.e. an unstable system. Any protrusion of the spinning disc causes temporary decrease of the spark gap distance and machining efficiency what leads to lower local erosion rate compared to the rest of the disc. It gradually increases the height of the defect compared to its surrounding. If the defect dimension reaches the spark gap distance the erosion is completely stopped by short circuit. This effect occurred both in axial and radial machining feed directions. The minimal intensity of this effect was observed in the tangential feed direction.

Fig. 2 Scheme of the complete design of the EDM plate grinder system

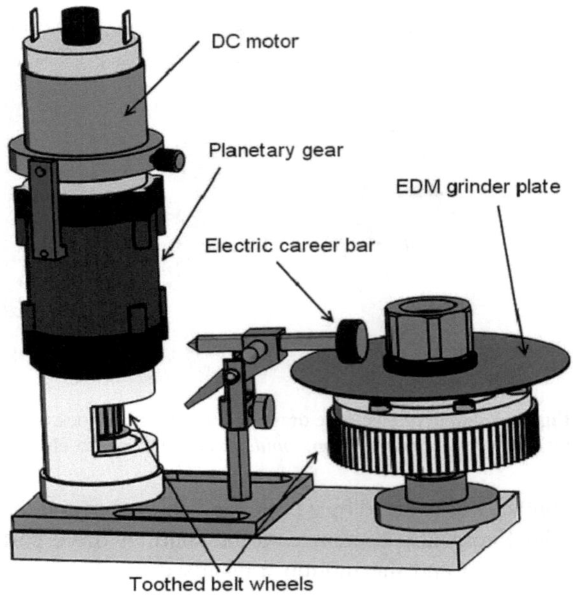

5 Conclusion

We designed and tested an EDM disc grinder for EDM electrodes dressing. We found a significant difference in EDM disc wear characteristics compared to common grinding process. A positive feedback in EDM disc erosion can be overcome by tangential feed direction or minimizing the disc wobbling below the spark gap distance values. We will continue in optimization of this system.

Acknowledgments The research work reported here was made possible by grand No. SGS14/055/OHK2/1T/12.

References

1. H.S. Lim, Y.S. Wong, M. Rahman, M.K. Edwin Lee, A study on the machining of high-aspect ratio micro-structures using micro-EDM. J. Mater. Process. Technol. **140**, 318–325 (2003)
2. M.P. Jahan, M. Rahman, Y.S. Wong, A review on the conventional and micro-electro discharge machining of tungsten carbide. Int. J. Mach. Tools Manuf. **51**, 837–858 (2011)
3. L. Bert, Micro EDM: State-of-the-art, applications and integrations with micro-milling, in: *Precisiebeurs* (KU Leuven, 2011), Presentation_Precbeurs_Micro_EDM_Lauwers.pdf. https://lirias.kuleuven.be/handle/123456789/332819
4. M. Yamazaki, T. Suzuki, N. Mori, M. Kunieda, EDM of micro-rods by self-drilled holes. J. Mater. Process. Technol. **149**, 134–138 (2004)
5. J. Fleischer, T. Masuzawa, J. Schmidt, M. Knoll, New applications for micro-EDM. J. Mater. Process. Technol. **149**, 246–249 (2004)

Optimized Design of the Wind Turbine's Composite Blade to Flatten the Stress Distribution in the Mounting Areas

P. Oganesyan, I. Zhilyaev, S. Shevtsov, and J.-K. Wu

Abstract This paper presents the numerical method and results of the structural optimization of the wind turbine blade's mounting zones to diminish and flatten the stress distribution at the action of the wind load on the blade. The blade, which has the stiff carbon/epoxy composite skin and less stiff lightweight core body, is mounted by two spokes to the rotor hub, and due to the wind load the areas of the blade near mounting of the spokes are extremely stressed. To minimize these stresses concentration the guess model of the blade was created using NX CAD tool, and finite element analysis was provided using COMSOL Multiphysics. Our modeling of the distributed air pressure load on the blade surfaces at the different wind orientation used the turbulent k-ε flow model, and allowed to find most loaded angular position of the blade at the rotor rotation. In order to minimize the peak von Mises stress around most stressed mounting hole the founded pressure field was imported to the Structural Mechanics/Optimization module, which solved the coupled mechanics/optimization problem. The optimized parts were the multilayered polymeric composite spacers with the hardened metallic folies and hole for the spoke fixing. As the design variables we used two parameters that describe the axial and radial dependencies of varied Young module and also the wall thickness of the spacer. The objective function was assumed in the form of the weighted sum of von Mises stress averaged over the area around the mounting holes. Finally we demonstrate some optimization results.

P. Oganesyan (✉)
Don State Technical University, Rostov on Don, Russia

South Center of Russian Academy, Rostov on Don, Russia
e-mail: gwolwer1ne@gmail.com

I. Zhilyaev • S. Shevtsov
South Center of Russian Academy, Rostov on Don, Russia

Southern Federal University, Rostov on Don, Russia
e-mail: zhilyaev@mail.com; sergnshevtsov@gmail.com

J.-K. Wu
National Kaohsiung Marine University, Kaohsiung, Taiwan, R.O.C.
e-mail: jiingkae.wu@gmail.com

V. Dynybyl et al. (eds.), *The Latest Methods of Construction Design*,
DOI 10.1007/978-3-319-22762-7_50

Keywords Wind turbine • Composite blade • Structural optimization • Stress reducing

1 Introduction

The problem of structural optimization is very important for many high loaded structures, especially in the cases when the weight and (or) size of such structures are constrained, or in the cases of high stress local concentration, and we need to flatten this stress distribution. Most often this problem can be solved by using the methods of the structural topology optimization [1, 2], by the parameterized geometry [3] or mechanical properties [4]. But there are structures whose geometry cannot be changed due to some requirements, e.g., constraints caused by the nearest parts of whole machine, by the aerodynamic shapes or other reasons. Such example that we consider here is the rotor blade of the wind turbine [5, 6]. The designed turbine is installed on a pontoon floating in the sea surface, and produces the electric energy for the fishnets. It consists of five blades, each mounted to the hub by two stiff spokes. The shape of the blade is the part of a helix (with diameter ~4 m and pitch 36 deg/m) whose axis coincides with the vertical axis of turbine rotation (see Fig. 1). The cross-section of the blade is the aerofoil with the chord ~0.75 m, and full length of the blade is ~3.5 m. The blade has the stiff carbon/epoxy composite skin and less stiff lightweight core body. Because of very high wind load with the different orientation the areas of the blade near mounting of two spokes are extremely stressed. Due to the turbine rotation the spatial distribution of the wind load changes at the different azimuth position of each blade, hence the stress intensity and distribution around the mounting points also changed. To minimize and flatten these stresses the guess model of the blade was created using NX CAD tool, and the following finite element (FE) analysis, which includes the optimization procedure, was provided using COMSOL Multiphysics.

In order to optimize the studied rotor blade we used the objective function in the form of the weighted sum of von Mises stress averaged over the area around the mounting holes, and the design variables were the diameter of this cylinder area and two parameters that determine the axial and radial dependencies of varied Young module of material.

2 Modeling of Wind Loading on the Blade

On the first stage of investigation the pressure load on the blade surface at the different wind orientation has been calculated using k-ε turbulent flow model with the parameters $C_{e1} = 1.44$, $C_{e2} = 1.92$, $C_{\mu} = 0.09$, $\sigma_k = 1.0$, $\sigma_e = 1.3$. $\kappa_v = 0.41$, $B = 5.2$. Such two steps approach to the optimization problem is similar to used

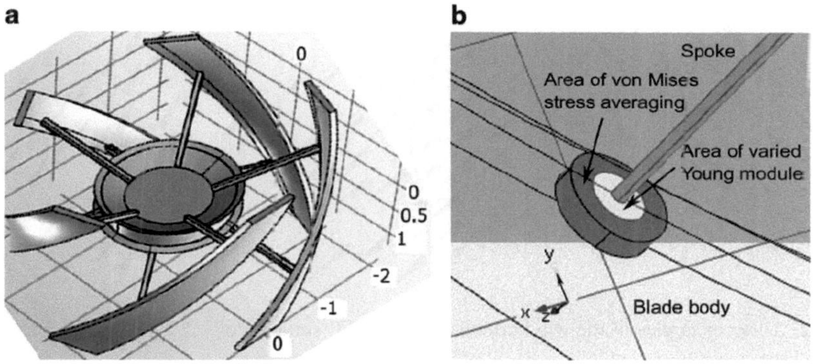

Fig. 1 CAD model of the studied wind turbine rotor (**a**) and mounting of spoke to the blade (**b**)

in [7]. We studied 12 wind directions (see, eg. Fig. 2) relative to one blade in the static state, and the velocity vector has the components according to the formula

$$[u_x \quad u_y \quad u_z] = [10 \cdot \cos(\phi_i) \quad 0 \quad 10 \cdot \sin(\phi_i)]; \quad \phi_i = \pi i/6; \quad i = 0...11 . \tag{1}$$

After these numerical simulations, two pressure fields, which act on the blade surface and match to the most stress state of material around the mounting zones have been selected for the following optimization procedure, and exported to the FE tool as the text tables.

3 Optimization of the Mounting Areas of the Blade

In order to eliminate the stress concentration near the mounting zones where rotor blade transmits the torque to the hub through the spokes made from aluminium alloy we use the reinforcing spacers, whose radial dimension, radial and axial stiffness distributions are the design variables. Such spacers can be manufactured by the alternating lay up of the impregnated fabric and titanium foil. The above mentioned axial and radial dependencies of the spacer's mechanical properties have been determined in consideration of mechanical properties of aluminium alloy (Young modulus $Y_{Al} = 70$ GPa and Poisson ratio $\nu_{Al} = 0.3$) and used polymeric composite ($Y_{comp} = 20$ GPa, $\nu_{comp} = 0.3$). These dependencies

$$Y_r(\mathbf{x}) = Y_{Al} - (Y_{Al} - Y_{comp}) \cdot r_n(\mathbf{x})^{k_1}, \tag{2}$$

$$Y_h(\mathbf{x}) = Y_r(\mathbf{x}) - (Y_r(\mathbf{x}) - Y_{comp}) \cdot h(\mathbf{x})^{k_2} \tag{3}$$

are expressed through dimensionless distances $r_n(x), h(x)$

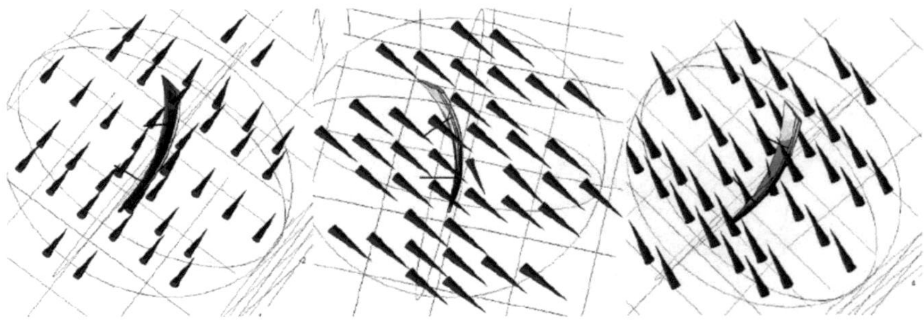

Fig. 2 Three examples of the wind orientation at the FE simulation

$$r_n(\mathbf{x}) = \begin{cases} [r(\mathbf{x}) - r_{\text{int}}]/r_{ext} & at \quad r(\mathbf{x}) > r_{\text{int}} \\ 0 & otherwise \end{cases}, \tag{4}$$

$$h(\mathbf{x}) = 1 - 2 \cdot |d/2 - l(\mathbf{x})|/d \tag{5}$$

where $\mathbf{x}$ is a coordinate of the arbitrary point of the reinforcing spacer, $r(\mathbf{x})$ is the distance between this point and axis of the hole, whereas $r_n(\mathbf{x})$ is its spacing to the hole surface, r_{ext} and r_{int} are the outer and inner radii of the spacer, d is the blade thickness in the hole location, l is the distance between point $\mathbf{x}$ of spacer and the nearest blade's skin. In these relationships the outer spacer's size $r_{ext} \in [2; 4]$ cm and two powers $k_1, k_2 \in [1; 4]$ that determine the spatial behavior of the spacer's Young module (see Fig. 3) are the design variables, which should be found by the optimization procedure.

The objective function was adopted in the form of weighted sum of averaged von Mises stress into the cylindrical subdomains of composite. The averaging procedure was the integration of Miss stress over these subdomains, which have the external radii 10 cm, and surround the spacers (see Fig. 1a).

The used optimization technique included static analysis performed under the FE model of the single blade with two fixed surfaces of the hole, and the pressure field that acts on the blade's aerodynamic surfaces, and also Nelder-Mead optimization algorithm, which call this FE model on each iteration step. Due to ability of Comsol Multiphysics FE tool to provide a collective work of these two applications, a whole problem has been solved in the common environment. In our numerical experiments the best performance of the Nelder-Mead algorithm has been established and proved in the different problem statements.

The histories of the problem convergence are presented in Fig. 4.

The diagrams in Fig. 4 demonstrate a fast convergence of the problem solution. The design variables tend to their limiting values, which is understandable. It is obvious that larger size of the spacers can provide more flatten stress distribution. In fact, their diameter is constrained by the dimensions of the blade because the mounting hole position is close to the leading edge of aerofoil.

The resulting values that characterise the stress concentration are sufficiently better comparing to their values before optimization. So, the peak stress around

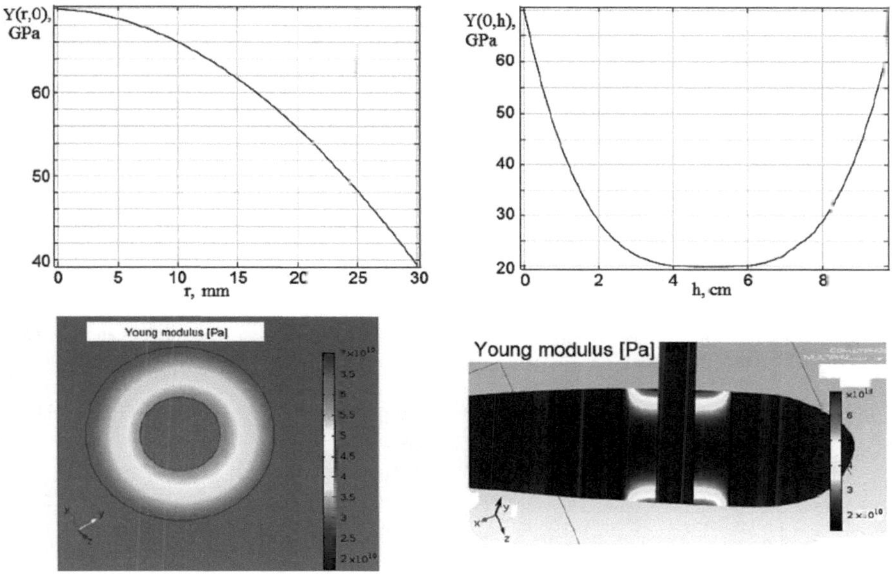

Fig. 3 Spatial distributions of the Young modulus in the radial (*left*) and axial directions (*right*). *Left plots*, which depict the module's distribution on the blade skin, and *right plot*, which relate to the cross-section of the blade, are built for $k_1 = k_2 = 3$

holes was diminished to 17 %, the averaged stress – up to 12 %, and the most stressed volume of composite blade was decreased up to 35 %.

4 Conclusion

The problem of the wind turbine blade optimization was formulated and solved by the sequential determination of the stress field, which acts on the blade surfaces at the different blade's azimuth position, with the following solving the coupled structural mechanics / optimization problem. The guess model of the blade was created using NX CAD tool, and exported for the finite element analysis to COMSOL Multiphysics. The size and spatially varied mechanical properties of the reinforcing spacers were used as the design variables, whereas the von Mises stresses averaged in the vicinity of mounting holes were used as the objective function. The optimized design was demonstrated the lowering of the peack stress concentration up to 17 % and reduction of overstressed volume up to 35 %. The presented example demonstrates that the proposed formulation is successful in incorporating the effect of wind loading into the structural optimization problem.

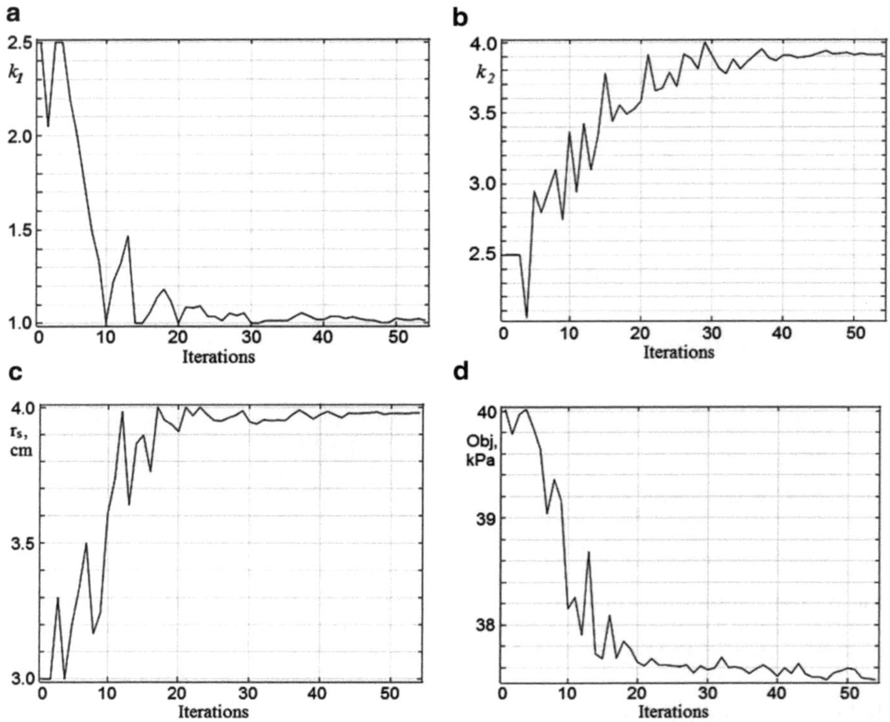

Fig. 4 The dependencies of the design variables (**a**–**c**) and objective function (**d**) on the iteration number at the solving of optimization problem

Acknowledgments This work has been financially supported by the Russian Foundation for the Basic Research (Grants 14-08-31612, 15-08-00849). The authors wish to acknowledge the valuable support from the National Kaohsiung Marine University (Taiwan, R.O.C.).

References

1. M.P. Bendsoe, O. Sigmund, *Topology Optimization. Theory, Methods and Applications*, 2nd edn. (Springer, Berlin, Heidelberg, New York, 2004), 370 p
2. R. Meske, J. Sauter, G. Zeynel, Recent improvements in topology and shape optimization and the integration into the virtual product development process, in *International Workshop on Advances in Shape and Topology Optimization*. Graz, Austria, 2008, 12 p
3. S. Shevtsov, M. Flek, I. Zhilyaev, Modeling and optimal design of power high stroke piezoelectric actuators for rotorcraft applications, in *Physics and Mechanics of New Materials and their Applications* (NOVA Publishers, New York, 2013), pp. 259–274
4. G.W. Reich, B. Sanders, Development of skins for morphing aircraft applications via topology optimization, in *Book of Proceedings of the 48th AIAA/ASME/ASCE/AHS/ASC Structures, Structural Dynamics, and Materials Conference* (Honolulu, Hawaii, 2007). 13 p

5. O. Guillermin, Going digital is key for composite wind blade manufacturers. *VISTAGY Wind Blade eBook Series*, vol. 2, 7 p. [online]. Available http://www.siemens.com/2014/energy/wind-power
6. S. Yoshioka, Development of a new blade profile for a vertical axis wind turbine, in *Book of Proceedings of the 2013 Comsol Conference* Boston, 2013, 6 p
7. R. Zakhama, M.M. Abdalla, Z. Gurdal, H. Smaoui, Wind load effect in topology optimization problems. J. Phys. Conf. Ser. **75**, 9 (2007)

Research of New Composites for Lightweight Construction with Low Impact on the Environment

M. Petrů, P. Lepsik, and Ondřej Novák

Abstract New types of composite materials are currently very important for use in lightweight structures with very high strength and resistance to external influences. Therefore, it is increasingly used for their properties in all areas of industrial production. Their application offers a significant weight savings and reduction of a manufacturing energy consumption. But current composite properties are not at maximum level, thus development efforts are focused on their improvements. One of the problems of new composites is a study and description of an interface between the phases. The paper describes experimental and FEM analysis dealing with mechanical properties at the micro level Analyses were focused on a cohesion and the strain at the interface of the system fiber-matrix-core. These composites are used for components of car parts such as doors, diffusers, spoilers and more detailed description can be applied for the improving of mechanical properties of composite structures.

Keywords Lightweight constructions • Composite parts • Environment • Cars

1 Introduction

Composite materials are now increasingly used for their properties in all areas of industrial production for high specific strength. Their use offers a significant weight savings and reduced energy consumption. But current composite properties are not at maximum level, and therefore the efforts of development on further improving are focused. The large amounts of energy required for the production of steel structures and their susceptibility to corrosion caused by climatic conditions and the external environment are well known. The above mentioned disadvantages of applied materials especially when used for structures in seawater could be addressed by using low-energy light composite materials with reduced

M. Petrů (✉) • P. Lepsik
Faculty of Mechanical Engineering, Technical University of Liberec, Liberec, Czech Republic
e-mail: michal.petru@tul.cz; petr.lepsik@tul.cz

O. Novák
Textile Faculty, Technical University of Liberec, Liberec, Czech Republic
e-mail: novak.ondra1@seznam.cz

V. Dynybyl et al. (eds.), *The Latest Methods of Construction Design*,
DOI 10.1007/978-3-319-22762-7_51

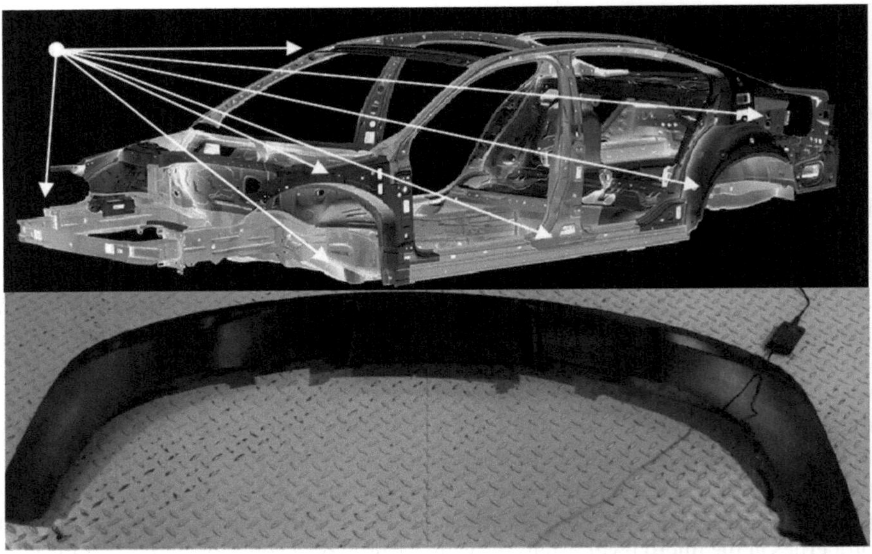

Fig. 1 Composites for lightweight constructions with low impact on the environment: a design of car frame (*above*), composite part – rear diffuser (*below*)

environmental impact [1–4]. These materials are suitable for a variety of reasons, including in particular their high specific strength, light weight, toughness, corrosion resistance, and thermal stability, etc. in comparison with the aluminium [4, 5] or with steel [6]. Composite materials consist of a matrix and reinforcement. A majority of materials used for this purpose is highly resistant to salt and thereby increase the efficiency of components and parts of vehicles (Fig. 1). The study and description of the interface among phases represents one of the problems of new composite structures. The paper describes experimental and numerical analysis dealing with mechanical properties at a micro level. Analyses were focused on a cohesion and strain at the interface of system fiber-matrix-core. The knowledge about behaviour on the micro-level can help to improve properties of new type of composites.

2 Materials and Methods – Theory

A damage mechanism of composite with continuous endless fiber *f*, which are wound around the geometry of the core connected by a matrix m is a complex analytical problem difficult to describe. This composite has strength χ_S (Eqs. 1 and 2) that is done with synergy depending on the properties of individual components, which form a compact structural system. Scattering of the strength of the fibers and the influence of microstructural defects can be roughly defined by introducing the strength of the fiber bundle χ_f as well as an efficiency coefficient of matrix ψ. Properties of fiber composites also significantly affect by the fiber orientation relative to the loading direction. If the angle of winding of fibers is greater than 10°, strength is reduced. Therefore, the strength is proportional to the winding angle of fibers. The strength of the matrix, the adhesion of the matrix to the fibers and a strength of the interface are crucial parameters of the composite. An interphase is affected by a transverse tensile strength (perpendicular to the fiber direction), which is highly dependent on the strength of the fibers in contact with the matrix and the tensile strength of the matrix. The structure of the composite can fail in the interphace by a shear type of damage χ_S^I – type I or out-phase damage χ_S^{II} – typ II (Eq. 3). The strength of the composite significantly reduces also the presence of pores a, which can be identified in many cases only microscopically (Eq. 4).

$$\chi_S = \psi \chi_f V_f + (\sigma_m)_{\varepsilon_f \lim} V_m \Big|_{\psi=\langle 1 \div 2 \rangle} \tag{1}$$

$$\chi_S^{\perp} = (\sigma_m)_{\varepsilon_f \lim} \chi_{fm} \left[1/\chi_{fm} + \sqrt{4V_f/\pi} - 1/\chi_{fm} \sqrt{4V_f/\pi} \right] \tag{2}$$

$$\chi_S^I = \frac{G_m}{V_m + V_f G_m / G_f} \text{ or } \chi_S^{II} = 2V_f \sqrt{\frac{E_f E_m V_f}{3V_m}} \tag{3}$$

$$\chi_S = \chi_S^{I,II} (1 - a + a^2)/(1 + a) \tag{4}$$

where V_f, V_m is the volume of the fibers and the matrix, E_f, E_m is the elastic modulus of the fibers and the matrix, G_f, G_m is shear modulus of fibers and matrix, σ_m is the stress in the matrix, χ_{fm} is the tensile strength of fibers in the contact with the matrix, $a = \upsilon / (1 - V_f)$, υ is the pore volume in the composite.

3 Experimental and Numerical Analysis

With the help of new technology [3], four types (see Fig. 2a) of composite structures with reduced environmental impact were produced. Composites were reinforced with carbon fibers that were wound on a PU tubular core. The winding angle α was 0°, 20°, 45° and 60°. Mechanical properties of composites through tensile tests were studied. For the study of an interfacial stress and interphace damage a numerical model using composite Chamis model [7–9] was built. Physical parameters of the model are put in Table 1 (Fig. 3).

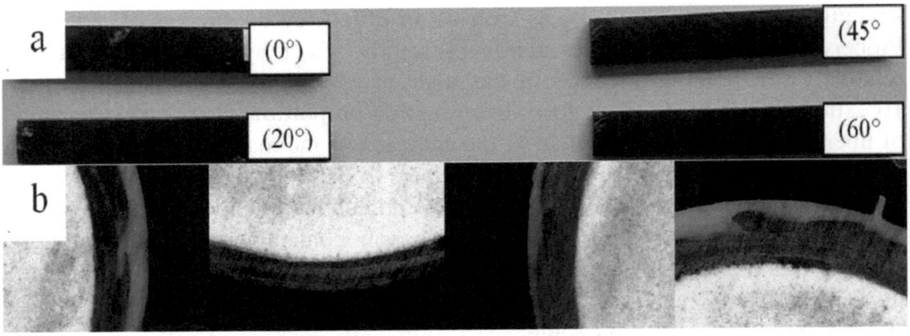

Fig. 2 (**a**) Samples of new environmental low-energy composites, (**b**) detail of cross-section

Table 1 Physical parameters of the material applied in numerical simulation [3, 4]

Material	Density [kg·m^{-3}]	Module of elasticity [GPa]		Module of shear [GPa]		Tensile strength [GPa]
		Long. E	Trans. E	Long. G	Trans. G	
Fibers of carbon	1750 ± 150	250	14	23	5.4	2.3
Epoxy resin (matrix)	1150 ± 370	3.2	3.2	1.24	1.24	0.067
Core	50 ± 0.36	0.0026		–	–	–

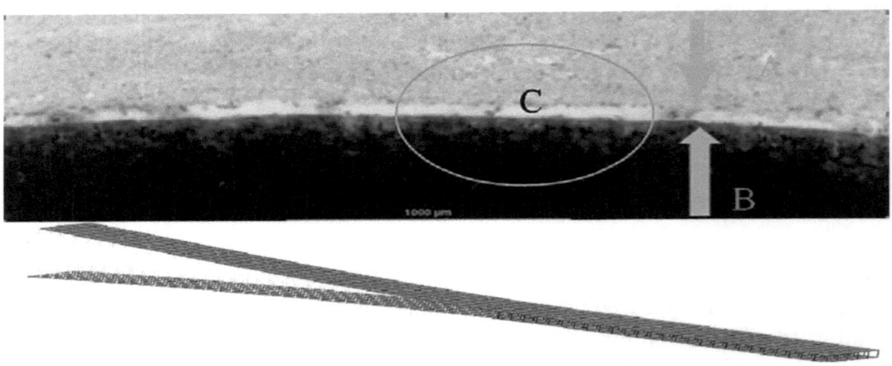

Fig. 3 Image analysis of interphase (*above*), FEM model of interphase (*below*)

4 Result and Discussion

From tensile tests it was determined that new types of composites with reduced environmental impact have quasi-isotropic course of a strain-stress dependence. The orientation of fibers affects not only the connections of fibers and matrix, but also a way of damage of resulting delamination of the composite structure

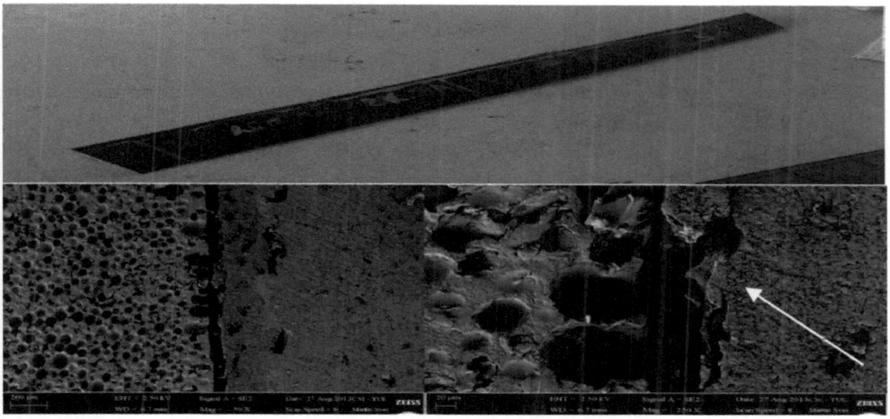

Fig. 4 Composite part (*above*), study delamination through the experimental analysis (*below*)

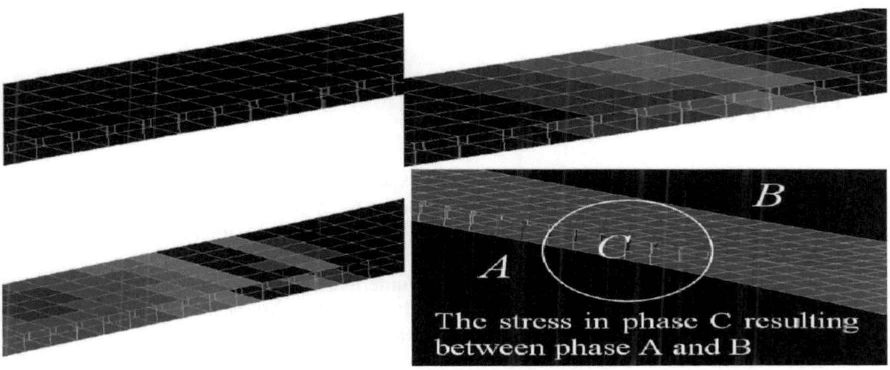

Fig. 5 Study delamination through the FEM model

(see Figs. 4 and 5). Experimental and numerical analysis has shown that the reinforcement oriented in the direction of the load (angle between load and fiber direction is 0°) exhibits the elastic modulus E 57.9 ± 7.3 MPa. If the angle between load and fiber direction is 20° the elastic modulus is approximately 46.3 ± 5.67 MPa. If the reinforcement is oriented at larger angles, then moduli report lower values (for 45° is $E = 19.6 \pm 2.1$ and for 60° is $E = 11.2 \pm 1.9$).

5 Conclusions

Experimental and numerical stress analyses of new types of composites with reduced environmental impact were carried out. These composites are developed for the construction placed in the sea and car bodies. From the results was determined that the direction of the fiber orientation also has an effect on the integrity and the creation of interphace and also significantly affects the size V_f, where the fiber volume increases significantly with winding angle α. On the contrary higher fiber volume does not necessarily mean higher strength, because for the tensile stress has been established that winding angle 20° shows E about 15 % lower compare to $\alpha = 0°$ [3, 4].

Acknowledgments The results of this project LO1201 were obtained through the financial support of the Ministry of Education, Youth and Sports in the framework of the targeted support of the "National Programme for Sustainability I" and the OPR&DI project Centre for Nanomaterials, Advanced Technologies and Innovation CZ.1.05/2.1.00/01.0005 and Project OP VaVpI Centre for Nanomaterials, Advanced Technologies and Innovation CZ.1.05/2.1.00/01.0005 and by the Project Development of Research Teams of R&D Projects at the Technical university of Liberec CZ.1.07/2.3.00/30.0024.

References

1. M. Petru, O. Novak, P. Lepsik, Numerical and experimental analysis of composite reinforcements applicable for a car seat innovation. Technolog **6**, 25–30 (2014)
2. L. Sevcik, P. Tuma, M. Petru, T. Martinec, R. Kovar, Composite reinforcement, Patent Application No. 2013–1066 (2013)
3. M. Petru, O. Novak, P. Lepsik, D. Mysakova, Numerical and experimental analysis of interphase of new environmental low-energy composites, *52nd conference on experimental stress analysis* (EAN 2014), pp. 101–104. ISBN: 978-80-231-0377-6
4. A. Lufinka, M. Petru, Determination of the composite tube mechanical properties, *52nd conference on experimental stress analysis* (EAN 2014), pp. 69–73. ISBN: 978-80-231-0377-6
5. J. Broncek, M. Dzimko, B. Hadzima, J. Takeichi, Experimental investigations of aluminium alloys 2024–t 3 form in terms of tribocorrosion characteristics. Acta Metall. Slovaca **20**(1), 97–104 (2014)
6. K. Petr, V. Dynybyl, J. CHMELAR, Test rig for gearboxes of rail vehicles, in *Modern Methods of Construction Design (ICMD 2013)*, Lecture Notes in Mechanical Engineering (Springer, 2014), pp. 169–175. ISBN: 978-3-319-05202-1
7. Y. Swolfs et al., Tensile behaviour of intralayer hybrid composites of carbon fibre and self-reinforced polypropylene. Compos. Part A **59**, 78–84 (2014)
8. Mehrabadi, F. A. Experimental and numerical failure analysis of adhesive composite joints, Int. J. Aerosp. Eng. 1–10 (2012)
9. H. Qing, L. Mishnaevsky, Unidirectional high fibre content composites automatic 3D FE model generation and damage simulation. Comput. Mater. Sci. **47**, 548–55 (2009)

System of Low Profile Roller Conveyors

F. Stary and V. Hanus

Abstract The article describes design of modular low profile roller conveyor for pallet transport. The article compares tangential and chain to chain variant. Further describes modifications of roller conveyor to reduce price of conveyor.

Keywords Roller conveyor • Chain arrangement • Modular system

1 Introduction

In the factories the goods is often transported by roller conveyors. Pallets loaded with goods is sometimes transported by pallet truck and then loaded on roller conveyor for further operations as pallet wrapping. For easy loading of pallet on conveyor without further devices the low profile roller conveyor is needed.

2 Demand on Design of Roller Conveyor

The roller conveyor should have low height about 150 mm to achieve loading of pallet with pallet truck without further devices. The design of conveyor must allow variable transport height from 150 mm to 800 mm. The conveyor should have modular design so the length of conveyor was 1500 mm, 2000 mm, 3000 mm and 4000 mm. Speed of conveyor is 0.2 m/s.

3 Concept Designs of Roller Conveyor

The chain was chosen as a driving element of rollers. Further were compared variants with tangential chain and chain to chain. The standard design of tangential roller conveyor is shown in the Fig. 1a. The standard design of chain to chain roller

F. Stary (✉) • V. Hanus
Czech Technical University, Prague, Czech Republic
e-mail: frantisek.stary@fs.cvut.cz; vit.hanus@fs.cvut.cz

V. Dynybyl et al. (eds.), *The Latest Methods of Construction Design*,
DOI 10.1007/978-3-319-22762-7_52

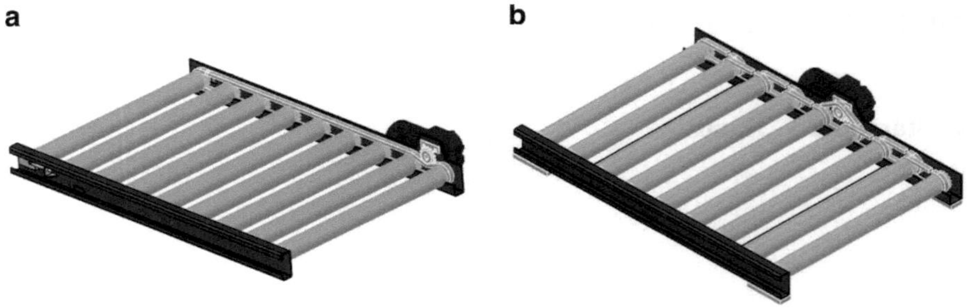

Fig. 1 Standard roller conveyor (**a**) tangential and (**b**) chain to chain

conveyor is shown in the Fig. 1b. Design of these conveyors was modified to meet requirements on minimal height of 150 mm. Used gearbox with electromotor is higher than demanded minimal height, but pallet on rollers is in demanded height.

The advantage of tangential chain is higher efficiency, simple construction (it is possible to choose any distances between rollers) and simple assembly. The disadvantage is demand on precise chain guiding on the lower or upper side of chain. The chain has to be also preloaded by stretching mechanism.

The advantage of chain to chain variant is maintenance-free run, higher chain engagement protection and no need of chain guiding. The disadvantage is demand on precise tolerance of distance between rollers, more complicated assembly and demand of two sprockets on one roller.

4 Modifications to Reduce the Price of Roller Conveyor

In the modification was used the fact that gravity roller (roller without sprocket) is cheaper than roller with sprocket or with two sprockets. The rollers in the conveyor are placed so that every second roller is driven. The arrangement for tangential variant is shown in Fig. 2a and for chain to chain variant in Fig. 2b. The arrangement in model of real conveyor is for tangential variant shown in the Fig. 3a and for chain to chain variant in the Fig. 3b.

From Fig. 2a, b is evident that each variant has different optimal number of rollers. For tangential variant with odd number of rollers are not both outer rollers driven and with even number of rollers is one outer roller driven and second is not driven. For chain to chain variant with odd number of rollers are both outer rollers driven and with even number of rollers is one outer roller driven and second is not driven.

If there is demand so that both outer rollers are driven, it is not possible to use tangential variant.

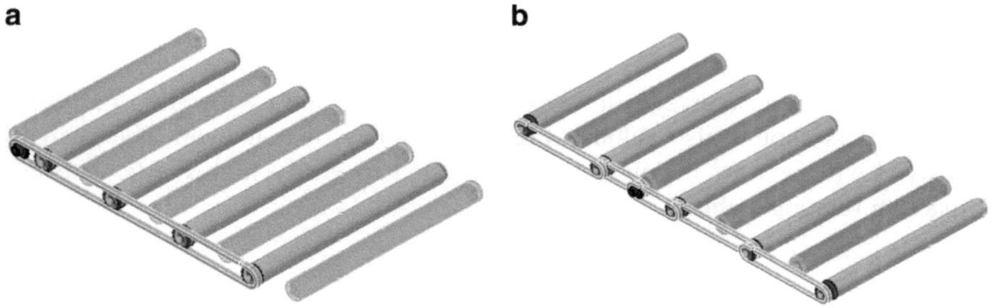

Fig. 2 Modifications reducing the prize off conveyor (**a**) tangential and (**b**) chain to chain

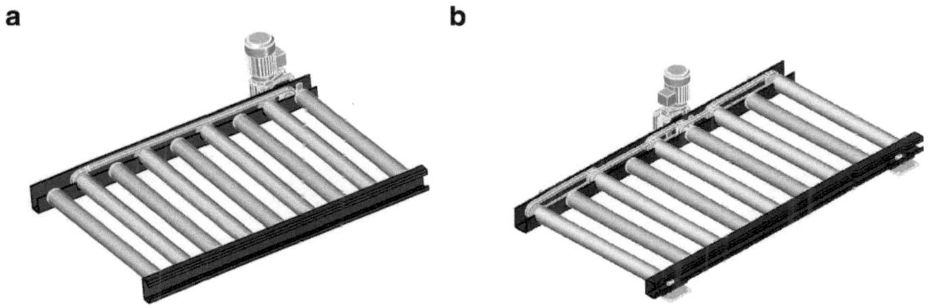

Fig. 3 Arrangement on real conveyor (**a**) tangential and (**b**) chain to chain

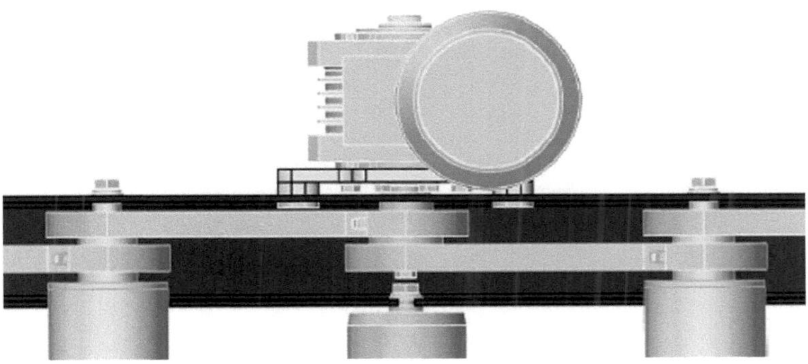

Fig. 4 Detail of driving sprocket

5 Chosen Variant of Roller Conveyor

On base of information mentioned in chapter "Designing of Mixing Equipment, Reactors and Bioreactors" was chosen chain to chain variant shown in Fig. 3b. Detail of driving sprocket is shown in the Fig. 4. The cost on rollers for this variant is from 7 to 25 % smaller to chain to chain variant with all rollers driven. The cost differs on base of pitch between rollers and size of roller.

6 Conclusion

The design of roller conveyor lowered cost of rollers and by lowering number of driven rollers should also achieve higher efficiency of conveyor. The designed conveyor will be now manufactured and tested to verify higher efficiency and impact on running stability.

Fiber Driven Tilting Mechanism

P. Svatoš, Z. Šika, M. Valášek, J. Zicha, V. Bauma, and V. Rada

Abstract The paper deals with the optimization of the experimental functional model of the fibre driven tilting mechanism with three degrees of freedom and four drives. Measured orientation of the platform is realized by means of incremental sensors in central joint. The optimization of the calibrability and the sensor placement shows, that the measurement of the platform position (orientation in the spherical case) is necessary both for the good calibrability and for its accurate position control.

Keywords Fibre driven mechanism • Parallel kinematical structure • Anti-blacklash control • Optimization

1 Introduction

Fibre-driven mechanisms, which are based on parallel kinematical structures, are a newly developing area in the field of light and smart mechanisms structures [1, 2]. Our research [3] of the fibre-driven redundant parallel kinematical mechanisms (PKM) has found certain appropriate kinematical structures with optional dimensions and optional geometrical configuration to achieve smooth motion between all positions together with the increase of all mechanical properties offered by parallel kinematical structures.

The parameters were optimized in order to achieve maximum workspace with anti-backlash properties, without collisions and with maximum dexterity and stiffness and dynamics with respect to the real components. Computational tools were used for optimization process via genetic algorithm, for seeking the optimum variant of the geometrical layout [4]. The most important part, which is based on anti-backlash control principle, ensures that all discrete positions over whole workspace are realised by fibres loaded by tension force.

P. Svatoš • Z. Šika (✉) • M. Valášek • J. Zicha • V. Bauma • V. Rada
Faculty of Mechanical Engineering, Czech Technical University, Technická 4, Prague 166 07, Czech Republic
e-mail: zbynek.sika@fs.cvut.cz

V. Dynybyl et al. (eds.), *The Latest Methods of Construction Design*,
DOI 10.1007/978-3-319-22762-7_53

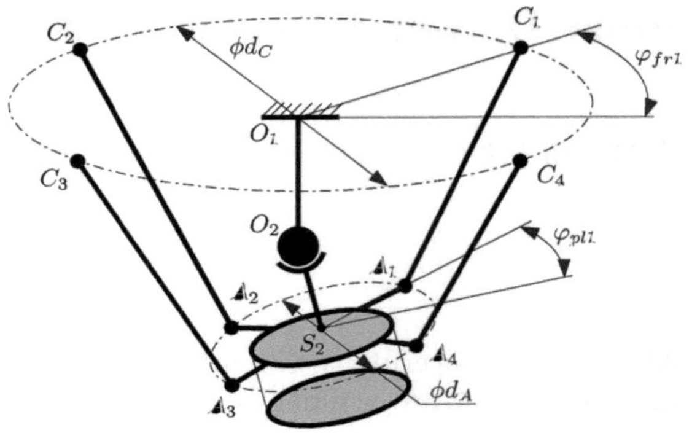

Fig. 1 Kinematical structure of the fibre-driven spherical mechanism

This paper presents the promising candidate of fibre-driven redundant PKM, a spherical mechanism with three degrees of freedom (DOFs) and four drives.

2 Fibre-Driven Spherical Tilting Mechanism

The Fig. 1 illustrates kinematical structure of the redundant fibre-driven spherical mechanism. The structure arises from the HexaSphere mechanism [5], but the platform is driven by only four fibres instead of six rods or fibres.

2.1 The Kinematical Structure

Now the moving platform is driven only by four fibres and it performs a spherical motion around central point O_2 with three DOFs. Fibres lead from the platform over pulleys (fixed to the base frame) to sliders which are moving along linear guidance, ball screw axis with AC servo motors.

Fibres are connected in the platform in points $A_{1\text{–}4}$ described by the diameter d_A and the angle $\varphi_{pl1\text{–}4}$. Four pulleys (points $C_{1\text{–}4}$) lie in a circle with the diameter d_C and they are described by the angle $\varphi_{fr1\text{–}4}$. Next parameters are lengths of bars O_1O_2 and S_2O_2. The kinematical structure is then determined by six optimized kinematical parameters.

2.2 *The Central Spherical Joint*

The central joint is equipped with triple of rotational sensors in order to ensure the absolute measurement of platform orientation without collisions of joint's components and to improve possibility of the on-line calibration.

All rotations are measured by RENISHAW RESM Signum angular encoder system, which consists of a ring, a read-head and an interpolator.

2.3 *Drives, Carbon Fibres and Pulleys*

The carbon-fibre composite wire is used for the proposed parallel kinematic mechanisms instead of steel wire. The production methodology for the selected type of carbon composite fibres was established via special-purposed template.

The fibre goes through the cylindrical hole in bearing from the pulley to the slider on the linear guidance. The linear motion of slider is controlled by the AC servomotor through the ball screw. This pulley has, in addition to the rotation of pulleys wheel, also other axis of rotation, which allows inclining the whole pulley.

3 Optimization Results

Optimization process had to consider real components, their dimensions, also movement limits with collisions. Results of the optimization are optimal dimensions of the components and position of the pulleys with fixed positions of fibres on the moving platform. These positions were not chosen arbitrarily but rectangularly.

The optimized variant of the spherical mechanism (Fig. 2) has the following parameters in the Table 1.

Details of the spherical central joint and the pulleys are shown in the Fig. 3.

4 Conclusion

The spherical fibre-driven redundant mechanism has been proposed and optimized from the point of view of the workspace, dexterity, collisions, calibrability and stiffness. The functional laboratory model, which includes four drives and three rotational sensors in the central spherical joint, has been built.

a

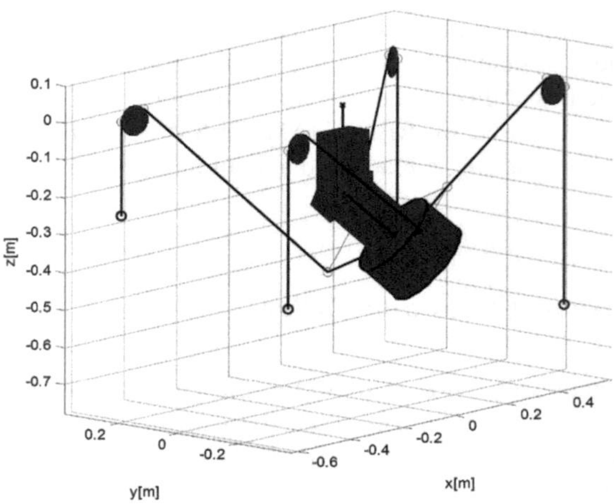

b

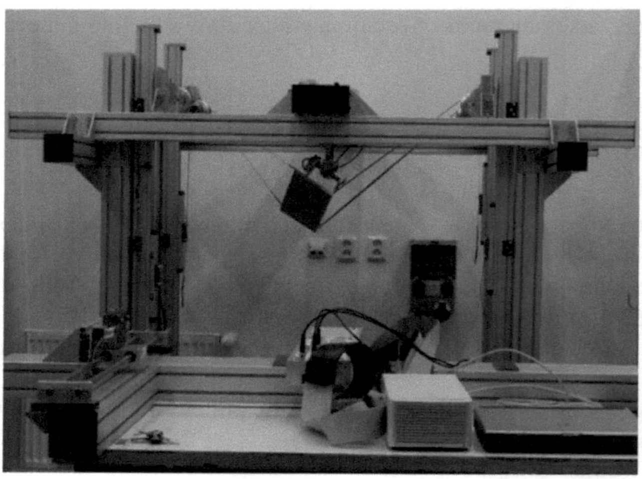

Fig. 2 Optimized kinematical structure: (**a**) MATLAB and (**b**) experimental functional model

Table 1 Results of optimization

d_C	φ_{fr}	d_A	φ_{pl}	O_1O_2	S_2O_2
1.199 m	27.78°	0.438 m	69.28°	−0.231 m	−0.180 m

a b

Fig. 3 Details of the components: (**a**) central joint and (**b**) pulleys

Acknowledgments The authors appreciate the kind support by the grant GA101/11/1627 “Tilting Mechanisms Based on Fiber Parallel Kinematical Structure with Antiblacklash Control” of the Czech Science Foundation and project SGS13/177/OHK2/3T/12 “Mechatronics a adaptronics 2013” of CTU in Prague.

References

1. X. Duan, Y. Qiu, Q. Duan, J. Du, Calibration and motion control of a cable-driven parallel manipulator based triple-level spatial positioner. *Advances in Mechanical Engineering* vol. 2014, (2014), ID 368018
2. Q. Chen, W. Chen, R. Liu, J. Zhang, Error analysis and flexibility compensation of a cable-driven humanoid-arm manipulator. *IEEE International Conference on Robotics and Automation*. (Shanghai, s.n., 2011), pp. 988–993
3. M. Valášek, Z. Šika, V. Bauma, P. Svatoš, Způsob a zařízení pro změnu tuhosti sériového nebo paralelního základního pohyblivého mechanismu, zvláště průmyslových robotů a obráběcích strojů. *Patent application PV 2012–621 CZ*, 10. 09 2012
4. P. Svatoš, M. Valášek, V. Bauma, P. Polach, Optimization of anti-backlash fibre driven parallel kinematical structures. Bull. Appl. Mech. **8**(31), 40–44 (2012). ISSN 1801–1217
5. M. Valášek, M. Karásek, HexaSphere with cable actuation. *Recent Advances in Mechatronics*. (2010), pp. 239–244

Part VII
Product Innovation

P. Baran, M. Bochinsky, R. Grega, S. Hosnedl, M. Kopecky, I. Mazinova, and M. Sasiadek

Pneumatic Dual Mass Flywheel – Damper Concept for Downspeeding

R. Grega and P. Baran

Abstract Modern drivetrain, new types of engines and the effort to reduce the weight of the system improves the efficiency of powertrain. But also it is the way that the given system much more sensitive to the vibration. Today's damper system, damped clutch disc and dual mass flywheel cannot remove or shift the resonance field from working area of engine in all cases which powertrain can have. Therefore we proposed a new type of the dual mass flywheel in the framework of our workplace in order to improve behaviour of powertrain in vibration field.

Keywords Downspeeding • Vibration • Torsional damper • Pneumatic dual mass flywheel

1 Introduction

The stringent worldwide CO_2 limit values is the primary requirement for future mobility. Environmentally vehicle's drivetrain are element of energy efficiency worldwide mobility. Hence, the development of efficient drivetrain is the top priority. That we can follow the optimization of existing system and introduction of new system. Most experts agree that "today's" powertrains with internal combustion engine and transmission will dominate of the world's mobility. The hybrid and electric drives need political provisions for increase their market share [7].

Trends utilizing in engine development pose individual claims on damper systems. With downsizing we reduce internal engine losses, but it resulting in higher torsional excitation due to lower number of cylinder and lower excitation frequencies. With downspeeding we achieve high torque, by high pressure turbocharging, at very low engine speed. This leads to lower excitation frequencies and high pressure peak increase excitation amplitudes. The cylinder deactivation when engine running in partial load pose specific claims for damper system to ensure good isolation of vibration [6, 8].

Lower CO_2 emissions is task also for transmissions and chassis not only for engine. There is a trend of higher ratio and higher number of gears. These concepts

R. Grega (✉) • P. Baran
Technical University of Košice, Košice, Slovak Republic
e-mail: robert.grega@tuke.sk; peter.baran@tuke.sk

V. Dynybyl et al. (eds.), *The Latest Methods of Construction Design*,
DOI 10.1007/978-3-319-22762-7_54

permit the engine be operated with low consumption. Further reducing of consumption by longer gear ratio included with lower operating engine speeds increased engine excitation and requires that the damper should have specific characteristic. According to [1], when we think about the operating speeds of a current 2.0 litter diesel engine and reduced it by 10 %, it is possible to consume 5.6 % less fuel. To ensure standard driving dynamic and comfort the same engine output must be achieved at a 10 % lower engine speed, which mean that engine torque must be increased. We can suppose that the engine operating speed will be closer to idle speed and then excitation frequency will be more closely to the natural frequency of the drivetrain.

Figure 1 summarized downspeeding effect on the rotational irregularity. The rotational oscillation range at the "downspeeding" engine is double than today's engine. Nevertheless, the irregularity at the input shaft of transmission must be the same. In order to take advantage of lower consumption reached by downspeeding and achieve good driving comfort we need a suitable damper [1].

Development of damper system in drivetrain starts with torsional damped clutch disc and continued by dual mass flywheel with very low torsional stiffness (Fig. 2).

The aim of this article is to show a new way of reducing vibration in the drivetrain. Introduce a new damper concept – the pneumatic dual mass flywheel and its effect on the vehicle powertrain.

2 Today's Damper System and Their Effect on Powertrain

The effect of torsional dampers on vehicle drivetrain we can show at dynamical analysis of simple model. Our dynamical analysis of vehicle powertrain consist of 3 main parts, specifically a combustion engine, a gear box and rest of drivetrain, that can be replaced with a three-mass mechanical system which torsionally oscillating. Such system has two natural frequencies.

To describe the behavior of this mechanical system is suitable to use a Campbell diagram. Figure 3 [2] show a Campbell diagram of three-mass system with damped clutch disc. Figure 4 show a diagram of this system with a dual mass flywheel.

Operating range of the mechanical system is shown by revolutions n_v and n_{max} in both diagrams. Thin straight lines marked from 0.5 to 6 represent harmonic excitation order of torque. The second order is the main harmonic order and it's marked by thick line. Thin lines marked from 1G to 5G represent the first natural frequencies for speed gear 1 till 5. Thick lines marked from 1G to 5G represent the second natural frequencies for speed gear.

In Campbell diagram in Fig. 3 we show two resonance field with the main harmonic excitation order. First filed is resonance with the first natural frequency and is under of working revolution of engine. This field is quickly break through while engine starting. Second resonance field by second natural frequency is directly in engine working area. This fact is cause of an increase the noise and vibration in the entire of powertrain.

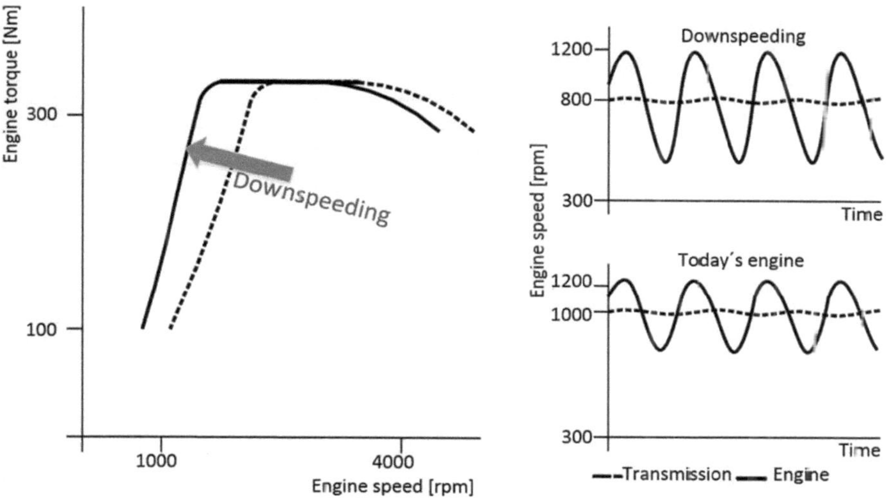

Fig. 1 Rotational irregularity at engine and transmission for today's and downspeeding engine

Fig. 2 Torsional damped clutch disc and dual mass flywheel

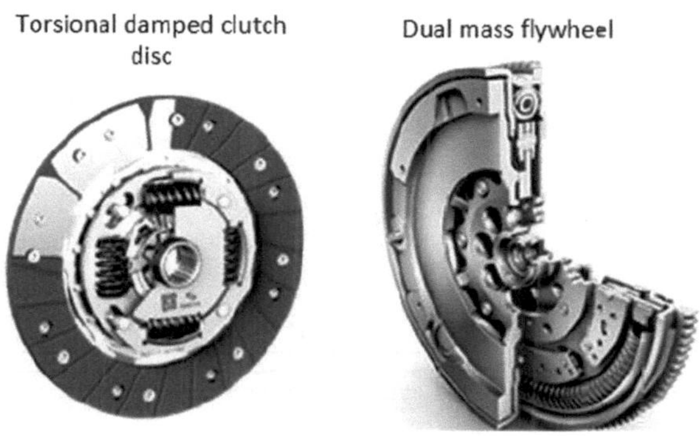

With dual mass flywheel the second natural frequency is shift to lower speed, therefore the second resonance field is also under of engine working area (Fig. 4). This is the main goal of the application of dual mass flywheel in powertrain.

3 Characteristics of Pneumatic Dual Mass Flywheel

Our research team obtained a great amount of experiences in the development area of machineries and systems projected at our workplaces for reduction of torsional vibrations [4]. Therefore we developed a new pneumatic dual mass flywheel (Fig. 5).

It's consists of the primary mass (1) and the secondary mass (2). The secondary mass is pressured with the pneumatic flexible chambers that are shaped like

Fig. 3 Campbell diagram of system with damped clutch disc

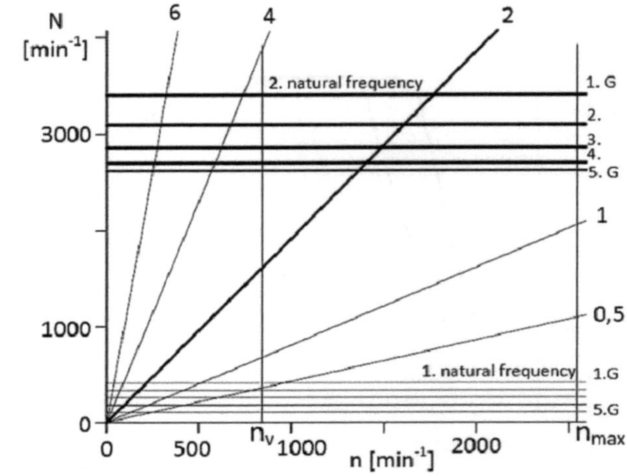

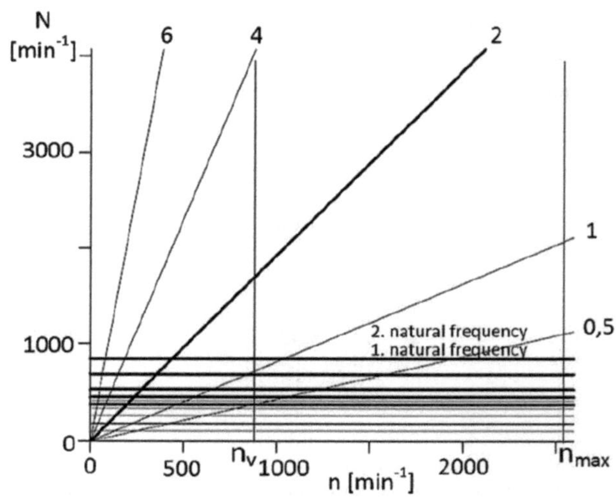

Fig. 4 Campbell diagram of system with dual mass flywheel

half-moons and are filled with the air (3). The primary mass is joined to the carrier (4), which is equipped with the compression pistons (5). The pistons are linked with the pneumatic flexible chambers. The chambers are compressed towards the pistons when they are loaded. The pneumatic dual mass flywheel is attached to the pneumatic accumulator situated out of the combustion engine. The main task of the pneumatic accumulator is keeping of a constant air pressure in the pneumatic flexible chambers. The air pressure in the pneumatic dual mass flywheel can be changed from the 100 kPa up to 800 kPa. The behaviour signed with the letters from *a* to *h* in the (Fig. 6a) are the loading characteristics and in the (Fig. 6b) are

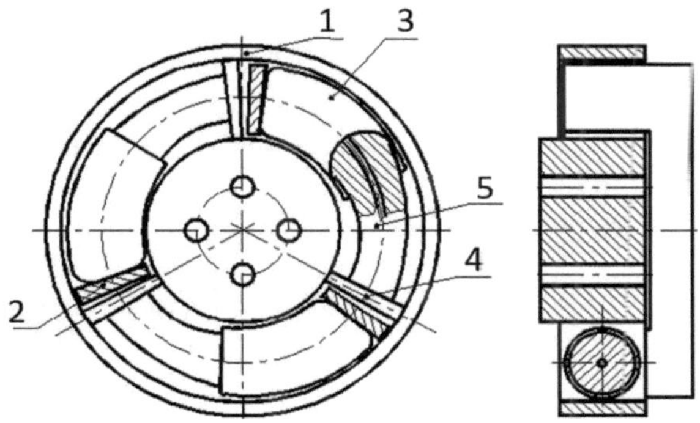

Fig. 5 Pneumatic dual mass flywheel

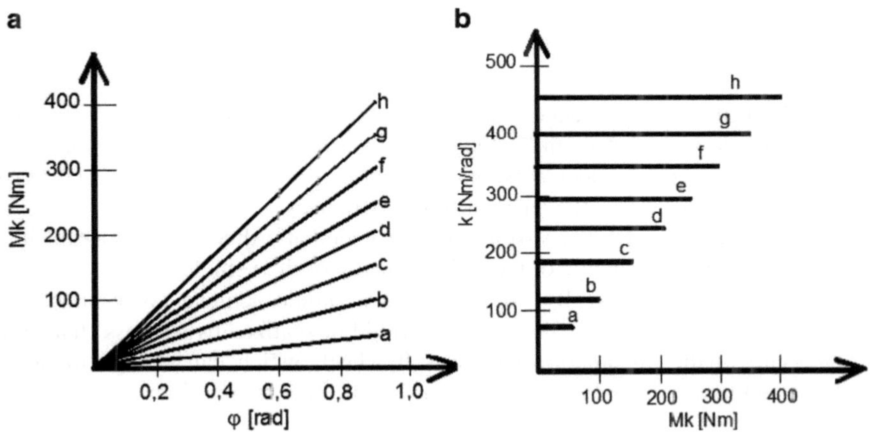

Fig. 6 Static characteristic of the pneumatic dual mass flywheel

presented the characteristic of static torsional stiffness of the pneumatic dual mass flywheel operating at the pressure levels from the 100 kPa up to 800 kPa.

Application of the suggested pneumatic dual mass flywheel ensures elimination of inception of resonance with all harmonic excitation order. Effect of pneumatic dual mass flywheel on vehicle powertrain for illustration shows Campbell diagram in (Fig. 7). Thick straight lines mark second natural frequency for speed gear 1 till 5. We see that the rays of natural frequencies are located between the rays of excitation harmonic orders and the resonance not occur [3].

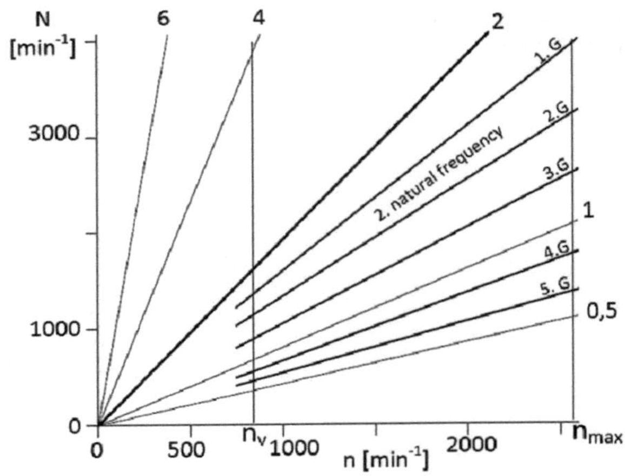

Fig. 7 Campbell diagram of system with pneumatic dual mass flywheel

4 Conclusion

On our department we are successfully devote the new design of flexible couplings with pneumatic spring elements, which provide great opportunities in a tuning of mechanical system to reduce torsional vibration [5]. The pneumatic dual mass flywheel was projected on the basis of this experiences.

By changing the torsional stiffness of the pneumatic dual mass flywheel it is possible to ensure the change of natural frequency of the vehicle drivetrain during the running of the engine. By this change, we should achieve no resonance in working area of powertrain. This condition could be realized by stepless change of air pressure in the pneumatic dual mass flywheel, which can change torsional stiffness and thus change the natural frequency of the vehicle drivetrain.

Acknowledgments The research work reported here was made possible by Grant Project VEGA: 1/0688/12 – Research and application of universal regulation system in order to master the source of mechanical systems excitation.

References

1. J. Kroll, M. Hausner, R. Seebacher, Mission CO_2 reduction, in *10th Schaeffler Symposium: Solving the Powertrain Puzzle.* (Schaeffler Technologies Gmbh & Co. KG, 2014), pp. 56–77
2. W. Reik, Torsional vibrations in the drive train of motor vehicle, principle consideration, in *4th Symposium LUK: Torsional Vibrations in the Drive Train* (LuK GmbH & Co. KG, Baden-Baden, 1990), pp. 5–28
3. R. Grega, J. Krajňák, P. Baran, The reduction of vibrations in a car – The principle of pneumatic dual mass flywheel. Zesz. Nauk. Politechniki Śląskiej. z.84 Nr.(1907), 21–28 (2014). ISSN 0209–3324

4. R. Grega, J. Krajňák, The pneumatic dual-mass flywheel. Zesz. Nauk. Politechniki Śląskiej **76** (1865), 19–24 (2012). ISSN 0209–3324
5. J. Homišin, Spôsoby ladenia mechanických sústav aplikáciou ladičov torzných kmitov. Acta Mechanica Slovaca, roč. 9, č. 3-b, (2005), pp. 5–12
6. P. Němeček, Diagnostika strojů s vibračním principem činnosti. Acta Mechanica Slovaca, roč. 12, č. 3-c, (2008)
7. L. Jakubovičová, M. Sága, P. Kopas, M. Handrik, M. Vaško, Some notes on analysis of bending and torsion combined loading. Mach. Dyn. Res. **34**(4), 97–105 (2010). ISSN 2080–9948
8. B. Łazarz, G. Wojnar, H. Madej, P. Czech, Influence of meshing performance deviations on crack diagnostics possibility, Transactions of the Universities of Košice: research reports from the Universities of Košice, (Košice, 3 – 2009), pp. 5–8, ISSN 1335–2334

Rationalising the Use of Design for Assembly as DfX Knowledge to Increase Competitiveness of Designed Technical Products

S. Hosnedl, M. Kopecky, and M. Sasiadek

Abstract Interdisciplinary Design for X knowledge and methods belong among powerful methodological ways for increasing effectiveness and efficiency of engineering design processes and consequently the competitiveness of the designed technical products. The letter X mostly indicates a specific property of technical products or more often their professionally related set. Such sets of knowledge and methods are then accordingly called for example: Design for Cost, Design for Assembly, Design for Safety etc. It is generally known that there are a lot of such DfX sets developed spontaneously including their names. A rational concept of a systematic taxonomy for DfX knowledge and methods is introduced in the paper. The core of this paper is focused on the general principles of DfX taxonomy and methodology applied on the Design for Assembly (DfA) field. This methodology is focused on reducing cost for assembly and curtailment assembly time of designed technical product while keeping its delivered quality. Validation of the introduced DfA methodology is presented on cases from industrial praxis.

Keywords Design for X • Design for assembly • Taxonomy of properties

1 Introduction

Design for X is one of the most spread and used methodological approaches for improvement and streamline of development process of new or innovative technical products. Their main goal is to contribute rise of delivered quality, reduce expended cost and curtail delivered terms to achieve increased competitiveness of the designed technical products. Their common feature is use of different specific professional often interdisciplinary knowledge and methods focused on

S. Hosnedl (✉) • M. Kopecky
University of West Bohemia, Pilsen, Czech Republic
e-mail: hosnedl@kks.zcu.cz, kopecky@kks.zcu.cz

M. Sasiadek
University of Zielona Góra, Zielona Góra, Poland
e-mail: m.sasiadek@iizp.uz.zgora.pl

V. Dynybyl et al. (eds.), *The Latest Methods of Construction Design*,
DOI 10.1007/978-3-319-22762-7_55

achievement of relative groups of inherent (i.e. not assigned) characteristics, attributes etc., mostly generally called properties of designed technical products.

Design for X methodologies have been thus mostly spontaneously established and developed by a lot of (groups of) authors all over the world based on their respective knowledge and experience [1, 2]. Thus their scopes, names and so on are often redundant, partially overlapping or on the other hand even not yet named. Consequently their classification and storage, effectiveness of their use and further development and maintenance also corresponds to this 'fuzzy' situation.

Our effort was therefore focused at first on rational establishment of a systematic nevertheless flexible and open taxonomy for both "all" existing and potential DfX methodologies. The traditional reason was to enable efficient survey and collection of existing knowledge and methods in this extremely large field, and then much effectively develop necessary generalisations followed by improvements and/or development of both general and concrete topical DfX methodologies as usual. We have verified and validated that such taxonomy system for DfX methodologies can be effectively overtaken from taxonomy of properties (incl. behaviours) of technical products, which we have already developed with use of the Theory of Technical Systems and largely validated.

Here we outline an "additional methodological value" of the introduced taxonomy system. It is rational identification of those concrete "X", which are critical for a currently designed technical product. These "weak X" rationally identified with use of the developed SW tool for support of "Specification of requirements and evaluation of their fulfilment" for any designed technical product can be then rationally improved with use of both available and/or newly developed DfX methodologies as also outlined.

2 Taxonomy of Properties of Technical Products

Theoretical knowledge on technical systems (TS) provide us with knowledge about technical products understood as technical systems, for example about their life cycle, structure and other properties, and so on. This knowledge is comprehensively described e.g. in [3–5]. Here we will focus especially on taxonomy of TS properties. TS Properties can be and are classified in a number of ways. Our developed consistent and transparent theory based hierarchical taxonomy stems from our research supported by inquiries and by our large both national and international university and industry related knowledge and experience.

All inherent (i.e. "implemented by designing") TS properties [6] are split in the first hierarchical level into the following three Domains:

(1) Descriptive properties comprising a description of TS itself including its descriptive features.
(2) Reactive (behavioural) properties comprising reactions/responses of TS to its external and/or internal load of any kind during TS Life Cycle.

(3) Reflected properties comprising TS reflections of external bodies on TS, i.e. on its Descriptive, Reactive and other Reflected properties.

These three Domains are then systematically and clearly split only into 10 classes, and their respective two or three logical subclasses of properties in their lower hierarchical levels. This developed comprehensive system can simply and unambiguously include any of the 'infinite number' of properties accompanying any TS during its whole Life Cycle. In addition it can also serve as a direct basis for taxonomy of 'Design for X' (DfX) as mentioned above as well as for large bulk of 'Prediction of X' (PoX) knowledge and methods [4, 7] not mentioned here.

3 Specification of Requirements and Analyses of their Fulfilment

Continuous specification of requirements and analyses and evaluations of their fulfilment enables a rational property driven management of engineering design processes. Software support of Specification of Requirements and of their Evaluation & Analyses for any designed technical product was developed based on the framework of the above introduced taxonomy of properties. It is being routinelly used in educational engineering design projects and also in research and industrial projects at Department of Machine Design of UWB.

The stated, obligatory and/or generally implied requirements on the designed technical product are completed into a input sheet (Fig. 1). It is clearly structured according to the taxonomy of TS properties outlined in the previous chapter. Each property is specified by selected number of property indicators and their values. After entering requirements a reference assessment of their fulfilment by current own and selected comparable/competitive TSs follows. The degree of fulfilment is evaluated scaled from 0 to 4 points. The results of evaluation are represented in a form of diagrams clearly depicting the respective evaluations and compare them for the current and competitive TS. This evaluations and analyses are finally completed by designed out technical product(s) as depicted in Fig. 2.

This comprehensive assessment of the required set of properties clearly indicates critical "weak" (and of course also strong) properties that should be enhanced during the following development of the designed technical product. For example it may happen that after evaluation of fulfilment of properties it is found out on graphs that assembly-ability is not satisfactory and thus designer is navigated to enhance it. Subsequently DfX knowledge can be used to help to solve this problem, especially Design for Assembly in this case. The way how knowledge and principles in Design for Assembly can be effectively used will be demonstrated in the following part of the paper.

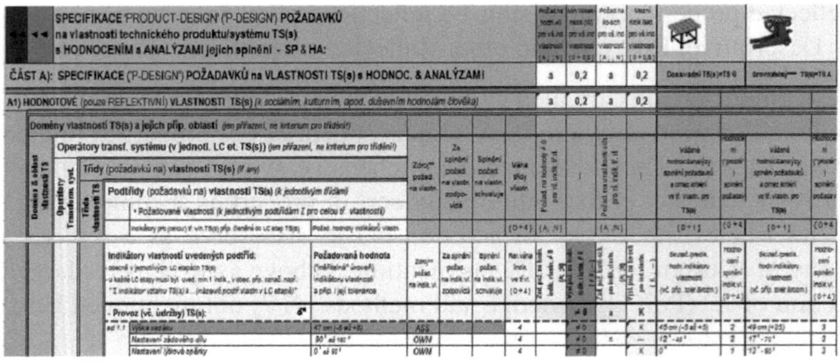

Fig. 1 SW support of specification of requirements and evaluation and analyses of their fulfilment for a designed technical product developed in MS EXCEL [5] – screen of an input section

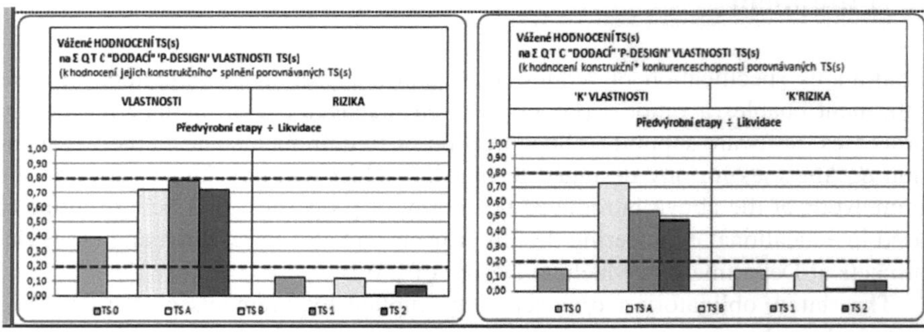

Fig. 2 SW support of specification of requirements and evaluation and analyses of their fulfilment for a designed technical product developed in MS EXCEL [5] – screen of diagrams [5]

4 Design for Assembly

Design for Assembly involves taking into account the assembly process requirements over the course of the product design cycle, so that the TS construction would be primarily adjusted to efficient, in terms of speed and cost, installation. Taking into account all the requirements of the assembly process at the design stage is a complicated task. The practical application of the principles of design for assembly is often limited to the intuitive consideration of these requirements during the development of the project. It is commonly known that manufacturing costs constitute the major part of the implementation process, while the installation costs can be even 30–50 % of the cost of manufacturing the product. Therefore, design for assembly is one of the most important methods in the large field covered by a large 'umbrella' of DFX.

In literature can be found many methods for assembly evaluation. Boothroyd Dewhurst Method, Hitachi Assembly Evaluation Method and Lucas Hull Method are for example described [8–11].

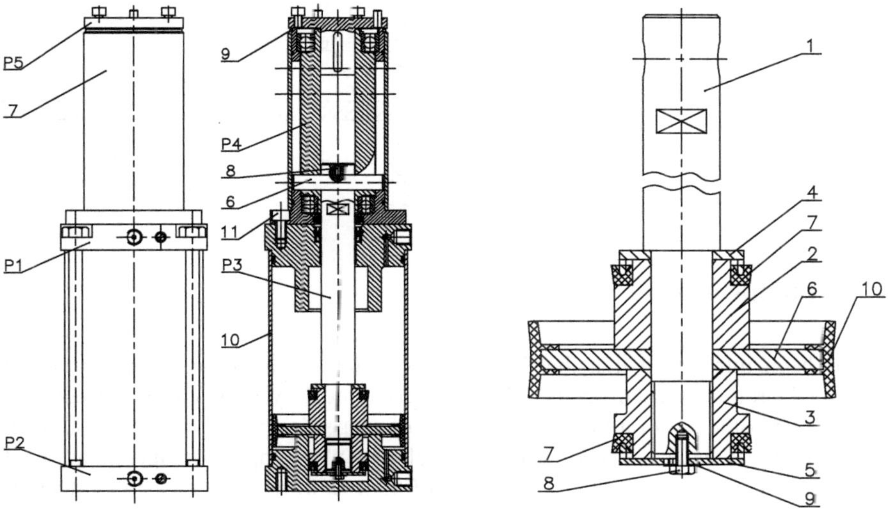

Fig. 3 Servo-motor structure and piston-rod unit

5 Application of DfA Principles

The result of computer implementation of the method of assembly sequence generation according to the DfA principles is the developed EASY ASSEMBLY computer programme, which was presented at the conference. The conducted analysis concerned a hydraulic servo-motor (Fig. 3). This article, for reasons of space, is limited on analysis of the only one subassembly. The servo-motor structure and the analysed piston-rod unit are depicted in Fig. 3. Piston-rod unit consists of 11 constituent parts specified on the figure: *1* – Piston-rod, *2* – Muffler front sleeve, *3* – Muffler rear sleeve, *4* – Front gland ring, *5* – Rear gland ring, *6* – Piston, *7.1* – U1 sealing, *7.2* – U1 sealing, *8* – M4x10 screw, *9* – Tab washer, *10* – Piston sealing.

As a result of analysis of the obtained solutions of the servo-motor subassembly, construction changes of the constituent parts have been proposed. These changes simplified the construction due to the assembly process requirements, which leads to increase its efficiency. Parts reduction has been suggested in the piston-rod unit. Instead of parts 2, 6 and 3, a solution aggregating the replaced parts into one has been proposed as depicted in Fig. 4. Modified construction of the piston-rod unit decreased the number of parts from 11 to 9.

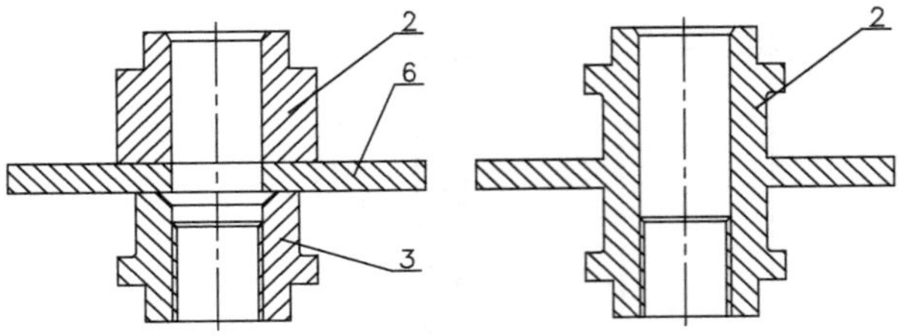

Fig. 4 Construction change of part no. *2*, *3* and *6* of the piston-rod unit

6 Conclusions

It has been introduced that the developed systematic taxonomy of properties of technical products based on the Theory of Technical systems can be effectively used among others for rational identification of properties of any designed technical product. It has been outlined that the developed SW support of Specification of requirements and evaluation and analyses of their fulfilment for a designed technical product developed in MS EXCEL based on this taxonomy is a powerful vehicle to perform it. Use of Design for X (DfX) knowledge and methods is efficient to support the folowing design improvements of the identified weak properties. For example use of Design for Assembly for a servo-motor is presented in this paper. Reduction of the number of constituent parts of the piston-rod unit by replacing three parts by the only one has been achieved in this case. This led to the decrease of the generated technological sequences from 8 to 1 as well as to the improvement of the sequence evaluation from 16.1 to 13.6 as evaluated with use of the developed EASY ASSEMBLY Software.

Acknowledgments This paper includes partial results of the Project SGS-2013-050 "Complex support of design engineering of technical products to improve their properties and competitiveness II".

References

1. M. Chiu, G. Okudan, Evolution of design for X tools applicable to design stages: A literature review, in *Proceeding of ASME 2010 – IDETC/CIE 2010*, Quebec 2010
2. J. Clarkson, C. Eckert, *Design Process Improvement: A Review of Current Practice* (Springer, London, 2005)
3. V. Hubka, W.E. Eder, *Theory of Technical Systems* (Springer, Berlin, 1988)
4. W.E. Eder, S. Hosnedl, *Design Engineering – A Manual for Enhanced Creativity* (CRC Press/Taylor & Francis Group, Boca Raton/Florida USA, 2008)
5. S. Hosnedl, *System Designing of Technical Products* (University of West Bohemia, Pilsen, 2012) (electronic version)
6. CSN EN ISO 9000 Czech v. of the European Standard Quality management systems – Fundamentals and vocabulary. (Czech Institute for Standardization, Praha, 2006)

7. G. Pahl, W. Beitz, J. Feldhusen, K.H. Grote in *Engineering Design, A Systematic Approach*. 3rd English edn. (Springer, London, 2007). (Trans. K. Wallace, L. Blessing from: Pahl, G. et al from Konstruktionslehre, Berlin: Springer-Verlag, 2003)
8. J.D. Booker, M. Raines, K.G. Swift, in *Designing Capable and Reliable Products*, (Springer-Verlag, 2001)
9. D.M. Anderson, in *Design for Manufacturability & Concurrent Engineering* (CIM Press, California, 2010)
10. G. Boothroyd, P. Dewhurst, *Design for Assembly* (Boothroyd Dewhurst Inc., Wakefield, 1991)
11. D.E. Whitney, in *Mechanical Assemblies: Their Design, Manufacture, and Role in Product Development* (Oxford University Press, 2004)

Materials and Sustainable Development

I. Mazinova and M. Bochinsky

Abstract This conference paper deals with ideas of sustainable development. How to introduce students to this? There is no completely "right" answer to questions of sustainable development, there is a thoughtful, well-researched response that recognizes the concerns of stakeholders, the conflicting priorities and the economic, legal and social aspects of technology as well as it environmental legacy.

Keywords Sustainability • Triple bottom line accounting • Analysing sustainable developments • Corporate social responsibility

1 Introduction

The concept of sustainable development is understood in different ways by different people, often depending on their needs for pursuing their own interests. Some politicians abuse the term for political purposes, and some large corporations use the term 'sustainability' only for purposes of growing their business. On the other hand oil companies understand the term sustainable development in relation to their future as a 'time after oil'. Generally it can be said that the majority of people think of sustainability as the unlimited use of all available technologies for the growth of gross domestic product whilst reducing the undesirable effects on the environment (for example, reduction of carbon emissions). And then there are those who do not trust this approach, who see the free market as the cause of environmental problems, which is the continuous driving force behind the growing threat to an unsustainable future. Sustainable development is seen as equilibrium, not growth. Between these two points of view there are many other opinions. We are living at a time when neither conclusive proof – nor one side of the argument, can be obtained – only time will tell.

I. Mazinova (✉) • M. Bochinsky
University of West Bohemia, Bohemia, Czech Republic
e-mail: mazini@kks.zcu.cz; bochinsk@students.zcu.cz

V. Dynybyl et al. (eds.), *The Latest Methods of Construction Design*,
DOI 10.1007/978-3-319-22762-7_56

2 Sustainability

So what then is sustainability? It depends on your point of view – the scale, whether it is local or global, and the time frame – whether it is weeks, years, or centuries.

Sustainable development is a means of developing human society which is carried out in agreement with economic and social progress whilst fully preserving the environment. The main objective of sustainable development includes the preservation of the environment for future generations with as little change as possible. The principles of sustainability are set out in standard CSN 01 0391, which was released in October 2013, entitled 'Corporate social responsibility management system – Requirements' [1]. The introduction of this standard states:

'Specifically, the integration of social responsibility manifests itself by integration of positive attitudes, practices and programmes into an organization's strategy at the level of top management. It requires a shift in the view of their social role from the level of "profit only" to a broader perspective in the context of the often-mentioned three pillars "P" – People – Planet – Profit'.

Unfortunately many firms see social responsibility as a fashionable trend a means for self-promotion.

The Fig. 1 shows sustainability represented as the 'Triple Bottom Line'

This apparently simple diagram is very hard to implement in practice. It demands a detailed analysis of single actions ('articulations') in a system and a subsequent synthesis of the facts and, most significantly, a large dose of common sense.

How should engineers approach the issue of sustainability? What and how should we teach students? Thanks to a certain Czech president (and not only him), the overwhelming opinion in the Czech Republic is that there is no threat. Technical specialists tend to see this issue as something for philosophers. Questions such as the following are regarded by technical specialists as, at the very least, strange:

How long can economic growth, which is always exponential, be sustainable?

Is strong economic growth compatible with the needs of today's and future population?

Is sustainable development (in the context of the continuously widening gap between rich and poor countries) actually achievable?

And is it really in our interests to worry about what comes after us?

3 Sustainability from the Viewpoint of Engineers

We would like to share with you a rational analysis of sustainability issues, conducted by Professor Ashby et al. from the perspective of material selection [2]. How is it done? By separating the rings and developing their meaning.

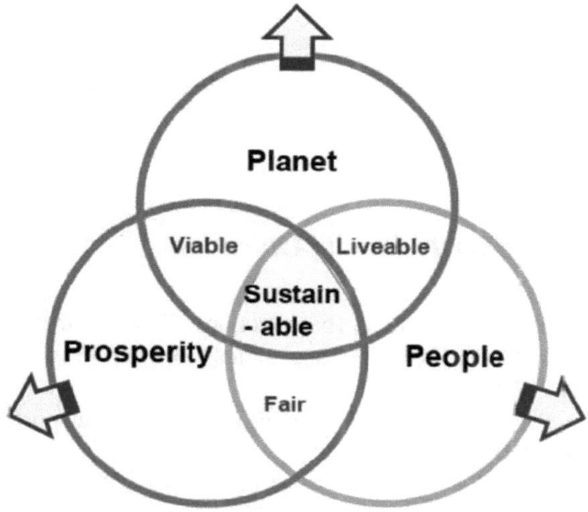

Fig. 1 Triple bottom line [4]

There are many different proposals ("articulations") that are used to promote a "Sustainable Development". Proposals are submitted from various stakeholders. Among the stakeholders include governments, the public, local communities, owners, manufacturers, suppliers, trade unions, customers, lobbyists, investors, national press, managers. They have their interests and their power of influence. Thereby get their proposals logically into conflict. Due to the length of the contribution skip lists the main proposals of individual stakeholders. Their interests can be derived following main categories (meaning still in terms of material selection):

- Materials (manufacture supply chain, life cycle)
- Environment (carbon release, air, water, land)
- Design-energy (energy efficiency, function, performance, safety)
- Legislation (awareness, compliance, reporting)
- Society (social equity, education, health)
- Economics (cost, benefit).

Individual proposals ("articulations") of stakeholders must be systematically explored. Generally, you can define a method to assess the sustainability of these five steps:

Step 1: Clarify objective
Step 2: Stakeholders analysis
Step 3: Fact finding (objective)
Step 4: Debate impact (subjective)
Step 5: Reflection

4 Case Study: The Electric Car

Practical application of the method is demonstrated on the electric car. Is the electric car sustainable? [5]

4.1 Step 1: Clarify Objective

Based on these facts (background): Global car production is 60 million units per year; 15 % of CO_2 release comes from car; Governments targets are 10 % electric by 2020.

Prime objective is decarbonize road transport. Scale is 8 million cars/year by 2020.

4.2 Step 2: Stakeholders and Concerns

- National and local government – carbon targets
- Car makers and distributors – sales
- Battery makers – supply chain, recycling
- Labour Unions – employment, rights
- Drivers, Automobile Association – range anxiety, cost
- Environmental campaigners – carbon footprint

Battery makers have strongest influence and greatest interest.

4.3 Step 3: Fact Finding

The strongest stakeholder interests split into major categories stakeholder interests, which have been described above:

- Material supply chain – Lithium
- Energy – batteries
- Legislation – battery directive
- Environment – CO_2 footprint
- Society – range anxiety
- Economics – acceptance of cost

Material Consumption per car is 7.3 kg Lithium, if the goal of 8 million cars per year, consumption is 58,400 t per year Lithium. Annual world production of Lithium in 2011 was 34,000 t. From these facts it follows that the requirement for 2020 would be 160 % of current world production.

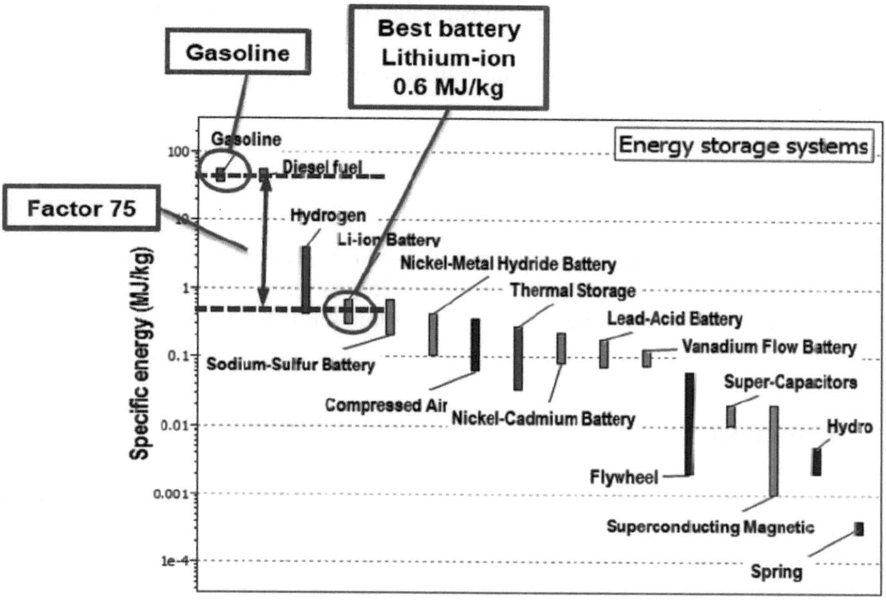

Fig. 2 Bar chart – energy storage systems [3]

Energy What are alternative batteries? Bar chart in Fig. 2 created using the CES software (link) shows the energy density of materials. The graph shows that the Lithium-ion battery has the highest energy capacity. Energy density of batteries is much smaller than gasoline.

Regulation EU battery directive is no batteries to landfill.

Society and stakeholders concerns–The travelling distance is limited by the battery weight and cost. 500 km range requires more than half tonne Li-on battery. It follows that battery weight is proximately 1 kg/km. Cost at todays prices $42,000 http://evobsession.com/electric-cars-2014-list/

Environment Can prime objective be met? Will be decarbonizes road transport?

The battery must be recharged from the National Grid (gas/coal fired) – the carbon footprint resulting in the production of electricity from gas is 140 g/MJ.

Delivered energy to propel small car is 0.6 MJ/km.

Efficiency of battery – electric motor set is 85 %.

It follows that carbon footprint of electric car is $140 \times 0.6/0.85 \approx 100$ g/km.

4.4 Step 4: Impact on the Three Capitals

Planet – the prime objective will not be met without grid decarbonized.

Prosperity – now is inadequate infrastructure for the production and recycling of Lithium.

People – their expectations of range anxiety are not currently met.

4.5 Step 5: Reflection

The issue of electric vehicles can be seen in the short-term (7 years), in this case, the electric car are not sustainable. In long-term (25 years), it is necessary to establish the infrastructure. Alternative strategy (for sustainability) is rethink car use.

5 Conclusion

Students should not think about the design of the product only from a technical point of view, but in a broader context, as indicated in this contribution.

Acknowledgments This lecture is co-financed by the European Social Fund and the state budget of the Czech Republic for project no. CZ.1.07/2.2.00/28.0206.

european social fund in the czech republic
EUROPEAN UNION
MINISTRY OF EDUCATION, YOUTH AND SPORTS
OP Education for Competitiveness

INVESTMENTS IN EDUCATION DEVELOPMENT

References

1. ČSN 01 0391. *Systém managementu společenské odpovědnosti organizací – Požadavky*. (Úřad pro technickou normalizaci, metrologii a státní zkušebnictví, Praha, 2013)
2. M. Ashby, *Analysing Sustainable Developments and the CES EduPack Sustainability Database* (Webinar, Cambridge, 2014)
3. CES EduPack software (Granta Design Limiting, Cambridge, UK, 2012) www.grantadesign.com
4. M. Ashby, *Materials and the Environment: Eco-informed Material Choice*, 2nd edn. (Elsevier/Butterworth-Heinemann, Boston, 2013) 616 p. ISBN 978-012-3859-716
5. *Udržitelný rozvoj* [online]. 13.4.2014. 2014 [cit. 2014-06-08]. Dostupné z:http://cs.wikipedia.org/wiki/Trvale_udr%C5%BEiteln%C3%BD_rozvoj

Part VIII
Experimental Methods and Measuring

G. Achtenova, O. Berka, M. Burian, M. Dub, V. Dynybyl, V. Fliegel, Z. Folta, I. Gajdac, M. Golda, P. Herajn, D. Herak, P. Hrabe, M. Hraska, M. Janda, A. Kabutey, M. Kobza, O. Kohl, J. Kolar, N. Kovalova, R. Kovar, M. Kral, P. Kribala, J. Kricka, P. Kulhavý, J. Kuzel, F. Lopot, A. Lufinka, K. Macuchova, P. Maly, R. Martonka, M. Mazac, O. Milacek, P. Mossoczy, M. Muller, S. Nemcova, J. Petrik, S. Poljak, R. Sigalingging, P. Srb, M. Svoboda, P. Syrovatka, M. Trochta, R. Uhlir, D. Vejrych, and J. Zicha

Design of Experiment for Load Gear Solution in Planetary Transmission

O. Berka, F. Lopot, and M. Dub

Abstract This paper is focused on examination of teeth loading in rotating gears during standard operation and describes the applied approach and device developed. One of the closely examined items in planetary transmission testing is the gearing. The tooth load of each wheel in planetary transmission is monitored individually. In large devices, also the load across the width of the tooth is assessed by the $K_{h\beta}$ coefficient.

Keywords Planetary transmission • Strain gauge • Stress state • Gear meshing

1 Introduction

One of the examined place during testing the planetary transmission are gears. The tooth load of each individual component in planetary transmission is monitored. By large devices is also monitored load across the width of the tooth. It is coefficient $K_{h\beta}$. The actual experiment would not be so difficult if it was not planetary transmission. It was necessary to determine the method of measurement on rotating parts. In case of determining the load on stationary gear, it is a common strain gauge measuring. For this measurement it is possible to use simple devices supplied by a number of manufacturers and implement data transfer over the standard wires. More complicated situation occurs if the gear is rotating around its own axis. In that event the wires can't be used for data transfer from strain gauge. There are used a Wi-Fi panels for instance. A planet is the most complicated piece of transmission to be measured. The planet is rotating around axis of other gear and at the same time around its own axis. Planet performs compound movement composed of multiple rotations and moreover it is closed in gearbox often. Any kind of using Wi-Fi panel is very hard for this gear. Therefor it was necessary to develop a special device. It is an autonomous record device. This device can be placed inside or outside of gear, due to its small dimensions. Moreover the device is supplemented by batteries and amplifiers for signal magnifying.

O. Berka (✉) • F. Lopot • M. Dub
Faculty of Mechanical Engineering, Czech Technical University of Prague,
Prague, Czech Republic
e-mail: ondrej.berka@fs.cvut.cz; frantisek.lopot@fs.cvut.cz; martin.dub@fs.cvut.cz

V. Dynybyl et al. (eds.), *The Latest Methods of Construction Design*,
DOI 10.1007/978-3-319-22762-7_57

How it was implied before, this is the exacting strain gauge measurement. In the introduction the demandingness of measurement due to kinematic planetary transfers was mentioned. Another limiting parameter may for example be a space between teeth of gears during joint meshing, the temperature in the toothing under load and aggression of lubricating oil. In compiling the experiment it must be all these boundary conditions taken into account. Space between gears tooth during joint meshing meanwhile the installation of strain gauge and wires on gears tooth must be respected. Toothing and lubrication oil temperature must be respected during choice of adhesive for strain gauge sticking, during choice of wires for strain gauge and during choice of wires and strain gauges protection. Equally the temperature influence electronic device in gearbox. Furthermore all of components must be protected from aggression of lubrication oil. That includes strain gauges, wires and all electronic devices, which are coming into contact with lubrication oil.

Above several conditions were described, which must be complied during designing of experiment. Large limitation are dimensions of space in gearbox. For this purpose the recording device was created. It can be placed in small places in gearbox due to its size. Whereas, that the experimental measurement is done in gearbox, it is not possible to perform more design modifications of parts, because transmissions are distributing to the end customers. With this recording device is not required.

Next point of experiment is the placement and connection the strain gauges on tooth. Location must be such that it will be the maximum measured value of stress in the teeth. Further, temperature in toothing must be compensated. Last point of experiment is that the location and the connection must be for both directions of rotation. Thus meshing by first or second side of the teeth.

2 Method of Measurement

Due to the complexity of the measurement it is necessary to be very strictly adhere the following procedure.

1. Assignment
2. Determine the input parameters of experiment
3. Preparation of the experiment
4. Installation of measuring apparatus
5. Experimental measurement
6. Visual check after measurement
7. Evaluation of measurement

Individual points will be commentated in follow chapters. In case of compliance of established method we prevent complications, which may come and which could result in unsuccessful experiment.

3 Conducted Experimental Measurement

Measurement was conducted in testing room for transmission in producer factory. Measurement was made during running-in and testing of developed transmission for rubber industry. The purpose was monitoring of new construction effect for following smoothness of gear loading along the width of teeth during meshing with other gear. Measuring apparatus could not affect structure of transmission without arranged places of parts. For safety securing relevant applicable data the measurement was made synchronously with three autonomous systems on the ring gear (Fig. 1), planet (Fig. 2) and pinion (Fig. 3). For measurement on the ring gear was used four channel stable measuring apparatus National Instruments. For measurement on planet was developed unique self-sufficient measuring apparatus with own recording device (Fig. 4). Data from pinion was recorded by four channel apparatus National Instrument with Wi-Fi module for in-time data transfer to the computer (Fig. 5). Program for data recording and processing was created in software LabView. For final evaluation was used Matlab.

The position of strain gauges in toothing on gears of transmission is evident from Fig. 6. Measuring places on gears are intentional aligned. It is for comparing value between gears.

Transmit frequency of record was set on 2000 Hz with respect to time of gears meshing.

Standard strain gauges was glued on tooth gap and connected to the bridge for following bending effect of loading in tooth bottom. Strain gauges and wires were protected from extraneous influence by layer of epoxide synthetic resin.

Partial results are represented graphically for better illustration and with relevant comments. Identification of meshing with individual planet gears is set up for better orientation in diagrams. Planet gears are symbolized by symbols S1, S2 and S3 (Fig. 7).

Progress of teeth loading is represented at Fig. 8. It is record form strain gauges placed on the one teeth. Main teeth loading is zone II. It is a tensile load of teeth during meshing with other gear. In zone I there is the tensile load of teeth too. But this load is produced by meshing previous tooth before investigated teeth with other gears. It is total deformation of gears. After meshing investigated teeth with other gears come up to unloading of teeth and teeth is changing over to pressure load. It is caused by meshing with the following tooth of the gear. This phenomenon is in zone III. It is total deformation of the gears again.

4 Conclusions

For verification the measurement it is compared with a gear-testing using colour. Tooth are painted by colour along their whole length. Colour is wipe off after contact with another tooth. Positioning of loading along length of teeth is evaluated

Fig. 1 Ring gear with strain gauges

Fig. 2 Pinion with strain gauges

Fig. 3 Planet gear with strain gauges and epoxide layer

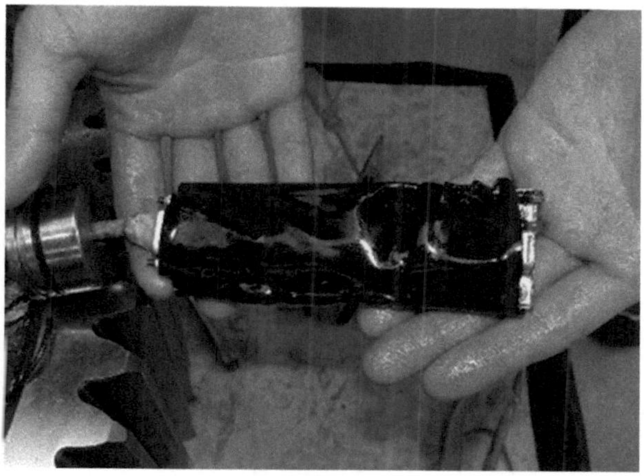

Fig. 4 Recording device

Fig. 5 Systems for data recording

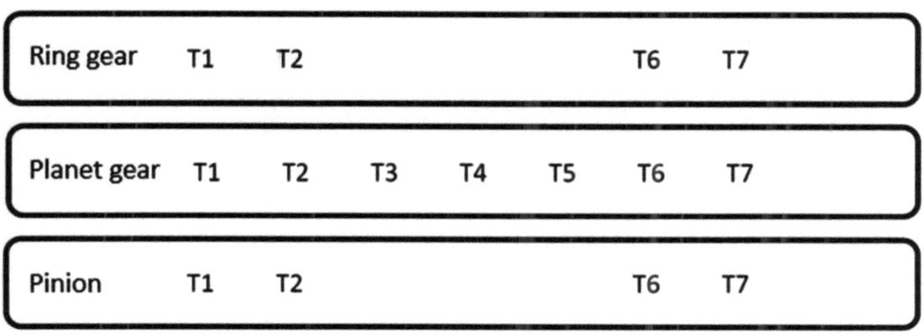

Fig. 6 Position of strain gauges

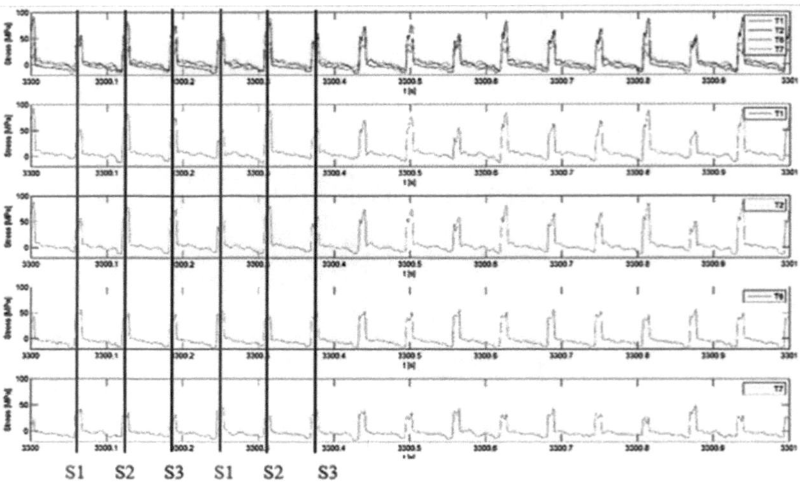

Fig. 7 Identification meshing with individual planet gears

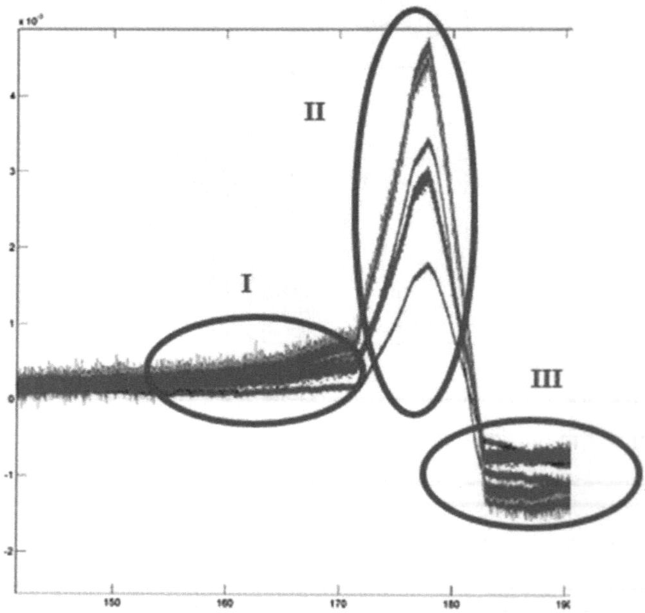

Fig. 8 Progress of recording value from strain gauges

by this test. 3D diagram is made for comparing. Diagram describes measured loading along length teeth (Fig. 9). Verification is done by visual comparing Figs. 9 and 10.

Designed method was used for this experiment. Measurement brought several additional piece of knowledge. These are information how to make better the next

Fig. 9 Toothing with testing colour

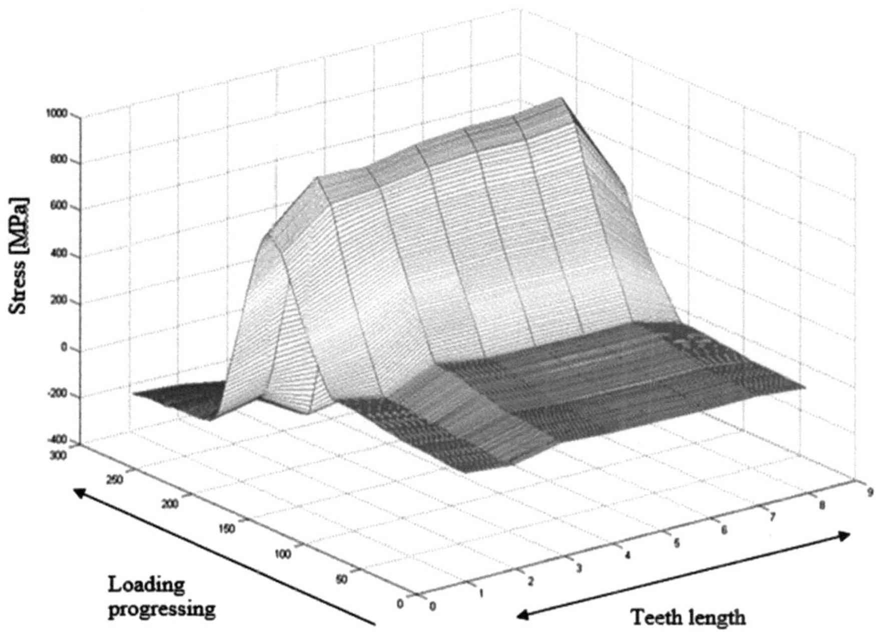

Fig. 10 Loading processing along teeth length

measurement. For example using another adhesive and epoxide, better location of strain gauge, etc. Results were compared with impression of teeth which were coloured. Resulting values agree. Experiment and measurement can be declared as successful.

Investigation of Tram Gearbox Hinge Behaviour

M. Dub, J. Kolar, F. Lopot, V. Dynybyl, and O. Berka

Abstract During the operation of trams in cities the chassis are dynamically loaded through acceleration, deceleration, passage curves and evidently by the roughness of the rail track. These added dynamic loads can significantly increase wearing of the chassis components and tram drive. The aim of this analysis is the investigation of stress state and motions of gearbox hinge. For this purpose, strain gauges and optical system Qualisys is used. Special type of strain gauges arrangement for measuring tension and bend is designed. Acquired data are processed in software Matlab and in user interface of the Qualisys software.

Keywords Tram drive • Gearbox hinge • Strain gauges • Qualisys • Matlab

1 Introduction

During working of tram the chassis are dynamically loaded through acceleration, deceleration, passage curves and evidently by the roughness of the rail track. These added dynamic loads can significantly increase wearing of components of the chassis and drive or it can cause damage of some parts. Thus all the components have to be designed with respect to given conditions. In this case, problems with bearing mounting into gear box occurred after relatively little amount of driven kilometres in almost all gearboxes. One of the hypothesis, which can explain this problem, is too rigid suspension of the gearbox. The hinge doesn't allow sufficient movements of the gearbox and generated reaction forces are transferred to the bearings of the output shaft as an additional load. Therefore testing of the gearbox hinge was suggested. The aim of the measurement is to analyse behaviour of tram gearbox and its hinge during the real driving conditions.

M. Dub (✉) • F. Lopot • V. Dynybyl • O. Berka
Faculty of Mechanical Engineering, Czech Technical University of Prague, Prague, Czech Republic
e-mail: martin.dub@fs.cvut.cz; flopot@seznam.cz; vojtech.dynybyl@fs.cvut.cz; ondrej.berka@fs.cvut.cz

J. Kolar
Charles University, Prague, Czech Republic
e-mail: josef.kolar@fs.cvut.cz

V. Dynybyl et al. (eds.), *The Latest Methods of Construction Design*,
DOI 10.1007/978-3-319-22762-7_58

1.1 Tram and its Drive

Tested tram is partially low-storeyed tram. It has three tractive chassis and it can reach the maximum speed of 60 km/h. Each chassis has two tractive mounted wheels, which are driven by partially spring-mounted drive. That consists of asynchronous motor of power 90 kW and two-gear reduction gearbox connected with claw coupling. Big gear of the gearbox is pressed on the axle of the mounted wheels, which is imbedded into bearing units. Their connection with the frame provide four rubber-metal cone springs, that create the primary springing of the tram chassis.

2 Hypothesis

The gearbox is furthermore connected to the chassis frame through the vertical gearbox hinge. The hinge catches reaction force generated by the rotation moment of the mounted wheels, but it has to allow some specific movements of the gearbox. The maximal tensile force in the hinge for the two-gear gearbox for quasi-static (without respect to the inertial forces of gearbox) can be calculated as

$$S = \frac{M_M(i_c - 1)}{n} \tag{1}$$

Where M_M is maximal torque of the asynchronous motor, i_c is gear ratio of the gearbox and n is distance between the output shaft and the hinge axis.

The hinge of tested tram is designed as a round bar of diameter 28 mm imbedded into rubber supports as it is shown in Fig. 1. These should behave nearly as an ideal spherical joints so the hinge can catch only tension or pressure force. Any bending stress in the hinge is parasitic because it is caused by lateral forces which produce undesirable additional radial and axial loading forces for bearings of the output gearbox shaft [1].

3 Instrumentation

The hinge should transfer only tension or pressure forces. It is desirable to investigate its stress state. Specifically tension or pressure which are expected and bending stress which is undesirable but it can confirm the above mentioned hypothesis. For this purpose strain gauges are used. For the verification and better description of gearbox suspension behaviour the system Qualisys for capturing and evaluating motion is also used.

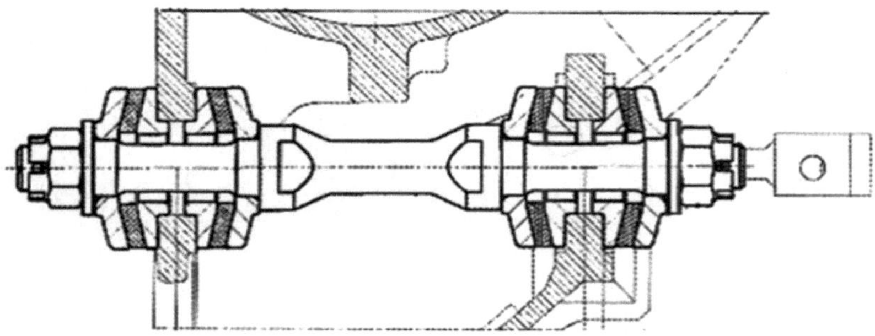

Fig. 1 Gearbox hinge design (rotated view), [2]

3.1 Strain Gauges

Foil strain gauges are used to investigate stress state of the hinge. It is desirable to measure tension – pressure and bend simultaneously. Moreover the bend has to be find out in two perpendicular planes to obtain complete knowledge of the bend and the resultant including its orientation. The first plane is coincident with the tram drive direction (longitudinal plane) and the second is perpendicular and lies in the lateral plane of the tram. For measuring tension-pressure Wheatstone half-bridge arrangement has to be used due to bend compensation (i.e. two strain gauges). For measuring bend also Wheatstone half-bridge arrangement has to be used due to tension-pressure compensation (i.e. for measuring in two planes four strain gauges). To decrease quantity of strain gauges new arrangement is designed (Fig. 2). Only four strain gauges are connected into four independent Wheatstone quarter-bridges with time synchronization.

This arrangement is very convenient because it is possible to measure tension-pressure and simultaneously measure bending in two perpendicular planes. Thanks to this improvement only four strain gauges has to be used, which spares not only the number of used strain gauges but also space and makes the connection of strain gauges easier. Acquired signals are obtained as strain from each strain gauge and they can be assembled after measurement according to following formulas. Tension-pressure stress can be calculated from formula (2). Bending stresses in two planes are a consequence of formulas (3), (4).

$$\sigma_t = \frac{E}{4}(\varepsilon_1 + \varepsilon_2 + \varepsilon_3 + \varepsilon_4) \tag{2}$$

$$\sigma_{b1} = \frac{E}{2}(\varepsilon_1 - \varepsilon_3) \tag{3}$$

$$\sigma_{b2} = \frac{E}{2}(\varepsilon_2 - \varepsilon_4) \tag{4}$$

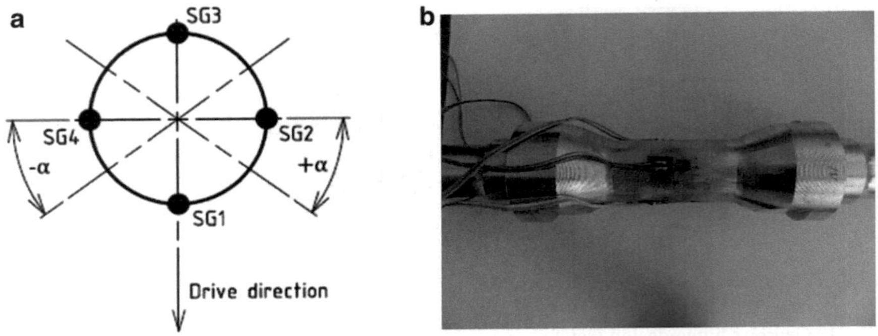

Fig. 2 Measuring arrangement of the strain gauges on the gearbox hinge

3.2 Qualisys System

In addition to measuring stress state of the hinge it is convenient to monitor relative motions of the gearbox connected with the mounted wheels to the frame of the chassis and to the asynchronous motor. Then it is possible to evaluate relative shifts and rotations of the output shaft of the asynchronous electromotor to the input shaft of the gearbox and required range of motions of claw coupling. For this purpose optical system for continual capturing and evaluating motions Qualisys is used. For tracking motions of the object it is necessary to use three cameras and three markers creating the reference plane on monitoring object. In this case six spherical markers are used, three are placed on the gearbox and remaining three are placed on the asynchronous motor. In each object (gearbox and asynchronous motor) is thus created local Cartesian coordinate system (see Fig. 3). Furthermore another marker is placed on the axle of the mounted wheels to sample revolutions so the covered distance and current speed can be calculated.

Altogether six cameras are used, they are divided into two subsystems – moving subsystem and stationary one. The first one is mounted on the special holder which is bolted to the tram frame so it moves with the tram. The second one is placed in the floor pit. The stationary system of cameras is there to verify accuracy of the moving system which could be influenced by the vibration of the tram frame. Both systems are calibrated together and time synchronized.

4 Data Acquisition, Processing and Evaluation

The test drive was divided into three regimes – maximal acceleration, maximal deceleration and common drive. The hinge with installed strain gauges and thermal sensor was mounted into rear mounted wheel of the rear chassis of tested tram. In addition to stress and motion measurement, the torque of two asynchronous motors mounted to the chassis was recorded from the control unit of the tram. Data

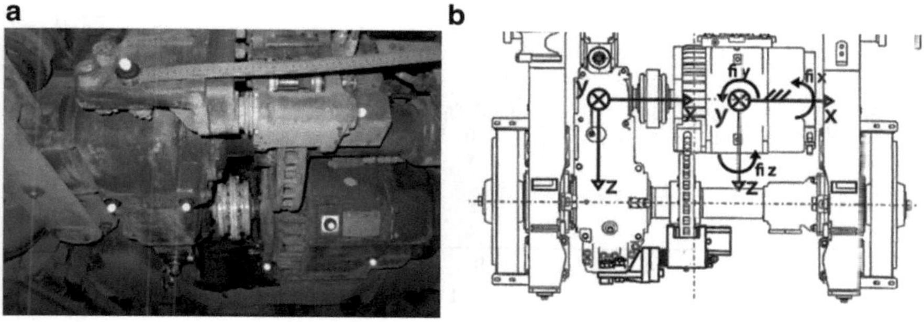

Fig. 3 Markers of the system Qualisys and Cartesian coordinate systems of the gearbox and the asynchronous motor

acquired by strain gauges contained lot of noise. That's why low-pass linear filter with Hamming's window was applied to reduce it. Script calculates tension, bend in two perpendicular planes and resultant with its orientation from data acquired from four independent Wheatstone quarter-bridges. Data captured by the Qualisys system were processed in user interface of Qualisys software. The main aim of this experiment is to detect presence of a bend in the hinge during test drive regimes. Tensile stress can be used as a verification with analytical tensile stress in the hinge which can be calculated according to the formula (1).

Presented graphs proved the presence of the bend in the hinge. Measured tensile stress is in very good agreement with analytically calculated value (Fig. 4). In Fig. 5 there is shown the rotation of the gearbox round the longitudinal axis of the tram during the acceleration regime. For better transparency black line is added into raw signal. It is a representation of mean value during the time. The average deviation of the angle ϕ_z during the acceleration (time about 2–9 s) is 5°.

5 Results and Conclusion

The measurement proved the hypothesis that the gearbox hinge doesn't catch only tensile and pressure forces, vice versa it is also bending stressed. But the measured values of bending stress are quite low. This can be caused by low dynamic effect. The tested tram was empty and the quality of the rail track in the depot was quite good. Another reason could be that the gearbox hinge was mounted exactly with specified pretension so the rubber bedding isn't so rigid and the hinge allows bigger movements. This hypothesis is proved by data measured by system Qualisys. Presented graph shows that movements are higher than admissible values for the claw coupling.

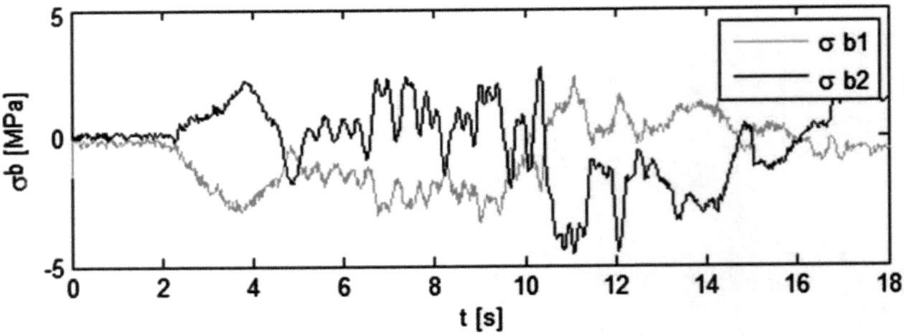

Fig. 4 Bending stress in two perpendicular planes for the acceleration regime

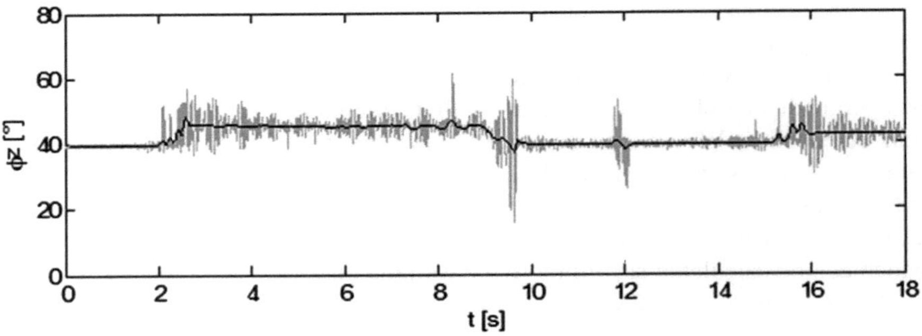

Fig. 5 Relative rotation of the gearbox to the asynchronous motor round the longitudinal axis *z*

References

1. J. Kolář, Teoretické základy konstrukce kolejových vozidel, (ČVUT v Praze, 2009)
2. J. Kolář, F. Lopot, M. Dub, Vyhodnocení měření namáhání závěsky u nápravové převodovky tramvaje, Výzkumná zpráva č. Z 14–01, ČVUT v Praze, Fakulta strojní (2014)

Mechanical Properties of Fasteners

V. Fliegel, A. Lufinka, and R. Martonka

Abstract The basic fasteners include bolts and nuts. These parts have a very wide range of applications not only in engineering but also in other related fields. With this type of connections we encounter at every step. Manufacturers and distributors of these components are required to guarantee the quality of goods sold. For this reason, they are forced to perform independent quality control by independent laboratories. Characteristic test of fittings include tensile, torsional (bolt) and the ability to disassemble (nut).

Keywords Experiment • Fastener • Bolt • Nut

1 Introduction

The basic mechanical connections belong to include the bolted connection. The screw connection is made up from a minimum, bolt and nut. The application is completed by a pair of pads. Screw connections are used by depend of the value of the load bearing joints such loaded (unloaded). In the case of bearing joints is important to choose the right size and holding power of the bolt and nut. For the production of screws is used a lot of kind of steels, which have from small to high and very high tensile characteristic.

In case of damage to well-designed screw joints we are looking for causes of damage. One cause of failure can be the mechanical properties of used material of the screw connection. Producers of parts screw connection regularly checks in cycles the mechanical properties of the material on the selected parts. These are usually put under control of destructive tests, defined by the relevant standards [1–3].

V. Fliegel (✉) • A. Lufinka • R. Martonka
Technical University of Liberec, Liberec, Czech Republic
e-mail: vitezslav.fliegel@tul.cz; ales.lufinka@tul.cz; rudolf.martonka@tul.cz

V. Dynybyl et al. (eds.), *The Latest Methods of Construction Design*,
DOI 10.1007/978-3-319-22762-7_59

2 Mechanical Tests

Mechanical properties of the bolts are measured in two ways of loading [1, 2]. For the first case is tension of defined bolt dimensions derived from the size of the thread. For the second case loads bolt by the torque. In both cases, determines the value of the relevant stress (tension, torsion), which is then compared with the minimum permissible stress for the material.

By loading of tensile is measured dependence tensile force in the bolt and the extension of the bolt. The test continues until to breaking bolt. Correctly broken bolt is that, which breaks on free part of the thread.

By torsional loading is measured the value of torque and the characteristic value is the maximum value. The test runs to twisting the bolt. Correctly twisted bolt is that, which breaks on free part of the thread.

The mechanical properties of nuts are determined only from the axial load. There is not important direction of the loading – tension or compression, only the nut is loaded with correspond force for the minimum time period [3, 4]. In case when after unloading is possible turn the nut by hand without any tools, material properties correspond to that material which must be used.

3 Experiments

3.1 Tensile Stress

The testing device contains an upper anchor plate, a removable flanges, a hardened nut, and a lower anchor plate. The upper and lower anchor plates were mounted on a movable jaw of the hydraulic stool. Test device was supplemented by a set of removable flanges and hardened nuts, to which are anchored to the appropriate test bolt (Fig. 1).

Tensile load force on the bolt is measured by a force transducer with adequate measuring range. In our case, we use a force transducer by company GTM with a measuring range to 100 kN. Measurement deformation on the bolt was measured by immediate place of the bolt, thereby to eliminate any unwanted deformation of parts of the measuring device. In our case, was used sensor WA 10 mm by company HBM.

3.2 Torsional Stress

The testing device consists of a pulley with a rope, torsion shafts, interchangeable adapters, hardened nut, split nuts. As the source of the torque was used linear hydraulic motor, which was pulling with a rope connected to the pulley. When we

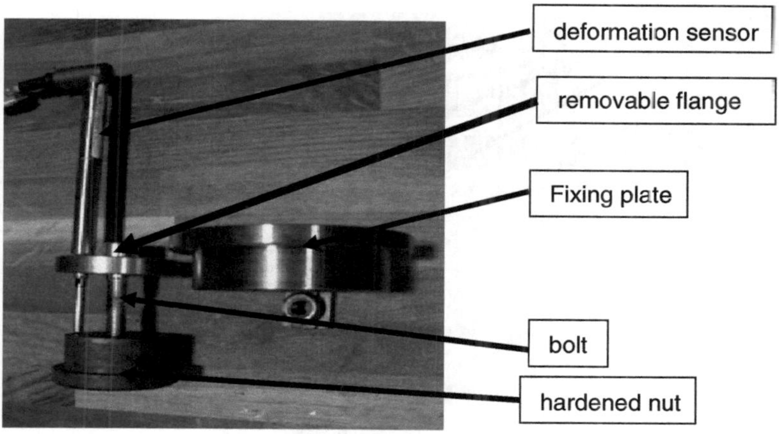

Fig. 1 The sensor, removable flange, hardened nut

know size of the radius (diameter) of the pulley is a simple recalculate (1) the value measured force in the rope to the value of load torque on the bolt. Between the cable and the motor was inserted the force transducer by company GTM with a measuring range to 2.5 kN. Measuring device has been added by precision angular potentiometer RSS 50 by company Megatron (Fig. 2).

$$Mk = F \cdot rp \tag{1}$$

There is

Mk – torque moment
F – measured force
rp – radius of pulley = 65 mm

4 Measured Data

4.1 Tensile Stress

For load bearing screw connections are used primarily materials with higher strength, it is a steel with a yield strength of the bland. Therefore, these materials introduces the yield strength – Rp0, 2nd in the sample case was used bolt M10 × 50 ISO 4762, property class 80 with a minimum value according to Rp0, 2 = 600 MPa (Fig. 3).

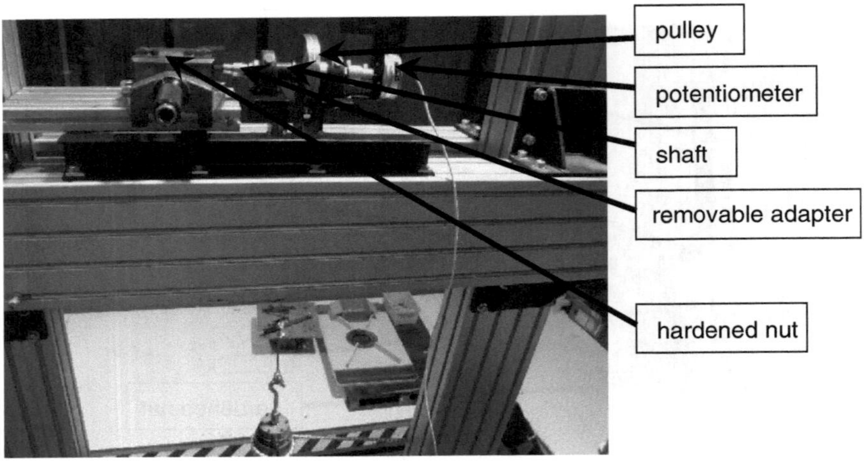

Fig. 2 Experimental device – torsion

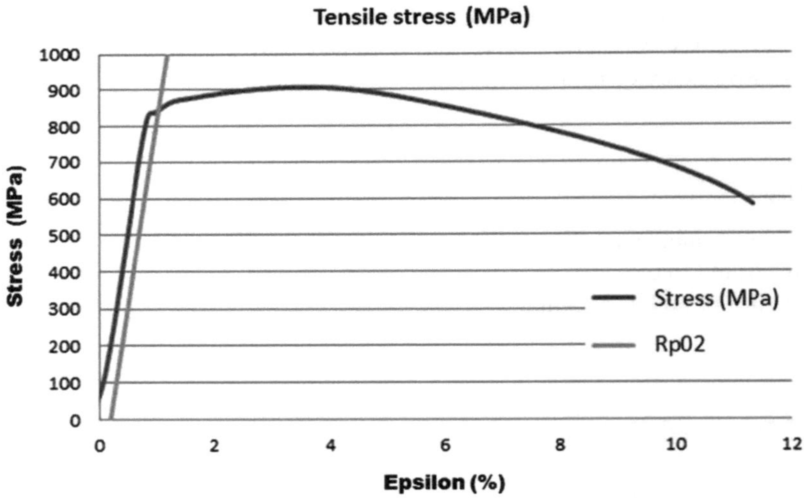

Fig. 3 Tensile stress – M10 × 50 ISO 4762

4.2 *Torsional Stress*

According to standard was defined minimum value of torque that bolt has able to withstand. To sample the value bolt 74 Nm. The graph indicated by the horizontal limit value. After conversion to torsional stress by (2) the minimum value is 740 MPa (Figs. 4 and 5).

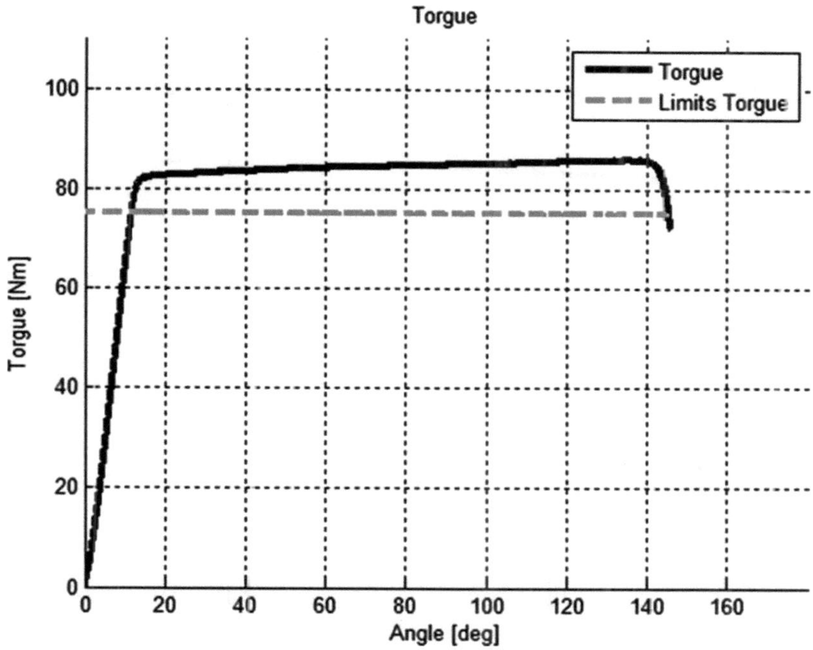

Fig. 4 Torque – M10 × 50 ISO 4762

$$\tau k = \frac{Mk}{Wk} = \frac{Mk}{\frac{\pi \cdot d^3}{16}} \quad (2)$$

There is

Tk – torsional stress
Mk – torque moment
Wk – section modulus in torsion

5 Conclusion

From the mentioned measured data and graphs implies that the bolt of strength steels can be used to approximate the safety k = 1.5 of tensile, k = 1.2 of torsion.

For both types of loading is observed equivalent characteristic, there is a steep increase of tension – elastic deformation, moderate yield point and then a slight increase up to the limit of strength.

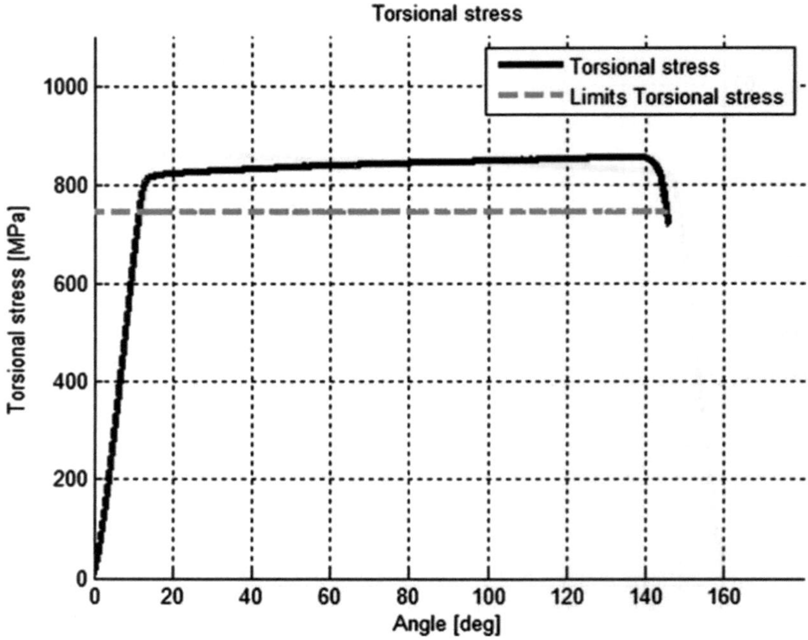

Fig. 5 Torsional stress – M10 × 50 ISO 4762

References

1. EN 20898–1: Mechanical properties of fasteners – part 1: bolts, screws and studs
2. ČSN EN ISO 3506 – 1: Mechanical properties of corrosion-resistant stainless-steel fasteners – bolts, screws and studs
3. EN 20898–2: Mechanical properties of fasteners – part 2: nuts
4. ČSN EN ISO 3506 – 2: Mechanical properties of corrosion-resistant stainless-steel fasteners – nuts

Mathematical Description of Non-linear Mechanical Behaviour of Materials Under Compression Loading, Case Study: Spruce Bulk Wood Chips

D. Herak, A. Kabutey, and R. Sigalingging

Abstract This communication discusses the mathematical description of mechanical behaviour of bulk spruce wood chips under compression loading. The dependency between compressive force and deformation curve characteristic was fitted by reciprocal slope transformation (RST) and tangent curve (TC) methods. The mathematical model coefficients of both methods were respectively determined by the least squares method and Marquardt Levenberg approximation process, all applicable in MathCAD software. It is confirmed here that both methods can suitably be used for the illustration of deformation curve characteristic of bulk spruce wood chips.

Keywords Tangent curve • Reciprocal slope transformation • Compressive force • Deformation • Marquardt • Levenberg

1 Introduction

In engineering applications they are involved the use of different methods for the identification of nonlinear mechanical behaviour of agricultural products such as wood, rubber, coconut, banana, oil palm and other plastic materials [1–4]. The general problem related to their applications is impossibility to apply the basic Hooke's law for defining the stress strain relationship. Few methods are used for mathematical description of nonlinear behaviour of industrial or agricultural materials. These methods include the finite element methods (FEM) [5], rheological methods (RHM) [6], which are based on the merits of understanding the inner structure of the materials; tangent curve method (TCM) [7] and reciprocal slope transformation method (RST) [8], are also based on the comprehension of materials

D. Herak (✉) • A. Kabutey • R. Sigalingging
Czech University of Life Sciences Prague, Prague, Czech Republic
e-mail: herak@tf.czu.cz; kabutey@tf.czu.cz; sigalingging@tf.czu.cz

V. Dynybyl et al. (eds.), *The Latest Methods of Construction Design*,
DOI 10.1007/978-3-319-22762-7_60

as a unit. Each method has its own pros and cons and it is not possible to simply define which method is most suitable for nonlinear description. Based on literature search, FEM and RHM are appropriate for detailed physical analyzes whiles TCM and RST are applicable in industrial engineering. The aim of this study is to analyse TCM and RST methods for mechanical description of bulk spruce wood chips under compression loading.

2 Materials and Methods

2.1 Deformation Curve Characteristic

The information of deformation curve characteristic of bulk spruce wood chips under compression loading presented here has been reported in previously published study [9]. The coefficient of variation was 5 % (Fig. 1). Dependency between compressive force and deformation was recorded using force 150 kN pressing vessel diameter 76 mm and initial height of bulk spruce wood chips 80 mm (Fig. 1) The experiment was observed under temperature 20 °C and pressing rate 1 mm s^{-1}. Weight of bulk spruce wood chips was (38.4 ± 3.5) g, porosity (70.01 ± 1.38) % and moisture content (6.62 ± 0.69) % d.b.

2.2 Reciprocal Slope Transformation Method

Dependency between compressive force F_{RST} (N) and corresponding deformation $\underline{x}$ (mm) Eq. 2 was transformed using reciprocal slope transformation method (RST) [8, 10] in form given by Eq. 1, where T (mm N^{-1}) is amount of reciprocal

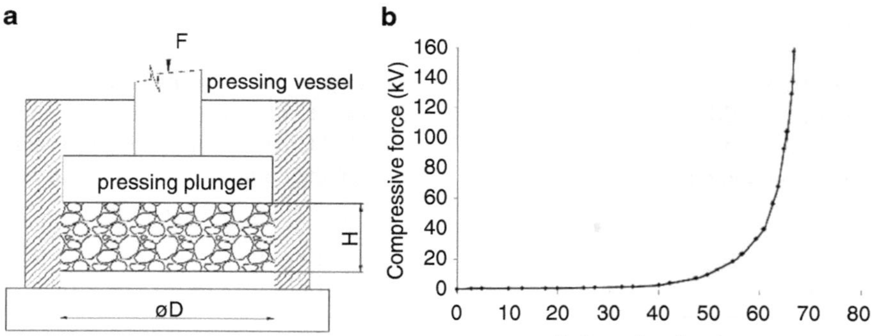

Fig. 1 Experimental illustration [9] of bulk spruce wood chips showing force deformation dependency

transformation, a (N^{-1} mm^{-2}), b (N^{-1} mm^{-1}), c (N^{-1}) and d (N^{-1} mm) are coefficients of the RST method.

$$T(x) = \frac{x}{F_{RST}} = ax^3 + bx^2 + cx + d \tag{1}$$

$$F_{RST}(x) = \frac{x}{ax^3 + bx^2 + cx + d} \tag{2}$$

The coefficients were determined by the least squares method applicable in MathCAD software.

2.3 Tangent Curve Method

Dependency between compressive force F_{TCM} (N) and corresponding deformation x (mm) was described by tangent curve method (TCM) [7] which is given by Eq. 3.

$$F_{TCM}(x) = A \cdot [\tan(B \cdot x)]^n \tag{3}$$

Where A (N) is force coefficient of mechanical behaviour, B (mm^{-1}) is coefficient of mechanical deformation behaviour and n (−) is curvature of deformation curve. These coefficients were determined by Marquardt Levenberg approximation process [11] using MathCAD software.

3 Results and Discussion

Determined coefficients of RST and TCM are shown in Table 1 and their mathematical description presented in Fig. 2. From the statistical analysis using ANOVA (Table 2) it follows that both methods can be used for description of deformation curve characteristic of bulk spruce wood chips under compression loading.

Single ANOVA analysis of the measured data from bulk spruce wood chips were statistically significant at significance level 0.05, that is, the values of F_{crit} (critical value comparing a pair of models) being higher than the F_{rat} values (value of the F – test) and P_{value} (significant level where it can be rejected the hypothesis of equality of models) (Table 2) also higher than 0.05 with very high coefficient of determination R^2 (Table 2).

From the graphical observation (Fig. 2) it is evident that RST clearly described the beginning of the deformation curve characteristic but the course of compressive force at large deformation was not accurately defined by the method.

The physical background of RST mathematical description is based on the compliance theory [8]. In Fig. 2 shows that TCM method described precisely the whole course of deformation curve characteristic except the origin of deformation

Table 1 Determined coefficients of RST and TCM method

RST				TCM		
a (N^{-1} mm^{-2})	b (N^{-1} mm^{-1})	c (N^{-1})	d (N^{-1} mm)	A (N)	B (mm^{-1})	n (−)
$4.333 \cdot 10^{-4}$	$5.673 \cdot 10^{-3}$	1.865	0.00	2100	0.021	4

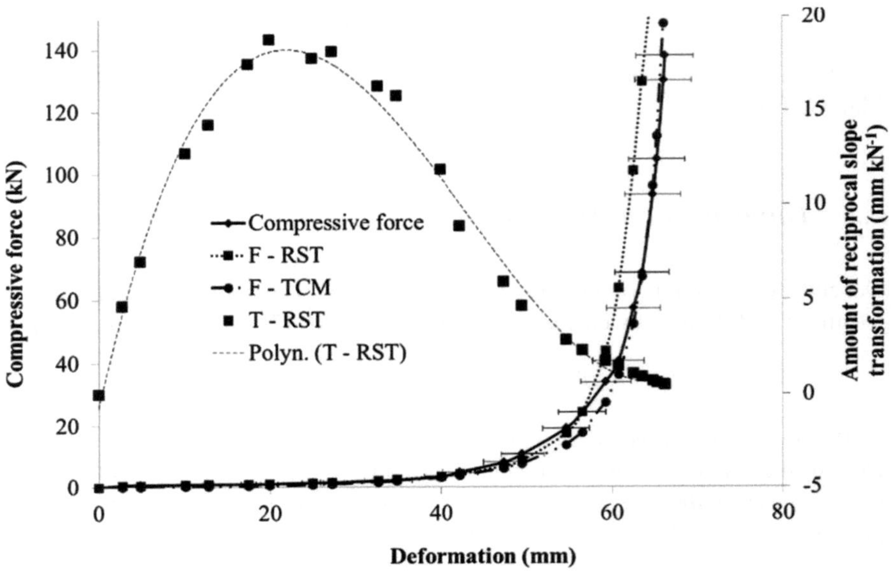

Fig. 2 Deformation characteristic of spruce bulk wood chips, marks in the graph indicates standard deviation

Table 2 Statistical analysis ANOVA

RST			
F_{crit} (−)	F_{rat} (−)	P_{value} (−)	R^2 (−)
4.042	0.007	0.978	0.995
TCM			
F_{crit} (−)	F_{rat} (−)	P_{value} (−)	R^2 (−)
4.042	0.671	0.416	0.97

curve due to the fact that the bulk material was supposedly to be pre stressed and thus shows different characteristic at very small deformation which usually has no practical implication in industrial engineering. For coefficient determination, it is necessary to use nonlinear fitting system for example, Marquardt Levenberg algorithm [11]. The physical understanding of this theory is based on the theory of system rigidity [7, 9]. It is also evident that using simple mathematical operations such as integration and derivation then the deformation energy and rigidity can be determined. Obviously, both methods can be used not only for description of various types of materials with nonlinear mechanical behaviour but also for different kinds of loading. The RST and TCM method can be utilized as a background for more complex models of materials with nonlinear mechanical behaviour and they can also help to understand in detail their deformation characteristic with minimum energy demand.

4 Conclusion

The dependency between compressive force and deformation curve characteristic of bulk spruce wood chips was described by two different methods namely the RST – reciprocal slope transformation and tangent curve (TCM). Based on statistical analysis of measured data both methods can be significantly used for accurate illustration of deformation curve characteristic and mechanical behaviour of bulk spruce wood chips under compression loading.

References

1. A.K. Bledzki, A.A. Mamun, J. Volk, Barley husk and coconut shell reinforced polypropylene composites: The effect of fibre physical, chemical and surface properties. Compos. Sci. Technol. **70**(5), 840–846 (2010)
2. S. Joseph, A comparison of the mechanical properties of phenol formaldehyde composites reinforced with banana fibres and glass fibres. Compos. Sci. Technol. **62**(14), 1857–1868 (2002)
3. K.M.M. Rao, K.M. Rao, Extraction and tensile properties of natural fibers: Vakka, date and bamboo. Compos. Struct. **77**(3), 288–295 (2007)
4. F.A. de Silva, N. Chawla, R.D.T. Filho, Tensile behavior of high performance natural (sisal) fibers. Compos. Sci. Technol. **68**(15–16), 3438–3443 (2008)
5. M. Petrů, O. Novák, D. Herák, S. Simanjuntak, Finite element method model of the mechanical behaviour of *Jatropha curcas L.* seed under compression loading. Biosyst. Eng. **111**(4), 412–421 (2012)
6. J. Ocenasek, J. Voldrich, Mathematical modeling of a biogenous filter cake and identification of oilseed material parameters. Appl. Comput. Mech. **3**, 1–12 (2009)
7. D. Herak, A. Kabutey, M. Divisova, S. Simanjuntak, Mathematical model of mechanical behaviour of *Jatropha curcas L.* seeds under compression loading. Biosyst. Eng. **114**(3), 279–288 (2013)
8. D. Herák, J. Blahovec, A. Kabutey, Analysis of the axial pressing of bulk *Jatropha curcas L.* seeds using reciprocal slope transformation. Biosyst. Eng. **121**, 67–76 (2014)
9. D. Herák, A. Kabutey, A. Sedláček, G. Gürdil, Tangent curve utilization for description of mechanical behaviour of pressed mixture. Res. Agric. Eng. **57**(1), 13–18 (2011)
10. J. Blahovec, Reciprocal slope transformation applied to physical problems. Appl. Math. Lett. **24**(12), 2038–2041 (2011)
11. D.W. Marquardt, An algorithm for the least-squares estimation of nonlinear parameters, SIAM. J. Appl. Math. **11**(2), 431–441 (1963)

Influence of Steel Sheet Width on Bearing Capacity of Resistance Spot Welding

P. Hrabe and M. Muller

Abstract A research was focused on an evaluation of a influence of a steel sheet S235J0 width on bearing capacity of a resistance spot welding. Different widths 15, 30 and 45 mm of 1 mm thick sheet of the steel were the tested parameters. A change of the bearing capacity, a deformation and an angle of the deformation were evaluated.

Keywords Bearing capacity • Deformation • Angle of deformation • Mechanical properties

1 Introduction

A choice of a suitable bonding method is one of basic steps which are needed for ensuring competitiveness. A prospective method of metal materials bonding is a spot welding. The spot welding belongs among a resistance welding that means a metallurgical bond is created. The production of bonds by the resistance spot welding technology belongs among simple that means technologically less demanding [1].

A bonding method "Clinching" can be classified among significant other methods. This method uses a shaping for the bond creation. It serves for bonding of materials using one or multi-degree production process. This principle can be characterized by a transfer of the material combined with a local cut or a plastic deformation and subsequent cold pressing [2–4].

Resistance spot welding is a very effective method [5] to join two or more metal sheets together with heat produced by welding current and it is widely used in automotive industry. The stiffness, strength and integrity of the vehicle constructions depend in part on the quality of the resistance spot welds.

A choice of a suitable bonding method is one of basic steps which are needed for ensuring a competitiveness [6–9]. It is possible to characterize three basic ways of bonding – mechanical, heat and chemical one [7]. When applying a particular

P. Hrabe (✉) • M. Muller
Czech University of Life Sciences Prague, Prague, Czech Republic
e-mail: hrabe@tf.czu.cz; muller@tf.czu.cz

V. Dynybyl et al. (eds.), *The Latest Methods of Construction Design*,
DOI 10.1007/978-3-319-22762-7_61

bonding technology a knowledge of technologic principles which influence qualitative features of the final bond is important [10].

The aim of the research was to set the influence of the pattern width on mechanical properties.

2 Materials and Methods

Test samples made from the constructional steel S235J0 were used for the experimental research. Sizes of the test sample which was welded were 1 × (15, 30 and 45) × 105 mm (made according to standard ČSN EN ISO 14273).

A surface of test samples was treated by grit blasting by a corundum of a fraction size F80. A mechanical treatment of the surface belongs among significant factors. Many authors deal with a research of the surface treatments.

A resistance electrode holder BV2 5 21 was used for the resistance spot welding, a maximum welding current is 6.4 kA, a maximum force among electrodes is 2 kN. It is declared a nugget of a size 3–3.5 mm. Soft resistance spot welding mode (that means low currents and long welding time intervals) was used at the samples production.

A principle of the bond rise is following:

- The mechanical treatment of the surface,
- Setting of parameters on a regulator that means the welding time 0.2 s for materials of the thickness 1 + 1 mm,
- Putting welded materials (overlapped) among welding electrodes,
- Pressing of welding electrodes – a transit of the electric current for the stated time.

A destructive testing that means a tensile shear strength test was performed on a universal testing machine LABTest 5.50ST (a sensing unit AST type KAF 50 kN, an evaluating software Test&Motion). A speed of the deformation corresponded to 6 mm*min^{-1}. An evaluating criterium was a set loading force (N) and a type of the weld failure. A size of the nugget was evaluated on the basis of a picture analysis.

3 Results

Comparing tests focused on different overlapped width (15, 30 and 45 mm) confirmed a decreasing of the bearing capacity of bonds created by the resistance spot welding. The result of measuring is visible in Fig. 1.

For the correct evaluation it is also important to determine the determination index R^2. It is the problem of the correlation analysis. The values of the determination index can be from 0 to 1. The functions presented in Fig. 1 are determined by equations in Table 1.

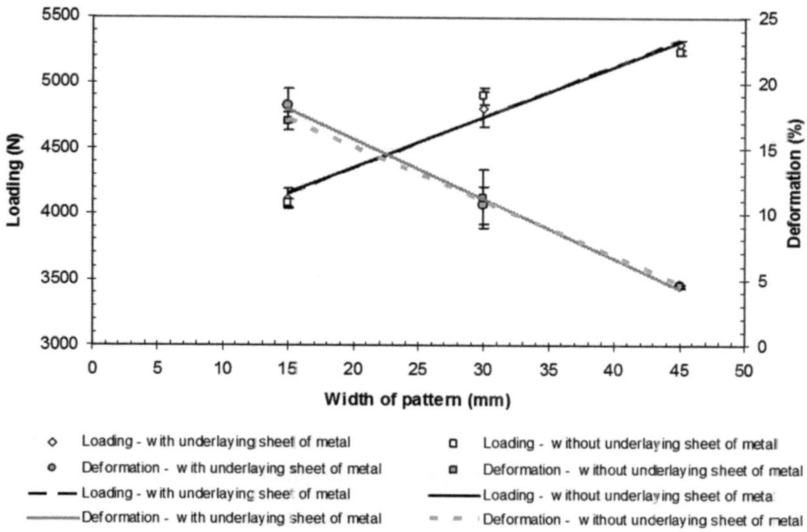

Fig. 1 Influence of sample width on loading and deformation of resistance spot welded bonds

Table 1 Equations of linear functions

Description	Functional equations	R^2
Loading – with underlaying sheet of metal	$y = 39.463x + 3554.6$	0.99
Loading – without underlaying sheet of metal	$y = 38.777x + 3579.8$	0.95
Deformation – with underlaying sheet of metal	$y = -0.4573x + 24.88$	0.99
Deformation – without underlaying sheet of metal	$y = -0.417x + 23.437$	0.99

It is visible form the results of the measuring that the deformation angle of the spot welded bonds also decreases in a linear way depending on the sample width (Fig. 2).

The functions presented in Fig. 2 are determined by equations in Table 2.

It is obvious from the tests results that the pattern width significantly influences the deformation / rotation of the spot welded bonds (Fig. 3). Thus, the bending moment leads to the tensile stress, which is perpendicular to the nugget, as shown in Fig. 3. A basic influence of packing pieces depending against a misalignment (an increase of the bending moment) of test samples at loading was not certified the results. Also Hongqiang Zang et al. came to similar results [5]. They tested 30 mm thick test samples and they ascertained the maximum angle of the rotation 12.8°. It is obvious from above mentioned tests that the angle was set 10.9–12.5° at the width 30 mm.

The results of the bearing capacity present an arithmetical mean of data ascertained in the reliability interval α 0.05 (Table 3). Tukey's HSD test was used for the statistical comparison of the mean value. It is possible to say on the basis of the results of Tukey's HSD test that they are statistically non-homogeneous

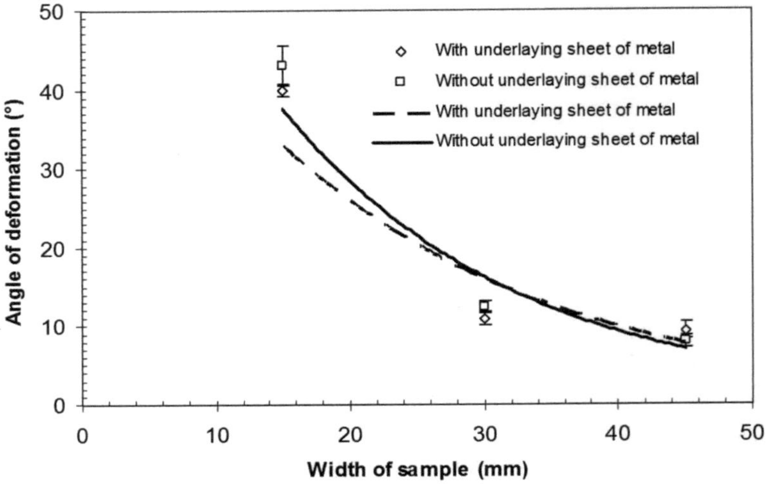

Fig. 2 Influence of sample width on deformation angle of bonded materials

Table 2 Equations of linear functions

Description	Functional equations	R^2
With underlaying sheet of metal	$y = 68.271e^{-0.0484x}$	0.83
Without underlaying sheet of metal	$y = 87.449e^{-0.0562x}$	0.93

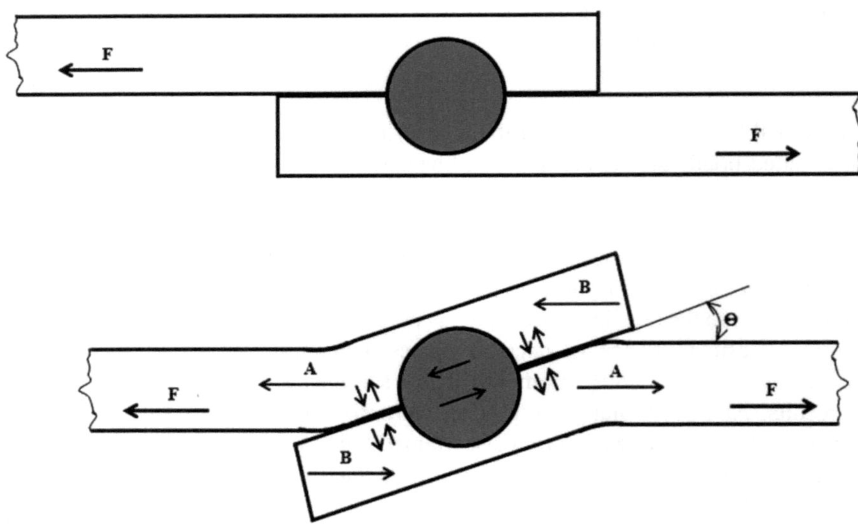

Fig. 3 Scheme of weld bond rotation at loading (shear tensile test)

Table 3 Statistical comparison of mean values – Tukey's HSD test

			Agreement				
With (mm)	Characteristic of tested set	Angle (°)	1	2	3	4	5
45	Without underlaying sheet of metal	7.97	*				
45	With underlaying sheet of metal	9.35	*	*			
30	With underlaying sheet of metal	10.91		*	*		
30	Without underlaying sheet of metal	12.46			*		
15	With underlaying sheet of metal	39.97				*	
15	Without underlaying sheet of metal	42.97					*

groups (Table 3). It followed from the results that the tested sets are distinguished for five different groups from the homogeneity point of view.

The bond failure occurred always in the boundary of the nugget and bonded materials.

4 Conclusion

Following conclusions concluded from performed experiments:

- The welded bond was always failed in the interface of the heat affected zone and the bonded material.
- The loading capacity of the resistance spot welded bonds ranged in the interval 4000–5300 N – linear increase.
- The essential influence is the width of test samples.
- The width of test samples depends of the deformation angle. It came to equalizing of the deformation / rotation angle since the width 30 mm.

Acknowledgments Supported by Internal grant agency of Faculty of Engineering, Czech University of Life Sciences in Prague [nr. 31140/1312/313133].

References

1. M. Černý, J. Filípek, Corrosion damage of rivet joints (Acta univ. agric. et silvic. Mendel. Brun., 2008), LVI, No. 4, pp. 37—46
2. Böllhoff [online]. 2010, [2011-26-08]. Available at: http://www.boellhoff.com/en/in/assembly_systems/rivclinch_clinching.php
3. Y. Abe, M. Kishimoto, T. Kato, K. Mori, Joining of hot-dip coated steel sheets by mechanical clinching. Int. J. Mater. Form. **2**(1), 291–294 (2009). ISSN: 1960–6206
4. J. Mucha, L. Kaščák, E. Spišák, Joining the car-body sheets using clinching process with various thickness and mechanical property arrangements. Arch. Civ. Mech. Eng. **11**(1), 135–148 (2011). ISSN: 1644–9665

5. Z. Hongqiang, Q. Xiaoming, X. Fei, B. Jie, C. Jianhe, Failure analysis of dissimilar thickness resistance spot welded joints in dual-phase steels during tensile shear test. Mater. Des. **55**, 366–372 (2013). ISSN: 366–372
6. M. Müller, Influence of adhesives storing temperature on adhesive bond strength. Manuf. Technol. **14**(1), 71–75 (2014)
7. M. Müller, D. Herák, Application possibilities of adhesive bonds – Europe, Indonesia. Sci. Agric. Bohem. **44**(3), 167–171 (2013)
8. M. Müller, D. Herák, P. Valášek, Degradation limits of bonding technology depending on destinations Europe, Indonesia. Tehnicki Vjesnik-Technical Gazette **20**(4), 571–575 (2013)
9. M. Müller, P. Valášek, Degradation medium of agrokomplex – adhesive bonded joints interaction. Res. Agric. Eng. **58**(3), 83–91 (2012)
10. M. Müller, P. Valášek, Assessment of bonding quality for several commercially available adhesives. Agron. Res. **11**(1), 155–162 (2013)

Calculation of Fuel Consumption on the Vehicle Simulation Model

M. Kobza

Abstract This paper describes the methodology for fuel consumption calculation on the vehicle simulation model with hydrostatic drive in the program MATLAB/Simulink. The consumed fuel quantity is calculated from the engine relative fuel consumption. This calculation is based on digital approximation of the map supplied by the engine producer. The amount of consumed fuel obtained on the simulation model is confronted with measurement on the real vehicle. The methodology of fuel consumption calculation will be used in the criterion function for the optimization kinetic energy recovery process.

Keywords Combustion engine • Approximation • Vehicle • Fuel consumption • Hybrid

1 Introduction

The calculation of fuel consumption has been developed to compare the individual results of drive cycle simulation. Instantaneous fuel consumption is determined by the product of mechanical power and the relative fuel consumption of combustion engine. Actual fuel consumption is obtained by using digital approximation of the map of combustion engine relative fuel consumption.

2 Problem Formulation

2.1 Digitization of Relative Fuel Consumption Map

The map of combustion engine relative fuel consumption, supplied by the engine manufacturer, was digitized into MATLAB by using *ginput* function, which records the cursor coordinates. Scanned points were approximated by *v4* method, based on

M. Kobza (✉)
Brno University of Technology, Brno, Czech Republic
e-mail: y126106@stud.fme.vutbr.cz

V. Dynybyl et al. (eds.), *The Latest Methods of Construction Design*,
DOI 10.1007/978-3-319-22762-7_62

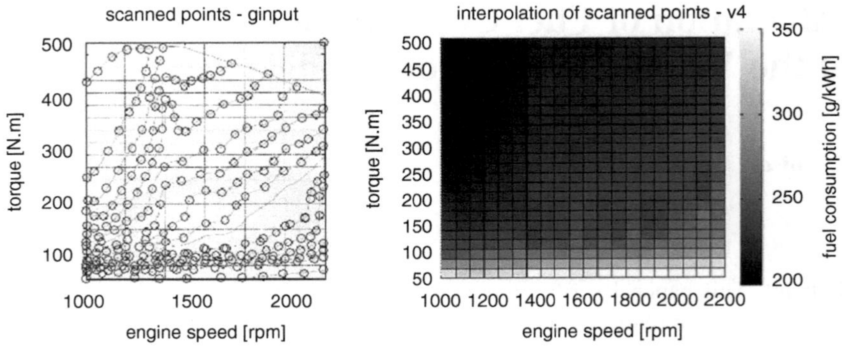

Fig. 1 Digital approximation of map of combustion engine relative fuel consumption

Delaunay triangulation (Fig. 1). The advantage of this method is obtaining a surface over the whole set *(speed, torque)*.

2.2 *The Model*

A vehicle simulation model with hydrostatic drive was created as part of the project dealing with the development of the regenerative braking system for heavy industrial vehicles, see diagram in Fig. 2. This vehicle simulation model is based on the regenerative device model [1].

The vehicle simulation model was equipped with the subsystem (Fig. 3), which calculates the total amount of consumed fuel during a driving cycle. The subsystem uses the *2-D Lookup Table* block, which performs a linear approximation of interpolated map of relative fuel consumption (Fig. 1) [2].

The inputs for *2-D Lookup Table* block are the instantaneous speed and torque of the combustion engine output shaft. The output of the block receives the corresponding value of relative fuel consumption. *One step delay* blocks are included to ensure the numerical stability of calculation. Mass of consumed fuel m is given by Eq. (1). (H = relative fuel consumption, ω = engine speed, T = torque, P = mechanical power)

$$m = \int H \cdot P \cdot dt = \int H \cdot \omega \cdot T \cdot dt \tag{1}$$

3 Comparison with Measurement

The calculation methodology was verified by comparing the amount of consumed fuel during the simulation and measurement on the real vehicle. The comparison was carried out on four separate records of driving cycle. The simulation model was

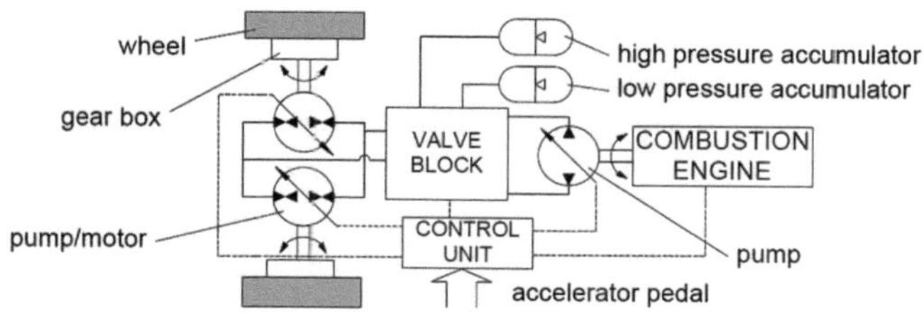

Fig. 2 Simplified diagram of the vehicle with hydrostatic regenerative drive

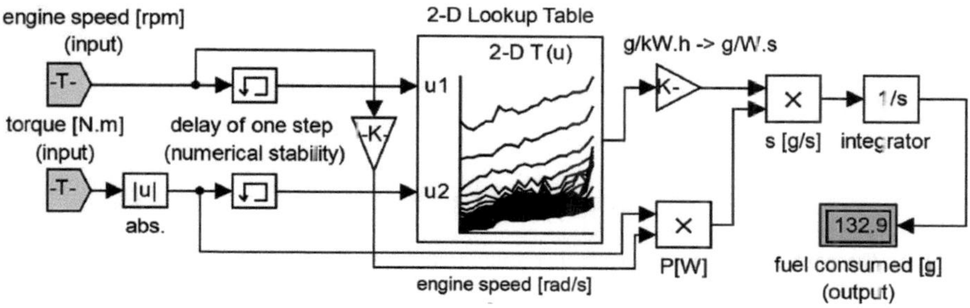

Fig. 3 Subsystem for the fuel consumption calculation in the vehicle simulation model (Simulink)

set to accurately imitate the previously measured driving cycle (a reverse working cycle when the vehicle cyclically changed the driving direction from forward to backward. The comparison of measured and simulated mass of consumed fuel is shown in Table 1. The comparison of courses of instantaneous fuel consumption during calculation and during simulation is shown in Fig. 4.

The average deviation between the mass of consumed fuel during simulated and real driving cycle reached 7.9 %. Measurements № 1 and 2 showed a more dynamic drive compared to measurements № 3 and 4. The engine was more often found in unstable operating conditions, when the map of relative fuel consumption is not sufficiently accurate [3].

4 Conclusion

The subsystem for calculation of consumed fuel mass will be used in the simulation model of hybrid vehicle with regenerative drive. Further development of the vehicle simulation model will offer an optimized process of kinetic energy recovery using the genetic algorithm. The mass of consumed fuel will be one of the inputs for the criterion function to evaluate the process of kinetic energy recovery.

Table 1 Comparison of consumed fuel mass during driving cycles

№.	Calculation time	Consumed fuel mass		Deviation of simulation
		Simulation	Measurement	
1.	298 s	1094 g	957 g	12.5 %
2.	228 s	877 g	751 g	14.3 %
3.	361 s	1114 g	1079 g	3.1 %
4.	341 s	1140 g	1122 g	1.6 %
			Average deviation:	**7.9 %**

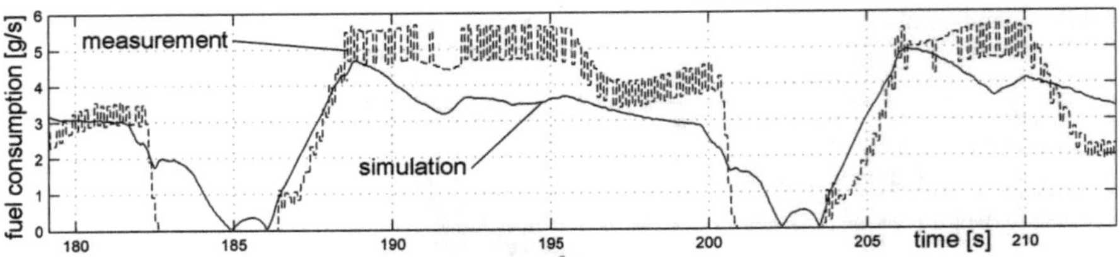

Fig. 4 The course of instantaneous fuel consumption during the reverse driving cycle

Acknowledgments This work is an output of research project Eureka (LF12029) "Hydrostatic system to energy recovery for commercial vehicles "supported by the Ministry of Education, Youth and Sports of the Czech Republic and NETME Centre, regional R&D centre built with the financial support from the Operational Programme Research and Development for Innovations within the project NETME Centre (New Technologies for Mechanical Engineering), Reg. No. CZ.1.05/2.1.00/01.0002 and, in the follow-up sustainability stage, supported through NETME CENTRE PLUS (LO1202) by financial means from the Ministry of Education, Youth and Sports under the "National Sustainability Programme I".

References

1. Z. Němec, Model cyklické rekuperace kinetické a tlakové energie – příklad 1. Dílčí pracovní zpráva k výzkumnému záměru. VUT v Brně, FSI. Prosinec 2012
2. I.N.C. CUMMINS, *Engine performance data* (Columbus, Indiana, 2007)
3. M. Kobza, Determination of efficiency of hydraulic drive with recuperation. Brno University of Technology, Faculty of Mechanical Engineering, 2014. Master thesis, 68 p., supervisor prof. RNDr. Ing. Josef Nevrlý, CSc

Identification of Parameters of Stiffness and Damping Based on the Amplitude Frequency Characteristics

O. Kohl

Abstract This text deals with the determination of stiffness and damping parameters of a dynamic system. These parameters are determined mathematically from the amplitude-frequency characteristics. In practice, this method is applied to automobile seat, which is equipped with a system of air springs without dampers.

Keywords Stiffness • Damping • Air spring

1 Introduction

In this text is described a method with which one can be determinate the parameters of stiffness and damping based on the knowledge of the amplitude-frequency characteristic, weight of mass placed on the sprung part of the support and displacement excitation. It is a linear vibration system with one degree of freedom, which is excited kinematically with forced harmonic oscillations.

The obtained stiffness and damping parameters are substituted into the motion an equation of which is formed amplitude-frequency characteristics, which is compared with measurements.

Load characteristics seat support is finally physically measured and then is verificated correctness of the result.

2 Dynamic Parameters of the System

Weights of mass m is placed in the vertical direction on the frame (base) with elastic-viscous coupling superseded by spring with stiffness k and by viscous damper with damping b. The frame moves in the vertical direction in harmoniously with the displacement $y(t) = Y \sin \omega t$ and causes the vertical vibrations of the body with the displacement $x(t)$ from the static equilibrium position when the spring is compressed by x_0 and transmits the own weight of the sprung mass (Fig. 1).

O. Kohl (✉)
Technical University of Liberec, Liberec, Czech Republic
e-mail: ondrej.kohl@tul.cz

V. Dynybyl et al. (eds.), *The Latest Methods of Construction Design*,
DOI 10.1007/978-3-319-22762-7_63

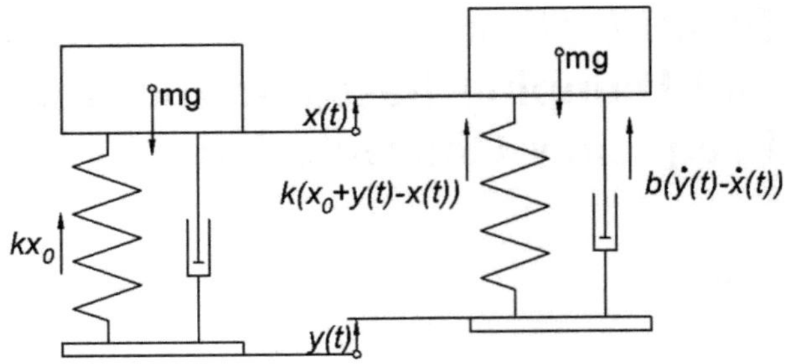

Fig. 1 Kinematically excited system

Motion equation of the sprung mass has the form

$$m\ddot{x}(t) = k(x_0 + y(t) - x(t)) + b(\dot{y}(t) - \dot{x}(t)) - mg \tag{1}$$

And can be adjusted:

(a) For absolute displacement *x(t)* to form

$$\ddot{x}(t) + 2D\Omega\dot{x}(t) + \Omega^2 x(t) = \Omega^2 y(t) + 2D\Omega\dot{y}(t) \tag{2}$$

(b) For the relative displacement *z(t)* = *x(t)*-*y(t)* of the body relative to the frame

$$\ddot{z}(t) + 2D\Omega\dot{z}(t) + \Omega^2 z(t) = -\ddot{y}(t) \tag{3}$$

The amplitude of the steady oscillation is

$$Z = \omega^2 Y \frac{1}{\sqrt{\left(\Omega^2 - \omega^2\right)^2 + 4D^2\Omega^2\omega^2}} \tag{4}$$

After adjustment

$$\frac{Z}{Y} = \frac{\eta^2}{\sqrt{\left(1 - \eta^2\right)^2 + 4D^2\eta^2}} \tag{5}$$

Where damping ratio is

$$D = \frac{b}{2m\Omega} = \frac{b}{2\sqrt{km}} \tag{6}$$

Own frequency is

$$\Omega = \sqrt{\frac{k}{m}} \tag{7}$$

Tuning

$$\eta = \sqrt{\frac{\omega}{\Omega}} \tag{8}$$

[1]

3 Getting Parameters of Stiffness and Damping

Into the above equation for getting the amplitude of steady oscillations was supplemented specified parameters and the equation was adjusted.

$$z = y\frac{(\sin(2\pi ft))^2}{\sqrt{\left(1-\left(\frac{2\pi f}{\Omega}\right)^2\right)^2 + 4\left(\frac{b}{2\sqrt{km}}\right)^2\left(\frac{2\pi f}{\Omega}\right)^2}} \tag{9}$$

The table in Microsoft Excel software was created replacement of amplitude-frequency characteristics. The course of the base stiffness was estimated together with a damping coefficient and it was gradually substituted into the equation. It was chosen constant damping coefficient and linear stiffness characteristic (see Fig. 2). Closest to the actual characteristic was linear progression stiffness characteristic,

$$k = 6,5x + 950N \cdot mm^{-1} \tag{10}$$

based on the ratio of displacement excitation and response (see Fig. 3). Damping coefficient was also elected and due to measured characteristic was its value chosen

$$b = 40N \cdot mm \cdot s^{-1} \tag{11}$$

The measured characteristics were incorporated into the amplitude-frequency characteristic (see Fig. 4), which was compared with the specified characteristics which was calculated from the measured load characteristics of the base (see Fig. 5).

4 Conclusion

Determination of the parameters of excitation transmission amplitude – frequency characteristics is possible, but because that it is necessary to gradually assign values stiffness and damping is very time consuming.

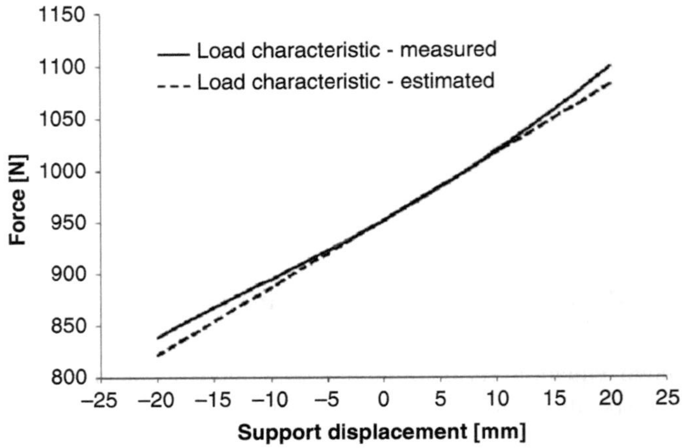

Fig. 2 Load characteristic

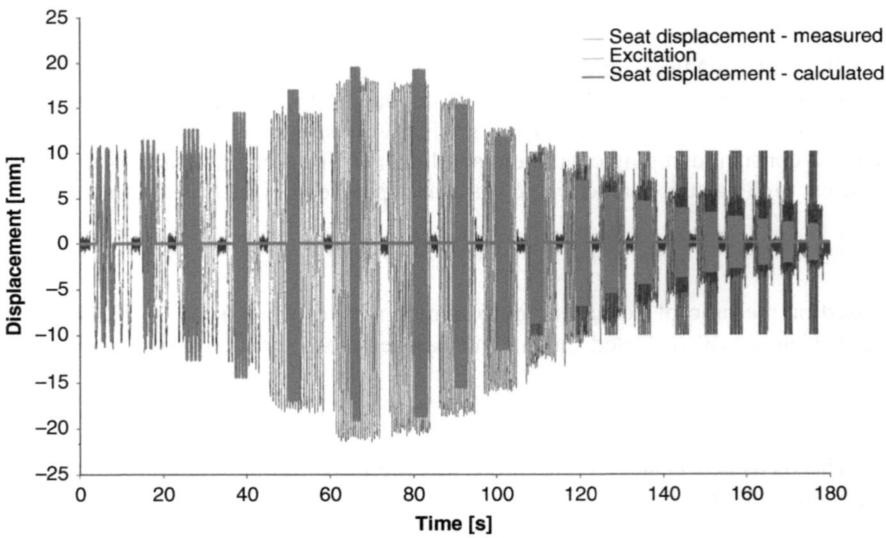

Fig. 3 Excitation, measured and calculated seat displacement

The coefficient of damping and stiffness are interdependent. The value of one variable affects the value of the other. By estimation was possible to have proper slope of the curve stiffness characteristic of the support. The initial stiffness in relation to the damping coefficient must have been verified by measurements.

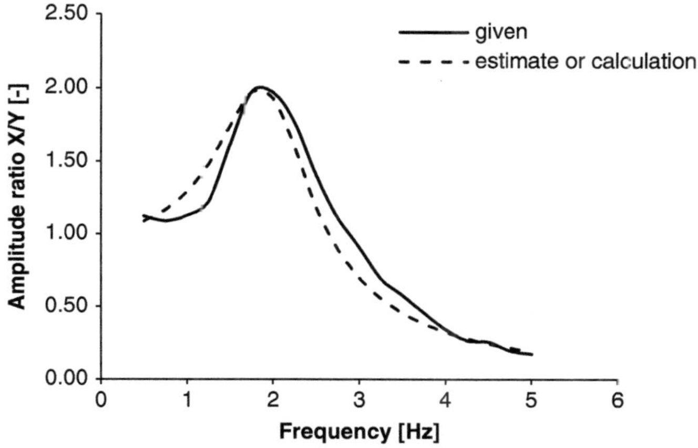

Fig. 4 Support transmission amplitude-frequency characteristic – given compared with the calculation

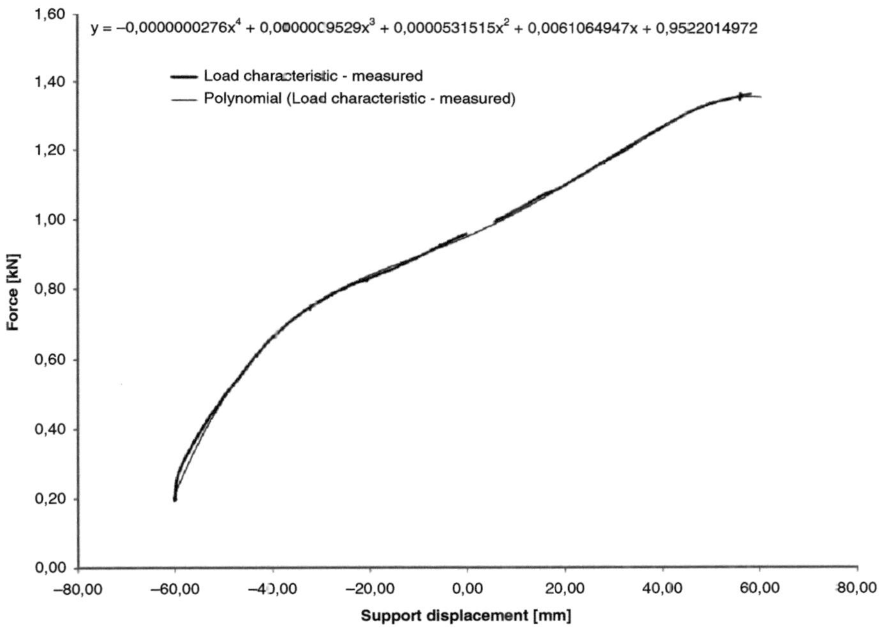

Fig. 5 Measured load characteristic of the support

References

1. V. Zeman, Z. Hlaváč, *Kmitání mechanických soustav* (Plzeň: Západočeská univerzita v Plzni, 2010). ISBN: 978-80-7043-337-9

Measurements of the Integral Characteristics of Twisted Flow Using a Five-Tube Pressure Probe

J. Kuzel, M. Golda, M. Kral, and M. Hraska

Abstract This document summarizes developments in the measurement of the integral characteristics of fans carried out at VZLÚ, a.s. in 2013–2015. This especially concerns the method of the measurement of fans without a stator stage, where the characteristics of the output flow do not satisfy the requirement of a steady axial flow, but fall within the area of a rotating 3D flow field.

Keywords Probe • Twisted • Pressure • Fan • Measurements

1 Introduction

Within the Fan Aerodynamics project dealing with methods for the aerodynamic design of axial fans and methods for the experimental validation of the designed processes, a method for the determination of the integral characteristics of fans was defined (pressure drop, flow rate), based on standard [1]. This method was validated using a measuring line fitted with 18 blade fans with a diameter of 630 mm, without a stator stage to balance the rotational component of the flow. For the measurement of total and static pressures necessary for the calculation of the integral characteristics of the fan, wall reading of the static pressures in front and behind the impeller was used, and the total pressure ridge probe was positioned as recommended by standard [1], at a distance of eight times the air duct diameter. At this point, axial flow is expected to be fully developed. However, with impellers not fitted with a stator stage, the blade geometry and the centrifugal forces generated in the air mass during the impeller rotation lead to such a high twisting of the flow that the condition of the fully developed axial flow is not satisfied, even at a distance significantly greater than eight times the diameter. The output flow has the

J. Kuzel (✉) • M. Golda • M. Hraska
Aerospace Research and Test Establishment, Prague, Czech Republic
e-mail: kuzel@vzlu.cz; golda@vzlu.cz; hraska@vzlu.cz

M. Kral
Aerospace Research and Test Establishment, Prague, Czech Republic

Czech Technical University in Prague, Prague, Czech Republic
e-mail: kral@vzlu.cz; michal.kral@fs.cvut.cz

V. Dynybyl et al. (eds.), *The Latest Methods of Construction Design*,
DOI 10.1007/978-3-319-22762-7_64

characteristics of a rotating vortex adhering to the air duct walls. For this reason, it is not possible to use the ridge probe to measure the total pressure along the flow air duct cross section, and as an alternative solution the measurement is performed using multiple-opening cylinder pressure probes positioned directly in front of and behind the impeller. This document also describes calibration method, see [2, 3], and the measurement using five-tube pressure probe.

2 Probe Calibration

The five-tube pressure probes are designed to measure both static and total pressures at any probe location and also make it possible to determine the flow velocity vector. Such measured values can be used – using integration along the air duct cross section – to determine the flow rates and the pressure drop at the fan. Both five-tube probes were first calibrated inside a wind tunnel. The probes were fitted and blown over in a wind tunnel, with gradual swivelling along two axes (Fig. 1); the pressures were measured and compared with the external atmospheric pressures at each probe opening. The obtained sets of measured values were used to calculate the non-dimensional coefficients of Cpa (1) and Cpb (2), which were then used to determine the flow velocity vector in the air duct measurement.

$$Cpa = \frac{\alpha_+ - \alpha_-}{P_T - \overline{P}} \tag{1}$$

$$Cpb = \frac{\beta_+ - \beta_-}{P_T - \overline{P}} \tag{2}$$

$$Cp_T = \frac{P_T - P_{DYN}}{P_T - \overline{P}} \tag{3}$$

$$Cp_S = \frac{\overline{P}}{P_T - \overline{P}} \tag{4}$$

$$\overline{P} = \frac{\alpha_+ + \alpha_- + \beta_+ + \beta_-}{4} \tag{5}$$

$$P_{DYN} = \frac{1}{2}\rho v^2 \tag{6}$$

The course of the Cpa and Cpb coefficients established during the probe calibration (Fig. 2) clearly showed in which air flow angle ranges α and β the probes can be used. The angle ranges are related, specifically probes AVE 1 and AVE 2 are useful within range α = (−35°, 35°) β = (−35°, 35°). However, if we restrict the range of values e.g. to α = (−5°, 5°), this increases the usable range for angle β = (−50°, 50°).

Angle β indicates the twist angle in the fan measurement, while angle α indicates the deviation from the air duct longitudinal axis, which is usually in the range of up

Fig. 1 Axis of probe

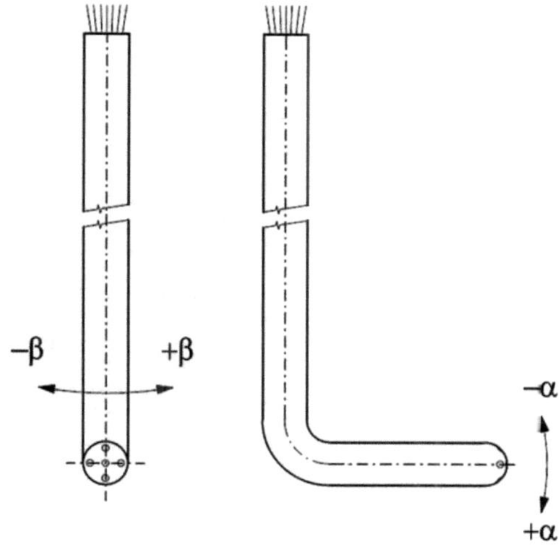

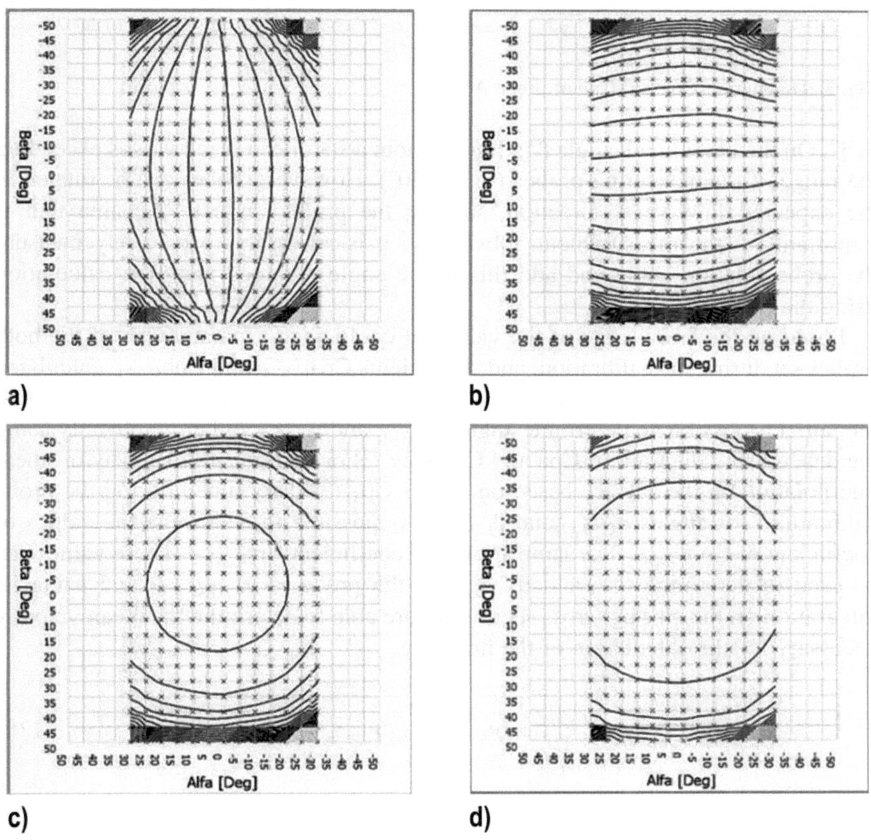

Fig. 2 Coefficients of probe AVE 1 (**a**) Cpa, (**b**) Cpb, (**c**) Cp_T and (**d**) Cp_S

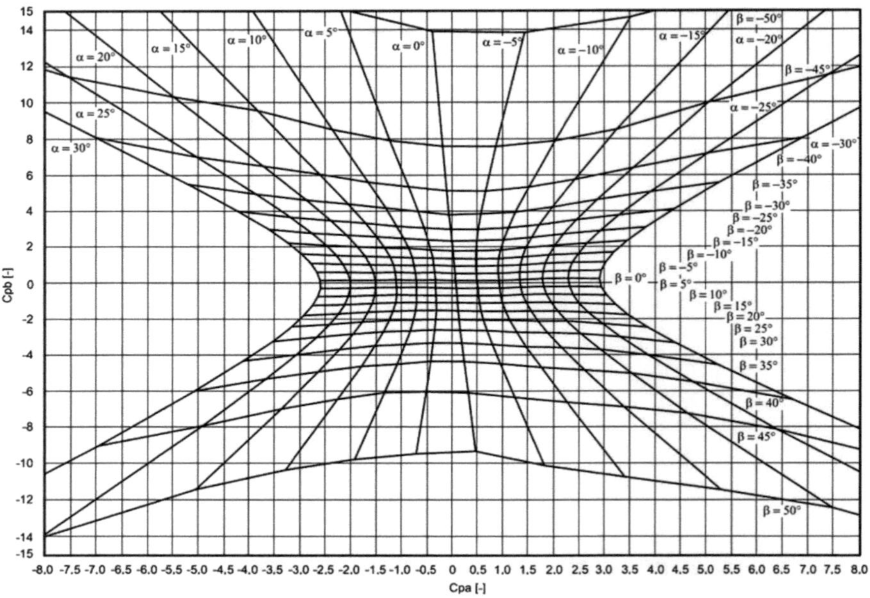

Fig. 3 Diagram of Cpa and Cpb of probe AVE 1

to 5°. On the other hand, angle β in fans without a stator can reach values exceeding the range of the measuring probes (−50°, 50°). However, probes can be rotated in the expected flow twist direction, shifting the probe range to the area that is important for the measurement. Obviously, it is necessary to precisely measure the probe rotation angle and add this to the angle obtained from the calculation using the probe calibration.

Based on the knowledge of the values of coefficients Cpa and Cpb of the both probes set during the calibration, and coefficients Cpa_{CALC} and Cpb_{CALC} calculated from the values measured in the probes built in the air duct in front of and behind the fan, it is possible to determine angles α_{CALC} and β_{CALC}, either graphically using the dependency diagram of Cpa and Cpb (Fig. 3) or numerically. By way of linear interpolation on the surface based on values Cp_T (3) determined during the probe calibration, for angles α_{CALC} and β_{CALC} it is possible to determine the values of coefficient of Cp_{TCALC}. The same applies to coefficient Cp_{SCALC}. These values can be used to determine the velocity against the probe in v_{CALC} (7) and using its integration on the air duct cross section we are able to determine the mean velocity necessary for the calculation of the flow rate.

$$v_{CALC} = \sqrt{\frac{2}{\rho}\cdot\left(P_T - \overline{P}\right)\cdot\left(1 - Cp_T - Cp_S\right)} \quad (7)$$

3 Measurements on the Fan

The velocity profile probing in the air duct was performed continuously along a rectilinear pathway by pushing both probes into the openings in the air duct wall. The probes were inserted together using a pair of cross members with 1 mm spacing. The first probe was located at the distance of 125 mm in front of the impeller, the second probe was fitted 130 mm behind the impeller, with the length of the horizontal part of the probes at 70 mm. At each measurement point, the local total and static pressures were determined using the established pressures, followed by the determination of the velocity vector at the given point of measurement. The velocity vector is determined using the dynamic pressure at the given air density, calculated from the barometric pressure and the ambient air temperature. The velocity vector was subsequently used to determine the axial component of the velocity, which is necessary for calculating the flow rate. Using the integration at the cylindrical coordinates of these partial measurements, the flow rate was determined together with the mean velocity and the mean total and static pressures in front of and behind the fan. The pressure drop at the fan was established from the difference in the mean static pressure in front of and behind the fan. Prior to the integration, edge conditions were added to the measured data, as these cannot be measured using a probe with final diameters. The assumption is zero velocity at the air duct wall, increasing linearly towards the value of the first measurement point. It is further assumed in the integration that all measured variables at the given annular ring defined with the radius and the dot pitch are constant. Due to the fact that this is a rotationally symmetrical task, this condition has been fulfilled. Another ignored value is the error caused by the final pitch of the measurement points, which, according to previous experience concerning similar measurement tasks, is negligible. Analytical integration of the measured values according to the following relationships and after transformation to a summation with step $\Delta r = 1$ mm:

$$v_m = \frac{2\pi \int_{R_1}^{R_2} v(r) \cdot r \cdot dr}{\pi \cdot \left(R_2^2 - R_1^2\right)} \approx \frac{2\pi \sum_{i=0}^{(R_2 - R_1)} v_i \cdot r_i \cdot \Delta r}{\pi \cdot \left(R_2^2 - R_1^2\right)} \tag{8}$$

$$p_{Sm} = \frac{2\pi \int_{R_1}^{R_2} p_S(r) \cdot r \cdot dr}{\pi \cdot \left(R_2^2 - R_1^2\right)} \approx \frac{2\pi \sum_{i=0}^{(R_2 - R_1)} p_{Si} \cdot r_i \cdot \Delta r}{\pi \cdot \left(R_2^2 - R_1^2\right)} \tag{9}$$

$$p_{Tm} = \frac{2\pi \int_{R_1}^{R_2} p_T(r) \cdot r \cdot dr}{\pi \cdot \left(R_2^2 - R_1^2\right)} \approx \frac{2\pi \sum_{i=0}^{(R_2 - R_1)} p_{Ti} \cdot r_i \cdot \Delta r}{\pi \cdot \left(R_2^2 - R_1^2\right)} \tag{10}$$

Where: R_2 – is the outer radius of the annular ring
R_1 – is the inner radius of the annular ring

This provided the values of the mean (integral) flow velocity v_m (8), the mean static pressure p_{Sm} (9) and the mean total pressure p_{Tm} (10), in front of and behind the fan. The mean flow velocity at a known cross section was then used to determine the flow rate. From the static pressure differences in front of and behind the fan, the pressure drop was determined.

4 Conclusion

Based on the measured and calculated values of the integral value of the fan and their comparison with the values predicted using the designed CFD calculation, the method of the measurement of twisted flow based on [2, 3], implemented with the use of five-tube probes, has proved to be reasonably accurate and an adequate alternative to the standard method of fan measurement according to [1]. In this manner, it is also possible to measure fans with a stator blade ring; compared to the standard method of measurement, this will also provide us with data describing the local parameters of the flow field.

Acknowledgments The research was performed with financial institutional support from government budget through the Ministry of Industrial and Trade of the Czech Republic.

References

1. ČSN 12 3061 Fans. Rules for testing
2. M.Y. GündoğDu, M.Ö. ÇARPINLIOĞLU, A multi-tube pressure probe calibration method for measurements of mean flow parameters in swirling flows. Flow. Meas. Instrum. [online]. 9(4), 243–248 (1998) [cit. 2015-05-27]. DOI: 10.1016/s0955-5986(98)00027-2
3. M. Matejka, Smerova kalibrace petiotvorove kuzelove sondy. XXVI. Setkani kateder mechaniky tekutin a termodynamiky (Herbertov, 2007), pp. 59–63. ISBN 80-86786-09-9

Method of Experimental Measurement of a Bucket in Laboratory Conditions

P. Kribala, V. Dynybyl, and J. Kricka

Abstract The paper deals with a method of measurement of a bucket. The bucket is an excavating component of the compact bucket wheel excavator K100 (by NOEN Inc.). The bucket wheel excavator is a machine for surface continuous mining. The tested bucket will be statically loaded and deformation of the bucket will be measured at defined spot. A stand was designed especially for this purpose.

Keywords Method of experimental measurement • Bucket • Compact bucket wheel excavator • Static load

1 Introduction

Bucket is excavating component of compact bucket wheel excavator K100, which was designed by NOEN Company. Main activities of the company include designing and engineering of machines and equipment for open-pit mining, long distance belt conveyor systems, material handling equipment at the dump and delivery of machines and equipment for bulk material stock-yards. The compact bucket wheel excavator will be used in chalk open-pit mine. It is equipped with 14 buckets. The buckets are attached around bucket wheel, which rotates and removes material from the excavation face. The bucket wheel is closed radially by a circular chute which only permits discharge of material from the bucket at the beginning of discharge opening. The material then falls onto a transfer chute that is mounted inside of the bucket wheel. The chute guides moving material to the side, out of the wheel, onto the conveyor in the bucket wheel boom. Compact bucket wheel excavator K100 is shown in Fig. 1.

P. Kribala (✉) • J. Kricka
Czech Technical University in Prague, Prague, Czech Republic
e-mail: petr.kribala@fs.cvut.cz; jaroslav.kricka@fs.cvut.cz

V. Dynybyl
Faculty of Mechanical Engineering, Czech Technical University of Prague, Prague, Czech Republic
e-mail: vojtech.dynybyl@fs.cvut.cz

V. Dynybyl et al. (eds.), *The Latest Methods of Construction Design*,
DOI 10.1007/978-3-319-22762-7_65

Fig. 1 Compact bucket wheel excavator K100 (NOEN Inc.) (*1* – bucket wheel, *2* – bucket)

1.1 Description of the Bucket

The bucket and its parts is shown in Fig. 2. Bucket consists of a body, six teeth, chain mat and a bucket attachment (front pin attachment and rear weld attachment). The body of bucket is welded and made of wear resistant plate. The tooth is inserted to a tooth adapter, witch is welded to bucket body. The chain mat presses material out of the bucket in the discharge region of the bucket wheel.

2 Experimental Measuring Method and Description of the Stand

The first measuring of the bucket will be done in lab. A stand will be designed especially for this purpose. The goal is to simulate conditions, which are similar to real conditions during excavation. The stand consists of five parts (frame for attachment of bucket, frame for attachment of bearing, arm, hydraulic cylinder and frame for attachment of hydraulic cylinder). The stand is shown in Fig. 3. Its parts is shown in Fig. 4. and in Fig. 5. The frame for attachment of bucket is welded from beams and plates. The bucket tooth will be loaded by the arm. The arm is attached to the bearing units. The bearing units are attached to the frame for attachment of bearing. The hydraulic cylinder will press the arm. The hydraulic cylinder is attached to the arm at one side and to the frame for attachment of hydraulic cylinder at other side. The frame for attachment of hydraulic cylinder will be welded from plates. Each individual measurement will be done for one tooth. Before measurement of other teeth the bolts will have to be disassembled and the arm with the hydraulic cylinder will have to be moved to the position against another tooth. A deformation of the bucket and a deformation of the frame for attachment of the bucket will be measured at defined spots.

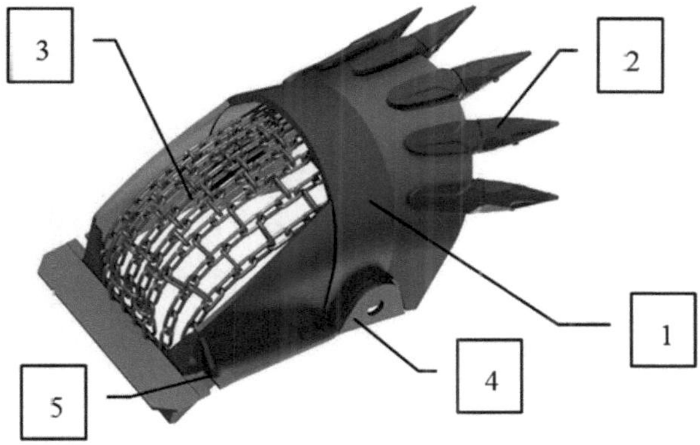

Fig. 2 Parts of the bucket (*1* – welded steel body, *2* – tooth with tooth adapter, *3* – chain mat, *4* – front pin attachment, *5* – rear weld attachment)

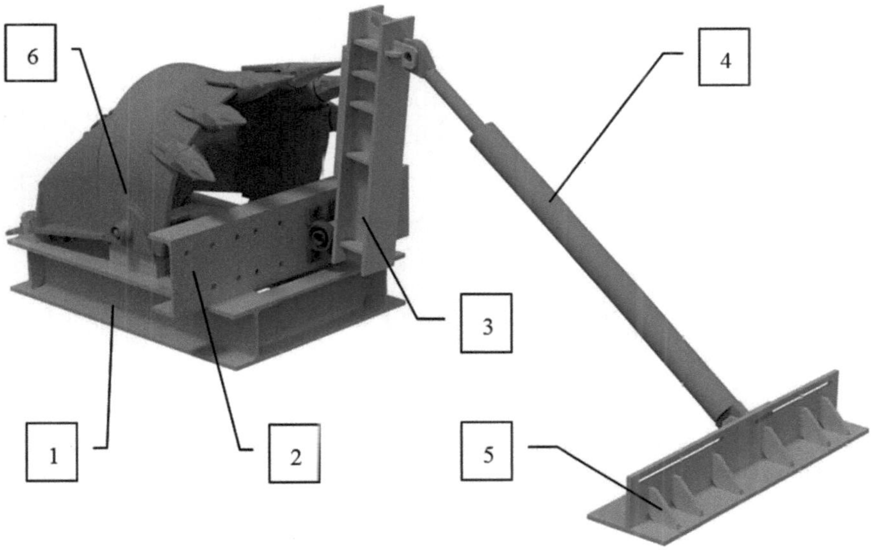

Fig. 3 Full test stand (*1* – the frame for attachment of bucket, *2* – the frame for attachment of bearing, *3* – the arm, *4* – the hydraulic cylinder, *5* – the frame for attachment of hydraulic cylinder, *6* – the bucket)

3 Conclusion

In this paper is described the testing stand for the bucket in laboratory conditions. The bucket will be tested statically.

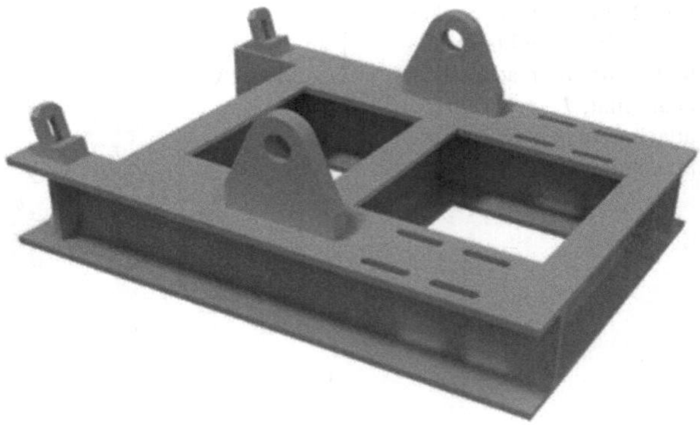

Fig. 4 The frame for attachment of bucket

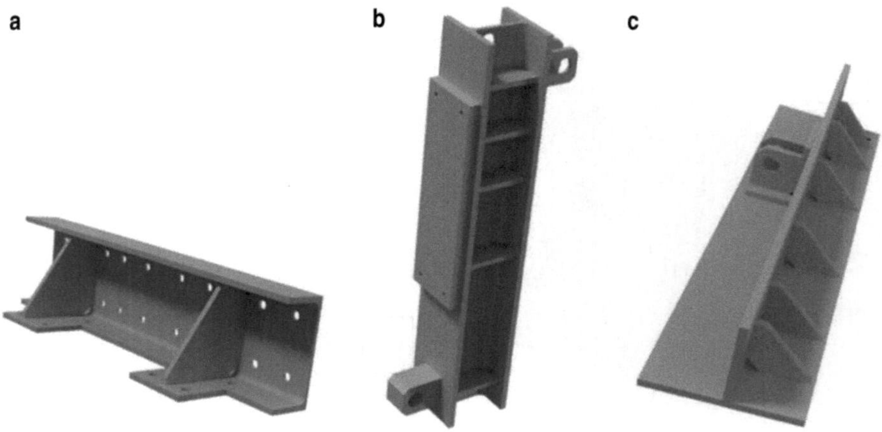

Fig. 5 (**a**) The frame for attachment of bearing; (**b**) the arm; (**c**) the frame for attachment for hydraulic cylinder

References

1. W. Durst, W. Vogt, Bucket Wheel Excavator (Clausthal-Zellerfeld [West Germany], Trans Tech Publications, 1988). ID: 19252305. ISBN 0878490752 9780878490752

Biaxial Stress of the Textile Car Seat Cover Tested in a Climatic Chamber

P. Kulhavý, N. Kovalova, R. Martonka, and J. Petrik

Abstract Knowing the mechanical and deformation properties of textile materials plays a major role in determining quality of the final product. While exploitation textile as a seat cover, influence doesn't have only tensile properties in one direction but also is transmitted to the adjacent yarns in the fabric. The role of a seat cover is not just to cover up the foam but also increase the rigidity of the seat cushion and have influence to a viscoelastic behaviour of the foams. In the vehicle there are also an important changes in microclimate conditions. Big fluctuations of temperature and local increasing of humidity at the same time. In order to assess the impacts all of this condition was executed an experiment for biaxial loading in a climatic chamber with variable temperature and humidity.

Keywords Seat cover • Textile • Climatic chamber • Biaxial loading

1 Introduction

The basic preconditions for a successful construction of a final product, is the knowledge mechanical properties of the used materials. For industrial textile materials, it is especially important to take into account the ambient conditions occurring in real operation. For textile materials which show considerable anisotropic or orthotropic us generally not sufficient testing in one axis, it is necessary to load two axes simultaneously. According to Quaglini [2] because of the small thickness of fabric is usually neglected component of the stress in the thickness direction of the material and technical textiles are affected by plane stress. In an effort be as close as possible to the real condition in the vehicle, where are a significant climatic stress, it was necessary to do test in a climatic chamber with variable temperature and humidity.

P. Kulhavý (✉) • N. Kovalova • R. Martonka • J. Petrik
Technical University of Liberec, Liberec, Czech Republic
e-mail: petr.kulhavy@tul.cz; natalia.kovalova@tul.cz; rudolf.martonka@tul.cz; jan.petrik@tul.cz

V. Dynybyl et al. (eds.), *The Latest Methods of Construction Design*,
DOI 10.1007/978-3-319-22762-7_66

2 Measurement of Textile Materials

As has Reinhardt [6] historically pointed already in his work aimed to methods of testing textiles, textile materials are almost always burdened biaxially. The simplest method used to testing the biaxial strain has been called bursting test: The bursting test using a circular specimen which is clamped along an edge. This flat membrane is deformed by lateral air or water pressure into a spherical shape and the pressure is increased till the specimen burst.

Anisotropic materials like a tested coating materials, according to Novak and Hanuš [1], due to the specific mechanical properties show a noticeable hysteresis and different deformation in according to a direction of loading. In tensile tests of textiles in one axis in a longitudinal direction, are formed the main stress concentrators and a subsequent rupture at the site near the clamping the jaws. In contrast, in the transverse direction is the tension more equally concentrated in the centre of the sample. When considering climate changes in the textile material are the most apparent a sorption process, which significantly affect their properties [3]. The sorption property means that the materials are able to absorb humidity, gases, chemicals, etc. (Fig. 1).

3 Measurement Method

According to Escárpita [4] is for a testing of biaxially loaded samples necessary to ensure that the test device meets the basic requirements. There must be strictly a tension or compression, avoiding spurious shear or bending loads. Must be guaranteed orthogonality between the axis of the load throughout the test [5]. Generally it is recommended to use hydraulic or mechanical system with linear motors, system of cables, pulleys and bearings. For climatic stress is according to the norm given standard for testing textiles. The air temperature must be 20 ± 2 ºC and humidity of 65 ± 2 %. The fabric should be dried before the test and then placed in a precisely air conditioned environment for 12–24 h.

For the experiment it was necessary to select and prepare samples according to ČSN EN 12751 Textiles – Sampling of fibers, yarns and fabrics for testing [7]. Then perform conditioning of samples according to standard number ČSN 80 0061 Conditioning of textile raw materials [8]. The samples were produced in company Johnson controls, s.r.o. The aim of this work is to find out the mechanical properties (tensile strength and extension at break) of the fabric related to different moisture and temperature (Fig. 2).

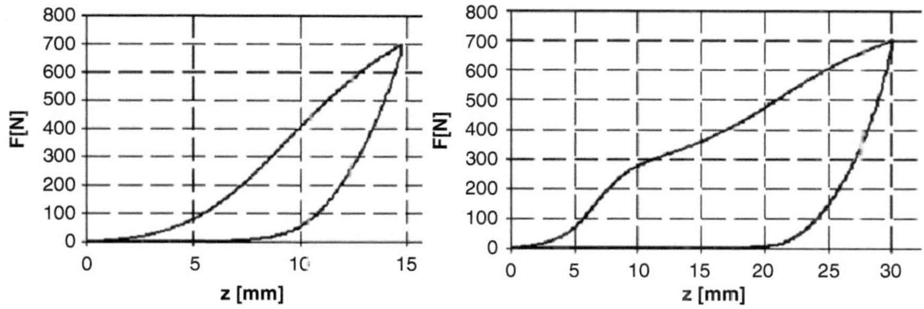

Fig. 1 Tension diagram of a cover textile material loaded in one axis (*left* weft, *right* warp) [10]

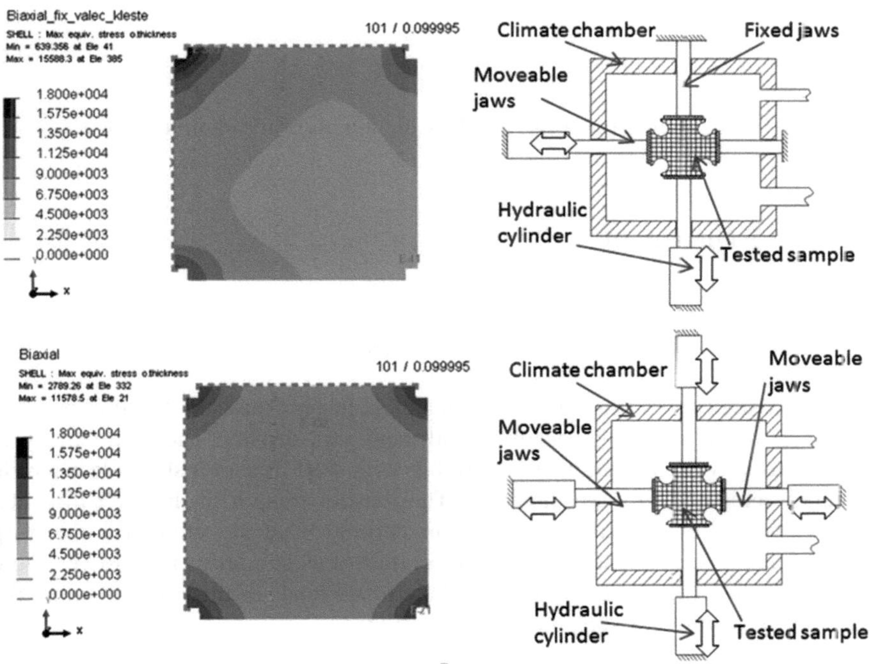

Fig. 2 Scheme of the experiment device and MKP simulation (Von Mises stress) with the example of loading with non guaranteed orthogonality between axis (*above*) and with ensured orthogonality (*below*)

3.1 Extension at Break

Extension at break in the warp or weft is defined as the elongation at maximum force (at break) to the original length of the fabric [9]. The elongation of fabric is dependent on the elongation of the yarns and the way of linking yarns in the fabric.

$$\varepsilon = \frac{L_p - L_o}{L_o} \cdot 100\% \tag{1}$$

where: ε is the extension at break [%], L_p is the length of the sample at the time of yarn breaking [mm] and L_o is the length of the original sample [mm].

3.2 Relative Strenght

The strength of the fabric depend on the force required to breach the fabric. The strength depends on the stability of the warp or weft yarns and the yarn density.

$$S = \frac{F}{L} \tag{2}$$

where: F is the applied force [N] and L is the length of the sample [mm].

4 The Experiment

The textile materials used for the experiment was the synthetic fabric with a twill weave of fabric (3/1), warp yarn density = 31 per cm and weft yarn density = 24 per cm (Fig. 3).

The tensile curve of the fabric has an almost linear character. The difference in the shape of the curves is because of different yarn density. The yarn density of warp is usually higher and that is why fabric in weft direction shows less tensile strength compared to warp direction. The tensile strength in the weft direction reaches for the type of material approximately 80 % of the strength in the warp direction. The dependence on the temperature and humidity, is however more obvious in the weft direction. During the climate loads in the warp direction the material still showed a very similar values (Fig. 4).

5 Conclusion

During the experiment were observed the changes in the mechanical properties of the fabric depending on humidity and temperature in climatic chamber. On the above mentioned surface diagram (Fig. 5) is quite clearly shown the dependence of the strength of the fabric within relation to the temperature and humidity in the climatic chamber. The exact values are shown in the Table 1. Strength of the fabric is generally decreases, when the humidity is increases. This may be caused by the moisture absorption of the fibers and also by changing the value of friction between the fibres.

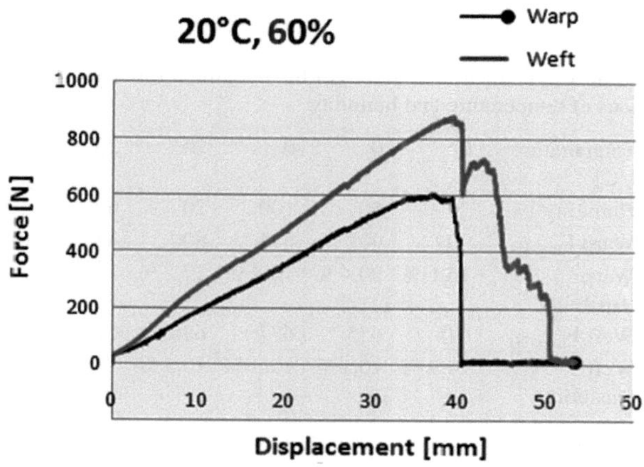

Fig. 3 The test of the sample accorded to the normalized climate conditions

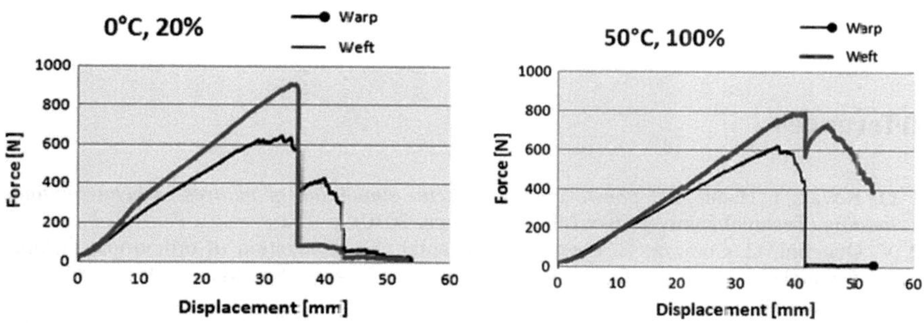

Fig. 4 The diagrams of the tensile stress showed at the two extreme points of the material strength. For the highest strength value (*left*) and the lowest value (*right*)

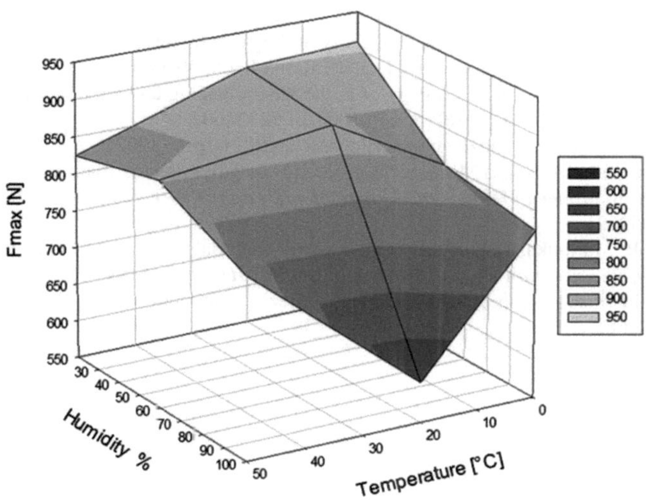

Fig. 5 Spatial diagram of depending the strength of the fabric on temperature and humidity

Table 1 The measured values of the maximum force and elongation measured for the combinations of temperature and humidity

Temperature [°C]	0	0	0	20	20	20	50	50	50
Humidity [%]	20	60	100	20	60	100	20	60	100
Warp $F_{max[N]}$	909	902	824	800	883	853	774	603	792
Warp tensibility	16.1 %	20.4 %	16.3 %	17.2 %	17.7 %	17.6 %	17.5 %	18.3 %	18.9 %
Weft $F_{max[N]}$	646	655	677	696	610	647	670	634	626
Weft tensibility	14.9 %	15.1 %	15.3 %	15.5 %	16.9 %	16.3 %	20.4 %	18.5 %	16.9 %

Acknowledgments This work was supported by the Ministry of Education of the Czech Republic within the SGS project no. 21011 on the Technical University of Liberec.

References

1. O. Novák, J. Hanuš, *3D nonwovens in medicine-simulation of mattresses behavior under loading* (Technická univerzita v Liberci, Liberec, 2010), p. 112
2. V. Quaglini, C. Corazza, C. Poggi, Experimental characterization of orthotropic technical textiles under uniaxial and biaxial loading. Compos. A: Appl. Sci. Manuf. **39**(8), 1331–1342 (2008). doi:10.1016/j.compositesa.2007.07.008
3. T. Kubeček, R. Nemčoková, *Identification of the Mechanical Parameters of Fabrics* (Technical University of Liberec, Liberec, 2013), p. 59
4. D. Escarpita, E.A.B. Koenders, D.B.F. Carvalho, in *Biaxial Tensile Strength Characterization of Textile Composite Materials. Composites and Their Properties* (InTech, 22 Aug 2012). DOI: 10.5772/48705
5. S. Kawabata, M. Niwa, in *Validity of the Linearizing Method for Describing the Biaxial Stress-Strain Relationship of Textile* (The University of Shiga Prefecture, 1984) Journal of the Textile Institute 92(3):38–52 · January 2001 DOI:10.1080/00405000108659613
6. H.W. Reinhardt, C. Corazza, C. Poggi, On the biaxial testing and strength of coated fabrics. Exp. Mech. **16**(2), 71–74 (1976). doi:10.1007/BF02328607
7. CSN 800 00 73. Pojmy používané při zkoušení textilií tahem (1992)
8. P. SRB, R. MARTONKA. Mechanical properties of polyurethane foam in different climate conditions, in *Modern Methods of Construction Design: Proceedings of ICMD 2013*. ISBN 9783319052021
9. R. Kovar, R. Martonka, Tensile test of the car parts materials in a climatic chamber, in *Modern Methods of Construction Design: Proceedings of ICMD 2013*. ISBN 9783319052021
10. J. Petřík, Interakce automobilové sedačky a zátěže. Disertační práce (TU v Liberci, KMO, Liberec, 2008), s. 125

Hexapod Control System Optimization

A. Lufinka and R. Martonka

Abstract Hexapod is a new testing device for generation of six degrees of freedom movement in the Laboratory of applied mechanics. Its control system is built from National Instruments components and control application is created in the software LabVIEW. Using the real-time system to run this application requires special programming procedures because the precise timing of the control loop has to be observed. The hexapod motion control is based on its kinematic model. This model calculates the engine control signals according to the input required by motion information. Its output data files are used as input data for complex 3D motion control. Equations for the six basic motions (3-axis movement and 3-axis rotation) were obtained from the kinematic model, too. These equations are implemented directly in the control program for basic movement and rotation control. The control system, program design and program optimization are described in this paper.

Keywords Hexapod • Control system • Real time application • Kinematic equations

1 Introduction

Hexapod is a testing device for generation of six degrees of freedom movement. It is used for testing of mechanical parts or assemblies properties (mostly car seats). Old equipment – "Six degrees of freedom platform" [1] was replaced by this new kinematic system. This solutions increased device stiffness so higher frequencies of testing signals can be used. Of course the control signals calculation is more complicated than in the old orthogonal system. The original control system could be moved to the new device because both devices have identical hardware scheme, six linear hydraulic engines are controlled by the position feedback. But a software application had to be completely reprogrammed because a feedback loop speed had to be doubled due to the device operating frequency increasing. New function for communication with the laboratory hydraulic power units control system had to be added to application, too. Old and new testing devices are shown in the Fig. 1.

A. Lufinka (✉) • R. Martonka
Technical University of Liberec, Liberec, Czech Republic
e-mail: ales.lufinka@tul.cz; rudolf.martonka@tul.cz

V. Dynybyl et al. (eds.), *The Latest Methods of Construction Design*,
DOI 10.1007/978-3-319-22762-7_67

Fig. 1 Old (*left*) and new (*right*) testing devices for 3D movement

2 Control System Hardware

National Instruments real-time system PXI1002 forms the control system core. The basic part of the control system was transferred from the old device. Piston position sensors are connected to the six channels amplifier and then to the PXI system by 16-bit A/D converter. Voltage signals from PXI output 16-bit D/A converters are transferred to the current signals for servo valves control. Galvanically isolated digital inputs and outputs control all hexapod auxiliary equipments (input hydraulic valves, signal lights, etc.). Digital lines are used for communication with the laboratory hydraulic power units control system [2], too. Direct hydraulic power units control is thus possible by the hexapod software application. The control system block diagram is shown in Fig. 2.

3 Software Application

Several parallel running loops can be created in a LabVIEW application. These loops can be very accurately timed by hardware counters if the real-time system is used for application running. All these possibilities are used in the hexapod control application. Six PID controllers are realized by the high priority hardware timed (2 kHz frequency) loop. Exact running of this loop must not be breached in any case so all data are transmitted to this loop by a data queue. Desired position values are read from a binary data file and stored to the queue by the next medium priority loop. Possible data reading irregularities are thus eliminated by the data reservoir in the queue. Third low priority loop is used for operator commands.

Execution time of all instructions into the HW-timed loop has to be less than its timing. So, the number and type of instructions have to be optimized to this

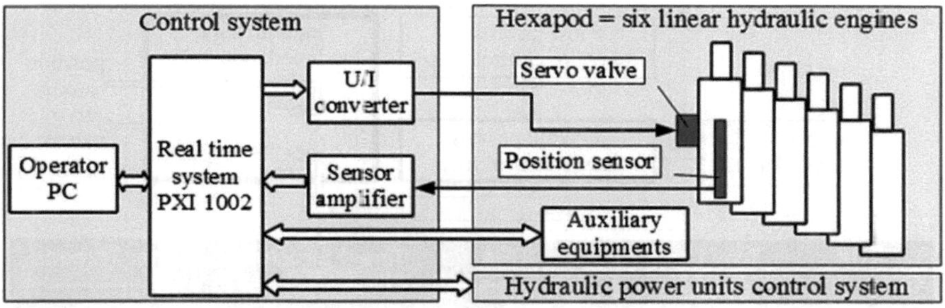

Fig. 2 The hexapod control system block diagram

condition fulfillment. This is a lengthy testing process in time optimal sequences selecting. The optimization results and application block diagram are shown in Fig. 3. HW-timed loop has approximately 30 % reserve and the CPU is used on 76 %. These are good results for the smooth application running.

4 Pre-processing

Timing analysis results show that real-time hexapod kinematics calculation is impossible in the current CPU. The special binary data file has to be used as the desired values source for position control. This file is created before start of the hexapod movement. First general 3D motion is recalculated to six engines movement by the hexapod kinematic model [3]. Its output data files are then converted to desired binary format by the pre-processing application (Fig. 4).

Whole testing process may be created as a sequence of several movements, these can be repeated and smooth transitions between the sequence blocks can be resolved using this application. Of course that this procedure is unnecessarily lengthy for generating simple one-axis movements and rotations. Therefore the conversion equations for each simple motion were created using the kinematic model and these equations were implemented directly to pre-processing application. So the kinematic model does not need to be used for simple motions generation and the test preparation time was shortened. An example of conversion equations for x-axis movement is shown in Fig. 5.

5 Conclusion

The described control system and software applications allow the hexapod to be used as a standard testing device. But the practical using showed one critical point. Conversion of 3D motion by the kinematic model is too slow for long input files.

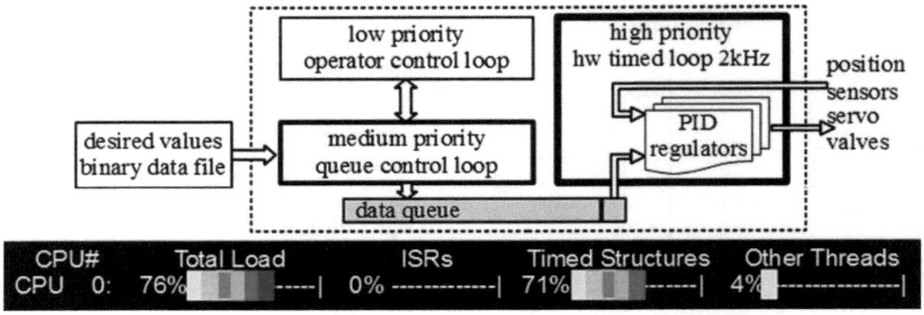

Fig. 3 Timing analysis results and the application block diagram

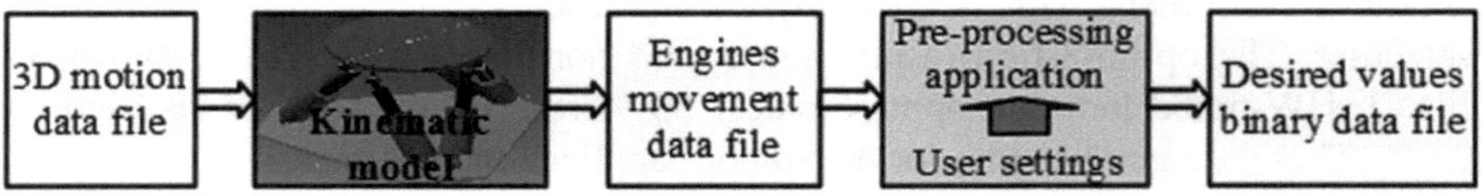

Fig. 4 The pre-processing block diagram

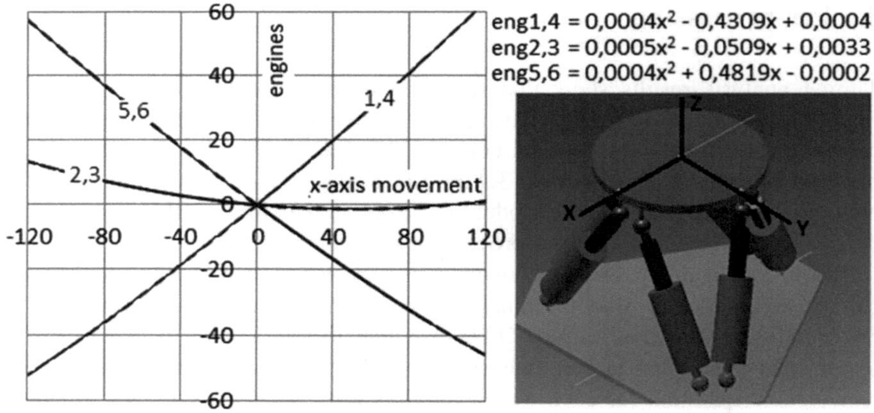

Fig. 5 The conversion equations for x-axis movement

Therefore, the future plan is to create a complex conversion kinematic matrix for 3D motion and to implement it into the pre-processing application. Computing time will be shortened and the pre-process will be much easier.

Acknowledgments The results of this project LO1201 were obtained through the financial support of the Ministry of Education, Youth and Sports in the framework of the targeted support of the "National Programme for Sustainability I" and the OPR&DI project Centre for Nanomaterials, Advanced Technologies and Innovation CZ.1.05/2.1.00/01.0005 and Project OP VaVpI Centre for Nanomaterials, Advanced Technologies and Innovation CZ.1.05/2.1.00/01.0005.

References

1. A. Lufinka, The six degrees of freedom platform, in *Book of Proceedings of 45th International Conference Experimental Stress Analysis* (University of West Bohemia in Pisen, Pilsen, 2007), pp. 71–72. ISBN 978-80-7043-552-6
2. A. Lufinka, Innovation of the Control System for Hydraulic Power Units, in *Modern Methods of Construction Design* (Liberec, Czech Republic, 2013), pp. 477–483. ISBN 978-3-319-05202-1
3. R. Martonka, V. Fliegel, Hexapod – the Platform with 6DOF, in *Modern Methods of Construction Design* (Liberec, Czech Republic, 2013), pp. 133–138. ISBN 978-3-319-05202-1

Integration of Light: A Way to Measure Spectral Characteristics

K. Macuchova

Abstract Measurement of spectral characteristics of various samples provides information about their chemical composition. Measurement of many samples is made difficult due to their scattering properties. Sometimes also the measured signal highly depends on the direction of test light reflection. Light integration and collection solves some of these difficulties and enables to get reliable data. Two sample use cases are presented to illustrate how light integrators may provide a solution.

Keywords Integrating components • Illumination • Light collector • Light integrator • Spectral measurement

1 Introduction

Spectroscopy studies relations and interactions between matter and radiated energy. Visible light is radiation with wavelength in $\lambda \in \langle 400; 760\rangle nm$ perceivable with human eye. Excitations of tested samples with infrared, visible, ultraviolet radiation are common measurement methods. Optical spectroscopy involves the excitation with ultraviolet A and B, visible and near-infrared radiation. The importance of spectroscopy as a measurement method is based on the unambiguous identification of the substance by its unique set of characteristic spectral lines. [1]

1.1 Theory of Scatter Light

Light scattering is caused when electromagnetic radiation interacts with matter. When the incident light beam strikes a surface – part of the light energy is reflected and part of it propagates further to the material. Mathematically, the reflection is described by bidirectional reflectance distribution function (BRDF). The

K. Macuchova (✉)
Czech Technical University in Prague, Prague, Czech Republic
e-mail: karolina.vosatkova@fs.cvut.cz

V. Dynybyl et al. (eds.), *The Latest Methods of Construction Design*,
DOI 10.1007/978-3-319-22762-7_68

propagation of wave into material describes bidirectional texture function (BTDF). [2] Different surfaces and materials are able to reflect the incident radiation differently.

Two boundary types of reflections are: specular reflection (from mirror-like surfaces) and diffusive reflection (from Lambertian surfaces).

1.2 Introduction to Light Integrators and Collectors

Integrating sphere (also known as an Ulbricht sphere) is an optical component with hollow spherical or hemi-spherical cavity covered with diffusive materials. It is used in photometric and radiometric measurements. Principally it integrates reflected light using the inner sphere cavity, where due to multiple scattering reflections the light flux is homogenized. [3] Integrating spheres might be characterized as preserving energy, but eliminating negative influences of polarization and directivity of incident light. The inner sphere is made of highly diffusive (Lambertian) surface. Detailed information about materials for light integrators is summarized in Sect. 2.1.

The integrator has several openings used to input incident light and output the integrated signal. Reflections from a glossy surface which may directly enter the spectrometer output aperture are absorbed by a light trap. A light trap is a barrier placed near the sample aperture.

During time, several standards were established. The CIE 2004a standard describes geometrical configuration of the integrating sphere used for colorimetric measurements. For example: the "8°: d" describes standardized geometrical configuration, where the number means the angle of incident light and *d* stands for integrating sphere. European standard of total integrating scattering measurement is ISO 13696. [4]

2 Selecting a Light Integrator

One of small typical dismounted integrating sphere can be seen in Fig. 1. It consists of white diffusion material with spherical cavity and black anodized chassis.

When choosing a proper integrating sphere, several features has to be taken in account: range of wavelengths of incident radiation, desired reflection rate, size of the integrator, or how many input/output apertures are needed. The first two features influence the choice of diffusive material – as further described in the next section. The size and proportions of the integrator has to be calculated as denoted in the Sect. 2.2.

Fig. 1 Integrating sphere

2.1 Diffusive Materials

Materials used for light scatterers and integrators have high diffusion reflectance in given wavelength interval. These surfaces are called Lambertian (according to Swiss mathematician and physician) Johann Heinrich Lambert (1728–1777) if they reflect uniformly the radiation in all direction. By definition the ideal Lambertian surface is the blackbody.

Ideal material for scatterers and integrators is therefore highly diffusive, highly reflective over wavelength range of interest, easily applicable and stable. [5] In the following Table 1 is an overview of current state-of-art materials used for integrating spheres.

2.2 Calculating Diffusion Properties

Sample calculation for an integrating sphere on Fig. 1, with inner diameter $D = 50.8\ mm$, one $D_i = 25.4\ mm$ opening aperture and one $D_o = 12.7\ mm$ output aperture, used for integrating visible light and measuring of $\phi_i = 100\ mW$ laser beam. This integrating sphere was used in two applications described in the next chapter.

Table 1 Overview of materials

Name (Manufacturer)	Wavelength range	Reflectance	Material	Description
Infragold (Labsphere)	700 nm–20 μm	94 %	Au	Durable, easily cleaned, easily adhere to metal substrates, vacuum stability, corrosive, sensitive to dents and scratches
Spectraflect (Labsphere)	300–2400 nm	97–98 % (99 %)	BaSo4	Stable, non-toxic, sprayable, easily recoatable, high laser damage threshold, thermally stable to 100 °C, water soluble, easily damaged
Spectralon (Labsphere)	400–1500 nm	99 %	PTFE	Machinable, inert, thermally stable
Permaflect (Labsphere)	350–1200 nm	94 %	PTFE	Durable, water resistant, pigmented for reflectance standards
Duraflect (Labsphere)	350–1200 nm	94–96 %	BaSo4	Sprayable, easily cleaned, durable, waterproof, for outdoor conditions
Optolon (Optronic Lab.)	400–900 nm	98 %	BaSo4	Sprayable, moldable, washable, durable
Avian D (Avian Tech.)	400–800 nm	98 %	Urethane	Durable, pigmented, water proof
Avian B (Avian Tech.)	350–850 nm	97 %	BaSo4	Machinable, sprayable
Fluorilon (Avian Tech.)	400–800 nm	99 %	PTFE	Chemically inert, thermally stable up to 300 °C
Op-Di-Ma (GigaHertz)	250–2500 nm	Max. 98 %	PTFE	Pigmented for reflectance standards, machinable
Optowhite (SphereOptics)	400–1000 nm	98 %	BaSo4	Sprayable, sensitive to contaminations, water soluble
Zenith (SphereOptics)	250–2500 nm	98 %	PTFE	Thermally stable up to 250 °C, inert, easily cleaned, humidity resistant, high radiation level threshold
ZenithGold (SphereOptics)	800–5000 nm	95 %	Au	Chemically inert, thermally stable up to 1000 °C

Material of diffusion surface: polytetrafluoroethylene (PTFE)

Experimentally measured reflectivity	$\rho = 0.97$
Area of diffusion surface	$A_s = \frac{\pi D^2}{4} = 2026.8\ \text{mm}^2$
Size of input aperture	$A_i = \frac{\pi D_i^2}{4} = 506.7\ \text{mm}^2$
Size of output aperture	$A_e = \frac{\pi D_e^2}{4} = 126.67\ \text{mm}^2$
Illuminated area	$A = A_s - (A_i + A_e) = 1393.4\ \text{mm}^2$
Port fraction	$PF = \frac{(A_i + A_e)}{A_s} = 0.3125$
Radiation of diffusion surface	$L = \frac{\phi_i \rho}{\pi A} = 22.16\ \text{W/m}^2\text{sr}$
Multiplying sphere factor	$M = \frac{\rho}{1 - \rho(1 - PF)} = 3,516$

The best multiplying factor is obtained with the smallest port fraction.

Time response $\tau = -\frac{2D}{3c\ ln\rho} = 3.7\ \mu s$ is significant for rapidly varying light signals.

3 Use Cases

3.1 Use Case 1: Colorimeter

The first use case describes design of a testing instrumentation that enables non-destructive spectroscopic and colorimetric method of examination of glossy microtraces of all kinds. [6] The image in Fig. 2 shows the view of the whole assembly. The integrating sphere is on position 4. It was used mainly because the measured samples are quite glossy and it was necessary to eliminate measuring of specular reflections. This way the specular reflection is converted into diffusion signal and measured.

3.2 Use Case 2: Laser Spectrometer

The second use case is a laser spectrometer. Spectrometric measurement of lasers with high power is problematic due to saturation threshold of the spectrometric detector. Therefore the sampling light flux has to be attenuated. An integrating sphere spreads the incident light on the whole inner diffusion surface. But only a small fraction of diffusion surface viewed under a narrow angle of view is sampled with the spectrometer.

For demonstration purposes, a spectrometer which was primarily designed to measure sparse astronomic events (like meteors, or bolids) was rebuilt. The spectrometer can be seen in Fig. 3.

The original purpose of the device was not fulfilled successfully, because the combination of lens and the matching size of integrating sphere resulted in bigger attenuation of the original signal then was desired. So this prototype was abandoned and rebuilt into a laser spectrometer. Future work on the project of bolide spectrometer leads to miniaturization of the lens with adjacent integrating sphere so the attenuation of the signal is kept as low as possible. This example stresses out the importance of illumination. The integrating sphere is very efficient with good illumination, where it shows off its best features, but might prove its weaknesses with low level signals.

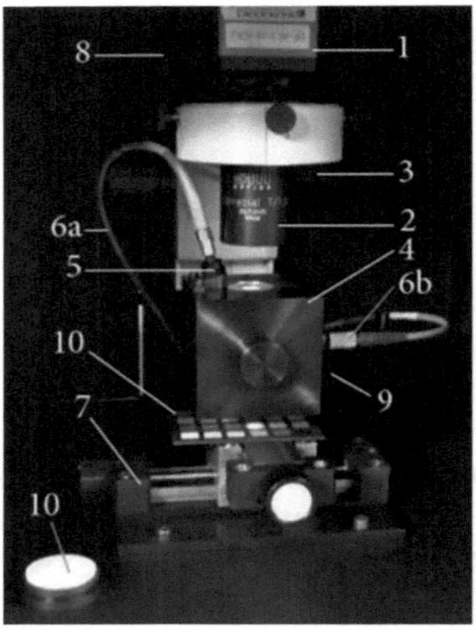

1 – camera	6b – spectrometer probe
2 – macro lens	7 – stage
3 – focuser	8 – stand
4 – integrating sphere	9 – cover
5 – collimating lens	10 – calibration standards
6a – illumination light guide	

Fig. 2 Colorimeter

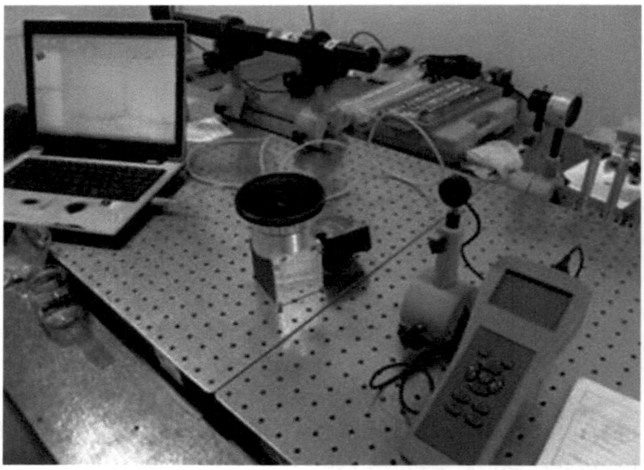

Fig. 3 Laser spectrometer

4 Conclusion

Two working use cases were presented where light collectors provide a solution. Presented method showed up stable, robust and proved good repeatability. Special care must be taken to applications which have to deal with weak light signals where the drop down of output intensity is the most significant.

Acknowledgments The research work reported here was partially supported by Czech Science Foundation under research programme GA14-19213S.

References

1. E. Fest, *Stray Light Analysis and Control* (SPIE Press, Bellingham, 2013)
2. J.C. Stover, *Optical Scattering: Measurement and Analysis* (SPIE Press, Bellingham, 2012)
3. N. Ohta, A.R. Robertson, *Colorimetry: Fundamentals and Application* (Wiley, Chichester, 2005)
4. B.E. Saleh, M.C. Teich, *Fundamentals of Photonics* (Wiley, Weinheim, 1991)
5. K. Macúchová, J. Zicha, Simulation, measurement and spectral analysis of night sky light background. Key. Eng. Mater. **2014**(607), 15–20 (2014). ISSN 1013-9826
6. K. Macúchová, J. Zicha, New approach to testing microtraces. Rom. Rev. Precis. Mech. Opt. Mechatron. **XIX**(41), 97–102 (2012). ISSN 1584-5982

Integration ... Light, AAW ... Measure...

References

1. ... Fish, ... Analysis ...
2. ... Optical ...
3. ... Robertson, ...
2005)
4. ... Teich, ...
5. ... Z. Zicha, ...
background ... Eng. Mater. ...
6. ... J. Zicha, ...
Mechanics XIX(4): ...

The Equipment for Temperature Measurements Near a Gear Tooth Flank

M. Mazac, P. Herajn, and M. Svoboda

Abstract This thesis deals with a design of equipment for measurements of temperature near the gear tooth flank. Temperature measurements are useful for understanding of processes occurring during gear operation. Measurements of temperature are necessary for application of other measuring systems (for example application of strain gauges). Temperature is parameter for application of right oil type too. The article describes choosing of temperature sensors and a method of data transport from the rotating shaft. Software for data transport and calibration is described here. The calibration of sensors is described too. Instruction for use are introduced in the text. The equipment for temperature measurements was tested on the little stand which was especially designed. The goal of this paper is design and testing of equipment for temperature measurements near a gear tooth flank.

Keywords Measurement • Temperature • Gear • Tooth flank

1 Introduction

Many different types of transportation vehicles are manufacturing and performed in all world. The vast majority of this vehicles use mechanical gears for transformation of engine power. Gears and gearboxes are very monitored parts of vehicles [1]. For the research of topics about gears a great emphasis is put every year. The right performance of a gearbox on many parameters of a used parts is depend. The main required parameters of gearboxes is low noise (f.e. [2]) and high efficiency (f.e. [3, 4]). The fuel consumption of vehicle is depend on efficiency of gearbox too. Efficiency of gearbox is very important parameter because the dissipated energy is mostly transformed to heat. The following temperature changes affects a physical parameters of gearbox parts. The gearbox can be destroyed during the extreme heat loading.

The main places of mechanical energy dissipation are contact surfaces where the tribological processes are ongoing. The contact places are in the bearing and shaft sealing for example. But the main place of friction processes is a contact area between tooth flanks. Many theoretical and practical information about the friction

M. Mazac (✉) • P. Herajn • M. Svoboda
Technical University of Liberec, Liberec, Czech Republic
e-mail: martin.mazac@tul.cz; pavel.herajn@tul.cz; miroslav.svoboda@tul.cz

V. Dynybyl et al. (eds.), *The Latest Methods of Construction Design*,
DOI 10.1007/978-3-319-22762-7_69

between tooth flanks are available in literature (f.e. a theoretical friction parameters as a load and a velocity in [5], practical information about temperature [6]).

From the informations above, results that the measuring of temperature near of a tooth flank is very useful for understanding of processes during gearbox operation. For the temperature measurements is necessary to have a special equipment. This short article describes the design of the special equipment for a temperature measurements realized near of a gear tooth flank of a real gearbox.

2 The Equipment for a Temperature Measurements

The next part of this article describes parts of equipment for temperature measurements near a tooth flank. The parts with optimal technical parameters were chosen from a wide range of commercial products.

2.1 Temperature Sensors

Temperature sensor is the main part of equipment for temperature measurement. Many type of sensors are used in technical praxis in present. Main used sensor types are resistance based sensors and voltage producing sensors- thermocouples. Thermocouples are famous with their small installation space. The thermocouple signal is relatively difficult to process. The resistance based sensor are more suitable for our application.

The two basic types of resistance based sensors are metal resistance type and semiconductor type. The resistance characteristic of metal resistance temperature sensor is linear. Linear resistance characteristic is better than non-linear. The metal resistance sensors are bigger than semiconductor sensors. The semiconductor temperature sensors is possible to subdivide as NTC and PTC thermistors. The resistance characteristic of thermistors is not linear but is possible to solve it by calibration. The resistance of PTC thermistor is increased with temperature and resistance of NTC thermistor is decreased with temperature. The modern types of thermistors are famous with their little dimensions, response time and high accuracy.

Finally we chose a miniature NTC sensor for our application. The type of chosen sensor is 2K7MCD1 by Telemeter Electronic GmbH (Fig. 1).

2.2 System for a Data Transportation

The main system of this equipment is electronic circuit for measuring of temperatures and data transport. Temperatures are measured on a rotating shaft and

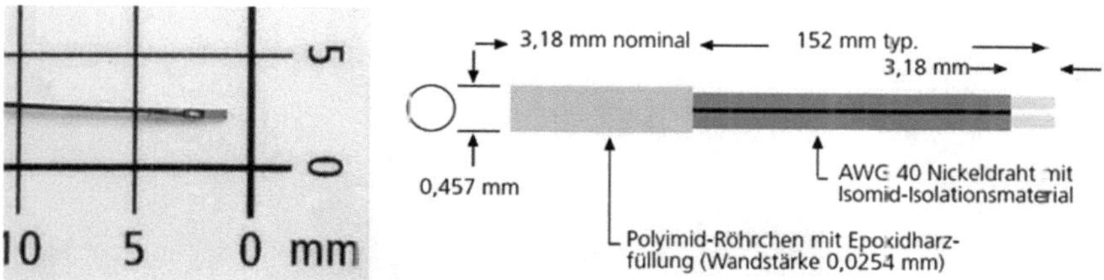

Fig. 1 Thermistor NTC 2K7MCD1

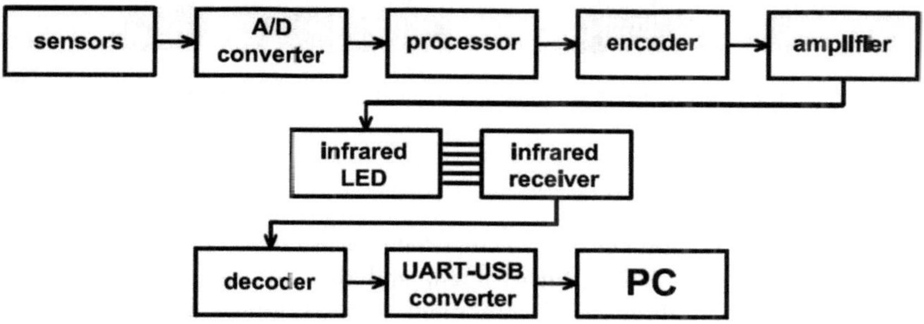

Fig. 2 The block scheme of measuring system

wireless transport to the stationary part is necessary. The first signal and data processing on the rotating shaft is performed already. The block scheme of the measuring system is on Fig. 2.

The sensor signal evaluation is the first operation. The measuring of sensor resistance is the first operation in fact. This first step is performed by the A/D converter. The output data from the converter are edited by processor. Processor is programmable chip which is necessary for the right cooperation of the A/D converter and the encoder. The data from processor are sending to the encoder which transform them for the last step performed on the rotor part. The last step is when the data from encoder are amplified and sending out by infrared diode. The rotating part of equipment is on Fig. 3a.

The infrared receiver is the first point where the data from rotor are received. The second step on the stator is decoding of signal from the infrared receiver. The operation of decoding is realized by IR decoder. The output data from decoder are in UART format. The UART-USB converter for conversion of signal is used. The output data from the converter are in USB format and save them to the computer is possible. The stationary part of equipment is on Fig. 3b.

Fig. 3 System for a data transport (**a**) rotating parts and (**b**) stationary parts

2.3 Calibration

Calibration of the equipment is very important for accuracy of measuring. Every temperature sensor is a little bit different and this differences is necessary to eliminate. For the calibration all equipment was assembled as on Fig. 4a. The calibration furnace Jofra ATC-650B was used. The eight connected sensors were put into the isothermal chamber of calibration furnace (Fig. 4b) and calibration cycle was performed. The calibration cycle was automatically controlled by the calibration furnace program.

The calibration cycle contained eleven temperature values, from 25 °C (77 °F) up to 125 °C (257 °F) therefore one step was 10 °C (18 °F). The data "representing" a temperature were stored into the computer. The stored data was assigned to temperatures during the postprocessing. The problem with the data recording in the next part of this paper is described.

2.4 Software for Data Processing

For the data processing, to create a computer program was necessary. The LabVIEW software was used for reading and evaluating of the data at first. The loading and displaying of the data was possible. This operation ran in real time, but the loading of the data by the LabVIEW application was problematic. For an other type of the data reading we decided. The easy program for the data loading in Delphi software was created. The format of the output data is a text file. It is not possible to show output data in real time, but the number of data errors is almost zero.

The output data are in text file as written. The LabVIEW software is possible to use as a smart tool for the postprocessing. The data are loaded into the program created in LabVIEW and the data processing is possible. The graphs of temperatures are created by the program finaly.

Fig. 4 Calibration (**a**) assembly of measuring system during calibration and (**b**) sensors of temperature in isothermal chamber

The special algorithm was used for the calibration of temperature sensors. The data stored during the calibration cycle (already described) were processed by the our created LabVIEW program and the constants of thermistors were calculated.

2.5 *Stand for Equipment Testing*

The small stand for the testing of the equipment for temperature measurements near a tooth flank was designed and manufactured. The testing stand is very easy equipment which is necessary for testing of the temperature measuring system not yet in his real operation condition but "in the office" only. This little stand is on Fig. 5.

The power supply is number one on the Fig. 5. The power supply gives 12 V for operation of a little electric motor. The electric motor is number 3 on the Fig. 5. The rotation speed of motor is possible to change by a regulator. The regulator is number 2 on the Fig. 5. The power of the electric motor is transmitted by the belt drive to the rotating shaft which is number 4 on Fig. 5. This shaft rotate with a similar frequency as in real gearbox. The rotating frequency of the shaft is about 5200 rpm. The rotating parts of the tested equipment for temperature measurements are applied on the shaft. The sensors are fixed in the middle of the rotating shaft.

The static parts (stator) of the measuring equipment on the small holder are installed, number 5 on the Fig. 5. The IR receiver is contained in the stator of

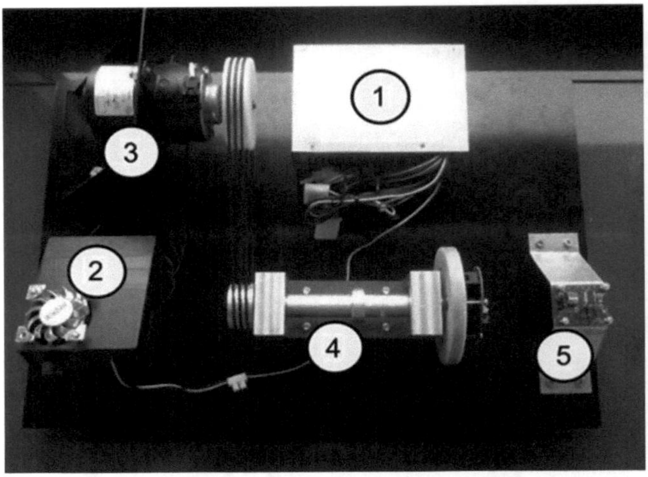

Fig. 5 Stand for equipment testing

equipment. The IR receiver have to be mounted in opposite orientation towards to infrared LED. The placement of the IR receiver in right distance and coaxially with the IR transmitter is necessary.

3 Conclusion

This short article deals with a design, manufacturing and testing of equipment for temperature measurements near of a tooth flank of gear in real gearbox. The article bring short information about used technologies, software and sensors. The calibration is shortly described here. The relative articles are [7, 8]. This kind of temperature measurements is very large issue and next outcomes will be presented in other publication.

Acknowledgments The research work reported here was made possible by the SGS 21009/115 project of Technical University of Liberec.

References

1. V. Moravec, Z. Dejl, M. Němček, Z. Folta, J. Havlík, *Čelní ozubená kola v převodovkách automobilů* (VŠB-TU Ostrava, 2009). ISBN: 978-80-7225-304-3
2. A. Dočekal, *Design optimization of earbox considering reduction of noise emission*. Ph.D. thesis, VUT Brno, 2003
3. T.T. Petry-Johnson, A. Kahraman, N.E. Anderson, D.R. Chase, Experimental investigation of spur gear efficiency, in *Proceedings of the ASME 2007 International Design Engineering*

Technical Conferences & Computers and Information in Engineering, Conference IDETC/CIE 2007, Las Vegas, DETC2007-35045, 4–7 Sept 2007
4. B.-R. Höhn, K. Michaelis, M. Hinterstoißer, Optimization of gearbox efficiency. GOMABN **48** (4), 441–480, December 2009, ISSN 0350-350X
5. V. Moravec, *Konstrukce strojů a zařízení II- čelní ozubená kola*, 2001. ISBN-80-7225-051-5
6. S.B. Rao, D.R. McPherson, *Gear tooth temperature measurements*, Gear Solutions, Aug 2009
7. M. Mazač, Stand for measuring temperatures of the main gears of automobile differential, in *54th International Conference of Machine Design Departments*, Hejnice. 10–12 Sept 2013. ISBN 978-80-7372-986-8.
8. M. Mazač, P. Herajn, Zařízení pro měření teplot v ozubených kolech, konference Konstruování-Green Engineering, Liberec, 26 Nov 2013

Determination of Power Losses Based on the Measurement in Closed Loop Test Bench for Automotive Gearboxes

O. Milacek and G. Achtenova

Abstract This article deals with estimation of power losses on the closed loop test stand dedicated for endurance tests of automotive gearboxes. This test bench has got two specific properties. Firstly, the gearbox is connected to the bench in the same manner as in the vehicle which means that the gearbox is rigidly connected with the bolts to the engine cast and this assembly is than hung using rubber mounts to the test rig frame. Second specific property of this test bench is that it can be used either for one- or more-load level tests. The main target of this contribution is to describe the power flow in the closed circuit which means to find out power losses of its all parts based on the measured load. The measurement will be done for both pretensioning mechanisms which means for manually operated worm gear as well as planetary unit. The magnitude of power losses will be evaluated for different revolutions, load levels and speed stages.

Keywords Gearbox test bench • Closed loop • Power losses • Efficiency • Power circulation

1 Introduction

The principle of the closed loop (back-to-back) test rig is well known and widely used for endurance testing of gear sets. Its advantage lies in low energy demands caused by power circulation in pretensioned closed testing circuit. In our case there is possible to choose of two kinds of pretensioning mechanisms. In case of one load level tests a standard worm gear set and frictional clutch are used. In case of more-load level tests a special unit with two planetary gear sets is used. This unit allows changing loading torque while running the test.

The description and comparison of power demands of the test rig using these two mechanisms is the main aim of this paper.

O. Milacek (✉) • G. Achtenova
Czech Technical University in Prague, Prague, Czech Republic
e-mail: ondrej.milacek@fs.cvut.cz; gabriela.achtenova@fs.cvut.cz

V. Dynybyl et al. (eds.), *The Latest Methods of Construction Design*,
DOI 10.1007/978-3-319-22762-7_70

2 Description of the Test Bench

The test bench was designed and made by Technical University of Ostrava. The supporting frame of the test bench is made of standard steel profiles. Both gearboxes (tested and technological) are connected to the frame in the same manner as in the vehicle which means that the gearbox is rigidly connected with the bolts to the engine cast and this assembly is than hung up to the test rig frame using rubber mounts including reaction rod. The advantage of this mounting is that all gear sets are forced in the same conditions as in the real car e.g. shafts displacement, lubrication, etc. Both gearboxes are connected with joint shafts. Between both input shafts there are inserted pretensioning mechanism and a pulley of the belt drive [3].

According to the direction of rotation and loading torque at the input shaft it is determined which gearbox is the tested one and which one is the technological. Maximal rpm and torque at the input shafts is 4000 min^{-1} and 200 Nm. Real design of this test bench is depicted in the Fig. 1.

The functional scheme of this closed loop test bench including descriptions and efficiencies of all parts can be seen in the Fig. 2.

2.1 Pretensioning Mechanisms

As it was mentioned, this test bench can be used either for one- or more-load level endurance tests. In the Fig. 3 [1] there is depicted a pretensioning mechanism which is used for one-load level tests. Its principle is following. Firstly the right shaft is locked to the frame using the pin. After that the worm is by inclining fitted and fixed to the worm wheel and is manually by its rotating preset desired torque in the circuit. Then are both shafts clamped using friction clutch. Finally is removed the worm and the pin and the test rig is ready to start the test.

Second used pretensioning mechanism is designed for more-load level tests, which means that during the test it must be possible to change the torque in the circuit. This is achieved by a special pretensioning unit, which is depicted in the Fig. 4 [2]. This unit consists of two identical planetary gear sets with common planet carrier. Both suns are connected with input shafts of gearboxes. One annular gear is rigidly connected to the frame; second one is able to rotate and managed by a worm gear. By using this unit it is possible to emulate a load spectrum of a vehicle.

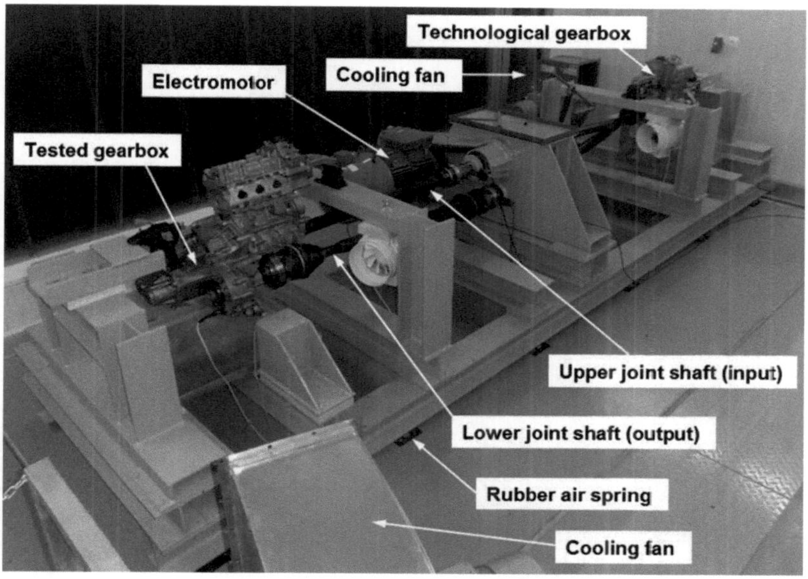

Fig. 1 Real design of the closed loop test bench

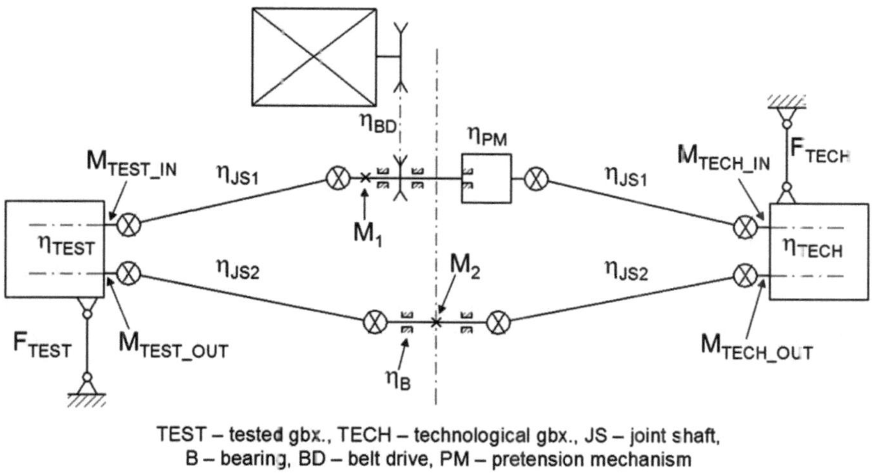

Fig. 2 Functional scheme of the closed loop test bench

3 Power Losses at the Test Bench

In the Fig. 5 there is depicted the diagram of power flow in the testing circuit. Used direction of loading and rotation determines the left gearbox as the tested one and the right gearbox as technological. The sum of all power losses is equal to a power

Fig. 3 Hand operated worm gear and friction clutch "ETP Techno" for one-load level tests

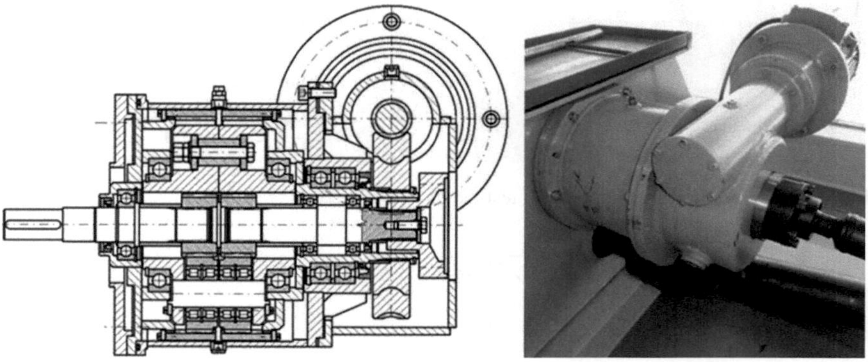

Fig. 4 Scheme and photo of the planetary pretensioning mechanism for more-load level tests

of the driving electromotor. The greatest losses are in gearboxes, joint shafts and in planetary pretensioning mechanism.

The torque is measured at two points (1 and 2) marked with crosses. Only in these two points is exactly known the power magnitude and can be counted the efficiency of the circuit's part between them. At the side of the tested gearbox is total efficiency the multiplication of efficiencies of input and output shaft and of a gearbox. At the technological side is furthermore a pretensioning mechanism (see Fig. 2).

$$\begin{aligned} \eta_{TESTpart} &= \eta_{JS1} \cdot \eta_{TEST} \cdot \eta_{JS2} \\ \eta_{TECHpart} &= \eta_{JS2} \cdot \eta_{TECH} \cdot \eta_{JS1} \cdot \eta_{PM} \end{aligned} \tag{1}$$

To determine individual efficiencies it is possible to use the equation for reaction torque of the gearbox which is calculated from a measured force in reaction rod.

$$M_R = M_{OUT} - M_{IN} = F_R \cdot l \tag{2}$$

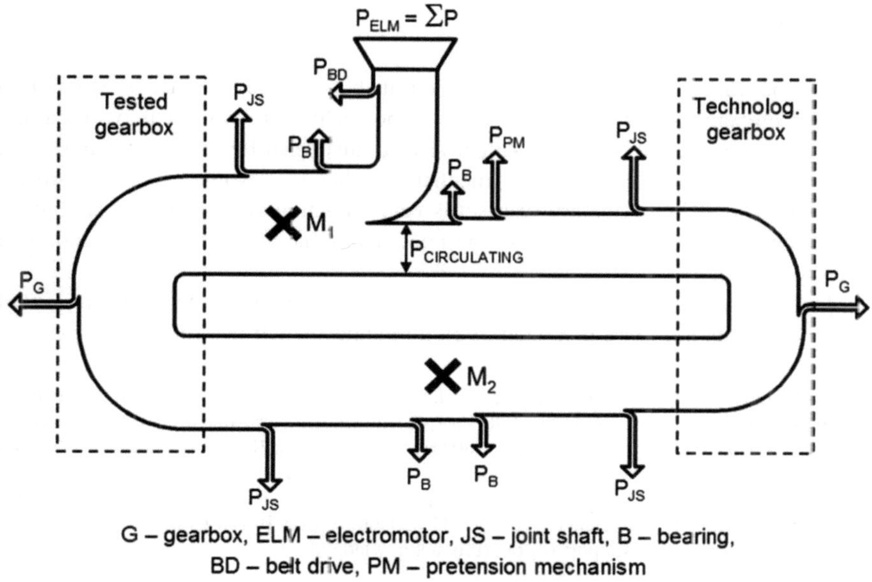

Fig. 5 Power flow diagram with depicted losses in the testing circuit

The distance $l = 0{,}382$ m. The sum of all losses in the circuit is equal to the power delivered from the electromotor reduced due to the efficiency of the belt drive. The subtraction of the power in the point 1 and the power from the electromotor determines the circulating power in the circuit – see Fig. 5.

In case of one-load level tests can be the efficiency of pretensioning mechanism counted as the efficiency of two ball bearings which support the worm wheel.

$$\eta_{PM} = \eta_{BB}^2 = 0{,}995^2 = 0{,}99 \tag{3}$$

In case of more-load level tests can be this efficiency counted as the efficiency of two serial planetary gear sets with common planet carrier. This value neglects efficiencies of bearings inside planetary gear sets.

$$\eta_{PM} = \eta_{PG1} \cdot \eta_{PG2} = 0{,}97^2 = 0{,}94 \tag{4}$$

Unfortunately it wasn't possible to provide all needed measurements till now and so the results of the comparison between computed and measured values of efficiencies will be presented only at the conference.

4 Conclusion

During endurance tests which are provided at this test bench must be very often changed engaged speed stage in both gearboxes because of local overheating. Standard testing period is 20 min. Afterwards it must be shifted in other speed stage.

Each of these two pretensioning mechanisms has its advantages and disadvantages. The way using planetary gear set is very advantageous because it enables to provide endurance tests continuously without any attendant. At the other hand there is worse efficiency.

Manually controlled worm gear is very advantageous from the point of view of energy demands but there is a need of an attendant, which is quite expensive.

Acknowledgments This research has been realized using the support of EU Regional Development Fund in OP R&D for Innovations (OP VaVpI) and Ministry for Education, Czech Republic, project # CZ.1.05/2.1.00/03.0125 –ED2.1.00/03.0125 Josef Božek Research Centre for Vehicles of Sustainable Mobility. This support is gratefully acknowledged.

This research has been realized using the support of Technological Agency, Czech Republic, programme Centres of Competence, project # TE01020020 Josef Božek Competence Centre for Automotive Industry. This support is gratefully acknowledged.

References

1. http://www.etp.se/en/products/power-transmission/etp-techno
2. Havlík, Jiří – Čelní ozubená kola v převodovkách automobilů, s. 154
3. V. Moravec, M. Němček, J. Havlík, Analýza toku výkonu v uzavřených zkušebních stavech pro životnostní zkoušky převodových skříní, TU of Ostrava, 2001, report number: D19-272 3000 9/2001

Experimental Measurements and Development of Agricultural Trailer

P. Mossoczy, R. Uhlir, and P. Maly

Abstract Article describes partial section of complex development of agricultural trailer. Article is focused on measuring equipment. In the early stages of development of trailer, the laboratory measurements were performed to determine the calibration constants of sensors. In the next stage of development of trailer, the experimental measurements were performed in the real operating conditions. Based on the proposed methodology for measuring some parts of trailer were selected to monitor the load and deformation. The results of measurement were used for further analysis. Furthermore, comparison of results of FEM analysis and experimental measurement was made.

Keywords Experimental measurement • Data acquisition system • Real-time controller • Agricultural trailer

1 Introduction

Experimental measurements are an important stage in the development of prototypes of agricultural machinery. These measurements verify the operability of prototypes and verify the strength calculations for individual components. In the first stage, measurements usually held in a laboratory environment. Simulations operating load of individual components of prototype run on the test stands there (Fig. 1). Operational tests of prototype on a test polygon, possibly in real operating conditions followed in the next stage [2], [3].

2 Design of Measuring System for Laboratory Tests

Selecting the right data acquisition (DAQ) hardware for design of measuring system is very important. For common laboratory tests are often used multifunction DAQ devices connected via suitable PC bus (PCI, USB, LAN). Multifunction DAQ

P. Mossoczy (✉) • R. Uhlir • P. Maly
Czech Technical University in Prague, Prague, Czech Republic
e-mail: pavel.mossoczy@fs.cvut.cz; roman.uhlir@fs.cvut.cz; pavel.maly@fs.cvut.cz

V. Dynybyl et al. (eds.), *The Latest Methods of Construction Design*,
DOI 10.1007/978-3-319-22762-7_71

Fig. 1 Laboratory tests of lifting mechanism

devices have a combination of analog inputs, analog outputs, digital inputs/outputs, and counters. Common low-cost DAQ devices do not have built-in signal conditioning, which most sensors require. External devices are therefore used for signal conditioning [1].

Another option to choose of suitable DAQ devices is a modular platform, which consists of a chassis to control timing and synchronization and a variety of I/O modules. More different modules configurations are possible (Fig. 2).

3 Design of Measuring System for Operational Tests

For operational tests, it is right to use portable and rugged measuring technique. Standard PCs are not designed to withstand the conditions of industrial environments. For instance, most laptops may not be rated for dusty or moist environments.

Operational tests have great demands on the reliability of the measuring system. The selecting right computer operating system for data acquisition is very important. For example, the common PC operating systems are not deterministic and cannot guarantee 100 % reliability of the measurement process. The ability to operate deterministically gives the real-time operating system (RTOS). Test applications can be execute according to precise timing requirements with high reliability.

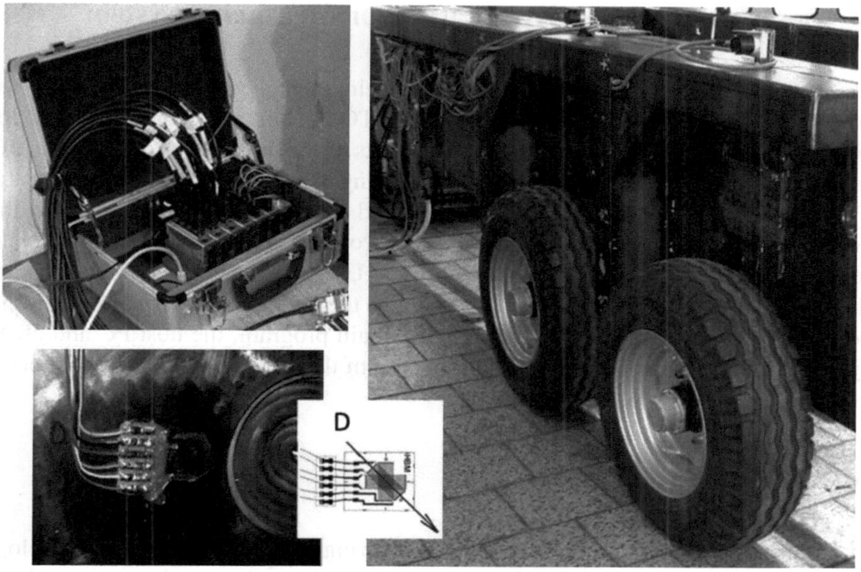

Fig. 2 Operational tests of prototype

4 Description of Measuring System for Tests of Prototype of Agriculture Trailer

Integrated measuring cRIO system (National Instruments) is system that combines an industrial real-time controller and reconfigurable field-programmable gate array (FPGA) chassis. This system allows executing high-volume applications for industrial machine control and monitoring. The chassis, which was used to the tests of prototype, has eight slots for different I/O plug-in modules connecting.

The plug-in I/O modules have been selected based on the draft measurement methodology. For the strain gauge measurement, three simultaneous bridge modules were used. Each bridge module contains all the signal conditioning required to power and measure up to four bridge based sensors simultaneously. Furthermore, the module with analog inputs, module with analog outputs, the module for thermocouple measurements and module with a digital input were used.

To monitor the load and deformation of the steel frame and lifting mechanism the foil resistive strain gauges (HBM) were used. To monitor the trajectory and position of the lifting mechanism the draw-wire encoders with the multi-turn potentiometer and linear encoders with TTL output (Larm) were used. To measure the pressure in the hydraulic system the pressure transmitter with analog output (Hydac) was used. To measure the temperature of steel frame the magnet mount thermocouple (Greisinger) was used (Fig. 2).

5 Description of Test Application for Measuring System

Test application programmed in graphical development system LabVIEW has two parts. They are the main program that runs in RTOS of controller and subroutine for FPGA chassis controlling. The subroutine sets the configuration of plug-in I/O modules and provides synchronous data reading. Data values are read from the FPGA in integer format and are then converted to scaled engineering units in the controller. The data are stored in internal nonvolatile memory. The application is loaded into the cRIO system from the host PC. User interface (GUI) of application that is displays on the host PC monitor contains the controls and displays to current acquired data monitoring. After running the main program, the host PC and cRIO system disconnect is possible. Measuring system then works in standalone mode.

6 Conclusions

This article describes the design of the measuring system for prototype development of agricultural trailer. Performed static and dynamic tests have verified the proposed concept of the measuring system in laboratory and operating conditions. The results of measurement for further analysis were used. Furthermore, comparison of results of FEM analysis and experimental measurement was made.

Acknowledgments This work has been supported by project MPO: FR-TI1/119.

References

1. V. Dynybyl, Z. Češpíro, J. Kanaval, P. Mossóczy, *Výkonová metrologie – Experimentální podpora vývoje a inovací mechanismů ve strojírenství* (ČVUT, Praha, 2009). ISBN 978-80-01-04325-7
2. R. Uhlir, P. Mossoczy, P. Maly, Experimental testing of agricultural trailer, in *Book of Proceedings of 54th International Conference of Machine Departments*, Technical University of Liberec, . Liberec, 2013, pp. 353–358. ISBN: 978-80-7372-986-8
3. R. Uhlir, P. Mossoczy, P. Maly, Evaluation of operational testing of agricultural trailer. Appl. Mech. Mater. **486**, 245–248 (2014)

Deformable Mirror for High Power Lasers

S. Nemcova and J. Zicha

Abstract The deformable mirror under development is a prototype of a large aperture adaptive optics for high power lasers. Its bimorph and lightweight structure consists of a composite core, electrodes, a piezo-ceramic layer and a reflective layer. A double exposure method and a real time holographic interferometry method are used for measuring deformations of the mirror sample induced by the active layer. Various piezo materials have been tested for its suitability, namely their hysteresis.

Keywords Deformable mirror • Piezoceramics • Hysteresis • Interferometry

1 Introduction

With high power lasers, it is possible to create extreme states of mass. Lasers are now used to generate a hot dense mass with a temperature of several million degrees and a density many times exceeding that of the solid state. Lasers under construction, such as EU project ERIC-ELI, aim to reach ultrarelativistic limits of electromagnetic field density leading to a polarization of vacuum or to create conditions close to those at a black hole horizon.

The main requirement is to focus the laser beam to the target place and its deposition to a small area with the highest possible energy concentration. However, without special care, lasers emit a beam that is amplitude modulated and deformed by phase shifts. In practice, the quality of a laser beam is affected by mechanical errors and vibrations, gradual degradation of elements, thermally induced inhomogeneous distribution of a refractive index, an unstable active and transport media and other random processes.

To maintain the high quality of a laser beam, the forming of a wavefront is critical, in other words, a phase control. A deformable mirror is considered a suitable space phase corrector. This method of wavefront correction is nowadays successfully used in

S. Nemcova (✉)
Czech Technical University in Prague, Prague, Czech Republic
e-mail: sarka.nemcova@fs.cvut.cz

J. Zicha
Faculty of Mechanical Engineering, Czech Technical University, Technická 4, Prague 166 07, Czech Republic
e-mail: josef.zicha@fs.cvut.cz

V. Dynybyl et al. (eds.), *The Latest Methods of Construction Design*,
DOI 10.1007/978-3-319-22762-7_72

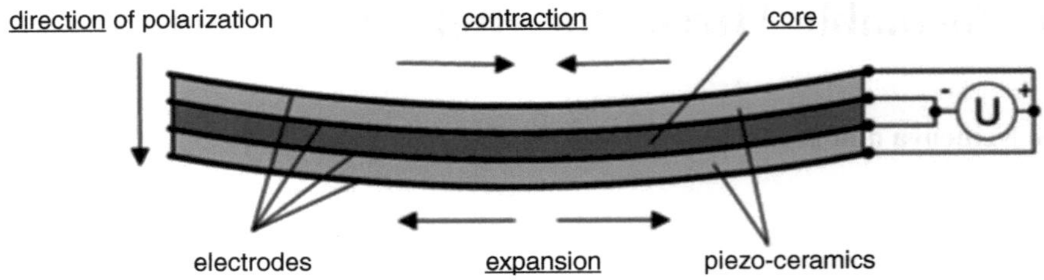

Fig. 1 The bimorph structure of deformable mirror

astronomy. Unfortunately, correctors for picture processing cannot be used for lasers, mainly because the high intensity of the laser beam would damage the mirror.

2 The Mirror Structure

We are developing a lightweight mirror with a bimorph structure [1]. The core is a composite layer. Both sides of the core will be covered in a mutual electrode (ground). The next layer (on both sides again) is made of piezo-ceramics. An electrode layer follows. The final arrangement of the electrodes is being designed now. The deformation of the mirror yields a transverse piezo effect (Fig. 1).

This is a different approach to the existing ones relying on separate piezo actuators [2, 3, 4, 5]. For high power active and adaptive optics it is possible to use a deformable mirror with separate segments of actuators or a continuous monomorph or bimorph layer. It is easier to achieve the parameters necessary for a beam correction with separate actuators. Unfortunately, with a larger aperture and therefore a higher number of actuators, the weight and the volume of the mirror grows rapidly.

3 The Measuring Methods

Two holographic methods have been used for evaluation of the samples' topography. A "system focus" method has been used for the measurement of hysteresis of various piezo-ceramics materials.

3.1 Double Exposure Method

An oblique illumination was used for the double exposure method (see Fig. 2). Firstly, we recorded a "zero state": no voltage applied to the mirror. Then, we applied a certain voltage at the electrodes and recorded the shifted state of the mirror.

The beam splitter consists of a birefringence prism and three half-wave plates. We can control the energy in the split beams by rotating the first $\lambda/2$ plate. In this way, the rate of intensities of the reference and the object beams can be set without

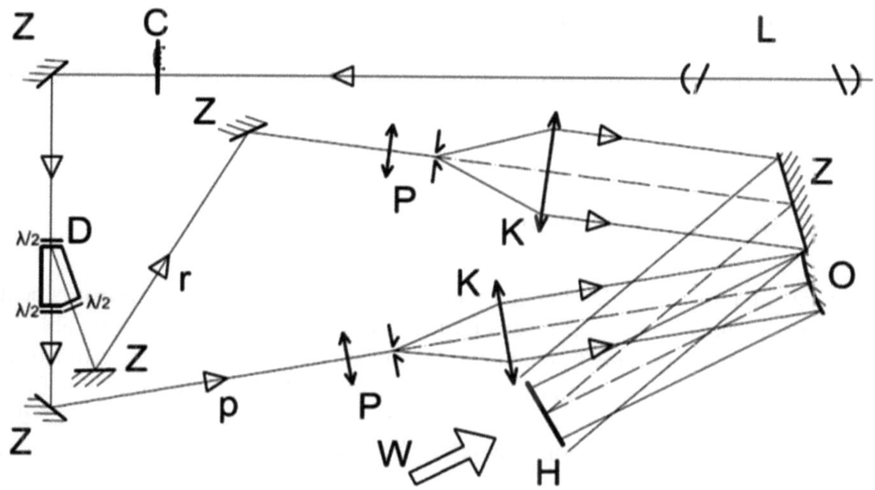

Fig. 2 Double exposure optical set-up. *L* – laser, *C* – shutter, *D* – beam splitter, *Z* – reference mirror, *O* – sample, *P* – space filter, *K* – collimator, *H* – holographic plate, *W* – direction of observation

energy losses. The two downstream λ/2 plates restore the polarization direction of the beams.

After the holographic plate is processed and illuminated by a reconstruction beam, we can analyse the interference pattern and evaluate the deformations of the mirror. The displacement Δl can be calculated as follows

$$\Delta l = \frac{N.\lambda}{2n\cos\left(\alpha_i/2\right)} \tag{1}$$

where N is the number of fringes, n is the refractive index and αi is the angle between the direction of illumination and observation.

To avoid possible errors and to make the calculations simpler, we also used a set-up with perpendicular illumination of the sample (i.e. $\alpha i = 0$). In this set-up, a plate beam splitter was used in front of the sample. This caused disturbing additional interference fringes created by multiple reflections at the splitter front and rear faces and successive interference of the reflected light. As this could not be overcome, we moved to the oblique illumination.

3.2 *Live Fringes Method*

The optical set-up is identical to the previous one. In this method, the "zero state" (unloaded) is recorded and the holographic plate is processed without removing it from the holder. Then the mirror is loaded and the interference pattern is observed through the hologram.

In contrast with the previous method, various loaded states can be observed through a single hologram.

3.3 System Focus Method

The measured mirror is illuminated by a beam of parallel rays. If the mirror is bent to a convex shape, we can measure its focal length directly. So as to be able to measure the concave shape too (a negative mirror), an additional positive mirror is placed downstream to form a positive system with the tested mirror (see Fig. 3).

As the shape of the tested mirror changes, the effective focal length of the two-mirror system changes consequently. The focal length of the tested mirror is determined by

$$f_1' = \frac{p_2' \cdot (d - f_2') - d \cdot f_2'}{p_2' - f_2'} \tag{2}$$

The sag h of the mirror is then calculated as follows

$$h = \frac{2 \cdot f_1' - \sqrt{4 \cdot f_1'^2 + D^2}}{2} \tag{3}$$

where D is the diameter of the tested mirror.

4 Experimental Results

4.1 Mirror Topography

Shown below are the results for the circular sample, real time method. In Fig. 4, the left picture is the "zero state". For a validation of our interference pattern analysis and the deformation calculations, we caused the deformation of the mirror by pushing a screw from the back of the mirror. In this way, we knew the deformation of the mirror and could cross-check this value with the one obtained from the analysis of the fringes.

Knowing the direction of illumination, we calculated the deformation from the interference pattern as 2.04 μm (6 fringes). Taking into account the fact that the mirror apparently changed its shape (most likely a relaxation after a previous load) – there is a residual interference fringe in the left picture, we can say the deformation was calculated closely enough.

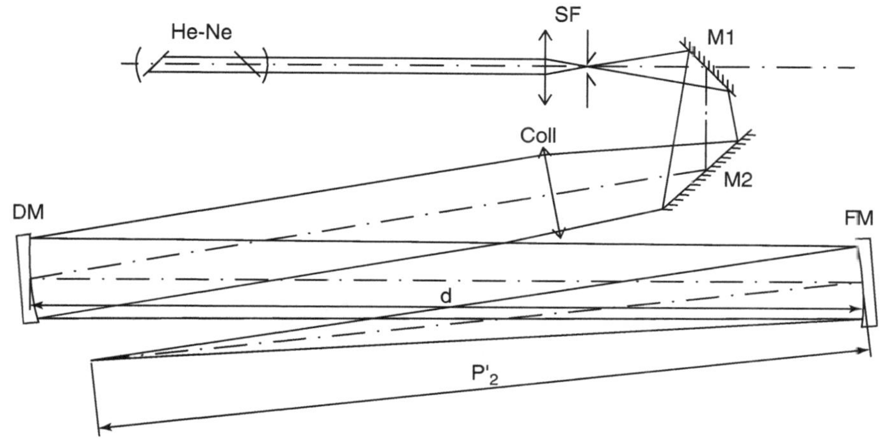

Fig. 3 Optical layout for the system focus method. *He-Ne* – laser, *SF* – space filter, *M1*, *M2* – folding mirrors, *coll* – collimating lens, *DM* – deformable mirror under test, *FM* – focusing mirror

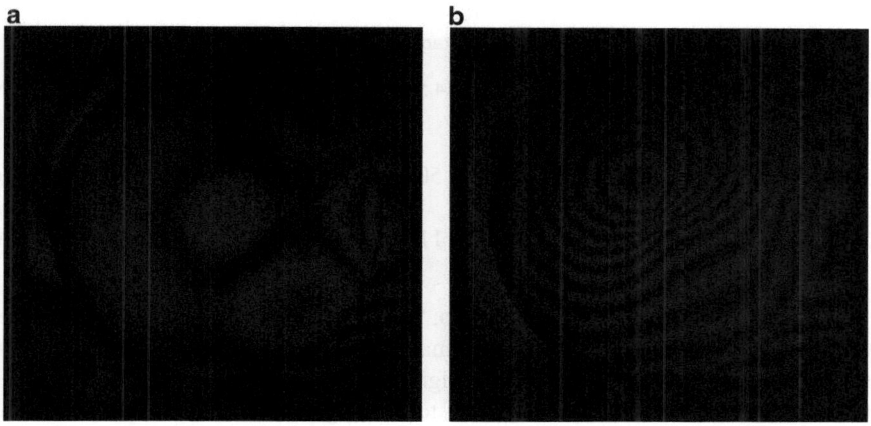

Fig. 4 Holographic interferometry: (**a**) unloaded mirror (**b**) interferogram of the mirror with a set deformation of 2.32 μm

We used Agfa 8E75 and Colour Holographic BB640 holographic plates, process JD2 (developed and bleached). Maximum voltage applied to the electrodes was 300 V.

4.2 *Hysteresis*

Three samples of piezo-ceramics were tested:

1. soft PZT ceramics, sample diameter 50 mm, thickness 0.2 mm, glass substrate diameter 70 mm, thickness 0.75 mm

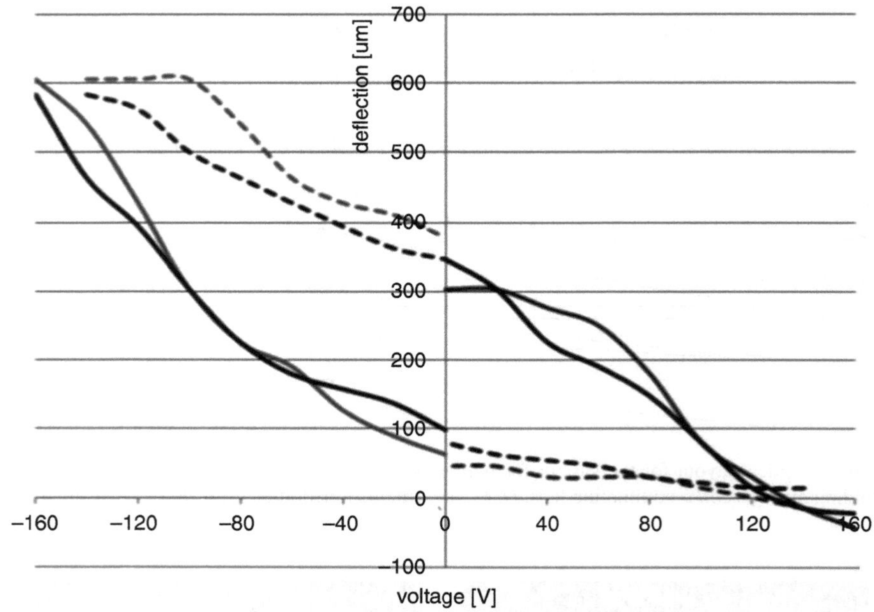

Fig. 5 Results for the first sample, hysteresis 44.5 % in the first cycle, 41.0 % in the second cycle

2. soft PZT ceramics, sample diameter 50 mm, thickness 0.3 mm, glass substrate diameter 70 mm, thickness 0.75 mm
3. hard PZT ceramics, sample diameter 110 mm, thickness 0.8 mm, glass substrate diameter 120 mm, thickness 2 mm

Measured cycles were always two, each with following voltage change: 0 V → max positive voltage → 0 V → max negative voltage → 0 V. In all cases, the measured hysteresis was slightly higher in the first cycle. Due to the higher thickness of the third sample, we had to use higher voltage to achieve similar bending of the mirror (Figs. 5, 6, and 7).

5 Conclusion

We have tested three different ceramics materials for their suitability as actuators for a deformable mirror with a bimorph structure. The holographic interferometry method revealed there is a noticeable relaxation time after a change of voltage load.

From the tested materials, the hard PZT shows least hysteresis, however, the voltage necessary to achieve the desired bending is too high – there is a danger of short circuit. We do not have a technology to cut a thinner piece of ceramics so as to

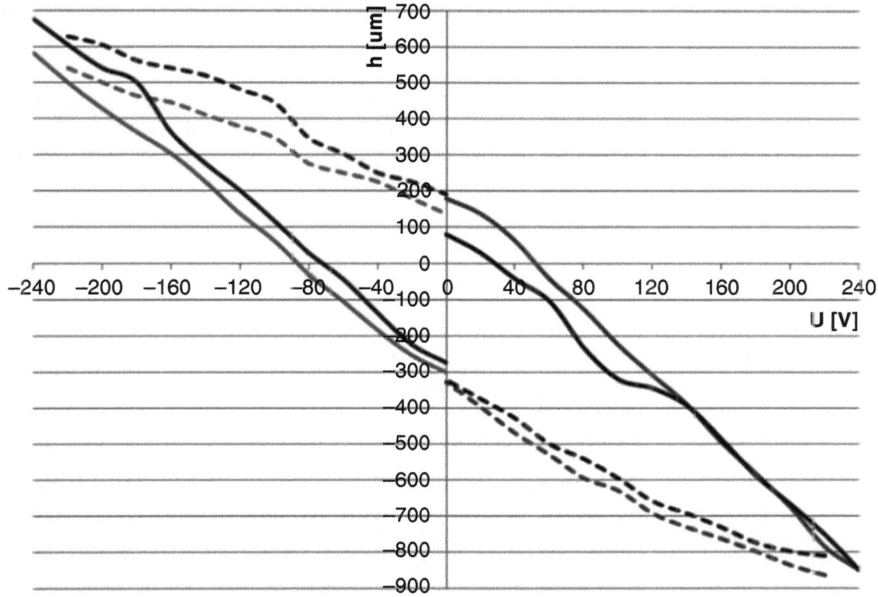

Fig. 6 Results for the second sample, hysteresis 36.7 % and 31.2 % in the first and second cycle, respectively

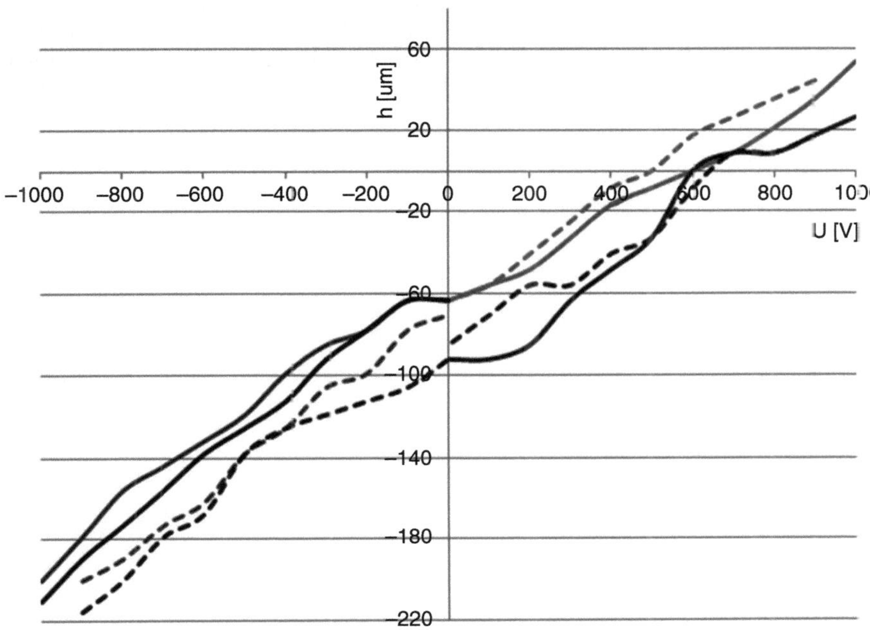

Fig. 7 Results for the third sample, hysteresis 16.1 % and 12 % in the first and second cycle, respectively

lower the voltage needed for the mirror actuation. Therefore the second sampleseems most promising.

Acknowledgments This work was performed with the support of the Technological Grant Agency of CR TA01010878. Large-aperture elements for laser active and adaptive optics are investigated in collaboration with the Institute of Plasma Physics, Academy of Science, CZ and 5 M Ltd, Kunovice, CZ.

References

1. V. Kmetik, S. Nemcova, L. Jiran, E. Stranakova, A. Inneman, Development of large aperture elements for active and adaptive optics, in *OaM 2012, Optics and Measurement International Conference*, Liberec, 16–18 Oct 2012
2. Z. Liao, Wavefront correction on high repetition rate, high energy laser system, in *The Advanced Maui Opticalvand Space Surveillance Technologies Conference*, Wailea, 2006, ed. by S. Ryan, , p. E32, 10–14 Sept 2006, The maui economic development board, M. Mansouri, Q. Ilham, H. Snoussi, C. Richard, Adaptive quantized target tracking in wireless sensors networks. Wirel. Netw. **17**(7), 1625–1639. ISSN 1022-0038
3. B. Wattellier, J. Fuchs, J.P. Zou, K. Abdeli, C. Haefner, H. Pépin, High-power short-pulse laser repetition rate improvement by adaptive wave front correction. Rev. Sci. Instrum. **75**(12), 5186–5192 (2004)
4. V.Y. Zavalova, A.G. Alexandrov, A.V. Kudryashov, A.L. Rukosuev, V.V. Samarkin, P.N. Romanov, The problems of adaptive optical system to correct the large aperture beams, in Proceedings of 7th International Workshop on Adaptive Optics for Industry and Medicine, 2009, p. 110. http://www.nightn.ru/files/publications/publications_ru.htm
5. S. Bonora, C.J. Hooker, S.J. Hawkes, J.L. Collier, C. Spindloe, in *Proceedings of the Sixth International Workshop on Adaptive Optics for Industry and Medecine*, ed. by C. Dainty, Imperial Coledge, pp. 106–111. ISBN 978-1-84816-110-8 (hardcover), ISBN 978-1-84816-111-5 (ebook)

Identification of Failures of Technical Equipment Using Acoustic Camera

S. Poljak and I. Gajdac

Abstract Article discusses the use of an acoustic camera for identification of noise sources in the technical equipment and identify sources of environmental noise. It is a device which can identify the sources of acoustic emission using a special microphone kit and it is possible to interpret the measured data graphically. Due to its composition, the application of acoustic camera to localize sound in various fields of human activity in the outer environment.

Keywords Acoustic camera • Identification of failures

1 Acoustic Camera

Acoustic imaging camera is a device used to illustrate the sound source and intensity. It uses a group of microphones, camera, and collector data along with the signal and image processing software.

System for use the acoustic cameras are usually made so that the digital camera field microphone is connected to the device for storing and processing data. Most often, the amount of data processing that uses powerful personal computer. The whole system is needed to create the final resulting image, which shows the sources of noise [1].

The principle of the device lies in the fact that each one microphone of the microphone array records an audio signal. Because each microphone in the array has a different position, the sound that comes in individual microphones is recorded with some delay. Microphone delay depends on the distance between microphones and source of noise. Due to the fact that the distance between the microphones in each array microphones are predetermined and each audio signal comes in different microphones with a delay, it is possible to calculate these noise sources using a variety of methods. Signal processing is used appropriate software that uses different methods can determine the exact location where it comes from the source sound propagation. The output of this calculation is a picture or 3D model, where individual pixels coloured picture associated values calculation (Fig. 1).

S. Poljak (✉) • I. Gajdac
University of Zilina, Žilina, Slovak republic
e-mail: silvester.poljak@fstroj.uniza.sk; igor.gajdac@fstroj.uniza.sk

V. Dynybyl et al. (eds.), *The Latest Methods of Construction Design*,
DOI 10.1007/978-3-319-22762-7_73

Fig. 1 Representation of the source of sound propagation from drilling machine

Most acoustic cameras use so-called two-dimensional acoustic mapping. This type of camera uses a unidirectional microphone array and one plot plane. Two-dimensional acoustic mapping works best when we assume that we are not dealing shape and surface properties. In this case, however, needs to be pursued object placed perpendicular to the acoustic camera.

2 2D Acoustic Camera

Two-dimensional acoustic mapping method introduces an error in the calculations right sound intensity in point. Two – dimensional mapping approximates the three-dimensional surface in the plane, allowing the specified distance between microphones and audio hotbed of action. However, this approximation ignores differences due to the distance of the surfaces to different depths at different points. In most applications of acoustic cameras this error is so small that their effects on the resulting measurement data is negligible. In confined spaces, this error becomes very important and can significantly affect the measurement result.

3 3D Acoustic Camera

Three-dimensional acoustic cameras can fix errors 2D cameras taking into account the depth of the surface and therefore correctly measure the distance between the microphone and each spatial point. Three-dimensional acoustic camera can produce a more accurate picture, but requires a 3D model of the measured object or area to be analysed. Three-dimensional acoustic cameras can be used to analyse in confined spaces or enclosed spaces. However, in order to carry out this measurement, it is necessary to use the omni-directional microphone (Fig. 2).

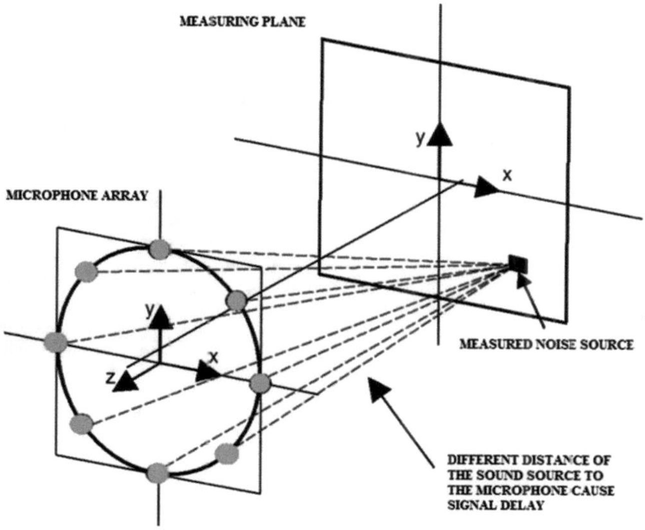

Fig. 2 The principle of acoustic camera

There are many applications for acoustic cameras, but most are designed to reduce noise. Cameras are often applied to reduce the noise of vehicles (cars, planes, trains) and technological units such as wind turbines. Other applications may be solutions to the problems of individual machines and mechanical parts [3]. Sound may itself be a major disruptive element in the environment, but in many cases there are much quieter noise sources that are psychologically important. Acoustic camera offers the user the choice of time and frequency and can also record called sound film. Use of acoustic cameras for these applications and the ability to sound analysis leads to better products and the environment [2].

4 The Basic Kit of Acoustic Camera

Basic acoustic camera assembly consists:

- Microphone array
- Digital camera, which is located in the middle of a microphone array
- A/D converter to convert the signal from the microphone
- PC control and evaluation software

The main condition for correct visualization of the sound is the right choice of a microphone array. Microphone arrays are selected mainly on the distance measured object and the reference frequency range (Fig. 3).

Fig. 3 Basic assembly of acoustic camera

5 Main Types of Microphone Arrays

Microphone array is an array composed of a number of free microphones. Typically, this field consists of all directional microphones distributed in the space of a microphone array, connected to a computer with A/D converter that records and interprets the uses of into a coherent shape. The distance between the microphones in the array microphone is directly proportional to the distance of monitored objects. The primary producers of acoustic cameras are offered in various configurations microphone arrays. In addition to standard sizes and shapes microphone arrays, you can also create an array of different shapes and sizes depending on customer requirements [4].

Circular acoustic field is characterized by the imposition of microphones in an annular shape. This arrangement is suitable microphone to measure the acoustic laboratories. Field structure is optimized so as to ensure the maximum spatial resolution and at the same time providing a high depth measurement display noise source. Due to the fact that in this field are used all shotgun microphone array should be directed to the source of noise and without any source of noise in the background. The recommended distance from the object being measured is 0.5–20 m.

Spherical microphone array is suitable for measuring in confined spaces. Field design is optimized to ensure the highest spatial resolution. The recommended distance from the object being measured is 0.5–25 m.

Radial array microphones are designed for external use. Recommended measuring range is 5–500 m from the noise source (Figs. 4, 5, 6, and 7).

Fig. 4 Ring microphone array

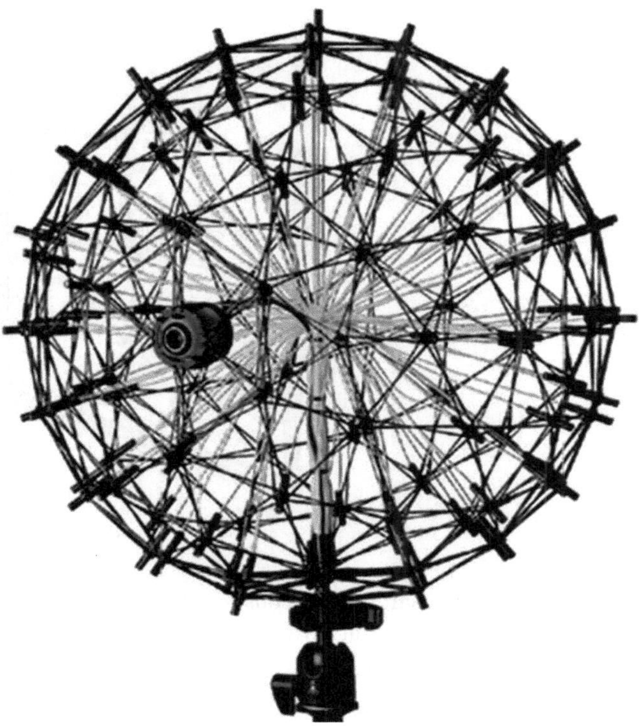

Fig. 5 Sphere microphone array

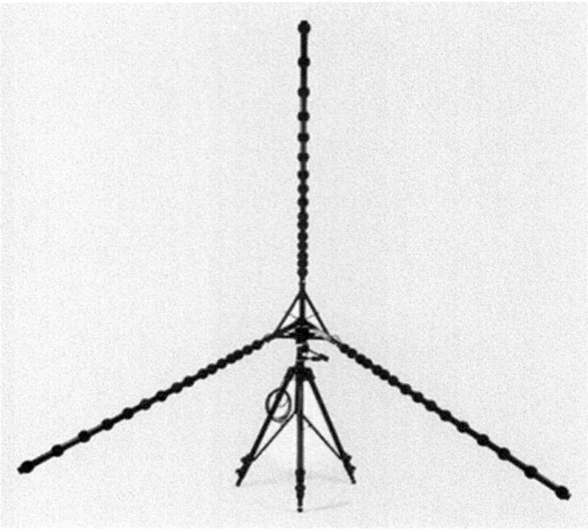

Fig. 6 Star microphone array

Fig. 7 Universal spiral microphone array

Microphones in the universal spiral acoustic array can be allocated freely through the grid. Can be measured at different distances from the source of the noise.

6 Software and Method of Calculation

Software evaluates the measured data using a variety of analytical methods. Most commonly used methods:

1. **NAH – Nearfield acoustic holography**

   The method is suitable for the measurement of noise sources in the smallest distances from the measuring system. Data processing to take advantage

2. **Beamforming**

   This method is used for measuring the sound field in the mid to väčšj distance from the noise source. It is also used for moving sources of noise or wind tunnel.

3. **SONAH – Statistically optimal near – field acoustic holography**

   Measurement method is similar to the method NAH, it is also designed to measure in the near acoustic field. The control software allows for data collection, analysis and subsequent export of measured values in a user-friendly interface.

7 Conclusion

Acoustic camera is a modular system designed to locate noise sources. Accurate, clear and quick display of noise contributes to solving problems of locating the source of the noise and problems associated with it.

This article discusses the use of an acoustic camera for spatial localization and identification of different sound emissions and their quantitative evaluation and frequency analysis in dynamic range. Acoustic camera allows perfect frequency analysis of noise sources at different distances.

Acknowledgments Tento článok vznikol vďaka podpore v rámci operačného programu Výskum a vývoj, pre projekt: Kompetenčné centrum znalostných technológií pre inovácie produkčných systémov v priemysle a službách, kód ITMS: 26220220155, spolufinancovaný zo zdrojov Európskeho fondu regionálneho rozvoja.

"This article has benefited from support under the Operational Programme Research and Development for the project: Competence Center for Knowledge technology innovation production systems in industry and services, ITMS code: 26220220155, co-funded by the European Regional Development Fund".

„Podporujeme výskumné aktivity na Slovensku/Projekt je spolufinancovaný zo zdrojov EÚ".

"We support research activities in Slovakia/Project is co-financed from EU funds".

Európska únia
Európsky fond regionálneho rozvoja

Agentúra
Ministerstva školstva, vedy, výskumu a športu SR
pre štrukturálne fondy EÚ

Operačný program
VÝSKUM a VÝVOJ

References

1. M. Badida, E. Lumnitzer, M. Moravec, P. Liptai, Uplatnenie akustickej kamery v priemysle, in *Strojárstvo, TUKE Košice*. Avaible via http://www.engineering.sk/index.php/clanky2/stroje-a-technologie/937--uplatnenie-akustickej-kamery-v-priemysle, 2012. ISBN 978-80-7372-986-8
2. *Acoustic Camera – Listening with Your Eyes*, gfai Tech GmbH. Avaible via http://www.acoustic-camera.com/fileadmin/acoustic-camera/support/downloads/AC_brochure_2014_EN.pdf, 2014
3. S. Hrček, R. Kohar, Š. Medvecky, Determination of the maximum roller bearing load with regards to durability thereof using FEM analysis. Commun. Sci. Lett. Univ. Žilina **14**(3), 55–61 (2012). ISSN 1335-4205
4. J. Halodova, Acoustic camera, in *Juniorstav 2008*, Pozemní komunikace, VUT Brno, 2008. Available via http://www.fce.vutbr.cz/veda/juniorstav2008_sekce/pdf/2_3/Halodova_Jana_CL.pdf

Assessment of the Mechanical Properties the Polyurethane Foam Under Multiaxial Loading in a Different Climatic Conditions

P. Srb, P. Kulhavý, and R. Martonka

Abstract Study of the mechanical properties of the polyurethane foams (such as comfortable filler of automobile seats) is important for a comprehensive understanding of the response of this nonlinear viscoelastic material. It is very essential for their comparison with other materials, because the current trend of development is pushing to find a new low-energy and low-density material. Two foams from the automobile seats with a different densities were assayed, under multiaxial loading for the several different climatic conditions.

Keywords Climatic chamber • Biaxial loading • Polyurethane foam • Car seat

1 Introduction

The polyurethane foam is a cellular material with specific mechanical characteristics and with low density (generally around 55 kg/m^3 for seat cushions and mattresses). The polyurethane foam is possible to characterize as a nonlinear viscoelastic material [1]. Assessment of its mechanical properties is very important for evaluating quality of seats and especially their safety. The importance is mainly in considering stress during deformation, which affects the pressure distribution in the contact zone and during interaction of the sample and the load [2]. This paper describes a static biaxial loading of the polyurethane foam samples which was taken on the assembled experimental device inside the climate chamber. Testing was performed under some different climatic conditions occurring on the Earth where are the cars ordinary used.

P. Srb (✉) • P. Kulhavý • R. Martonka
Technical University of Liberec, Liberec, Czech Republic
e-mail: pavel.srb@tul.cz; petr.kulhavy@tul.cz; rudolf.martonka@tul.cz

V. Dynybyl et al. (eds.), *The Latest Methods of Construction Design*,
DOI 10.1007/978-3-319-22762-7_74

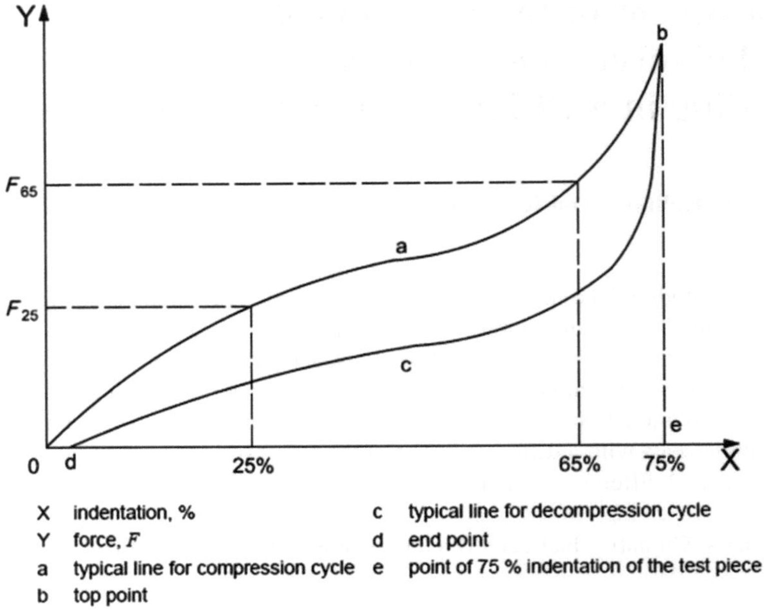

Fig. 1 Typical force-intendation curve [3]

2 Static Characteristic of the Polyurethane Foam

The Static characteristic of elastic materials are determined by static tests where is the tested sample exposed to a very low speed of deformation [3]. The resulting course of force (stress) and deformation indicate together the static characteristic.

The graph Fig. 1. shows that the polyurethane foam has a highly non-linear process of the force on the deformation. The loading and unloading curves form the hysteresis loop. The area bounded by this hysteresis loop represents a range of damping in the material. During loading the polyurethane foam, are the cells inside the material deformed and they are not able to return in the same time to their original form before the deformation.

The Compressive deflection coefficient, $\mathbf{S_f}$ is given by the equation:

$$S_f = \frac{F_{65}}{F_{25}} \tag{1}$$

where: F25 is the force at 25 % indentation in compression, stated in newtons and F65 is the force at 65 % indentation in compression, in newtons [kg.$m.s^{-2}$].

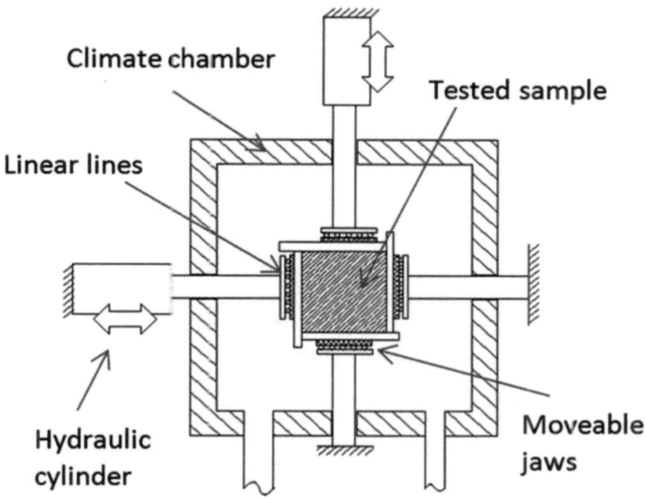

Fig. 2 Scheme of experimental device

3 Measurement Method

The measurements were taken in the external box of the climate chamber where some various climate conditions could be achieved. The size of the external box was approx. 1000 × 1000 × 500 mm. The temperatures inside the box was approximately from −70 °C up to +130 °C and the relative humidity from 10 % up to 90 % [4]. To the external box are connected two hydraulic cylinders showed in the Fig. 2. The hydraulic cylinders have the integrated position and a force sensor.

A specimens of the measured foam was placed in the horizontal position on the table between the four linear movable jaws (Fig. 3). There were used a specimens of the polyurethane foam of size 100 × 100 × 50 mm, the speed of pressure was chosen 60 mm/min up to 75 % deformation (50 mm) [5]. Four cycles of the measuring were conducted and the last one of them was evaluated. The measurement was performed in a five types of the climate condition. Each sample stayed 1 h in the chamber before the measurement started to stabilise conditions. For the measurements were used two types of polyurethane foams (TDI, MDI). For both types of foam were measured five specimens.

The measure results of MDI foam under temperature −10 °C and humidity 30 % are shown in Fig. 3, there are shown the forces in both (X, Y) loading directions. The values of force are almost the same course, which confirm the homogenous behaviour of measured material. The similar results were found for all specimens, so the next results are referred for one direction (Table 1). The measurement results of the course of the loading during the different climatic conditions are in Figs. 4, 5, and 6.

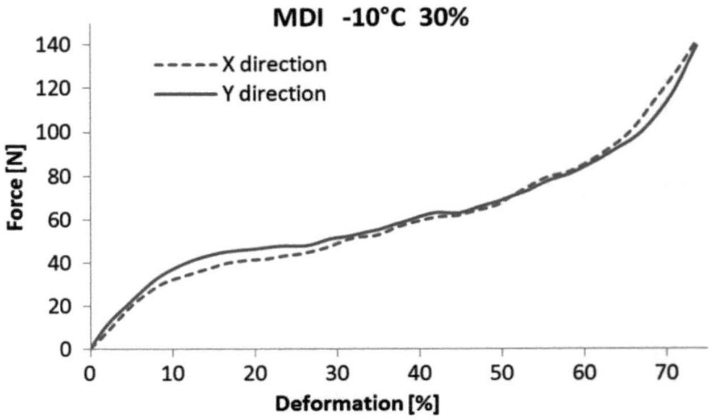

Fig. 3 Course of the loading force of MDI foam under −10 °C 30 %

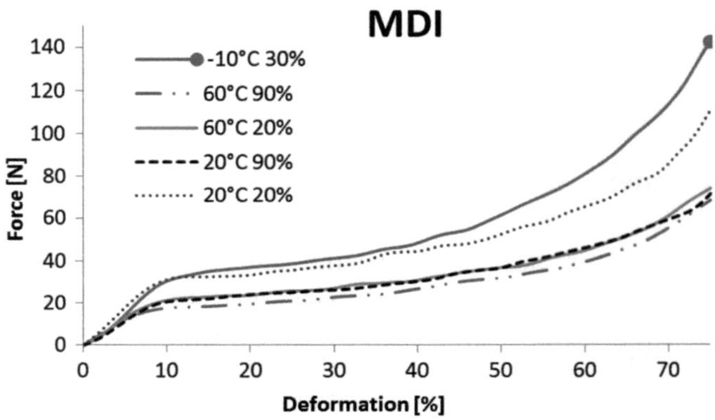

Fig. 4 Course of loading force MDI

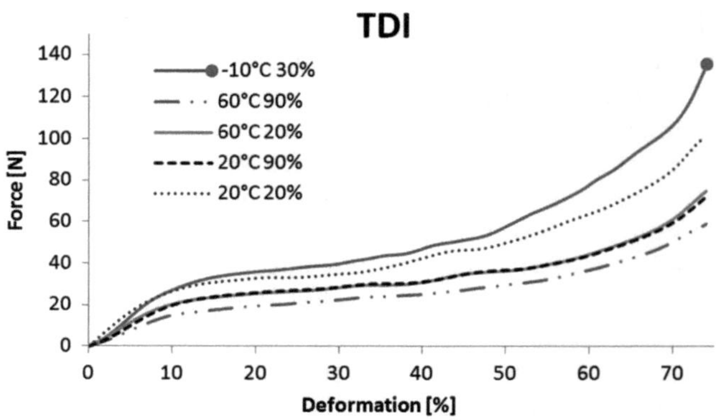

Fig. 5 Course of loading force TDI

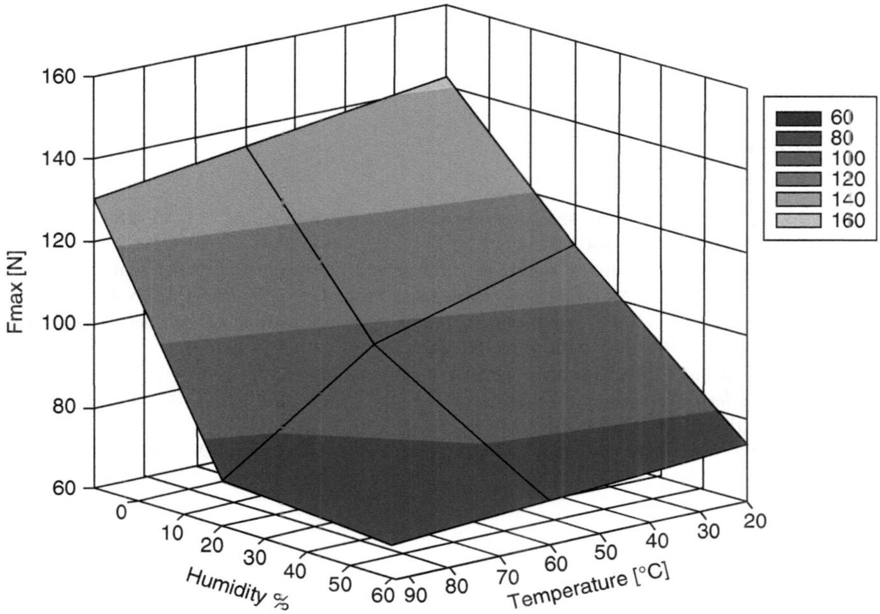

Fig. 6 The spatial scheme of an impact the climatic loads to the compressive force

Table 1 Results of the maximal force and the compressive deflection coefficient

Temperature [°C]	Humidity [%]	MDI max force [N]	MDI S_f [−]	TDI max force [N]	TDI S_f [−]
−10	30	142	2.19	135	2.42
20	20	110	2.06	100	2.18
20	90	71	2.04	72	1.89
60	20	74	2.02	75	1.92
60	90	68	2.19	59	2.05

4 Conclusion

This article described an experimental measurement of the biaxial loading of polyurethane foam specimens under different climate conditions. The values of maximum forces were evaluated which showed that with decreasing temperature the foam is stiffer. With increasing humidity maximum force decreases. Another evaluated parameter was compressive deflection coefficient, for optimal seats it should be between 2 and 4, which was achieved under most of conditions except TDI foam at 20 °C 90 % and 60 °C 20 %.

Acknowledgments This work was supported by the Ministry of Education of the Czech Republic within the SGS project no. 21011 on the Technical University of Liberec.

References

1. J. Petřík, P. Srb, Simulation of static loading of polyurethane foam. ACC J. **XVIII**(1/2012/Issue A), 85–91, Liberec. ISSN 1803-9752
2. M. Petrů, J. Petřík, Systems to optimize comfort and developments of car seat. Acta Tech. Corviniensis Bull. Eng. Ann. Fac. Eng. Hunedoara Fasc. **4**, 55–59 (2009). ISSN 1584-2665
3. ISO-2439, Flexible cellular polymeric materials – determination of hardness (indentation technique), 4. vyd. ISO copyright office (Geneva, 2008)
4. P. Srb, R. Martonka, Mechanical properties of polyurethane foam in different climate conditions, in *Modern Methods of Construction Design: Proceedings of ICMD*, Liberec, Czech Republic, 2013. ISBN 9783319052021
5. R. Martonka, Measuring properties of car seats and its inovation, Disertační práce, TU v Liberci, KST, Liberec, 2009, s. 117

Experimental Analysis of Prototype Rubber Suspension for Agricultural Machine

P. Syrovatka, F. Lopot, and M. Janda

Abstract This article deals about the experimental analysis of the effect on agricultural machine using rubber suspension and driving on testing track. Rubber suspensions with differential size of cross section and without rubber suspension are used during the analysis.

Keywords Rubber suspension • Testing • Qualisys system • Strain gauge

1 Introduction

There are many factors that influence design of rubber suspension like value of load, amplitude of loading, speed of loading, desired service life, and properties of used rubber.

Properties of rubber depend on blend of materials, vulcanization process, dimensions and shape of a rubber suspension. Because of many input variables, which have an impact on the resulting design of suspension, we decided to create prototype of suspension based on experience we have gained, and analysed result effects on a real machine.

There is a scheme of measuring set in Fig. 1. Tractor is connected with special construction, which simulate agricultural machine. Prototype of suspension is assembled in this machine. We could change load changing different weight on a special frame. This machine is in Fig. 2.

2 Description and Preparation of the Measurement

We decided to measure both loads which operate on the bearing unit connecting frame and steel roller, and loads which operate on the frame. Moreover compression of rubber suspension was measured. We used strain gauges for measuring loads and Qualisys system for measuring relative position. Strain gauges and

P. Syrovatka (✉) • M. Janda
Czech Technical University in Prague, Prague, Czech Republic
e-mail: pavel.syrovatka@fs.cvut.cz; martin.janda@fs.cvut.cz

F. Lopot
Faculty of Mechanical Engineering, Czech Technical University of Prague, Prague, Czech Republic
e-mail: frantisek.lopot@fs.cvut.cz

V. Dynybyl et al. (eds.), *The Latest Methods of Construction Design*,
DOI 10.1007/978-3-319-22762-7_75

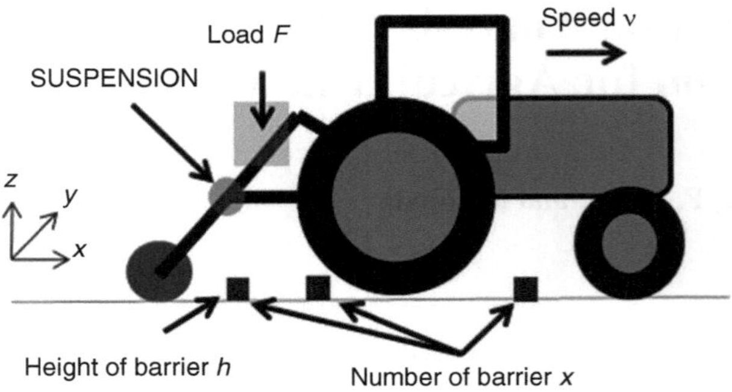

Fig. 1 Scheme of tractor and testing prototype

Fig. 2 Special machine for measuring rubber dynamic properties

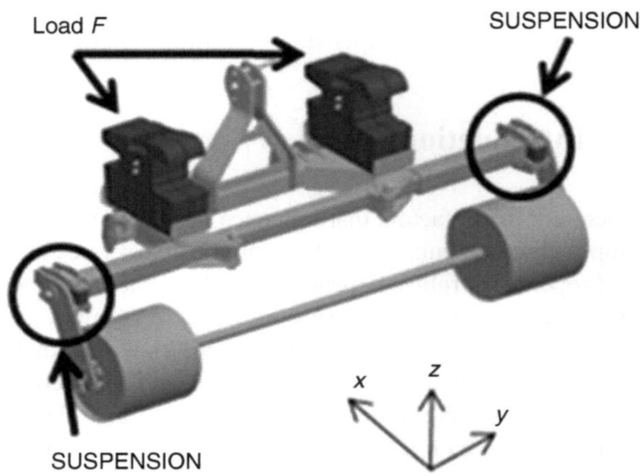

special markers for Qualisys system were installed on designated places of testing track (Fig. 3) [1,2].

3 Experimental Measuring on Testing Track

Testing track was hard-packed dirt road with steel barriers that were used to simulate real barriers like stones. We drove with this set with required speed over these special barriers (Fig. 4) [3].

Fig. 3 Special machine connect with tractor (*left*), detail with markers for Qualisys (*right*)

Fig. 4 Steel barrier (*left*) and static compressed rubber suspension (*right*)

4 Analysis of Measured Data and Results

Measured data from strain gauges were analysed in Matlab. Data from Qualisys were analysed in Qualisys postprocessor and in Matlab as well.

Negligible differences depending on the size of rubber suspension cross section are obvious in the Figs. 5 and 6. However, noticeable effect of rubber suspension occurrence is significant and obvious from the figures (WOS – without suspension) [3].

All samples were made from 50 % Butadien and 50 % natural rubber. Sample with mark * is made from 30 % Butadien and 70 % natural rubber.

Higher relative deformations at sample from high natural rubber are shown in Fig. 7.

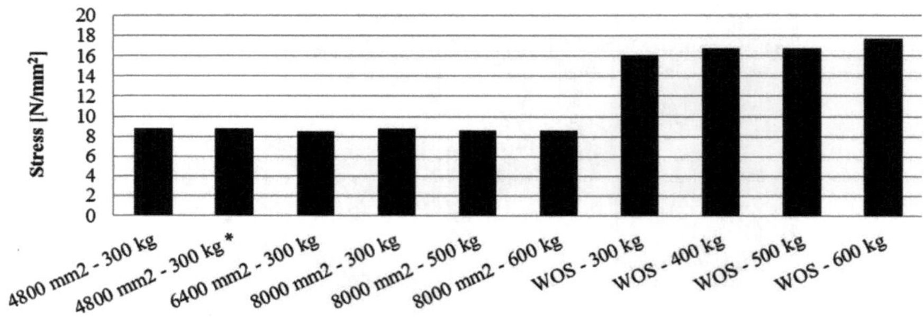

Fig. 5 Stress in frame of 15 kmph for load 300–600 kg

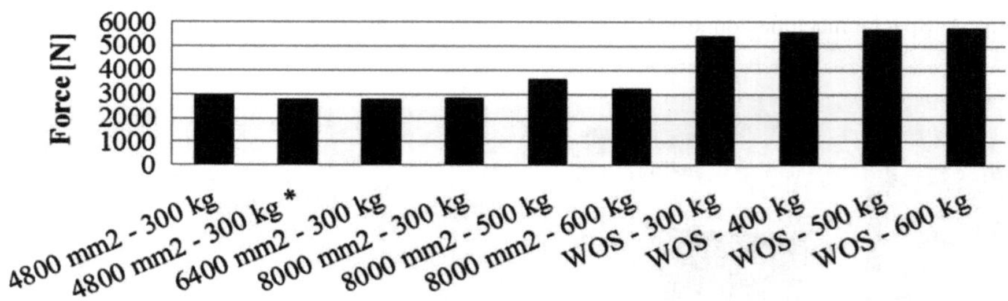

Fig. 6 Force on bearing of 15 kmph for load 300–600 kg

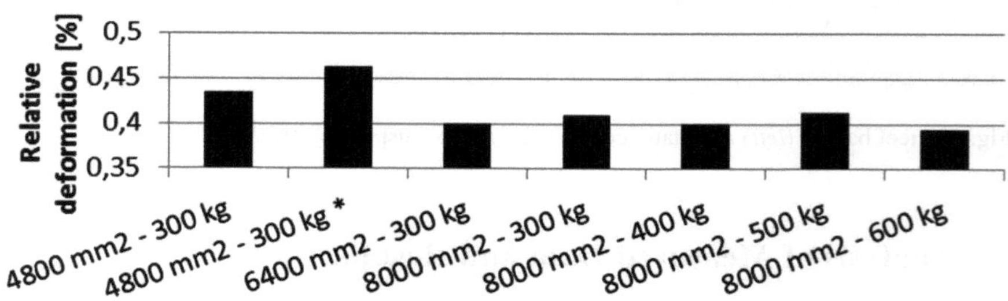

Fig. 7 Relative deformation of rubber suspension of 15 kmph for load 300–600 kg

5 Conclusion

This article is a concise description of comprehensive work which contains preparing and measuring in testing track. Results from the tests were used primarily for subsequent development of rubber suspension. Positive results significantly contributed for next development.

Acknowledgments This work has been supported by project TAČR Alfa: TA02010099.

References

1. P. Syrovátka, V. Dynybyl, Experimental analysis of the operational loads on agricultural machine, in *Experimental Stress Analysis 51*, ed. by K. Krish (Trans Tech, Uetikon-Zurich, 2014). ISSN 1660-9336. ISBN 978-3-03785-977-3
2. M. Mrňák, *Návrh odpružení ložiska pro zemědělské stroje Diplomová práce* (ČVUT v Praze, Praha, 2013)
3. M. Růžička, M. Hanke, *Dynamická pevnost a životnost, Fakulta strojní* (ČVUT v Praze, Praha, 1987)

Contribution to the Methodology of Measuring Transmission Error

M. Trochta, Z. Folta, and M. Burián

Abstract Measuring of transmission error is a complex problem, which includes many variables in process. This contribution describes problems which we encounter during preparation of methodology for measuring transmission error on gears made with different manufacturing precision. Mainly the problems of not repeatable measuring are concerned, which are caused by changes in temperature.

Keywords Transmission error • Measuring • Gear • Temperature

1 Introduction

The main topic of this contribution is methodology of measuring transmission error. In preparation state for measuring we encounter the problem with not repeatable results of measuring where the difference between measuring was up to 17 % (Fig. 2).

The testing apparatus (Fig. 1) for measuring transmission error is designed as a close circle testing apparatus. Testing circle includes testing gearbox, technological gearbox, preload device, shafts and motor. The device enables testing of the gears either in conformable sense of loading and rotating or in unconformable sense.

Measuring error up to 17 % was a big problem for us, so we start to search for reason of this error. After careful evaluation of result and check of incremental rotary encoder's accuracy we found the possibility that could cause the error of measurements. This possibility was the temperature change of oil in gear box during the measurement (Fig. 2).

M. Trochta (✉) • Z. Folta • M. Burián
VŠB-TU Ostrava, Ostrava, Czech Republic
e-mail: miroslav.trochta@vsb.cz; zdenek.folta@vsb.cz; miroslav.burian@vsb.cz

V. Dynybyl et al. (eds.), *The Latest Methods of Construction Design*,
DOI 10.1007/978-3-319-22762-7_76

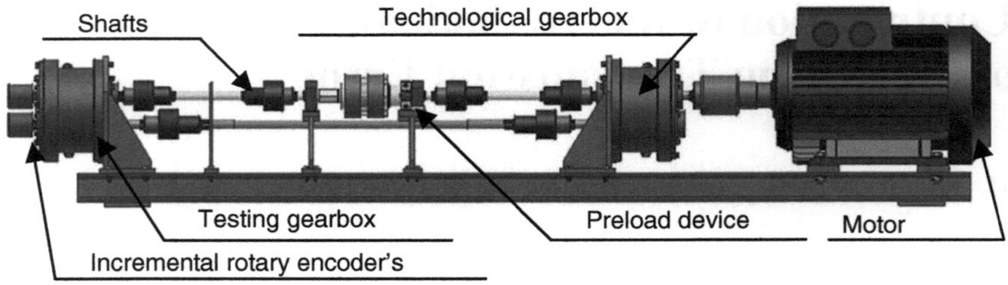

Fig. 1 Testing apparatus

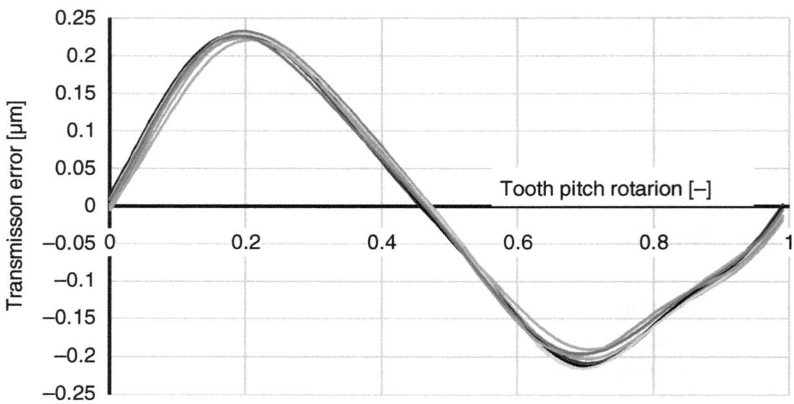

Fig. 2 Chart of measured transmission error

2 Temperature-Dependence of a Transmission Error

To evaluate the temperature-dependence of a transmission error we had to do some changes of the testing apparatus. The main change was inserting of a thermometer directly into the gearbox (Fig. 3), so that we could measure temperature of oil in gearbox.

Next we prepared measuring and during a 2-h-long test we monitored temperature of oil in gearbox. Measuring was done with torque 120 Nm and revolutions of motor 1500 min^{-1}. The test lasted for 2 h because after that time temperature was almost stable (Fig. 4).

During measuring we also measured the transmission error on different temperatures at the same torque (120 Nm) and 1500 RPM. At Fig. 5 you can see example of three curves showing a visible difference of transmission error on different temperatures.

Fig. 3 Photo of testing gearbox

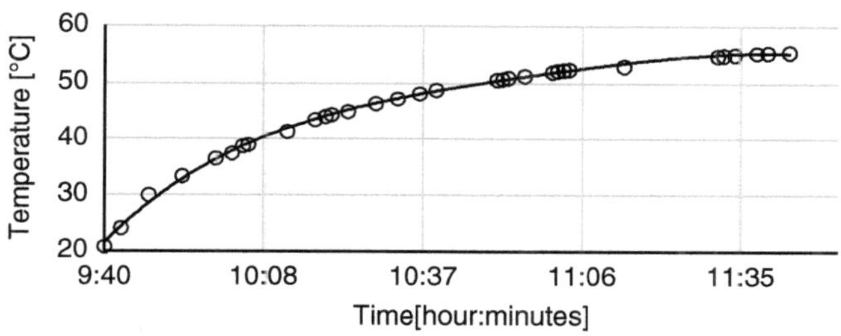

Fig. 4 Chart of temperature raising in gearbox

3 Conclusion

As of now, we have data from two of such measuring at our disposal. You can see linear dependence of transmission error on temperature of oil in testing gearbox (Fig. 6). This indicates that at least for gears with small modules we have tested we need to have stable environment for measuring of the transmission error. If it is not possible to provide a stable environment for measuring, this measuring will be affected with a non-acceptable error caused by the change of oil temperature in gear box.

When we observe stable temperature during measurement (50 ± 3 °C), a variance of results of repeated measurements is better than 1 % now.

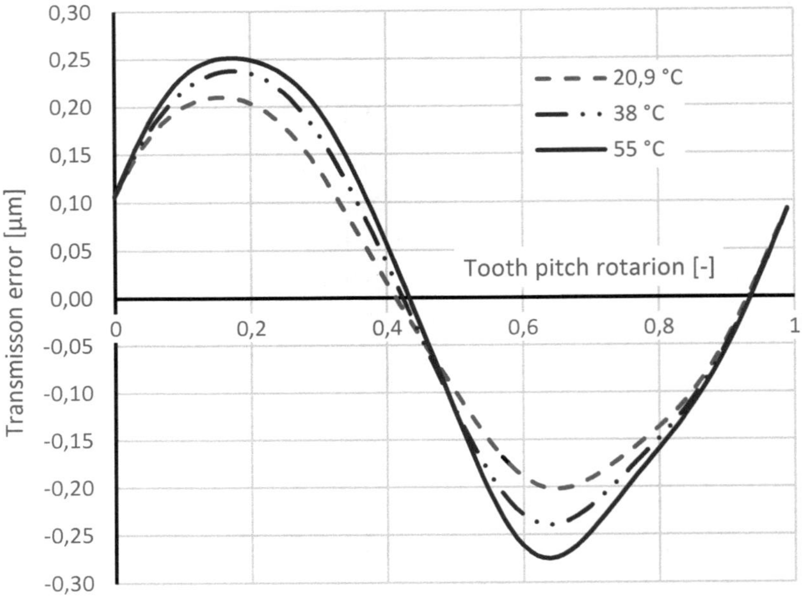

Fig. 5 Chart of transmission error dependence on temperature

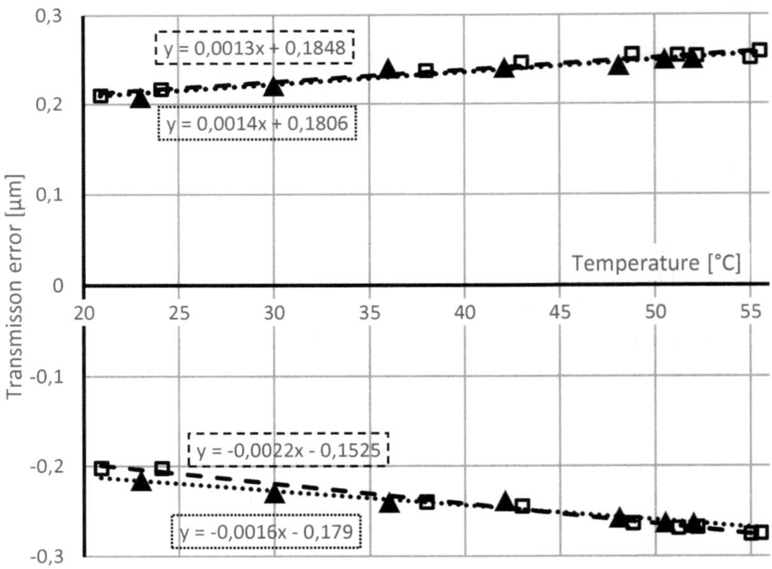

Fig. 6 Chart of temperature-dependence of an transmission error

Acknowledgment This research has been realized using the support of Technological Agency, Czech Republic, program Centers of Competence, project # TE01020020 Josef Božek Competence Centre for Automotive Industry. This support is gratefully acknowledged.

Contribution has been done in connection with project "Posouzení vlastností ozubených převodů z hlediska jejich geometrie a přesnosti výroby", reg. no. SP2014/24.

Climate Chamber for Testing of Multiaxial Loading

D. Vejrych, R. Martonka, and R. Kovář

Abstract This paper describes the equipment for testing materials in different climatic conditions with control of humidity environment. The described device allows testing of samples for tension, pressure and multiaxial loading. For static testing can be samples used in separate of clima camber or in external clima chamber with chance of fixture sample and additional members for the electronic transfer of accurate readings to an external computer with assesses software. In the described system are recorded maximum and minimum values for measurements with a demonstration of the possibility of measuring outcomes.

Keywords Temperature • Humidity • Chamber • Tension • Biaxial

1 Introduction

Tensile tests are at present most important studies, how to know mechanical properties of testing material. If a material is to be used as part of an engineering structure that will be subjected to a load, it is important to know that the material is strong enough and rigid enough to withstand the loads that it will experience in service. This paper is a continuation of uniaxial testing of material properties, which is describe in [1]. In this case same materials were used for biaxial testing. Each one of them was cut from a bumper of another type of vehicle. The test was carried out again in climatic chamber. By using the biaxial tensile test the properties of these materials were determined at different temperature and humidity. This paper leads to compare mechanical properties of all tested materials between uniaxial and biaxial loading at the same conditions.

The most common type of test used to measure the mechanical properties of a material is the Tension test. The major parameter that describes the stress–strain curve obtained during the tension test is the tensile strength.

D. Vejrych (✉) • R. Martonka • R. Kovář
Technical University of Liberec, Liberec, Czech Republic
e-mail: david.vejrych@tul.cz; rudolf.martonka@tul.cz; radovan.kovar@tul.cz

V. Dynybyl et al. (eds.), *The Latest Methods of Construction Design*,
DOI 10.1007/978-3-319-22762-7_77

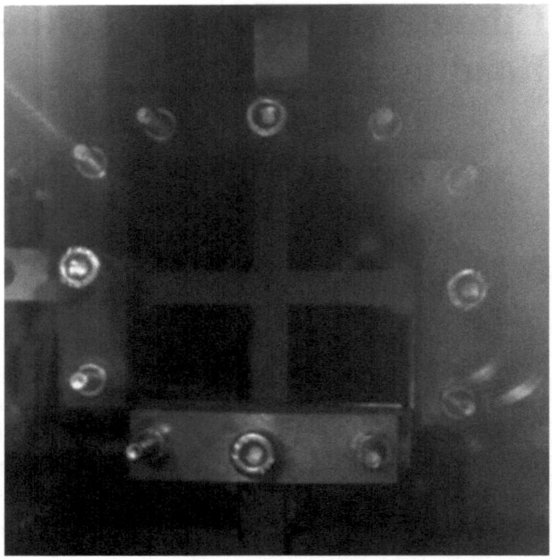

Fig. 1 Fixed specimen in the climatic chamber immediately before tests start

2 Biaxial Tensile Test

Biaxial tension is used for through deforming a specimen in two directions. In this case specimens are tested in a climatic chamber, where different condition of temperature and humidity are simulated. A tensile load is applied to the specimen until it fractures. The load is obtained by hydraulic cylinders. There are four jaws mounted on the end of each cylinder flange. With this jaws it is possible to strongly fix the specimen. The specimen, which is fixed and prepared for testing, is shown in Fig. 1.

During the test, the load required to make a certain elongation on the material is recorded. Engineering Stress is obtained by dividing the load by the original area of the cross section of the specimen. Elastic Region is placed on the part of the stress–strain curve up to the yielding point. Elastic deformation is recoverable. In the elastic region, stress and strain are related to each other linearly. That is described by Hooke's Law. Yield Strength is the stress level at which plastic deformation starts. The beginning of first plastic deformation is called yielding. It is an important parameter in design [2]. To generate useful strength data, a biaxial specimen must be capable of meeting a set of Requirements. The specimen should accept arbitrary biaxial load ratios. Specimens were obtained by cutting from a functional part, so their dimensions were not according the norm. Width of each specimen is 10 mm, but thickness depends on the thickness of the product.

Finite Element Analysis (FEA) is used extensively in the design of structures and sub-components. Measurements were made that demonstrated the biaxial nature of the strain in the testing specimen to find places, where the specimen will be firstly broken. FEA analysis of the test specimen also proved that the stress and strain

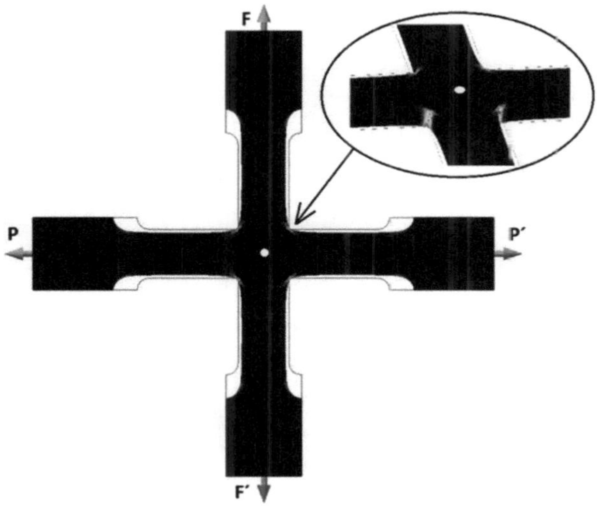

Fig. 2 FEM analyse of stress tension

distributions in the centre of the specimen were equi-biaxial. Simulation shows, that the main stress is in the corners. So in this place the specimen will be broken at first time. The result of simulation is shown in Fig. 2.

3 Experiment

For the experiment the same kinds of material were used, as in the case of uniaxial experiment. Each material comes from some part of a vehicle. Specifically, it comes from a front bumper. One test under same climatic conditions were carried out for each material. Each specimen stays ten minutes in the chamber before the test starts, to have a temperature of his vicinity. The goal was to compare the yield strength in relation to different temperatures. Each material is based on polypropylene and contains some specific components. Master batch is added to the base plastic to improve its quality. Additional ingredient to the plastic is talc which increases the dimensional stability of the product [3]. First tested material is labeled Sumika D320G02. It is made from basic Sumika and 15 % of talc. Second material is labeled 20MBTF. The plastic is composed of 87 % of natural Sabic and 13 % of master batch. Half of master batch consists of talc. So there is 6.5 % of talc in the entire volume. Last material is labeled 108MF10. That is the only natural Sabic plastic without any master batch or talc. All materials were tested at the value of temperature 10 °C and same value of humidity 10 %. Progress of the tests in compare with uniaxial loading is shown in the Fig. 3.

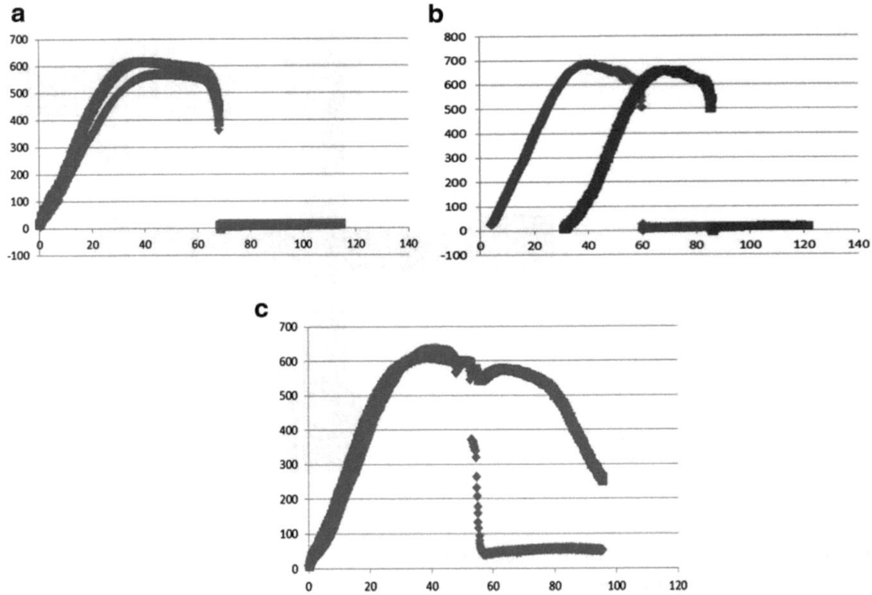

Fig. 3 Process graph of tensile tests (**a**) Sumika D320G02, (**b**) 108MF10 and (**c**) 20MBTF

4 Conclusion

During the tests were found progressions tensile tests of different materials at same value of temperature and humidity. An investigation of effect of temperature on the material was conducted and yield point was chosen as a measure. From the graphs we can read value of yield point. We need lower force and greater distance to tear the specimen. The material without any ingredients has the lowest value of yield point. Therefore, master batch and talc improve properties of the basic plastic material and make him more elastic.

Acknowledgments This work was supported by the Ministry of Education of the Czech Republic within the SGS project no. 21011 on the Technical University of Liberec.

References

1. L. Ševčík, P. Lepšík, M. Petrů, I. Mašín, R. Martonka, Modern method of construction design, in *Book of Proceedings of 54th International Conference of Machine Departments* (Technical University of Liberec, Liberec, 2014), pp. 357–366. ISBN 978-3-319-05203-8
2. ČSN EN ISO 527–1,2 (640604). *Plasty – Stanovení tahových vlastností.* (CNI Praha, Praha, 1997). 25 s
3. K. Karger, *Polypropylene: An A-Z Reference* (Kluwer, Dordrecht, 1999). ISBN 04-128-0200-7

Part IX
Nanotechnology

R. Kovář

System for High-Temperature Calcination of Nanofibers

R. Kovář

Abstract From past there is known a process for producing nanofibers. At present, there is the question of how to increase the productivity of their production. At the University of Liberec was developed principle of spinning from a roller. The system is made from spinning electrode and a collector, which are placed in an electric field. Using a modification of both parts is possible to change a value of intensity in the vicinity of both electrodes. This is able to leads to improve production of nanofibers. When correct settings are used, we can call this option the continuous process of nanofibers production. There is a theory that we can carry on more than one process of electrospinning in a closed space. After the process of spinning is necessary to subsequently modify the collected material. In this case, the material is composed by plastic parts, organic parts and titanium oxide. For the separation of titanium oxide is necessary to burn the material at the required temperature. After burning process it is possible to receive clear titanium oxide. For this task was build experimental furnace as a last part of production line.

Keywords Nanofibers • Electrospinning • Calcination • High-temperature

1 Introduction

At present, inorganic nanofibers have a lot of possibilities of using. That's why industry needs more nano material for its production. For this task was developed a production line. This production line is composed by a few parts. At the beginning is experimental lab. In this machine nanofibers are produced by using the principle of electrospinning from a "roller". The principle is based on the patent from Technical university of Liberec and the machine is also called Nanospider. In this machine there are two important equipments [1].

First of this is spinning electrode. It is a roller made from conductive material, which stays in a bath. The bath is filled by polymer solution and the surface of the roller is in contact with this solution. During the rotation of the spinning electrode the solution goes on the top and comes into contact with the electric field. On the

R. Kovář (✉)
Technical University of Liberec, Liberec, Czech Republic
e-mail: radovan.kovar@tul.cz

V. Dynybyl et al. (eds.), *The Latest Methods of Construction Design*,
DOI 10.1007/978-3-319-22762-7_78

other site of the machine is placed second electrode. This electrode is called collector and works as the second main part of the process. The collector is made also from conductive material. After beginning of the process of electrospinning it is possible to see spinned nanofibers on surface of the collector as shown [2].

For improving production of nanofibers this electrodes was modified. Productivity of nanofibers production is dependent on intensity around spinning electrode. The spinning electrode stays, but the material and geometrical shapes of the bath was changed. Previously was used plastic material. But when the material is changed, is proved, that the intensity around the spinning electrode increase. The second electrode (collector) has the shape of the rotary cylindrical brush. This shape has reason for continuous production. Charged spinning fiber electrodes are attracted to the grounded brush. Speed of rotation is one rotation in eight. During one rotation the one half of the brush is coated with fibers while on the other hand leads to entrainment of fibers from brush surface. At the beginning of the second rotation clean surface is getting in the area of spinning [3]. After this are nanofibers transported by using special mechanism to the ending part of the production line. Last operation of nanofibers production is calcination. For this task a calcination furnace was made.

2 Calcination Process

Calcination is the last step of nanofibers production. Nanofibers in inorganic form are in general prepared in two steps: electrospinning of polymeric solution to get fibrous morphology and calcination to remove organic compounds (polymer carrier) and to form inorganic material in specific crystal structure as shown [4]. Calcination of precursor fibers calcination is carried out at a temperature of 450–600 °C. At this temperature the polymer carrier is removed and TiO_2 stays. Crystalline phase and measured surface of TiO_2 are fixed at this temperature. Time of calcination is about 90 min to get required structure.

3 Experimental Furnace

For calcination process an experimental continuous furnace was made. As mentioned above, this machine represents last part of production line for continuous process of inorganic nanofibers production. The experimental furnace consists of two parts. First part is the furnace and the second part is support frame. The furnace forms a chamber in which occurs a process of calcination. Trough the chamber goes endless metal belt with nanomaterial on his surface. The belt is made from stainless steel. The chamber was separated to three single parts where different heating condition were used. First part is behind the entrance of the chamber and it is called preheating zone. Here the material is gradually heated. When the temperature

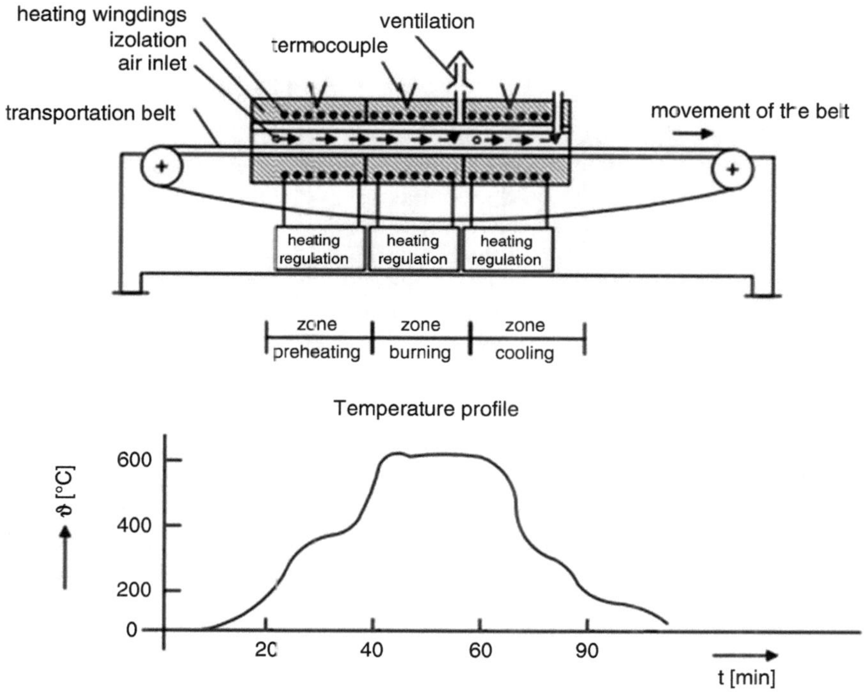

Fig. 1 Scheme of continuous furnace with temperature profile

reaches values higher than 400 °C, begins the process of burning. In this second zone all polymer carries are separated and TiO_2 gets new phase as mentioned above. In the end of the chamber there is the last zone. This zone is called cooling zone and here the burned material cools. All three zones with temperature profile are shown in Fig. 1.

3.1 *Support Frame*

The frame serves as support for the surface and for all of another parts. As a construction material were used aluminum profiles. Advantage of this solution is, that profiles are light and have a high level of variability. The frame consist of three parts. Main part is in the middle of the frame and hold the furnace. Under the furnace there is a cleaning mechanism composed by electric drive with flange. On this flange there is a rotation cylindrical brush with metal wires. As the brush rotates, wires clean the surface of the belt and removes waste, which was created during burning. On the right part there is drive of the belt made from electrical motor, worm gear and aluminum sheave. With this option it is value of speed of the

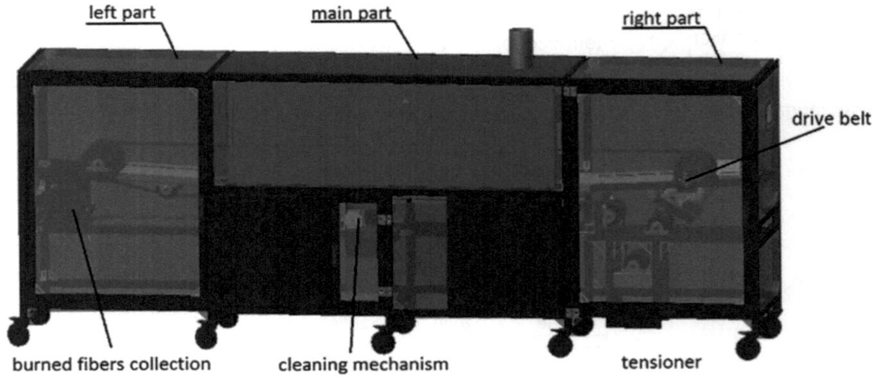

Fig. 2 Experimental furnace with the support frame

belt 1.5 m per hour. Under the drive is placed tensioner, which keeps tension in the belt and reduced effect of thermal expansion. On the left part there is second aluminum sheave. Both of sheaves represents main support for the endless belt. Under the sheave there is mechanism which collects burned fibers. The 3D model of furnace with the frame is shown in Fig. 2.

4 Conclusion

The work presented continual furnace for calcination TiO_2 nanofibers. This machine is the last part of continual production line for production inorganic nanofibers. After the end of calcination process fibers will have optimal structure without plastic carries. The machine is made from furnace and supporting aluminum frame. The furnace satisfies the conditions of calcination with the prescribed temperature profile.

Acknowledgments This work was supported by the Ministry of Education of the Czech Republic within the SGS project no. 28007 on the Technical University of Liberec.

References

1. O. Jirsák, Method of nanofibres production from a polymer solution using electrostatic spinning and a device for carrying out the method, Patent, WO 0127365, 2004
2. D. Lukáš, A. Sakar, L. Martinová, K. Vodsedálková, K. Lubasová, D. Chaloupek, J. Pokorný, P. Mikeš, J. Chvojka, M. Komárek, in *Physical Principles of Electrospinning (Electrospinning as a Nano-scale Technology of the Twenty-first Century)*, Textile Progress, No. 41, 2009. pp. 59–140

3. L. Ševčík, P. Lepšík, M. Petrů, I. Mašín, R. Martonka, Modern method of construction design, in *Book of Proceedings of 54th International Conference of Machine Departments* (Technical University of Liberec, Liberec, 2014), pp. 509–516. ISBN 978-3-319-05203-8
4. F. He, F. Ma, J. Li, T. Li, G. Li, Effect of calcination temperature on the structural properties and photocatalytic activities of solvothermal synthesized TiO_2 hollow nanoparticles. Ceram. Int. **40** (5), 6441–6446 (2014)

Author Index

A
Achtenova, G., 465–470
Andjelkovic, M., 173–182
Andrus, J., 203–206

B
Baran, P., 361–366
Barton, L., 3–6
Bauma, V., 353–357
Berka, O., 385–391, 393–398
Bochinsky, M., 377–382
Bosansky, M., 145–150
Bozca, M., 313–318
Broncek, J., 117–125
Burián, M., 7–13, 503–506

C
Chmelar, J., 215–220

D
Drahokoupil, J., 331–334
Dub, M., 15–18, 385–391, 393–398
Dvoracek, J., 319–323
Dvorak, J., 197–202
Dynybyl, V., 99–102, 203–206, 215–220, 269–274, 393–398, 433–436
Dzimko, M., 117–125

F
Fabian, P., 117–125
Fliegel, V., 399–404
Folta, Z., 503–506
Fryza, J., 127–132

G
Gajdac, I., 483–489
Galas, R., 133–137
Golda, M., 427–432
Grega, R., 361–366
Gregor, M., 81–86
Grznár, P., 81–86

H
Haluška, M., 81–86
Hanus, V., 349–352
Hartl, M., 153–157, 159–163
Hasan, M., 291–297
Havlik, J., 7–13
Hendrych, J., 15–18
Herajn, P., 457–462
Herak, D., 405–409
Hoidekr, J., 203–206
Homišin, J., 325–330
Hornak, S., 139–143
Hosek, J., 331–334
Hosnedl, S., 197–202, 369–374
Hrabe, P., 411–415
Hraska, M., 427–432
Hrcek, S., 207–212, 235–240

I
Ibrahim, T., 185–193

V. Dynybyl et al. (eds.), *The Latest Methods of Construction Design*,
DOI 10.1007/978-3-319-22762-7

J
Janda, M., 497–500
Jiroutova, D., 19–25
Jirout, T., 19–25

K
Kabutey, A., 405–409
Kaláb, K., 221–226
Khidhair, M., 291–297
Kloucek, V., 227–232
Kobza, M., 417–420
Kohar, R., 57–61, 207–212, 235–240
Kohl, O., 75–79, 421–425
Kolar, J., 27–35, 393–397
Komenda, L., 319–323
Kopecky, M., 369–374
Kopilakova, B., 145–150
Kostal, D., 153–157
Kovalova, N., 437–442
Kovář, R., 509–512, 515–518
Kral, M., 427–432
Kratky, L., 37–42
Krepela, J., 269–274
Kribala, P., 433–436
Kricka, J., 433–436
Križan, B., 45–49
Kronerova, E., 243–247
Křupka, I., 153–157, 159–163
Kubík, M., 51–56
Kulhavý, P., 437–442, 491–495
Kuzel, J., 427–432

L
Lepsik, P., 343–348
Lopot, F., 385–391, 393–398, 497–500
Lufinka, A., 399–404, 443–446

M
Machalek, R., 69–73
Macuchova, K., 111–114, 449–454
Madaj, R., 57–61, 81–86, 305–310
Maly, P., 471–474
Martonka, R., 399–404, 437–446, 491–495, 509–512
Matuszak, Z., 249–253, 255–261
Mazac, M., 457–462
Mazinova, I., 377–382
Mazůrek, I., 51–56
Milacek, O., 465–470
Moravič, M., 325–330
Mossoczy, P., 471–474
Mrazek, J., 99–102
Muller, M., 411–415

N
Nakashima, K., 159–163
Nečas, D., 159–163
Nemcek, M., 63–67
Nemcova, S., 475–482
Nemec, Z., 263–268
Nevrly, J., 263–268
Nohal, L., 319–323
Novák, O., 343–348

O
Oganesyan, P., 335–340
Omasta, M., 127–137
Ondra, M., 69–73

P
Pesik, L., 75–79
Petrik, J., 437–442
Petr, K., 215–220, 269–274
Petrovic, R., 173–182
Petrů, M., 343–348
Poljak, S., 483–489

R
Rada, V., 353–357
Radek, N., 117–125
Radosavljevic, M., 173–182

S
Sadek, S., 15–18
Sasiadek, M., 197–202, 369–374
Sásik, R., 81–86
Sawae, Y., 159–163
Semrad, D., 69–73
Sevcik, L., 87–91
Shevtsov, S., 335–340
Sigalingging, R., 405–409
Šika, Z., 353–357
Skarolek, A., 75–79
Skaroupka, D., 69–73
Slanina, R., 93–98
Smejkal, D., 165–170

Sniehotta, P., 93–98
Sojka, J., 299–303
Spacek, P., 15–18
Spaniel, M., 277–282
Šperka, P., 159–163
Spišák, P., 305–310
Srb, P., 491–495
Stary, F., 99–102, 349–352
Suransky, M., 103–110
Svatoš, P., 353–357
Svoboda, M., 457–462
Syrovatka, P., 497–500

T
Todic, N., 173–182
Trochta, M., 7–13, 503–506

U
Uhlir, F., 69–73
Uhlir, R., 471–474

V
Valášek, M., 353–357
Vejrych, D., 509–512
Vrbka, M., 159–163

W
Wahab, A., 185–193
Wu, J.-K., 335–340

Y
Yarimitsu, S., 159–163

Z
Zapotocny, J., 145–150
Zhilyaev, I., 335–340
Zicha, J., 111–114, 353–357, 475–482
Zlebek, M., 103–110
Zoul, V., 283–288

Subject Index

A
Acoustic camera, 483–489
Acoustic emission, 320, 321
Adhesion, 133, 165–170
Adhesion coefficient, 133
Air spring, 421
Air temperature, 185
Analysis, 197–202, 283–288
  sustainable developments, 377–382
Angle measurements, 112
Angle of deformation, 413–415
Anti-backlash transmissions, 227–232
Approximation, 417
Assembly, 88–89
Assembly set, 81–86
Automatic control circuit, 328–329
Axial bearing, 320
Axial impeller, 20–24
AXMAT, 319–323

B
Ballistic protection, 277
Bearing cage, 58, 305–310
Bearing capacity, 411–415
Bending stress, 313
Biaxial, 509
Biaxial loading, 438, 491
Biofuel, 37
Biotribology, 160
Blending, 23, 25
Bolt, 399
Bond graphs, 249–253

C
Calcination, 515–518
Calibration, 139–143
Carbonized depth, 269–274
Cardan joint, 111–114
Car seat, 491
Case hardened layer, 270–273
C–C gearing. *See* Convex–concave (C–C) gearing
Chain arrangement, 350
Chamber, 509–512
Chi-square, 217–219
Climatic chamber, 437–442, 491, 493
Clinched joint, 299–302
Clinching, 300–302, 411
Closed loop, 465–470
Combustion engine, 417
Compliant, 163
Composite blade, 335–340
Composites parts, 344, 347
Compressive force, 406, 407, 409
Compressor ratio, 185
Contact ratio, 314
Contact stress, 315–318
Contaminants, 166
Control algorithm of extremal regulation system, 329
Control system, 443–446
Convex–concave (C–C) gearing, 146–150
Conveyor belt, 221
Corporate social responsibility, 378
Corrosion, 291–297

V. Dynybyl et al. (eds.), *The Latest Methods of Construction Design*,
DOI 10.1007/978-3-319-22762-7

D
Damage mechanics, 277
Damping, 421–425
Deformable mirror, 475–482
Deformation, 406–409, 411–415
Dental laboratories, 93–98
Design(ing), 15–18, 87–91
of mechanical parts of traction drive, 27–28
project, 69–73
Design for assembly (DfA), 369–374
Design for X (DfX), 369
Diameter of cylinder, 174
Dimension, 245
Disc grinder, 331–334
Distribution of internal loading, 235–241
Downspeeding, 361–366
Driving shaft angle, 177
Ductile fracture, 277–282
Dynamic processes, 173
Dynamic viscosity, 176

E
Education, 153–157
Efficiency, 263, 466, 468–470
Elastic–plastic deformation, 270
Electrode machining, 331–334
Electro discharge machining (EDM), 331
Electrospinning, 515
End effector, 99–102
Energy, 263–268
Energy factors, 255–261
Environment, 343–348
Equations of motion, 203–206
Erosion, 333, 334
Evaluation, 69–73
Experiment, 173–183, 400–401
Experimental device, 104–106, 118
Experimental tests, 270

F
Falling body viscometer, 140
Fan, 427, 428, 430–432
Fastener, 399–404
Fatigue, 3–6, 221
Fatigue life, 207–212
Fatigue wear, 150
Filters, 87–91
Finite element method (FEM), 237, 269–274
Flow coefficient, 174, 175, 177
Flow of streams, 255–261
Fracture face, 221–226
Fracture locus, 278
Friction, 118, 133–137, 159–163
modifier, 133–137
Fuel consumption, 417–420
Fused deposition modelling, 104

G
Gas temperature, 185–193
Gas turbine, 185–193
Gearbox, 7–13, 215–220
hinge, 393–398
test bench, 465–470
Gears(ing), 7, 269–274, 313, 457–162, 503
mesh(ing), 269, 386, 387, 390
trains, 227–232
Geometrical flow section, 179, 180, 182
Grease, 127–132
Grip force, 100–102

H
HCR, 7–13
gearing, 147, 148, 150
Heat treatment, 291
Hexapod, 443–446
High-temperature, 515–518
Humidity, 509
Hybrid, 419
Hydraulic, 263–268
Hydrofoil impeller, 22–23
Hysteresis, 476, 479–480

I
Identification of failures, 483–489
Illumination, 453
Industrial, 7–13
Industrial design, 69–73
Influence, 104
Integrating components, 450
Interferometry, 153–157, 479
Involute profile, 63–67
Iron alloys, 291–297

K
Kinematic equations, 445
Kruskal–Wallis (KW), 218

L
Layer of nanofibers, 87
Levenberg, 407
Life cycle, 197–202

Light collector, 450
Light integrator, 450
Lightweight construction, 343–348
Lignocelluloses, 37–42
Liquid waste, 15
Logistic transport solution, 81–86
Low-floor trams, 27
Lundberg–Palmgren theory, 207

M
Machine element, 45–49
Magnetorheological (MR) fluid, 52
Magnetorheological (MR) valve, 51–56
Marquardt, 407
Masks, 87
Material analysis, 270
Measurement, 427–432
  validity, 255
Measuring, 503–506
Mechanical joining, 299–300
Mechanical properties, 412
Mechatronics, 112
Method of extremal regulation, 328
Methods of calculation, 283
Modern AC drives, 27
Modular system, 349
Module, 243–247
MR fluid. *See* Magnetorheological (MR) fluid
MR valve. *See* Magnetorheological (MR) valve
Multi-stage impeller, 23–24

N
Nanofibers, 515–518
New aspects, 283–288
Numbers of cylinders, 175
Nut, 399

O
Online assessment, 69–763

P
Parallel structure, 215–220
Parameters, 103–110
Particle suspension, 19
Pasty material, 15
Piezoceramics, 476, 479
Piston pump, 173–182
Pitting, 145–150
Planetary transmission, 385–391
Pneumatic dual mass flywheel, 361–366
Pneumatic flexible shaft coupling, 325
Pneumatic spring, 76
Polymer, 46
Polyurethane foam, 491–495
Polyvinyl alcohol (PVA) hydrogel, 159–163
Power circulation, 465
Power losses, 465–470
Precision position control, 227
Preload, 228
Pressure, 427–432
Probe, 427–432
Prototype, 82
PVA hydrogel. *See* Polyvinyl alcohol (PVA) hydrogel

Q
Qualisys system, 394, 497

R
Radial clearance, 175, 182
Rail vehicle, 27–35
Rapid prototyping, 82–83
Real time application, 444
Reciprocal slope transformation, 405
Recovery, 263–268
Resistance, 292
Risk, 197–202
Riveting, 300
Roller conveyor, 349–352
Rolling bearings, 207, 235–340
Rolling elements, 235
Rubber suspension, 497–500

S
Sanding, 165–170
Scale-up, 37–42
Seat cover, 437–442
Seizure, 128
Semi-active valve, 51–56
Separation, 15
Sheet-metal, 49, 299–300, 302
Solidification, 15
Spectral measurement, 449–454
Spur gear, 243–247
Squalane, 139
Stability, 331
Stabilization, 15

Stand, 3–6
Standard, 7
Start of involute, 63–64
Statistics, 215–220
Steel C45, 122
Stiffness, 421–425
Strain gauges, 101, 351, 385, 497
Stress reducing, 339
Stress state, 386
Structural optimization, 336
Sunflower seeds hulling machine, 203–206
Surface analysis, 221
Sustainability, 377

T
Tangent curve, 405, 407
Taxonomy of properties, 370
Teaching, 48–49
Technical systems, 199, 249–253
Temperature, 457–462, 503–505, 509–512
Tension, 509–511
Testing, 3–6, 498–500
Test rig, 215–220
Textbook, 46–49
Textile, 437–442
Thermal-expansionary treatment, 37–42
Tooth flank, 145–150, 457–462
Torsional damper, 362
Torsional vibration, 283, 286, 325–327, 330
Torsion-bar spring, 228, 230
TOX®-Joining System, 300, 301
Traction, 133–135, 137
Train electric units, 33–35
Tram drive, 394, 395
Tram pillars, 3
Transmission error, 503–506
Tribology, 117–125, 153–157
Triple bottom line accounting, 378, 379
Twin disk, 166
Twisted, 400, 427–432

U
Undercutting, 63–68

V
Vacuum casting, 60, 61, 305–309
Vacuum chamber, 307, 308
Vacuum cleaner, 93–98
Vehicles, 263–268, 307, 344, 417–420
Vibration, 15–18, 361–363, 366
Vibration isolation system, 75–79
Viscosity, 139–143

W
Wear, 118, 123, 127–134, 136, 137
Weight, 100–102
Wheat straw, 38, 39
Wheel flange lubrication, 132
Wheel–rail contact, 117, 127–137, 165–170
Wind turbine, 335–340